The Gut Microbiome in Health and Disease

Dirk Haller
Editor

The Gut Microbiome in Health and Disease

Springer

Editor
Dirk Haller
Chair of Nutrition and Immunology
Technical University Munich
Freising, Germany

ISBN 978-3-030-08039-6 ISBN 978-3-319-90545-7 (eBook)
https://doi.org/10.1007/978-3-319-90545-7

Softcover re-print of the Hardcover 1st edition 2018

Printed on acid-free paper

This Springer imprint is published by the registered company Springer Nature Switzerland AG.
The registered company address is: Gewerbestrasse 11, 6330 Cham, Switzerland

Preface

The German Research Foundation (DFG) started in 2013 to fund the Priority Program SPP 1656 entitled "Intestinal Microbiota" (www.intestinal-microbiota.de) and thereby consolidated a concerted action of the German Society of Hygiene and Medical Microbiology (DGHM) to support microbiome research in Germany. An interdisciplinary network of scientists, including microbiologists, gastroenterologists, immunologists, nutrition scientists, and physicians, worked together over the past few years to achieve novel insights into the role of the gut microbiome in health and diseases. In addition to numerous scientific accomplishments, the consortium made an effort to use their complementary expertise in educating the next generation of young researchers. In 2018, the members of the Priority Program SPP 1656 organized the 1st Summer School on "Microbiome in Health and Disease" within the frame of the annual Seeon Conference (www.seeon-conference.de), aiming to establish a continuous platform for education in this rapidly developing area of science. In addition, and complementary to the Summer School, this book provides a comprehensive review on the gut microbiome and its functions in health and a variety of intestinal as well as extraintestinal diseases, covering basic principles of the gut microbial ecosystem (composition, metabolic activities, and evolution over time and life stages), its reciprocal interaction with the immune system, and the clinical implementation related to diagnosis and therapy. We focus on bacteria as the dominant type of microorganism in the intestine, despite the fact that viruses, archaea, phages, and fungi emerge as relevant players in the regulation of the bacterial ecosystem and host functions. Considering the need for a continuous education process of students and health professionals, this book provides a structured overview about the methodologies applied as well as the scientific and clinical aspects of microbiome–host interactions, highlighting perspectives on historic developments and controversies in the field.

Munich, Germany — Dirk Haller
June 2018

Contents

1 Intestinal Microbiome in Health and Disease: Introduction

Dirk Haller

Abstract

At the end of the nineteenth century, Robert Koch and Louis Pasteur developed the concept that transmissible human diseases are caused by microbial infections and, thereby, revolutionized the view of physicians on how to prevent and treat epidemics. More than 100 years later, the next conceptual revolution implies that naturally occurring communities of "commensal" microbes, collectively called microbiome, in and on human body sites affect health and the development of numerous diseases. The intestine provides an explicitly large interface to the environment and is critically involved in immune and metabolic homeostasis, providing the conceptual basis that this spatially adapted communities of microorganisms affects human health. Immune, metabolic, and xenobiotic receptors sense and process microbial signals and thereby contribute to a mutualistic relationship between the microbiome and the host. It seems a plausible hypothesis that the microbiome, considered as the *forgotten organ*, coevolved with the mammalian host, leading to a symbiotic interdependence of this metaorganism. Increasing evidence suggests that "unfavorable or so-called dysbiotic" changes in the gut microbiome lead to a distortion of microbe–host homeostasis and potentially affect disease susceptibility. In this book, we discuss breakthroughs, challenges, and applications of microbiome research at a cutting-edge level.

D. Haller (✉)
Chair of Nutrition and Immunology, Technische Universität München, München, Germany
e-mail: dirk.haller@tum.de

At the end of the nineteenth century, Robert Koch and Louis Pasteur developed the concept that transmissible human diseases are caused by microbial infections and, thereby, revolutionized the view of physicians on how to prevent and treat epidemics. More than 100 years later, the next conceptual revolution implies that naturally occurring communities of "commensal" microbes in and on human body sites affect health and the development of numerous diseases. Over the past decade, large science consortia in Europe (MetaHIT; Metagenomics of the Human Intestinal Tract) and the USA (Human Microbiome Project) have started to acquire data on the genomic potentials, phylogenetic relationships, and functional properties of microbial communities, collectively called microbiome, in healthy and diseased human populations. The technical breakthroughs and affordability of next-generation sequencing (NGS) stimulated an enormous boost of scientific activities leading to almost 40,000 publications indexed under the search term "microbiome" in the database of the US National Library of Medicine (PubMed) (Fig. 1.1). A broad variety of disorders, including

D. Haller (ed.), *The Gut Microbiome in Health and Disease*,
https://doi.org/10.1007/978-3-319-90545-7_1

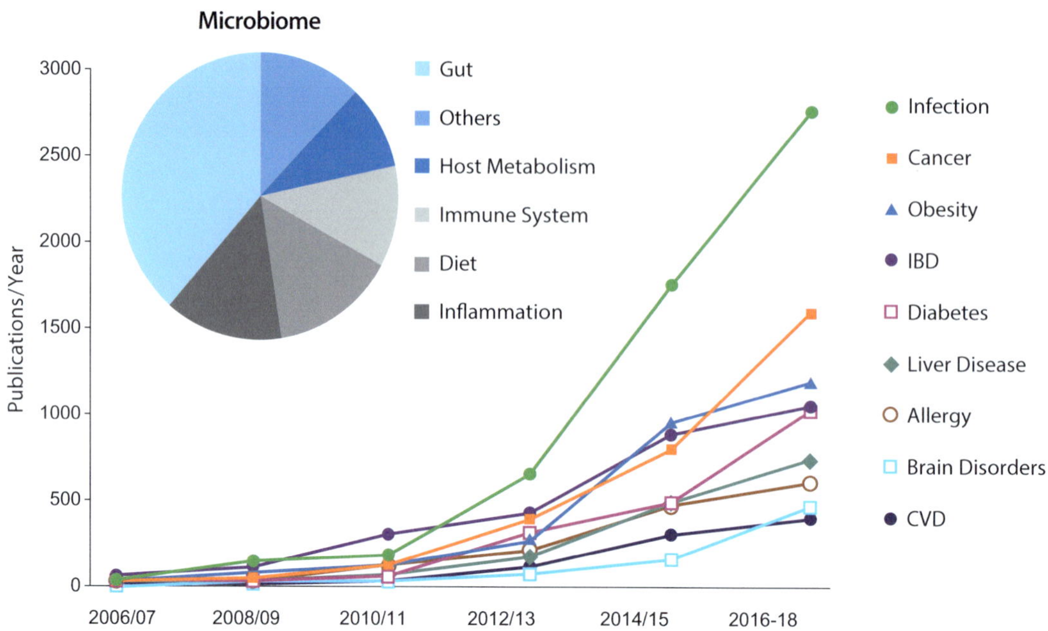

Fig. 1.1 Number of publications related to the search term "microbiome." Data were obtained by searching the database of the US National Library of Medicine (www.ncbi.nlm.nih.gov/pmc). The term "microbiome" retrieved a total of 39,592 publications (February 20, 2018). The pie chart illustrates the relative contribution of different aspects in microbiome research related to host organs (gut, total of 15,335 publications), host processes (inflammation, metabolism, immune system, total of 13,805 publications), or diet (total of 5709 publications). The annual contribution of publications related to disease categories is displayed between 2006 and 2018. Abbreviations: *IBD* Inflammatory Bowel Diseases, *CVD* Cardiovascular Disease

infectious as well as immune- and metabolically driven diseases, are associated with microbiome changes in the most densely colonized body site—the gut. Our digestive organ provides an explicitly large interface to the environment and is critically involved in immune and metabolic homeostasis, providing the conceptual basis that this spatially adapted community of microorganisms affects human health. Immune, metabolic, and xenobiotic receptors sense and process microbial signals and thereby contribute to a mutualistic relationship between the microbiome and the host. It seems a plausible hypothesis that the microbiome, considered as the *forgotten organ*, coevolved with the mammalian host, leading to a symbiotic interdependence of this metaorganism. Increasing evidence suggests that "unfavorable or the so-called dysbiotic" changes in the gut microbiome lead to a distortion of microbe-host homeostasis and potentially affect disease susceptibility. Nevertheless, the clinical relevance of microbiome changes remains speculative. Given the substantial interindividual variations in the microbiome of human populations and the pleiotropy of confounding factors, NGS-based analyses in cross-sectional studies are correlative and require validation in well-controlled replication studies using a careful selection of participants based on extensive phenotyping. The implementation of prospective (longitudinal) and treatment-naïve early-onset or birth cohorts may help to identify disease-relevant microbiome signatures in a progressive fashion and at very early stages. Disorders with low incidence require however prospective cohorts with probably unrealistic size in order to reach relevant numbers of cases. In addition to a better stratification of human phenotypes, the implementation of standardized protocols for sampling and analysis is needed to improve the reproducibility and comparability of microbiome signatures at a meaningful taxonomic resolution. An essential question arising from many human studies is whether microbiome alterations are the cause or simply the consequence of pathologies, exemplifying the need to better understand the functional relationships of microbial communities with their host at the mechanistic level. One has to accept the fact that knowledge in this area of research is still not consolidated, and the major challenge is to establish a causal understanding of microbiome-host interactions and to address the obvious knowledge gaps. First, sample preparation and NGS technologies are subject of constant refinement complicated by methodological limitations for data interpretation. Bioinformatic algorithms need to cope with the inherent complexity, and the implementation of machine-learning algorithms is a growing need. Second, sequencing-based knowledge gain requires biological backup leading to the obvious need for an expansion on the isolation of yet uncultured taxa and the development of large-scale bacterial strain repositories. Third, the generation of disease-relevant gnotobiotic animal models, being colonized by either simplified or complex microbial consortia, is a prerequisite to unravel the mechanistic basis of microbe-host interactions. Finally, and based on the total sum of microbiome research, the aim must be to develop therapeutic and prognostic tools for targeted clinical implementation.

Composition and Function of the Gut Microbiome

2

Michael Blaut

Abstract

The human gastrointestinal tract harbors a plethora of microorganisms, most of which belong to the domain Bacteria. Owing to manifold effects on host physiology and host health, there is a growing interest in better understanding the role and function of gut microbial communities. Microbiota composition changes along the gastrointestinal tract in response to changes in the physicochemical conditions and substrate availability. Moreover, large interindividual differences are observed. One major function of the gut microbiota lies in the conversion of indigestible dietary carbohydrates and host-derived glycans to short-chain fatty acids, which provide energy to the host and have regulatory functions. Microbiome analysis has led to the notion of a "core microbiome" which encodes functions shared by human individuals. Gut microbial community members interact with each other and with the host constituting a functional microbial ecosystem. However, there are still major gaps in our understanding of the molecular mechanisms underlying such interactions.

M. Blaut (✉)
Department of Gastrointestinal Microbiology, German Institute of Human Nutrition Potsdam-Rehbruecke (DIfE), Nuthetal, Germany
e-mail: blaut@dife.de

2.1 Introduction

Prokaryotic microorganisms (Bacteria and Archaea) have conquered essentially every habitat on earth and may therefore be considered ubiquitous. They occupy environments that differ profoundly in their physicochemical conditions and the substrates available for growth. Microbial habitats range from marine and sweet water environments, deep-sea hydrothermal vents, soil, and air to plants and animals. The microbes thriving in a given habitat are optimally adapted to the conditions prevailing therein. Some microbial communities withstand even harsh conditions such as high temperature, high salinity, and low or high pH. The ability of prokaryotes to colonize essentially all habitats on earth reflects 4 billion years of evolution. Depending on the environment, prokaryotic organisms may be phototrophic, chemotrophic, lithotrophic, autotrophic, heterotrophic, and combinations thereof, indicating a high metabolic variability. Besides playing essential roles in the global cycles of carbon, nitrogen, and sulfur, prokaryotes also occur in and on animals and humans. They occupy various body sites including the skin, nose, throat, as well as the urogenital and gastrointestinal tracts. These habitats differ with respect to the availability of substrates and oxygen, but, at least in mammals, they all provide a constant temperature favoring microbial growth. The intestinal tracts of herbivores differ from those of carnivores or omnivores not only in their anatomies

D. Haller (ed.), *The Gut Microbiome in Health and Disease*,
https://doi.org/10.1007/978-3-319-90545-7_2

but also in the microbial communities they harbor reflecting adaptations to the respective preferred food source. There is evidence that the intestinal microbial communities coevolved with their respective host (Ley et al. 2008).

2.2 Distribution of Microbial Communities in the Human Gastrointestinal Tract

Environmental conditions in the human gastrointestinal tract are not uniform but differ considerably between the stomach and colon. It's therefore not surprising that the microbial communities resident in the various sections of the digestive tract differ in several aspects including cell density, composition, and metabolic activity.

2.2.1 Stomach

Between meals, the pH in the stomach of healthy adults is typically 1–2 but increases following food ingestion. Transit time through the stomach determined in eight healthy subjects with a magnet tracking system ranged between 5 and 133 min, with a median of 56 min (Worsoe et al. 2011). Transit time is influenced by food consistency (shorter for fluids than for solid and un-chewed food), osmolarity (longer for monosaccharides compared to polysaccharides), nutrient composition (longer for fats and carbohydrates), and energy density (longer for high-energy diets). The low pH of gastric juice largely prevents the growth of ingested microbes explaining the low density of $<10^3$ microbial cells per ml of gastric content. However, a culture-independent survey of microbial 16S rRNA gene sequences in 23 gastric mucosa biopsy samples revealed a diverse community of 128 phylotypes belonging to the phyla *Firmicutes* (36 phylotypes), *Bacteroidetes* (35 phylotypes), *Proteobacteria* (32 phylotypes), *Actinobacteria* (12 phylotypes), *Fusobacteria* (10 phylotypes) (Fig. 2.1), and minor components of other phyla (Bik et al. 2006). A high proportion of the detected sequences were assigned to oral bacteria, such as *Streptococcus salivarius*, *Streptococcus mitis*, *Streptococcus parasanguinis*, various *Prevotella* and *Porphyromonas* spp., *Rothia dentocariosa*, *Atopobium parvulum*, and *Fusobacterium nucleatum*. It may be surmised that the main habitat of many of these species is the oral cavity, from where they get into to stomach by swallowing. Nineteen of the 23 subjects harbored *Helicobacter pylori*. This organism resides in the mucus layer of the stomach and is known to secure its survival in the gastric environment by the production of urease, which catalyzes the release of ammonia (and carbon dioxide) from urea resulting in an increase of the pH in the immediate environment of the cell. Bacteria isolated from gastric contents include *Lactobacillus* spp. and *Streptococcus* spp., which are capable of surviving at relatively low pH.

2.2.2 Small Intestine

The small intestine represents the longest part of the digestive tract with changing conditions and increasing bacterial cell densities along its course. The relatively short residence time of intestinal contents, namely, 209–391 min with a median of 255 min (Worsoe et al. 2011), limits the growth of microorganisms to high density, in particular in the duodenum. Bacterial cell counts increase from the duodenum to the terminal ileum from approximately 10^4 to 10^8 per ml of intestinal content (Booijink et al. 2010; Finegold et al. 1983), and also the number of taxa detectable with culture-independent methods increases (Hayashi et al. 2005). The ileal effluents of ileostomy patients were reported to contain species of the *Lactobacillales*, *Clostridiales*, and the *Veillonella* group as well as *Streptococcus bovis*-related species at relatively high abundance, while species related to *Ruminococcus gnavus*, *Ruminococcus obeum*, and *Bacteroides plebeius* were present at lower relative proportions (Fig. 2.1) (Booijink et al. 2010). The microbial taxa detected in ileostomy effluent were in part the same as those retrieved from the small intestine of four healthy subjects and in part similar to those detected in their feces (Zoetendal et al. 2012). In general, the small intestinal microbiota

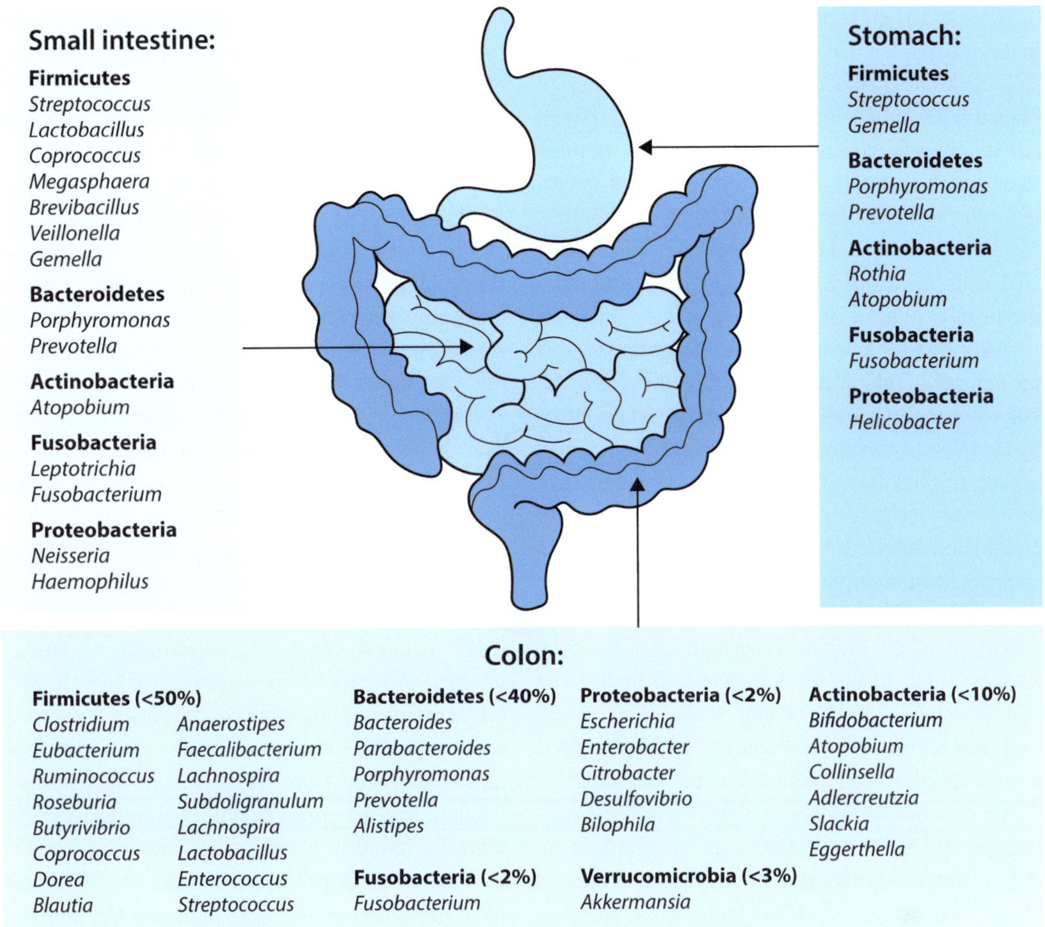

Fig. 2.1 Major bacterial genera encountered in the various sections of the gastrointestinal tract

composition was more variable among individuals and over time, when compared to the fecal microbiota. A more recent study, which compared the duodenal microbiota of 30 liver cirrhosis patients and 28 healthy subjects, reported the presence of the genera *Brevibacillus*, *Veillonella*, *Prevotella*, *Porphyromonas*, *Fusobacterium*, *Leptotrichia*, *Atopobium*, *Megasphaera*, *Gemella*, *Haemophilus*, and *Neisseria* (Chen et al. 2016). A large fraction of microorganisms in the small intestine are facultative anaerobes, but their proportion decreases from the duodenum to the terminal ileum because oxygen becomes more and more limited and the redox potential decreases. It is important to note that the number of studies investigating the small intestinal microbiota is considerably smaller than the number dealing with the colonic or fecal microbiota.

2.2.3 Colon and Feces

Owing to a relatively long mean colonic transit time of 35 h (Metcalf et al. 1987), colonic microorganisms have more time to proliferate. The absorption of water and ions during passage of colonic contents also contributes to an increase in bacterial density from cecum (10^8 ml^{-1}) to distal colon (10^{11} ml^{-1}) (Sender et al. 2016b). Clearly, the colon is the most densely populated body site. The total number of microbial cells inhabiting the human body has until previously been estimated to exceed the number of host cells by a factor of 10 (Savage 1977). A more recent publication is in conflict with this estimate. Based on thorough considerations, the total number of microbial cells harbored by a reference male person (20–30 years of age with a weight of 70 kg and a height of 170 cm) was estimated to be in the range of 4×10^{13} with more than 99% of these cells residing in the colon, while the number of human body cells is approximately 3×10^{13}, with red blood cells contributing 84% to this number (Sender et al. 2016a). Hence, the number of microbial cells in the human body is 1.3-fold higher than the number of body cells. However, if only nucleated cells are considered (0.3×10^{13}), this ratio increases to a factor of 10.

Although there is no dispute about the fact that the number of microbial species or phylotypes in the colon is quite high when compared with other microbial habitats, estimations of the number of microbial species or phylotypes present in colonic contents or feces vary significantly. While Eckburg and coworkers detected 395 bacterial phylotypes in samples from multiple colonic mucosal sites and in feces of three healthy subjects (Eckburg et al. 2005), other researchers estimated the number of bacterial species found in the human intestinal tract to be approximately 800 (Backhed et al. 2005), while 16S rRNA gene sequence analysis of 190 resected tissue samples from patients with inflammatory bowel diseases and control subjects led to the estimation of 15,000 to 36,000 species (Frank et al. 2007). These differences may in part be explained by the error-prone 16S rRNA amplicon sequencing which can result in the detection of false positives, which can be circumvented by using low-error amplicon sequencing (Faith et al. 2013). The application of this method to fecal samples from 37 healthy adults, who were sampled several times over up to 296 weeks, revealed that these individuals harbored 101 ± 27 species. These few examples show the large range of numbers of bacterial species estimated to be present in the human intestine. In this context, it is important to specify whether the phylotype or species numbers given refer to all human fecal 16S rRNA gene sequences available in database, to those obtained from a group of human subjects, or to one individual only. For example, using shotgun sequencing, Qin et al. clearly stated that a cohort of 124 European individuals harbored 1000 and 1150 bacterial species and each individual at least 160 species, which are also largely shared (Qin et al. 2010). It may be deduced that the number of species present in the intestinal tract of a given individual is rather in the range of hundreds than of thousands. While species richness in the human gut is high, the 16S rRNA gene sequences affiliate with only a small proportion of the 92 presently known bacterial phyla with cultured representatives: *Firmicutes*, *Bacteroidetes*, *Actinobacteria*, *Proteobacteria*, *Fusobacteria*, *Verrucomicrobia*, and *Cyanobacteria* (Hug et al. 2016). These phyla differ greatly in their relative contribution to bacterial cells in the microbiota. In one study involving 18 human subjects including monozygotic twins and their mothers, members of the *Firmicutes*, *Bacteroidetes*, *Actinobacteria*, and *Proteobacteria* were reported to account for approximately $\geq$95% of bacterial cells in the gut microbiota (Turnbaugh et al. 2009). While the study participants differed considerably in the relative abundance of the phyla, they displayed high similarity in the relative abundance of gene categories among the collected samples, suggesting that different taxa can exert identical functions. A number of studies reported 16S rRNA gene sequences to be indicative of the presence of members of the *Cyanobacteria* in fecal samples. However, so far no representative of this phylum has been isolated from the intestinal habitat. Instead, phototrophic *Cyanobacteria* are typically found in oceans, lakes, rivers, and ponds. However, whole-genome reconstruction of human fecal metagenomic samples indicated the presence of a new candidate phylum closely related to *Cyanobacteria*, for which the authors proposed the designation *Melainabacteria*. Genome analysis suggested that both lineages, *Cyanobacteria* and *Melainabacteria*, had a common ancestor, which was a non-photosynthetic, anaerobic, and obligately fermentative bacterium (Di Rienzi et al. 2013).

Other numerically minor components of the human gut microbiota include methanogenic Archaea and eukaryotic yeasts, whose abundance, based on the proportion of these organisms' genes in the intestinal metagenome, is in the range of 0.8% and 0.1%, respectively (Qin et al. 2010). The Archaea are represented by *Methanobrevibacter smithii*, which converts H_2 and CO_2 or formate to methane, and by *Methanosphaera stadtmanae*, which in addition is capable of reducing methyl groups to methane. The amount of methane excreted by humans in breath is variable; approximately every other person harbors detectable populations of methanogens (Florin et al. 2000). Among eukaryotic intestinal microorganisms, fungi are the most prominent members of the intestinal microbiota (Huffnagle and Noverr 2013). Intestinal fungi, referred to as gut mycobiome, have been much less studied than those of intestinal prokaryotes. Analysis of fecal samples from 98 healthy individuals led to the identification 66 fungal genera and an estimated number of 184 species (Hoffmann et al. 2013). *Saccharomyces*, *Candida*, and *Cladosporium* were the most prevalent genera, being found in 89%, 57%, and 42% of the samples, respectively. This investigation did not allow any conclusion on whether the detected fungi were resident or merely transient. Expert mycologists stated in a recent paper: "This diversity, while impressive, is illusory. If we examine gut fungi we will quickly observe a division between a small number of commonly detected species and a long tail of taxa that have been reported only once" (Suhr and Hallen-Adams 2015). A more recent study on the fecal mycobiota of healthy human vegetarians identified at least 46 distinct fungal OTUs affiliated with two phyla only: Ascomycota and Basidiomycota (Suhr et al. 2016). *Fusarium*, *Malassezia*, *Penicillium*, *Aspergillus*, and *Candida* (in decreasing order) were the most commonly detected genera. Even though fungi are considered important members of the microbial community in the human gut, knowledge about the role of these organisms in health and disease is still very limited in comparison to the bacterial members of the community. While viruses are not considered to be organisms, viruses can readily be detected in fecal samples. Sequencing of the metagenome and of DNA from virus-like particles in human fecal samples led to the identification of several thousand bacteriophage genomes referred to a virome (Minot et al. 2011). The role of the latter for the ecosystem is far from being understood.

2.3 Adaptation of Microbes to the Intestinal Environment

The term metagenome refers to the collective genome of all members of the microbiota in a given habitat, also referred to as microbiome. The proteins encoded in the human intestinal microbiome roughly reflect the functions required by intestinal bacteria to cope with the gut environment. A large proportion of the genes identified in the microbiome are related to fundamental functions required by every cell to grow and divide. Genes and proteins related to energy generation, synthesis of cellular components, and reproduction are found in every bacterial cell and are usually well conserved among bacteria. Examples include ribosomal proteins, RNA polymerase, and ATP synthase. Metabolic pathways including glycolysis and the tricarboxylic acid cycle are widely but not universally distributed among bacteria in general and intestinal bacteria in particular. For example, *Bifidobacterium* spp. degrade hexoses using the unique fructose-6-phosphate phosphoketolase pathway rather than by glycolysis. Moreover, in many strict anaerobes, the tricarbonic acid cycle is incomplete, and the remaining enzymes of the cycle preferentially fulfill anabolic functions. Physiology and the metabolic capacity of intestinal microorganisms are adapted to the conditions prevailing in the digestive tract.

2.3.1 Physicochemical Conditions in the Digestive Tract and Electron Transport

The intestine, especially the colon, is characterized by low oxygen partial pressure and highly reduced conditions with a redox potential (E_h) of approximately –215 mV, a value measured in the large intestine of pigs (Hornich and Chrastova 1981). Therefore, >99% of human fecal bacteria and also methanogenic Archaea are strict anaerobes. They cannot grow in the presence of oxygen because critical enzymes become inactivated under oxic conditions. Even though facultative aerobic or aerotolerant bacteria make up less than 1% of microbial cells in the human intestine, they play an important role. In particular facultative aerobes such as the *Enterobacteriaceae* are capable of utilizing oxygen as a terminal electron acceptor. For example, *Escherichia coli* is capable of expressing two ubiquinol-dependent oxidases, one of which has a low affinity for oxygen and a high turnover rate (cytochrome *o*-type oxidase), while the other one (cytochrome *d*-type oxidase) has a high affinity for oxygen and a low turnover rate. If no oxygen is available, *E. coli* is capable of gaining energy anaerobically either by anaerobic electron transport using nitrate, trimethylamine-*N*-oxide, dimethyl sulfoxide, or fumarate as terminal electron acceptors. If none of these is available, *E. coli* gains energy by mixed acid fermentation. The metabolism of this facultative anaerobe is regulated in such a way that the most efficient mode of energy generation is turned on while less efficient, alternative modes are turned off. Bacteria that tolerate oxygen but cannot gain energy by respiration are called aerotolerant. Lactic acid bacteria, for example, exclusively gain energy by substrate-level phosphorylation involving the conversion of carbohydrates such as lactose to lactic acid and some minor fermentation products. In spite of the scarcity of oxygen in the human intestinal tract, small amounts of oxygen become available by swallowing air during meals and by diffusion of oxygen from blood circulation to the mucosal surface. In addition to facultative aerobes, which tolerate high oxygen partial pressures, some gut bacteria previously considered strict anaerobes are capable of utilizing oxygen as long as nanomolar concentrations are not exceeded (Baughn and Malamy 2004). *Bacteroides fragilis* and other *Bacteroides* spp. harbor a high-affinity cytochrome *bd*-type oxidase allowing ATP generation by aerobic respiration. The term nanaerobes has been coined for such bacteria (Baughn and Malamy 2004). *Faecalibacterium prausnitzii* is an oxygen-sensitive organism which nevertheless grows near the mucosal surface where the oxygen partial pressure is relatively high. This is possible because *F. prausnitzii* transfers electrons to extracellular flavins and thiols, present in the gut, to reduce oxygen (Khan et al. 2012). However, even though this mechanism enables the organism to reoxidize reduced electron carriers, ATP generation is less efficient than aerobic respiration.

2.3.2 Alternative Electron Acceptors

Anaerobes capable of transferring electrons derived from oxidation reactions onto external electron acceptors do not have to use valuable intermediates such as pyruvate as electron acceptors. For example, under anoxic conditions *E. coli* and other *Enterobacteriaceae* are capable of using nitrate or trimethylamine-*N*-oxide as electron acceptor. Nitrate may be formed in the inflamed gut in which NO levels are increased because of an upregulated nitric oxide synthase (iNOS). NO reacts with reactive oxygen species to peroxynitrate ($ONOO^-$) which isomerizes to nitrate. Accordingly, nitrate produced under inflammatory conditions not only stimulates the growth of *Salmonella* but also that of *E. coli* (Lopez et al. 2012; Winter et al. 2013). Various sulfur compounds including sulfate, thiosulfate, and tetrathionate may also serve as electron acceptors. Sulfate reaching the colon may be of dietary origin, but the majority is derived from sulfated mucins by the action of bacterial sulfatases (Christl et al. 1992). Sulfate-reducing bacteria in the human gut include species of the genera *Desulfovibrio*, *Desulfobacter*, and *Desulfobulbus* (Nava et al. 2012). Since sulfate (SO_4^{2-}) is a poor electron acceptor, it first needs to be activated to adenosine-5′-phosphosulfate (APS) ($SO_4^{2-} + ATP \rightarrow APS + PP_i$). APS is subsequently reduced to sulfite (SO_3^{2-}) in a two-electron transfer reaction and thereafter to sulfide (S^{2-}) in a six-electron transfer reaction. Some bacteria including *Bilophila wadsworthia* are capable of gaining sulfite from sulfonates such as taurine (Carbonero et al. 2012). Therefore, bile acids conjugated with taurine stimulate the growth of this organism. Trimethylamine-*N*-oxide, another electron acceptor used by *Enterobacteriaceae*, is formed by the oxidation of trimethylamine as catalyzed by host monooxygenases (Bennett et al. 2013). Trimethylamine in turn originates from the bacterial degradation of choline or carnitine in the human intestine.

2.4 Metabolic Activities of the Intestinal Microbiota

Besides conferring colonization resistance on the host and protecting against pathogens, the gut microbiota primes the immune system and provides enzymes that expand the metabolic capacity of the host. A major function of the intestinal microbiota is the conversion of dietary and endogenous substrates that escape digestion including carbohydrates, proteins, secondary plant metabolites, and xenobiotics. The conversion of these substrates supports the growth of intestinal microbes by providing energy and metabolites for anabolic reactions. The gut microbiota's metabolic capacity has been proposed to rival that of the liver encompassing a wide range of reactions that reflect the low redox potential and the scarcity of oxygen available in most parts of the intestinal tract.

2.4.1 Substrates of the Intestinal Microbiota

Even though intestinal bacteria differ in how they generate energy, they share the same environment, i.e., the physicochemical conditions (pH, redox potential, temperature) at a given intestinal site, and they are dependent on the substrates coming from the diet (Table 2.1) or endogenous substrates provided by the host (Table 2.2). However, some bacterial population groups cross-feed other community members by converting these primary substrates into products that can be utilized further by bacteria that depend on these substrates; examples include lactate, formate, and hydrogen.

There is evidence that the intestinal microbiota coevolved with the respective animal host, suggesting that the bacteria resident in the digestive tract are optimally adapted to the specific environment and the nutritional habits of the host species. In humans, one of the main functions of the gut microbiota is the breakdown of dietary components that escape digestion by host enzymes. Nondigestible polysaccharides include resistant starch, plant cell wall components such as cellulose (β-[1→4] D-glucose), β-[1→3, 1→4] glucans, and pectins (α-[1→4]-linked D-galacturonic acid esterified by methanol to varying degree) as well as inulin (β-[2→1] fructose with a chain-terminating glucosyl moiety), which serves as a storage polysaccharide in various plants (Table 2.1). Some of these polymeric carbohydrates occur in conjunction with lignin and are referred to as dietary fibers, which represent the main substrate source for intestinal bacteria. However, the extent to which dietary fiber becomes utilized by intestinal bacteria depends on the physicochemical properties of the polymeric components, in particular on water solubility, water-binding capacity, and viscosity. These properties in turn depend on their chemical structure: type of carbohydrate units present, the way in which they are linked, and the degree of branching and polymerization. Dietary fiber may be categorized into structural polysaccharides originating from plant cell walls such as cellulose, pectin, xylan, mannan, and β-glucan and into storage carbohydrates such as inulin and starch.

Starch is the main carbohydrate source in a typical human diet. Humans are equipped with α-amylase, which is produced in salivary glands and pancreas and catalyzes the breakdown of starch to maltotriose and maltose. However, certain forms of starch, referred to as resistant starch, escape digestion because the glycosidic bonds cannot be accessed by human enzymes. Raw potatoes, green bananas, legumes, and unprocessed grains are typical sources of resistant starch. Moreover, heating and cooling of starch-containing foods such as potatoes and noodles may lead to the formation of retrograded starch, which represents one form of resistant starch. The intestinal microbiome has the capacity to depolymerize resistant starch and to utilize the cleavage products as sources of energy and carbon. Other indigestible dietary carbohydrates originate from whole-grain products, legumes, vegetables, fruits, and nuts. For the sake of completeness, it has to be mentioned that in addition to such complex carbohydrates, some mono- or oligosaccharides to a greater or lesser extent escape digestion in the small intestine and therefore become substrates of the intestinal microbiota. These include sugar alcohols such as sorbitol and xylitol, disaccharides such as lactulose (4-*O*-β-D-galactopyranosyl-D-fructofuranose), as well as oligosaccharides such as stachyose (α-D-galactopyranosyl-[1→6]-α-D-galactopyranosyl-[1→6]-α-D-glucopyranosyl-[1→2]-β-D-fructofuranoside), fructooligosaccharides, galactooligosaccharides,

Table 2.1 Substrates of dietary origin utilized by the intestinal microbiota

Dietary origin		
Category	Class	Composition or representative compounds
Polysaccharides	Resistant starch:	
	Amylose	α(1→4) Glucan
	Amylopectin	α(1→4), α(1→6) Glucan (branched)
	Cellulose	β(1→4) Glycan
	Pectins	α(1→4), α(1→6) Galacturonan (methylesters) α(1→4) Galactan and mixed linked arabinans
	Pentosans	β(1→4) Xylan with some arabinose and uronic side chains
	Hexosans	β(1→4) Glucomannan, β(1→3), β(1→4) Glycan (single or mixed)
	Xyloglycans	β(1→4) Glucan with β(1→6)-linked xylose side chains
	Galactomannans (Guar gum)	β(1→4) Mannans with α(1→6)-linked galactose side chains
	Chitin	β(1→4) *N*-Acetylglucosamine
	Laminarin	β(1→3) Glucans
	Inulin	β(1→2) Fructan
Oligosaccharides	Stachyose	α(1→6) Galactosyl raffinose
	Raffinose	α(1→6) Galactosyl sucrose
	Lactose	β(1→4) Galactosyl glucose
	Lactulose	β(1→4) Galactosyl fructose
Sugar alcohols	Sorbitol	
	Xylitol	
Secondary plant metabolites	Flavonoids	– Quercetin, Luteolin, Cyanidin, Daidzein
	Tannins	– Polymers of ellagic acid, gallic acid, pyrogallic acid
	Glucosinolates	– Glucoraphanin, Sinigrin, Sinalbin, Glucobrassicin
	Lignin	– Cross-linked macromolecule formed from paracoumaryl, coniferyl, and sinapyl alcohol
Proteins	Sarcoplasmatic and myofibrillar proteins	

and xylooligosaccharides. In human and animal studies, the latter three have been demonstrated to stimulate the growth of bacterial population groups considered to be beneficial and to have health-promoting properties. They are referred to as prebiotics. However, the original concept has recently been challenged and been revised (Bindels et al. 2015).

2.4.2 Breakdown of Complex Carbohydrates

Humans and other mammals lack the enzymes required for the breakdown of the large variety of complex dietary carbohydrates. However, the human gut microbiome provides a wide range of depolymerizing enzymes enabling the host to take advantage of dietary fiber by utilizing the bacterial degradation products. Metagenomic studies revealed that the human colonic microbiome in comparison to all sequenced microbial genomes is enriched in genes involved in the breakdown of dietary polysaccharides, whereas genes encoding other functions such as energy production and lipid metabolism are underrepresented. Genes representing more than 80 different glycoside hydrolase families, also referred to as carbohydrate-active enzymes (CAZymes), were identified in the distal human gut microbiome (Gill et al. 2006). High-throughput functional screens enabled the isolation of 310 clones exhibiting β-glucanase, hemicellulase, galactanase, amylase, or pectinase activities with 26 clones

Table 2.2 Substrates of the intestinal microbiota provided by the host

Host		
Category	Class	Composition or representative compounds
Glycoproteins	Mucus	Protein backbone with characteristic carbohydrates (fucose, sialic acid, *N*-acetyl-galactosamine, *N*-acetyl-galactosamine) preferentially linked to serine and threonine residues
	Hyaluronate	Polymer of glucuronic acid β(1→3) *N*-acetyl-galactosamine
	Chondroitin sulfate	Polymer of glucuronic acid β(1→3) *N*-acetyl-galactosamine, the latter being sulfated in C4 and/or C6
	Mucosal surface glycoproteins	Fucosylated proteins
Proteins	Digestive enzymes	Trypsin
		Chymotrypsin
		Leucine aminopeptidase
		Elastase
		Lipase
		Nucleic acid hydrolase
	Desquamated epithelial cells	
Bile acids	Primary bile salts	Taurocholate, glycocholate Taurochenodeoxycholate, glycochenodeoxycholate
	Secondary bile salts	Deoxycholate and its taurine or glycine conjugates Lithocholate and its taurine or glycine conjugates

being particularly efficient in the degradation of raw plant polysaccharides (Tasse et al. 2010). Seventy-three CAZymes from 35 different enzyme families were discovered, 32 of which were highly homologous to prevalent genes found in the gut microbiome of 20 human individuals. The results obtained in this study are consistent with the occurrence of horizontal gene transfer among intestinal bacteria (Tasse et al. 2010).

The first step in the utilization of nondigestible polymeric carbohydrates by intestinal bacteria requires their breakdown, which results in the formation of oligomeric and monomeric carbohydrates (Fig. 2.2). This process involves various enzyme families such as glycoside hydrolases, polysaccharide lyases, glycosyltransferases, and carbohydrate esterases (Cantarel et al. 2009; Flint et al. 2012). These CAZymes may also be categorized according to the type of substrates they act on, namely, plant cell wall components such as cellulose, pectins, xylans, β-glucans, and mannans or storage carbohydrates such as inulin and fructooligosaccharides. Another type of CAZymes acts on glycans produced by the host in the form of mucins and other glycoproteins (see further below in Sect. 2.4.3). The availability of an increasing number of draft genomes of human intestinal bacteria and metagenomic analyses has helped to identify gene clusters encoding putative CAZymes (http://www.cazy.org/). However, the catalytic features and the regulation of the majority of these proteins have not yet been investigated. It is also important to note that in addition to the enzymes catalyzing the depolymerization of glycans, auxiliary proteins are required for substrate binding, transport, and regulation. They act hand in hand and efficiently provide bacterial substrates for energy generation.

2.4.3 Degradation of Glycans by Members of the *Bacteroidetes*

Early on, it was recognized that *Bacteroides* species play an important role in the degradation of nondigestible carbohydrates in the human colon (Salyers et al. 1977a). As a representative of the genus, *Bacteroides thetaiotaomicron* has been used as a model organism to study carbohydrate

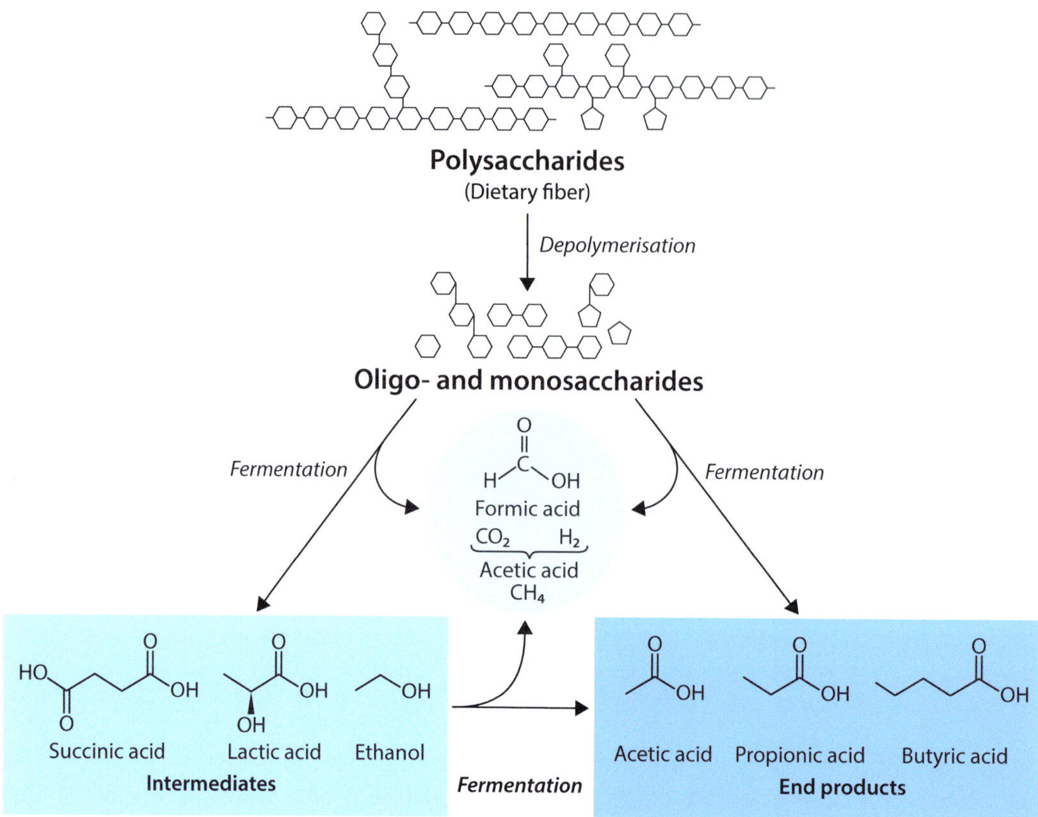

Fig. 2.2 Major steps in the breakdown of complex carbohydrates by the colonic microbiota

utilization in detail. Transcriptome analysis of *B. thetaiotaomicron* recovered from mono-associated mice fed either a polysaccharide-rich or a simple sugar diet revealed that the organism not only induces glycoside hydrolases in a diet-dependent manner but also expresses outer membrane proteins engaged in polysaccharide binding (Sonnenburg et al. 2005). *B. thetaiotaomicron* preferably utilizes simple carbohydrates or host glycans when dietary polysaccharides are not available, indicating a high degree of metabolic flexibility. Detailed studies of the starch utilization system of *B. thetaiotaomicron* led to a model widely used as a paradigm for the degradation of polysaccharides by *Bacteroides* species (Cho et al. 2001; Shipman et al. 2000). Two sets of proteins are involved in starch utilization by *B. thetaiotaomicron*. One set, which is referred to as starch utilization system (Sus), encompasses seven proteins contributing to starch degradation (SusABCDEFG) and a regulatory protein (SusR). The *mal* (maltose) regulon with the regulatory protein MalR also plays a role in starch utilization besides controlling the expression of *malA*, an α-glucosidase gene (Cho et al. 2001). Deletion of *malR* attenuates *sus* gene expression, indicating that SusR in conjunction with MalR control their expression. SusR activated by maltose, maltotriose, or longer glucose oligomers binds to the promoter region located upstream of *susB* and thereby activates the transcription of *susBCDEFG*. As *susA* is located upstream of *susB*, it has its own promoter (Reeves et al. 1997) (Fig. 2.3). SusDEF are lipoproteins anchored in the outer membrane, where they form a complex capable of binding starch (Shipman et al. 2000). SusG, which is also anchored in the outer membrane, is an α-amylase cleaving starch molecule bound to SusDEF (Shipman et al. 1999). The products formed by SusG are

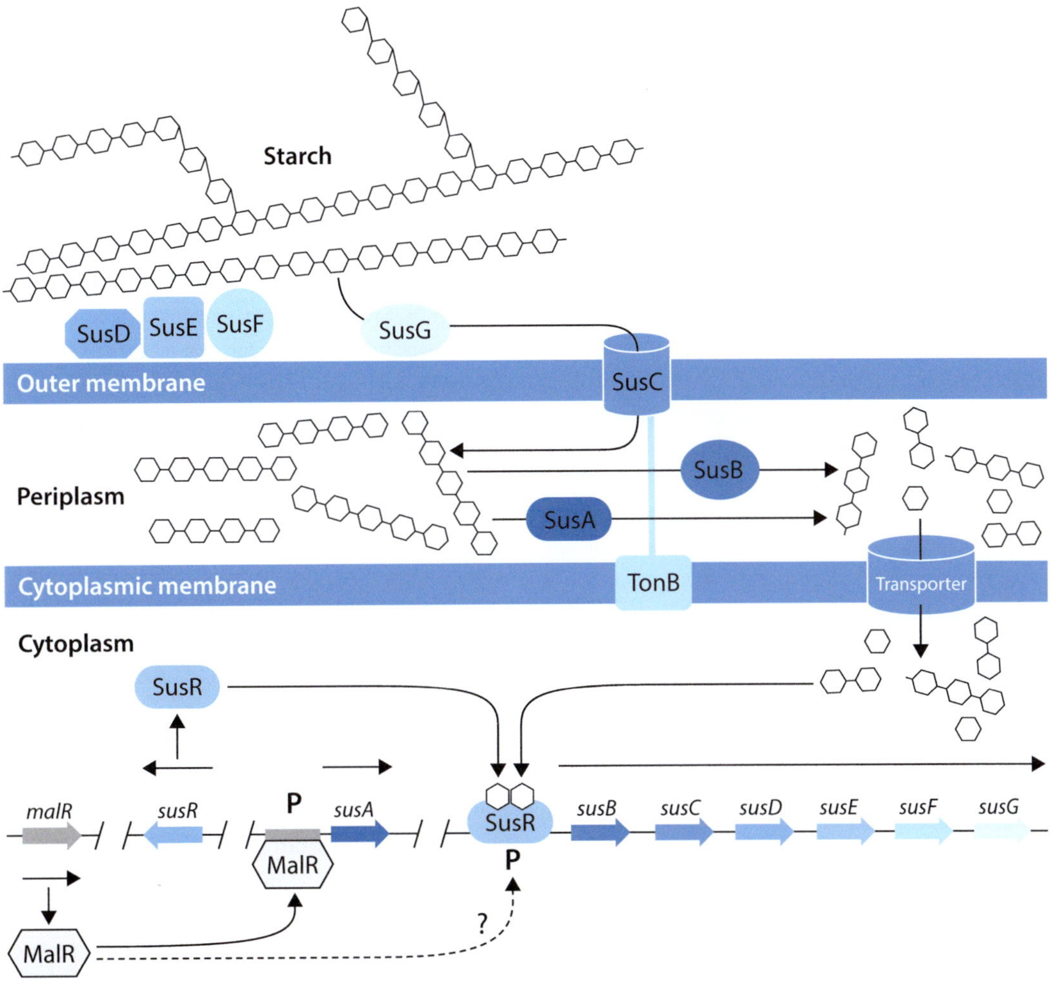

Fig. 2.3 Starch utilization by *Bacteroides thetaiotaomicron* as catalyzed by the starch utilization system (Sus). Mal refers to the maltose regulon

sufficiently small to reach the periplasm through the TonB-dependent transporter SusC located in the outer membrane. TonB is a complex in the cytoplasmic membrane of Gram-negative bacteria; it promotes the transport of various nutrients including complexed iron and cobalamin across the outer membrane (Schauer et al. 2008). Interestingly SusD is required for the binding and the subsequent uptake of starch through the TonB-dependent SusC complex. SusA and SusB, which are located in the periplasm, exhibit α-amylase and α-glucosidase activity, respectively, resulting in the release of mainly maltotriose and maltose, which are produced from the starch chunks released by SusG and transported into the periplasm. In turn, maltose and maltotriose are subsequently transported into the cytoplasm for further degradation and fermentation.

Inspection of the genome of *B. thetaiotaomicron* revealed the presence of approximately 90 *sus*-like gene loci, referred to as polysaccharide utilization loci (PULs) accounting for 18 % of this organism's genome (Martens et al. 2008; Sonnenburg et al. 2005). Approximately two thirds of these Sus-like PULs probably serve the degradation of plant-derived dietary polysaccharides, while one third plays a role in the degradation of host-derived glycans such as present in mucins and other glycoproteins. The gene clusters for proteoglycan degradation also contain genes encoding sulfatases and esterases that catalyze the removal of the corresponding functional groups from glycans. It is important to note that *sus*-like gene loci have not only been identified in the *B. thetaiotaomicron* genome but also in genomes of other *Bacteroidetes*. *Bacteroides* genomes are enriched in genes involved in glycan utilization compared with genomes from other bacterial groups of the gut microbiota. A comparison of the genomes of *B. thetaiotaomicron*, *Bacteroides fragilis*, *Bacteroides vulgatus*, and *Parabacteroides distasonis* (previously *Bacteroides distasonis*) with those of two non-gut *Bacteroidetes* species predicted that *B. thetaiotaomicron* has the largest number of glycoside hydrolases and polysaccharide lyases for the degradation of both plant and host glycans indicating that *B. thetaiotaomicron* is capable of utilizing a wide range of glycans. Based on these findings *B. thetaiotaomicron* was designated a generalist, whereas *P. distasonis*, which has the smallest genome among these species and the smallest repertoire of genes involved in carbohydrate degradation, environmental sensing, and gene regulation, has been considered a specialist (Xu et al. 2007). Only two classes of enzymes involved in glycan degradation are more abundant in *P. distasonis* than in the three *Bacteroides* species, namely, α-amylase-like proteins, *N*-acetylhexosaminidases, and polysaccharide deacetylases. The latter two are required for the degradation of epithelial glycans, which contain *O*-acetylated carbohydrates such as sialic acid. By deacetylating such glycans, *P. distasonis* not only provides substrates for itself but also for other members of the microbiota devoid of the corresponding enzymes. This is one example of cross-feeding, which is a characteristic of cooperative links among members of microbial communities in anoxic environments. The genome of *B. vulgatus* indicates that the repertoire of glycan-degrading enzymes is intermediate between that of *P. distasonis* and *B. thetaiotaomicron*, respectively, and that it possesses the most complete set of pectin-degrading enzymes including methyl esterase, acetyl esterase, and polygalacturonase (Xu et al. 2007). These analyses show that the degradation abilities of the four representative *Bacteroidetes* species overlap to some extent but they also reveal a certain degree of specialization enabling each species to occupy an ecological niche.

One well-studied example of host glycan utilization is *B. thetaiotaomicron*'s ability to cleave off fucose residues from the ileal epithelium decorated with this carbohydrate and to utilize it as a substrate (Bry et al. 1996). Fucosylation of ileal epithelium in germ-free mice starts 17 days after birth but comes to an end at approximately 28 days of age. This fucosylation program continues or restarts only when the mice are associated with *B. thetaiotaomicron* or with a complete mouse intestinal microbiota. Colonization induces fucosyltransferases in the host, which catalyze the decoration of the epithelial surface with fucose. *B. thetaiotaomicron* mutants, in which the fucose-utilization genes have been deleted, fail to induce the fucosylation program, and are also less efficient in colonizing the mouse intestine compared with wild-type mice (Bry et al. 1996). The transcriptional regulator FucR acts as a fucose sensor. FucR binds to the promoter of the fucose-utilization genes and thereby represses their transcription (Hooper et al. 1999). If fucose is present, it binds to FucR, which leads to the release of FucR from the promoter unblocking the transcription of the fucose-utilization genes. It has been proposed that the *B. thetaiotaomicron* chromosome harbors a second locus, called *csp* (control of signal production), encoding a protein that induces fucosylation in the host. In the proposed model, expression of Csp is also regulated by FucR, which in conjunction with fucose blocks csp transcription. If fucose is absent, csp transcription is no longer blocked (Hooper et al. 1999).

Studies investigating the regulation of glycan utilization by intestinal bacteria and the consequences for their growth are scarce. However, an investigation on the utilization of fructans shed some light on the principal mechanisms (Sonnenburg et al. 2010). Genetic and functional differences among *Bacteroides* species in PULs targeting various fructans were found to predict the competitiveness of these bacteria in the intestinal tract. In *B. thetaiotaomicron* regulation of fructan utilization involves a hybrid two-component signaling sensor that controls the expression of the corresponding gene cluster (Sonnenburg et al. 2010). The gene content of this fructan utilization locus differs among *Bacteroides* species and thereby determines the specificity and the type of fructans that can be utilized. For example, only *B. thetaiotaomicron*, which possesses an extracellular β-[2→6] endo-fructanase, is able to grow on levan.

The amount and type of carbohydrates accessible by community members have a major impact on the abundance of microbial population groups (McNulty et al. 2013). Therefore, the differences in fecal microbiota composition, in particular in the abundance of *Prevotella* and *Xylanibacter* species, observed between children from Burkina Faso and Italy, can be attributed to differences in the intake of fiber and starch (De Filippo et al. 2010). A recent mouse study suggests that certain community members may even get lost completely, if the substrates they require for growth are not available for a longer period of time, i.e., over generations (Sonnenburg et al. 2016).

2.4.4 Horizontal Gene Transfer of PULs

Several lines of evidence indicate that PULs genes can be horizontally transferred among members of the intestinal microbiota (Tasse et al. 2010) and between bacteria resident in the digestive tract and bacteria ingested with food (Hehemann et al. 2010). *Zobellia galactanivorans*, a marine member of the *Bacteroidetes*, is capable of degrading the sulfated polysaccharide porphyrin, which is present in marine red algae, and utilizing it as a growth substrate. There is evidence that the genes encoding the porphyranases, agarases, and accessory proteins required for porphyrin degradation have been transferred to the gut bacterium *Bacteroides plebeius*. Interestingly, this species was isolated exclusively from Japanese individuals (Kitahara et al. 2005) who consumed porphyrin-containing seaweed and thereby probably also ingested *B. plebeius* present on it. Intestinal microbiome analyses revealed that the genes encoding porphyranase and agarase are frequently found in Japanese subjects but not in North American individuals.

2.4.5 Degradation of Complex Carbohydrates by *Firmicutes*

Investigations into the breakdown of complex carbohydrates by intestinal bacteria have largely concentrated on *Bacteroidetes* even though it is clear that members of the *Firmicutes*, in particular *Ruminococcaceae* and *Lachnospiraceae*, also play an important role in polysaccharide degradation (Flint et al. 2012). For example, in human subjects the consumption of a diet rich in resistant starch led to an increase in the abundance of intestinal bacteria related to *Ruminococcus bromii* (Abell et al. 2008). *R. bromii* not only outperformed *B. thetaiotaomicron* in the degradation of resistant starch but also promoted its utilization by other starch-degrading bacteria including *Eubacterium rectale*, *Bifidobacterium adolescentis*, and *B. thetaiotaomicron* (Ze et al. 2012). Therefore, bacteria related to *R. bromii* have been proposed to play a critical role in the initial steps of resistant starch degradation. However, amylases have also been identified in *Roseburia inulinivorans* and other *Roseburia* species (Flint et al. 2012). While *Bacteroides* species employ several proteins anchored in the outer membrane to capture and cleave starch, amylolytic Gram-positive gut bacteria such as *R. inulinivorans* take advantage of amylases that are bound to the bacterial cell wall. In conjunction with a variable number of carbohydrate-binding modules on the cell surface, they effectively bind and cleave starch (Ramsay et al. 2006). Interestingly, starch also induces the formation of flagella in *R. inulinivorans*, which possibly help the organism to reach the substrates (Scott et al. 2011). Nine to 13 putative glycohydrolase genes were identified in the genomes of *Roseburia* and *E. rectale*, but their exact roles are not yet clear (Ze et al. 2012). A considerable number of intestinal *Firmicutes* play a role in the degradation of complex carbohydrates of plant origin. For example, human strains of *Ruminococcus albus*, *Roseburia intestinalis*, and *Faecalibacterium prausnitzii* utilize galactomannan, xylan, and pectin, respectively (Chassard et al. 2007; Lopez-Siles et al. 2012; Salyers et al. 1977b). A β-fructofuranosidase in *R. inulinivorans* catalyzes the depolymerization of fructans of different chain lengths. This activity is linked to an ATP-dependent sugar carrier, and expression of the corresponding genes was increased in the presence of inulin. Various mucin degraders including *Ruminococcus torques* have also been identified among the *Firmicutes* (Hoskins 1993). It may be concluded that *Firmicutes* play an important role in the degradation of both plant and host glycans, but in comparison with glycan-degrading *Bacteroidetes*, knowledge on mechanistic details is relatively limited.

2.4.6 Formation of Short-Chain Fatty Acids by Bacterial Fermentation in the Colon

The depolymerization may be considered the first step in the utilization of glycans (Fig. 2.2). Further steps include the fermentation of the cleavage products, i.e., monomeric and oligomeric saccharides. To some extent they become available to bacteria lacking enzymes for the breakdown of complex polysaccharides (cross-feeding). Bacterial population groups in the gut differ in the pathways they employ for the fermentation of these saccharides and in the respective spectrum of fermentation products. In the overall fermentation process in the colonic ecosystem, lactate, succinate, and ethanol merely represent intermediates (Fig. 2.2), which are converted further by other bacterial taxa. These activities give rise to the major end products of bacterial fermentation in the colon, namely, the short-chain fatty acids (SCFA) acetate, propionate, and butyrate, which are formed at an approximate molar ratio of 60:23:17. Total SCFA concentrations in the colon are in the range of 90–120 mM (Cummings et al. 1987). However, both the molar ratios and the concentrations of colonic SCFA are highly variable and depend on the type and amount of dietary fiber ingested. The majority (95%) of the SCFA formed by the colonic microbiota becomes absorbed (Topping and Clifton 2001). Following their absorption, acetate and butyrate may become oxidized in body tissues providing energy to the host. SCFA, preferentially butyrate, provide up to 70% of the energy required by colonic epithelial cells (Roediger 1980). Propionate may serve as a gluconeogenic substrate in the liver and acetate as a substrate for lipogenesis (Cummings 1995). However, SCFA also play various regulatory roles. For example, by inhibiting histone deacetylase, butyrate influences gene expression, which in colon cancer cells results in cell cycle arrest and activates apoptosis (Lazarova et al. 2013). Thus, SCFA play an important role in maintaining homeostasis of the colonic mucosa. SCFA have also been recognized as ligands of the G-protein-coupled receptors FFAR2 (free fatty acid receptor 2) and FFAR3, earlier referred to as GPR43 and GPR41 (Brown et al. 2003), which are expressed in ileal and colonic enteroendocrine L cells, adipocytes, and immune cells. FFAR2 activation triggers the release of leptin from adipocytes (Xiong et al. 2004) and the excretion of peptide YY (Tazoe et al. 2008) and glucagon-like peptide from enteroendocrine cells (Tolhurst et al. 2012). Since these molecules reduce appetite (Wren and Bloom 2007), they have been proposed to play a role in the control of appetite regulation. However, recent animal studies cast some doubts on such a role of intestinal FFAR (Lin et al. 2012; Tang et al. 2015).

2.4.7 Bacteria Involved in SCFA Formation

The majority of intestinal bacteria produce acetate, some in larger and others in smaller quantities. Homofermentative and heterofermentative lactic acid bacteria produce no or only small amounts of acetate. The third major group of lactate-producing intestinal bacteria, bifidobacteria, produces considerable amounts of acetate in addition to lactate (2 glucose → 2 lactate + 3 acetate). Major propionate producers include various *Bacteroides* species, *Veillonella parvula*, *Dialister succinatiphilus*, *Phascolarctobacterium succinatutens*, *Selenomonas ruminantium*, and *Megasphaera elsdenii*, many of which also produce succinate as a by-product or an intermediate that can be taken up again to be converted to propionate. Three different propionate formation pathways have been identified: the methylmalonyl-CoA pathway, the acrylate pathway, and the propanediol pathway. Primers targeting genes characteristic of either pathway were used to test the presence of the corresponding genes in representative human gut species (Reichardt et al. 2014). The majority of bacterial species were found to use the methylmalonyl-CoA pathway.

Butyrate-producing human fecal bacteria capable of utilizing lactate include strains related to *Eubacterium hallii*, *Anaerostipes caccae*, and distant relatives of *Clostridium indolis* (Duncan et al. 2004). Butyrate-forming human colonic bacteria include *R. intestinalis*, *E. rectale*, and *F. prausnitzii*, all of which convert glucose but not lactate to butyrate (Duncan et al. 2002).

Lactate is a major source for both propionate and butyrate. Some organisms are capable of producing both butyrate and propionate depending on the substrate. In the presence of acetate, *Coprococcus catus* converts lactate mainly to propionate; in contrast, mainly butyrate is formed from fructose with net consumption of acetate (Reichardt et al. 2014). Experiments in fecal slurries moreover suggest that the pH is a major factor influencing the conversion of lactate in the ecosystem and the propionate/butyrate ratio (Belenguer et al. 2007).

2.4.8 Utilization of Hydrogen and Formate

In addition to SCFA, the gut microbiota produces formic acid and the gases H_2 and CO_2 (Fig. 2.2), which are partly excreted and partly utilized. H_2 is produced by various bacterial population groups in the colon. For example, *Enterobacteriaceae* such as *E. coli* produce H_2 and CO_2 from formate catalyzed by formate-hydrogen lyase, while strict anaerobes such as *Clostridium* species and other *Firmicutes* may release H_2 in the course of pyruvate oxidation as catalyzed by pyruvate: ferredoxin oxidoreductase. The reduced ferredoxin produced in this reaction is used by hydrogenase for the reduction of two protons to produce H_2. The latter and formate play a role in methanogenesis and acetogenesis. Approximately 50% of humans excrete methane. The intestinal archaeon *Methanobrevibacter smithii* produces CH_4 from H_2 and CO_2 ($CO_2 + 4H_2 \rightarrow CH_4 + 2H_2O$) or formate ($4HCOOH \rightarrow 3CO_2 + CH_4 + 2H_2O$). Homoacetogenic bacteria such as *Blautia hydrogenotrophica* or *Blautia producta* may also take advantage of H_2 and CO_2 and/or formate (Bernalier et al. 1996; Liu et al. 2008). Using the Wood-Ljungdahl pathway, these bacteria catalyze the formation of acetate ($2CO_2 + 4H_2 \rightarrow CH_3\text{–}COOH + 2H_2O$ or $4HCOOH \rightarrow CH_3\text{–}COOH + 2CO_2 + 2H_2O$) (Ragsdale 2006). H_2 is an important product of anaerobic fermentation as it enables anaerobes to reoxidize electron carriers without the need to use intermediates such as pyruvate as electron acceptors, enabling a higher ATP gain per hexose metabolized compared to other fermentations.

2.4.9 Utilization of Proteins by Intestinal Bacteria

Based on measurements of nitrogen digestibility in ileostomized patients, it has been estimated that 5–10% of the ingested dietary protein reaches the colon (Darragh and Hodgkinson 2000). Endogenous proteins, in particular digestive host enzymes, are an additional source of protein for bacterial fermentation in the colon. The proteolytic activity of host proteases decreases from proximal to distal colon indicating that they become inactivated due to bacterial proteolysis (Gibson et al. 1989). Depending on diet, the total daily amount of protein entering the colon has been estimated to vary between 6 and 18 g (Cummings and Macfarlane 1991; Yao et al. 2016). The majority of colonic bacteria prefer carbohydrates over proteins for energy generation, but many species are also capable of utilizing proteins, peptides, and amino acids, alternatively or simultaneously. Protein can be used for energy generation and for biosynthetic purposes as they deliver both carbon and nitrogen for microbial growth. Protein degradation in the colon occurs in several steps and involves different bacteria. The first step in protein utilization is proteolysis, which results in the release of peptides and amino acids (Fig. 2.4). Strains of *B. fragilis* and *B. vulgatus* use cell-bound proteases for the cleavage of proteins, while strains of the genera *Clostridium*, *Propionibacterium*, and *Bacillus* take advantage of extracellular proteases. Fecal *Streptococcus* and *Staphylococcus* isolates possess both forms of proteases (Macfarlane et al. 1986). The majority of intestinal bacteria prefer ammonia over amino acids as a source of nitrogen. However, organisms such as bifidobacteria and clostridia, which utilize oligopeptides for anabolic purposes, retain only certain amino acids of absorbed peptides and excrete the remaining ones (Hespell and Smith 1983).

The pathways used by colonic bacteria for amino acid fermentation in the colon may differ among bacterial species. However, the initial steps of bacterial amino acid breakdown are restricted to only a few: oxidative deamination, reductive deamination, simple deamination, and decarboxylation resulting in α-oxo-acids, carboxylic acids, enoates, and (poly)amines, respectively (Fig. 2.4). Intestinal bacteria such as *Clostridium sticklandii* utilize pairs of different amino acids: one of the amino acids is subjected to oxidative deamination, while the other one undergoes reductive deamination, a process referred to as Stickland reaction. As a consequence of the release of ammonia, the luminal pH increases from proximal to distal colon indicating protein becoming more important as an energy source as carbohydrates become increasingly exhausted (Macfarlane et al. 1992). Amino acid fermentation results in the formation of SCFA, formate, H_2,

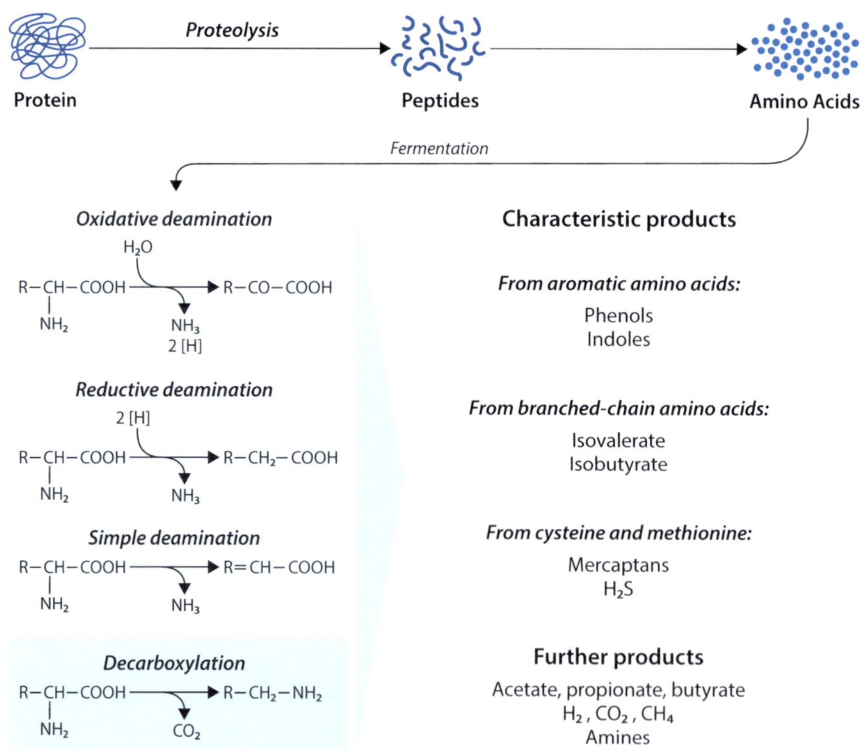

Fig. 2.4 Bacterial proteolysis in the large intestine

and CO_2 with the latter three serving as potential substrates for methanogenesis and homoacetogenesis, i.e., the same products also formed from carbohydrates. Products characteristic of amino acid fermentation include branched-chain fatty acids such as isovalerate and isobutyrate, which are formed from branched-chain amino acids; phenols and indoles, which stem from the fermentation of the aromatic amino acids tyrosine, phenylalanine, and tryptophan; as well as amines. Colonic fermentation of the sulfur-containing amino acids cysteine and methionine gives rise to hydrogen sulfide (H_2S) and mercaptans. A high protein intake is accompanied by increased bacterial sulfide generation in the human colon (Magee et al. 2000).

2.4.10 Conversion of Bile Acids

Even though bile acids are not a major source of energy for the gut microbiota, they may suppress the growth of bacteria sensitive to these detergents. They play a role in fat digestion and are ligands for the farnesoid X receptor (FXR), the liver X receptor (LXR), and the G-protein-coupled receptor TGR5; and they undergo conversion by the gut microbiota (Jones et al. 2008). They are synthesized in the liver from cholesterol, conjugated with glycine or taurine, and subsequently stored in the gall bladder. When required, they are excreted into the gut to solubilize dietary fat and support the formation of micelles. Many intestinal bacteria are able to deconjugate the primary bile acids to the corresponding unconjugated forms. Many intestinal bacteria including *Clostridium perfringens, Lactobacillus plantarum, Lactobacillus johnsonii, Bifidobacterium bifidum, B. longum,* and *B. adolescentis* harbor bile salt hydrolase genes (Ridlon et al. 2006). Metagenomic analysis revealed that bile salt hydrolases are enriched in the human gut microbiome and that they are present in all major bacterial divisions as well as in the archaeal methanogens *M. smithii* and *M. stadtmanae*, suggesting that this activity is relevant to survival in the mammalian gastrointestinal tract (Jones et al. 2008). The unconjugated bile acids may be further converted by bile acid dehydroxylases and hydroxysteroid dehydrogenases. Intestinal bacteria harboring enzymes involved in bile acid conversion include *E. lenta, C. perfringens, B. producta, B. fragilis, B. thetaiotaomicron, E. coli, Clostridium absonum, Clostridium sordellii, Clostridium innocuum, Clostridium scindens, Clostridium hylemonae, Clostridium bifermentans, Clostridium limosum, Clostridium leptum,* and *Clostridium paraputrificum* (Ridlon et al. 2006). A recent study revealed intestinal bacteria such as *Ruminococcus gnavus* favor the growth of *Bacteroides* spp. owing to their ability to detoxify deoxycholic acid by converting it to the corresponding 3-β-hydroxy bile acid epimer isodeoxycholic acid (Devlin and Fischbach 2015). Dehydroxylation of bile acids, which occurs in a position-specific and stereo-selective way, was studied in *Clostridium scindens* in detail. The eight genes required for bile acid dehydroxylation are organized in the *bai* (bile acid-inducible) operon (*baiBCDEAFGHI*) encoding the 27 kDa 3α-hydroxysteroid dehydrogenase (BaiA), the 58 kDa bile acid CoA ligase (BaiB), the 70 kDa 3-dehydro-4-chenodeoxycholic acid/cholic acid steroid oxidoreductase (BaiCD), the 72 kDa 3-dehydro-4-ursodeoxycholic acid/7-epi cholic acid steroid oxidoreductase (BaiH), the 19.5 kDa 7α-dehydratase (BaiE), a hypothetical 22 kDa 7-β-dehydratase (BaiI), the 47.5 kDa bile acid CoA hydrolase and a hypothetical bile acid CoA transferase (BaiF), and the 50 kDa transmembrane protein (BaiG), which catalyzes H^+-dependent bile acid transport (Ridlon et al. 2006). It is important to note that none of the bacterial enzymes acting on the bile acids cleaves the steroid ring structure.

Potential benefits for intestinal bacteria may arise from the utilization of the glycine or taurine moieties of bile acids as carbon or nitrogen source. Utilization of taurine by *B. longum* as a nitrogen source is in accordance with the finding that the bile salt hydrolase gene is co-transcribed with the glutamine synthetase adenylyltransferase gene (*glnE*), which is part of the nitrogen regulation cascade (Tanaka et al. 2000). There is evidence that taurine stimulates growth of the colitogenic *B. wadsworthia* by providing the electron acceptor for sulfite reduction (see Sect. 2.3.2).

2.4.11 Conversion of Secondary Plant Metabolites

In addition to carbohydrates and protein, diet may contain non-nutritive secondary plant metabolites such as polyphenols, which are found in grains, fruits, and vegetables. Polyphenols such as lignans and flavonoids have been reported to exert beneficial health effects. Therefore, their uptake, bioavailability, and biological activities in humans have been studied (Clavel et al. 2006b; Hollman and Katan 1999). The chemical structure of polyphenols and the composition of the intestinal microbiota affect the fate of these compounds in the digestive tract. Polyphenols are usually glycosylated, and, depending on the extent of absorption, they pass into the colon where they undergo conversion by intestinal bacteria. For example, intestinal bacteria convert the lignan secoisolariciresinol diglucoside (SDG) to enterolactone in several steps (Axelson et al. 1982). Various *Bacteroides* and *Clostridium* spp. are capable of deglycosylating SDG, but *Clostridium saccharogumia* turned out to be the most effective species of the strains tested (Clavel et al. 2006a, 2007). *Butyribacterium methylotrophicum*, *B. producta*, *Eubacterium callanderi*, and *E. limosum* are capable of catalyzing the second step, namely, the *O*-demethylation of matairesinol, whereas *C. scindens* and *E. lenta* dehydroxlyate the *O*-demethylated matairesinol to enterodiol (third step). The last step in this pathway, the conversion of enterodiol to enterolactone, is catalyzed by *Lactonifactor longoviformis* (Clavel et al. 2007). A defined consortium of four species, each catalyzing one of the four reactions, converts SDG to enterodiol and enterolactone. Gnotobiotic rats associated with this community excreted the two metabolites in urine and feces when fed a flaxseed diet, which is rich in SDG (Woting et al. 2010). The ability of humans to convert SDG to enterodiol and enterolactone is widely distributed among humans with women tending to harbor higher concentrations of enterolactone-producing intestinal bacteria (Clavel et al. 2005).

Isoflavones represent a subgroup of flavonoids, which like the lignans have been implicated in preventive effects against hormone-related cancers and cardiovascular disease as well as in alleviating menopausal symptoms. These effects have mainly been attributed to one of its bacterial transformation products, namely, equol, which undergoes urinary excretion (Setchell and Clerici 2010). Isoflavones mainly occur in their glycosylated form. Interestingly, some intestinal bacteria, e.g., the *Lachnospiraceae* strain CG19-1 and *Eubacterium cellulosolvens*, are capable of cleaving the more stable *C*-glycosides in addition to the more common *O*-glycosides (Braune and Blaut 2012; Braune et al. 2016). Daidzein and genistein are major isoflavones present in soy. They may undergo metabolization by intestinal bacteria such as *Adlercreutzia equolifaciens* and *Slackia isoflavoniconvertens*, which have been identified as equol formers (Maruo et al. 2008; Matthies et al. 2009). In *S. isoflavoniconvertens,* daidzein induces the expression of eight genes involved in its conversion, three of which were found to encode daidzein reductase, dihydrodaidzein reductase, and tetrahydrodaidzein reductase, respectively (Schroder et al. 2013). Heterologous expression of the latter two resulted in the reduction of dihydrodaidzein to equol. In the meantime, quite a number of intestinal bacteria metabolizing flavonoids including isoflavones and enzymes involved have been identified (Braune and Blaut 2016).

Following the consumption of soy, 33–50% of healthy subjects excreted equol, while 80–90% excreted the biologically less active *O*-desmethylangolensin, in addition or alternatively (Atkinson et al. 2004). One organism shown to convert daidzein to *O*-desmethylangolensin is *Eubacterium ramulus* (Schoefer et al. 2002). Which pathway dominates depends on the composition of the gut microbiota. These examples highlight the fact that intestinal bacteria convert a wide range of non-nutritive metabolites and that there may be alternative conversion pathways resulting in several intermediates and end products.

2.4.12 Core and Variable Microbiome and/or Microbiota

The intestinal tract environment favors bacteria that have the capacity to grow therein. Since bacteria in the gut share the same environment, it can be surmised that they have certain gene functions in common in addition to the so-called housekeeping functions required by all living cells. Therefore, it is not surprising that the microbiomes from different individuals share a high proportion of gene functions, including genes that encode enzymes required to degrade dietary fiber and host glycoproteins. These have been referred to as core microbiome because they represent metabolic activities that are found in every subject, while others are only present in some individuals but absent from others (Turnbaugh et al. 2009). The latter category, which has been termed the variable microbiome, includes methanogenesis, oxalate degradation, conversion of isoflavones, and the utilization of porphyrin from marine red algae (see Sect. 2.4.5) (Atkinson et al. 2004; Hehemann et al. 2010; Kumar et al. 2002; Wolin and Miller 1983). The idea of a core microbiome, which encompasses key functions and is shared by each individual, is a useful concept as it reflects distinct environmental influences in a given habitat. However, the value of defining a core microbiota (not microbiome!), which encompasses key species shared among humans, is questionable because their relative abundance is highly variable (Turnbaugh et al. 2009).

2.5 Conclusions

The microbial communities inhabiting the human intestinal tract play a major role in the breakdown of dietary components that cannot be utilized by the host and in the conversion of host metabolites. By expanding the metabolic capacity of the host and interacting with the host immune system, the gut microbiota profoundly affects host physiology. Even though metagenomics, transcriptomics, and proteomics have increased our knowledge about important functions of the intestinal microbiota, there still is a gap in our understanding of the exact molecular mechanisms underlying microbe-microbe or host-microbe interactions. Therefore, identifying the exact role of members of the gut microbiota and their competitive or cooperative links is of major importance. Considerable differences in microbiota composition among human individuals and populations impede the elucidation of the mechanisms underlying the role of intestinal bacteria in various diseases. Moreover, it often is not possible to find out whether disease-related changes in the gut microbiota are cause or consequence of the disease. We should strive to identify microorganisms that play critical roles in physiological and pathophysiological processes, with the ultimate goal to identify the bacterial molecules involved and their targets in the host.

▶ **Controversy**

Analysis of human fecal samples revealed differences in the relative abundance of key taxa. Three robust patterns or clusters, referred to as enterotypes, were identified (Arumugam et al. 2011). The three enterotypes are characterized by differences in the relative abundance of *Bacteroides*, *Prevotella*, or *Ruminococcus*. While the enterotypes did not correlate with gender, age, or body weight, long-term dietary habits were reported to influence the enterotype (Wu et al. 2011). The underlying concept was subsequently extended to other mammalian hosts including the mouse (Wang et al. 2014). Quite a number of scientists have found the enterotype concept appealing and conducted similar analyses. Several studies confirmed that the fecal microbiota of human subjects can be categorized into two of the three proposed enterotypes, namely, *Bacteroides* and *Prevotella*, while the *Ruminococcus* enterotype was usually not found. In a Korean study, 72% of the fecal samples collected from Korean monozygotic twin pairs belonged to the same enterotype and 2 years later the affiliation with either enterotype was still the same for 80% of the individuals (Lim et al. 2014). However, this also means that 20% of the subjects changed from one enterotype to the other during this time. Interestingly, a more recent study in Taiwanese adults, which was based on the analysis of 181 fecal samples, identified *Escherichia* rather than *Ruminococcus* as a representative genus of a third enterotype (Liang et al. 2017).

A recent in vitro study revealed that the fermentation of various fermentable polysaccharides is determined by the enterotype of the fecal donor (Chen et al. 2017). The inoculum with a dominance of *Prevotella* versus *Bacteroides* was dominated by fiber-fermenting bacteria. In agreement with this observation in children in Burkina Faso, who consumed a diet rich in fermentable fiber, *Prevotella* accounted for 53% of intestinal bacteria but were absent in age-matched Italian children (De Filippo et al. 2010). In spite of these interesting observations, the value of the enterotype concept has been challenged for several reasons (Knights et al. 2014): Dominant genera including *Ruminococcus* and *Bacteroides* are highly variable among individuals belonging to the same enterotype. Available datasets propose continuous abundance in gradients rather than discrete clusters despite the fact that the absence of *Prevotella* as observed in Italian children (De Filippo et al. 2010) inevitably results in discrete clustering. Most importantly, the affiliation of human subjects with a given enterotype may vary over time arguing against the notion that enterotypes are discrete states (Knights et al. 2014). Indeed, a large cohort study revealed that the microbiota profile of the majority of the study subjects corresponded to one of the enterotypes, while others had intermediate profiles, impeding a clear assignment to an enterotype (Huse et al. 2012). Based on these findings, it may be concluded that the enterotype concept, however appealing it may appear, does not really promote a better understanding of the gut microbiome.

History

The ubiquitous existence of microorganisms only became evident with the invention of the light microscope by Antonie van Leeuwenhoek (1632–1723) and the studies of Louis Pasteur (1822–1895) and Robert Koch (1843–1910), which revealed that bacteria catalyze reactions and may cause infections. Theodor Escherich (1857–1911) was one of the first researchers, who became interested in the role of intestinal bacteria in the digestive tract, in particular of infants. He isolated a fecal bacterium that later on was named after him, namely, *Escherichia coli*. For a long period of time, bacteria have primarily been perceived as culprits even though most bacteria known to date are nonpathogenic. This might explain that the intestinal fermentation was considered a detrimental process. The British surgeon William Arbuthnot-Lane (1856–1943) removed the colon from some of his patients because he assumed that the colonic fermentation led to an "autointoxication." For a long time, the investigation of the intestinal microbiota was impeded, because adequate methods for handling strict anaerobes were not yet available. So the exploration of the ecosystem only started after pioneers such as Robert E. Hungate (1906–2004) and Sydney M. Finegold (born 1921) developed methods for the isolation and handling of strict anaerobes (Hungate 1969; Sugihara et al. 1974). These early researchers and others laid the foundation for the field. They isolated and described a considerable number of bacterial species and tried for the first time to link the gut microbiota to health and disease (Finegold et al. 1975). The development of cultivation-independent methods (Amann et al. 1995) facilitated and accelerated the characterization of various microbial habitats, including that of human fecal samples (Suau et al. 1999). Steady improvements in sequencing methods and simultaneously decreasing costs have made metagenome sequencing a readily available tool, enabling researchers to assess all microbial gene sequences in an ecosystem and, in conjunction with transcriptomics and metabolomics, to characterize the metabolic potential of intestinal microbial communities (Dumas et al. 2006; Gill et al. 2006; Jiang et al. 2016). However, the prediction of gene functions depends on the correct annotation, which in turn is largely based on work of scientists who previously isolated bacteria and characterized their enzymes and genes. Isolation of new community members and identification of new gene functions can be tedious and usually receive little appreciation by the scientific community. This may be the reason that this important work is presently neglected even though a large proportion of gene functions have not yet been identified and gene functions predicted and annotated based on sequence similarity have not been experimentally verified. Therefore, it is still necessary to isolate as many bacteria as possible and to study their genes and enzymes.

Highlights

- The digestive tract of human and animals is colonized by microbial communities encompassing bacteria, archaea, and fungi, referred to as gastrointestinal microbiota.
- The gastrointestinal microbiota has coevolved with the host, and its members are well adapted to the different sections of the digestive tract, which differ in the physicochemical conditions and the availability of substrates. The composition of the intestinal microbiota at species level is highly variable among humans.
- The majority of intestinal microorganisms are strictly anaerobic bacteria, which gain energy by fermenting dietary fiber and endogenous substrates mainly to short-chain fatty acids, carbon dioxide, molecular hydrogen, and methane.
- The collective genome (metagenome) of all members of the intestinal microbiota represents the intestinal microbiome which encodes the functions of all community members. While a large proportion of functions are shared among the microbiomes of human individuals, some activities are only observed in certain human populations.
- Bacterial groups in the intestinal tract interact with each other and with the host. The majority of interactions are of mutualistic or commensal character. However, the mechanisms underlying such interactions are incompletely understood.

References

Abell, G. C., Cooke, C. M., Bennett, C. N., Conlon, M. A., & McOrist, A. L. (2008). Phylotypes related to *Ruminococcus bromii* are abundant in the large bowel of humans and increase in response to a diet high in resistant starch. *FEMS Microbiology Ecology, 66*, 505–515.

Amann, R. I., Ludwig, W., & Schleifer, K.-H. (1995). Phylogenetic identification and *in situ* detection of individual microbial cells without cultivation. *Microbiological Reviews, 59*, 143–169.

Arumugam, M., Raes, J., Pelletier, E., Le Paslier, D., Yamada, T., Mende, D. R., Fernandes, G. R., Tap, J., Bruls, T., Batto, J. M., et al. (2011). Enterotypes of the human gut microbiome. *Nature, 473*, 174–180.

Atkinson, C., Berman, S., Humbert, O., & Lampe, J. W. (2004). In vitro incubation of human feces with daidzein and antibiotics suggests interindividual differences in the bacteria responsible for equol production. *The Journal of Nutrition, 134*, 596–599.

Axelson, M., Sjövall, J., Gustafsson, B. E., & Setchell, K. D. R. (1982). Origin of lignans in mammals and identification of a precursor from plants. *Nature, 298*, 659–660.

Backhed, F., Ley, R. E., Sonnenburg, J. L., Peterson, D. A., & Gordon, J. I. (2005). Host-bacterial mutualism in the human intestine. *Science, 307*, 1915–1920.

Baughn, A. D., & Malamy, M. H. (2004). The strict anaerobe *Bacteroides fragilis* grows in and benefits from nanomolar concentrations of oxygen. *Nature, 427*, 441–444.

Belenguer, A., Duncan, S. H., Holtrop, G., Anderson, S. E., Lobley, G. E., & Flint, H. J. (2007). Impact of pH on lactate formation and utilization by human fecal microbial communities. *Applied and Environmental Microbiology, 73*, 6526–6533.

Bennett, B. J., de Aguiar Vallim, T. Q., Wang, Z., Shih, D. M., Meng, Y., Gregory, J., Allayee, H., Lee, R., Graham, M., Crooke, R., et al. (2013). Trimethylamine-N-oxide, a metabolite associated with atherosclerosis, exhibits complex genetic and dietary regulation. *Cell Metabolism, 17*, 49–60.

Bernalier, A., Willems, A., Leclerc, M., Rochet, V., & Collins, M. D. (1996). *Ruminococcus hydrogenotrophicus* sp. nov., a new H_2/CO_2-utilizing acetogenic bacterium isolated from human feces. *Archives of Microbiology, 166*, 176–183.

Bik, E. M., Eckburg, P. B., Gill, S. R., Nelson, K. E., Purdom, E. A., Francois, F., Perez-Perez, G., Blaser, M. J., & Relman, D. A. (2006). Molecular analysis of the bacterial microbiota in the human stomach. *Proceedings of the National Academy of Sciences of the United States of America, 103*, 732–737.

Bindels, L. B., Delzenne, N. M., Cani, P. D., & Walter, J. (2015). Towards a more comprehensive concept for prebiotics. *Nature Reviews. Gastroenterology & Hepatology, 12*, 303–310.

Booijink, C. C., El-Aidy, S., Rajilic-Stojanovic, M., Heilig, H. G., Troost, F. J., Smidt, H., Kleerebezem, M., De Vos, W. M., & Zoetendal, E. G. (2010). High temporal and inter-individual variation detected in the human ileal microbiota. *Environmental Microbiology, 12*, 3213–3227.

Braune, A., & Blaut, M. (2012). Intestinal bacterium *Eubacterium cellulosolvens* deglycosylates flavonoid C- and O-glucosides. *Applied and Environmental Microbiology, 78*, 8151–8153.

Braune, A., & Blaut, M. (2016). Bacterial species involved in the conversion of dietary flavonoids in the human gut. *Gut Microbes, 7*, 216–234.

Braune, A., Engst, W., & Blaut, M. (2016). Identification and functional expression of genes encoding flavonoid O- and C-glycosidases in intestinal bacteria. *Environmental Microbiology, 18*, 2117–2129.

Brown, A. J., Goldsworthy, S. M., Barnes, A. A., Eilert, M. M., Tcheang, L., Daniels, D., Muir, A. I., Wigglesworth, M. J., Kinghorn, I., Fraser, N. J., et al. (2003). The orphan G protein-coupled receptors GPR41 and GPR43 are activated by propionate and other short chain carboxylic acids. *The Journal of Biological Chemistry, 278*, 11312–11319.

Bry, L., Falk, P. G., Midtvedt, T., & Gordon, J. I. (1996). A model of host-microbial interactions in an open mammalian ecosystem. *Science, 273*, 1380–1383.

Cantarel, B. L., Coutinho, P. M., Rancurel, C., Bernard, T., Lombard, V., & Henrissat, B. (2009). The Carbohydrate-Active EnZymes database (CAZy): An expert resource for glycogenomics. *Nucleic Acids Research, 37*, D233–D238.

Carbonero, F., Benefiel, A. C., Alizadeh-Ghamsari, A. H., & Gaskins, H. R. (2012). Microbial pathways in colonic sulfur metabolism and links with health and disease. *Frontiers in Physiology, 3*, 448.

Chassard, C., Goumy, V., Leclerc, M., Del'homme, C., & Bernalier-Donadille, A. (2007). Characterization of the xylan-degrading microbial community from human faeces. *FEMS Microbiology Ecology, 61*, 121–131.

Chen, Y., Ji, F., Guo, J., Shi, D., Fang, D., & Li, L. (2016). Dysbiosis of small intestinal microbiota in liver cirrhosis and its association with etiology. *Scientific Reports, 6*, 34055.

Chen, T., Long, W., Zhang, C., Liu, S., Zhao, L., & Hamaker, B. R. (2017). Fiber-utilizing capacity varies in Prevotella- versus Bacteroides-dominated gut microbiota. *Scientific Reports, 7*, 2594.

Cho, K. H., Cho, D., Wang, G. R., & Salyers, A. A. (2001). New regulatory gene that contributes to control of *Bacteroides thetaiotaomicron* starch utilization genes. *Journal of Bacteriology, 183*, 7198–7205.

Christl, S. U., Gibson, G. R., & Cummings, J. H. (1992). Role of dietary sulphate in the regulation of methanogenesis in the human large intestine. *Gut, 33*, 1234–1238.

Clavel, T., Henderson, G., Alpert, C. A., Philippe, C., Rigottier-Gois, L., Dore, J., & Blaut, M. (2005). Intestinal bacterial communities that produce active estrogen-like compounds enterodiol and enterolactone in humans. *Applied and Environmental Microbiology, 71*, 6077–6085.

Clavel, T., Borrmann, D., Braune, A., Dore, J., & Blaut, M. (2006a). Occurrence and activity of human intestinal bacteria involved in the conversion of dietary lignans. *Anaerobe, 12*, 140–147.

Clavel, T., Dore, J., & Blaut, M. (2006b). Bioavailability of lignans in human subjects. *Nutrition Research Reviews, 19*, 187–196.

Clavel, T., Lippman, R., Gavini, F., Dore, J., & Blaut, M. (2007). *Clostridium saccharogumia* sp. nov. and *Lactonifactor longoviformis* gen. nov., sp. nov., two novel human faecal bacteria involved in the conversion of the dietary phytoestrogen secoisolariciresinol diglucoside. *Systematic and Applied Microbiology, 30*, 16–26.

Cummings, J. H. (1995). Short chain fatty acids. In G. R. Gibson & G. T. Macfarlane (Eds.), *Human colonic bacteria: Role in nutrition, physiology and pathology* (pp. 101–130). CRC Press: Boca Raton.

Cummings, J. H., & Macfarlane, G. T. (1991). The control and consequences of bacterial fermentation in the human colon. *The Journal of Applied Bacteriology, 70*, 443–459.

Cummings, J. H., Pomare, E. W., Branch, W. J., Naylor, C. P., & Macfarlane, G. T. (1987). Short chain fatty acids in human large intestine, portal, hepatic and venous blood. *Gut, 28*, 1221–1227.

Darragh, A. J., & Hodgkinson, S. M. (2000). Quantifying the digestibility of dietary protein. *The Journal of Nutrition, 130*, 1850S–1856S.

De Filippo, C., Cavalieri, D., Di Paola, M., Ramazzotti, M., Poullet, J. B., Massart, S., Collini, S., Pieraccini, G., & Lionetti, P. (2010). Impact of diet in shaping gut microbiota revealed by a comparative study in children from Europe and rural Africa. *Proceedings of the National Academy of Sciences of the United States of America, 107*, 14691–14696.

Devlin, A. S., & Fischbach, M. A. (2015). A biosynthetic pathway for a prominent class of microbiota-derived bile acids. *Nature Chemical Biology, 11*, 685–690.

Di Rienzi, S. C., Sharon, I., Wrighton, K. C., Koren, O., Hug, L. A., Thomas, B. C., Goodrich, J. K., Bell, J. T., Spector, T. D., Banfield, J. F., & Ley, R. E. (2013). The human gut and groundwater harbor non-photosynthetic bacteria belonging to a new candidate phylum sibling to Cyanobacteria. *eLife, 2*, e01102.

Dumas, M. E., Barton, R. H., Toye, A., Cloarec, O., Blancher, C., Rothwell, A., Fearnside, J., Tatoud, R., Blanc, V., Lindon, J. C., et al. (2006). Metabolic profiling reveals a contribution of gut microbiota to fatty liver phenotype in insulin-resistant mice. *Proceedings of the National Academy of Sciences of the United States of America, 103*, 12511–12516.

Duncan, S. H., Barcenilla, A., Stewart, C. S., Pryde, S. E., & Flint, H. J. (2002). Acetate utilization and butyryl coenzyme A (CoA): Acetate-CoA transferase in butyrate-producing bacteria from the human large intestine. *Applied and Environmental Microbiology, 68*, 5186–5190.

Duncan, S. H., Louis, P., & Flint, H. J. (2004). Lactate-utilizing bacteria, isolated from human feces, that produce butyrate as a major fermentation product. *Applied and Environmental Microbiology, 70*, 5810–5817.

Eckburg, P. B., Bik, E. M., Bernstein, C. N., Purdom, E., Dethlefsen, L., Sargent, M., Gill, S. R., Nelson, K. E., & Relman, D. A. (2005). Diversity of the human intestinal microbial flora. *Science, 308*, 1635–1638.

Faith, J. J., Guruge, J. L., Charbonneau, M., Subramanian, S., Seedorf, H., Goodman, A. L., Clemente, J. C., Knight, R., Heath, A. C., Leibel, R. L., et al. (2013). The long-term stability of the human gut microbiota. *Science, 341*, 1237439.

Finegold, S. M., Flora, D. J., Attebery, H. R., & Sutter, V. L. (1975). Fecal bacteriology of colonic polyp patients and control patients. *Cancer Research, 35*, 3407–3417.

Finegold, S. M., Sutter, V. L., & Mathisen, G. E. (1983). Normal indigenous intestinal flora. In D. J. Hentges (Ed.), *Human intestinal microflora in health and disease* (pp. 3–31). Academic Press: New York/London.

Flint, H. J., Scott, K. P., Duncan, S. H., Louis, P., & Forano, E. (2012). Microbial degradation of complex carbohydrates in the gut. *Gut Microbes, 3*, 289–306.

Florin, T. H., Zhu, G., Kirk, K. M., & Martin, N. G. (2000). Shared and unique environmental factors determine the ecology of methanogens in humans and rats. *The American Journal of Gastroenterology, 95*, 2872–2879.

Frank, D. N., St Amand, A. L., Feldman, R. A., Boedeker, E. C., Harpaz, N., & Pace, N. R. (2007). Molecular-phylogenetic characterization of microbial community imbalances in human inflammatory bowel diseases. *Proceedings of the National Academy of Sciences of the United States of America, 104*, 13780–13785.

Gibson, S. A., McFarlan, C., Hay, S., & MacFarlane, G. T. (1989). Significance of microflora in proteolysis in the colon. *Applied and Environmental Microbiology, 55*, 679–683.

Gill, S. R., Pop, M., Deboy, R. T., Eckburg, P. B., Turnbaugh, P. J., Samuel, B. S., Gordon, J. I., Relman, D. A., Fraser-Liggett, C. M., & Nelson, K. E. (2006). Metagenomic analysis of the human distal gut microbiome. *Science, 312*, 1355–1359.

Hayashi, H., Takahashi, R., Nishi, T., Sakamoto, M., & Benno, Y. (2005). Molecular analysis of jejunal, ileal, caecal and recto-sigmoidal human colonic microbiota using 16S rRNA gene libraries and terminal restriction fragment length polymorphism. *Journal of Medical Microbiology, 54*, 1093–1101.

Hehemann, J. H., Correc, G., Barbeyron, T., Helbert, W., Czjzek, M., & Michel, G. (2010). Transfer of carbohydrate-active enzymes from marine bacteria to Japanese gut microbiota. *Nature, 464*, 908–912.

Hespell, R. B., & Smith, C. J. (1983). Utilization of nitrogen sources by gastrointestinal tract bacteria. In D. J. Hentges (Ed.), *Human intestinal microflora in health and disease* (p. 21). New York, London: Academic Press.

Hoffmann, C., Dollive, S., Grunberg, S., Chen, J., Li, H., Wu, G. D., Lewis, J. D., & Bushman, F. D. (2013). Archaea and fungi of the human gut microbiome: Correlations with diet and bacterial residents. *PLoS One, 8*, e66019.

Hollman, P. C., & Katan, M. B. (1999). Dietary flavonoids: Intake, health effects and bioavailability. *Food and Chemical Toxicology, 37*, 937–942.

Hooper, L. V., Xu, J., Falk, P. G., Midtvedt, T., & Gordon, J. I. (1999). A molecular sensor that allows a gut commensal to control its nutrient foundation in a competitive ecosystem. *Proceedings of the National Academy of Sciences of the United States of America, 96*, 9833–9838.

Hornich, M., & Chrastova, V. (1981). The redox potential of the large intestine in swine in relation to swine dysentery. *Veterinary Medicine (Praha), 26*, 593–598.

Hoskins, L. C. (1993). Mucin degradation in the human gastrointestinal tract and its significance to enteric microbial ecology. *European Journal of Gastroenterology & Hepatology, 5*, 205–213.

Huffnagle, G. B., & Noverr, M. C. (2013). The emerging world of the fungal microbiome. *Trends in Microbiology, 21*, 334–341.

Hug, L. A., Baker, B. J., Anantharaman, K., Brown, C. T., Probst, A. J., Castelle, C. J., Butterfield, C. N., Hernsdorf, A. W., Amano, Y., Ise, K., et al. (2016).

A new view of the tree of life. *Nature Microbiology, 1*, 16048.

Hungate, R. E. (1969). A roll tube method for cultivation of strict anaerobes. In J. R. Norris & D. W. Ribbons (Eds.), *Methods in microbiology* (p. 117). Academic Press: New York.

Huse, S. M., Ye, Y., Zhou, Y., & Fodor, A. A. (2012). A core human microbiome as viewed through 16S rRNA sequence clusters. *PLoS One, 7*, e34242.

Jiang, Y., Xiong, X., Danska, J., & Parkinson, J. (2016). Metatranscriptomic analysis of diverse microbial communities reveals core metabolic pathways and microbiome-specific functionality. *Microbiome, 4*, 2.

Jones, B. V., Begley, M., Hill, C., Gahan, C. G., & Marchesi, J. R. (2008). Functional and comparative metagenomic analysis of bile salt hydrolase activity in the human gut microbiome. *Proceedings of the National Academy of Sciences of the United States of America, 105*, 13580–13585.

Khan, M. T., Duncan, S. H., Stams, A. J., van Dijl, J. M., Flint, H. J., & Harmsen, H. J. (2012). The gut anaerobe *Faecalibacterium prausnitzii* uses an extracellular electron shuttle to grow at oxic-anoxic interphases. *The ISME Journal, 6*, 1578–1585.

Kitahara, M., Sakamoto, M., Ike, M., Sakata, S., & Benno, Y. (2005). *Bacteroides plebeius* sp. nov. and *Bacteroides coprocola* sp. nov., isolated from human faeces. *International Journal of Systematic and Evolutionary Microbiology, 55*, 2143–2147.

Knights, D., Ward, T. L., McKinlay, C. E., Miller, H., Gonzalez, A., McDonald, D., & Knight, R. (2014). Rethinking "enterotypes". *Cell Host & Microbe, 16*, 433–437.

Kumar, R., Mukherjee, M., Bhandari, M., Kumar, A., Sidhu, H., & Mittal, R. D. (2002). Role of *Oxalobacter formigenes* in calcium oxalate stone disease: A study from North India. *European Urology, 41*, 318–322.

Lazarova, D. L., Chiaro, C., Wong, T., Drago, E., Rainey, A., O'Malley, S., & Bordonaro, M. (2013). CBP activity mediates effects of the histone deacetylase inhibitor butyrate on WNT activity and apoptosis in colon cancer cells. *Journal of Cancer, 4*, 481–490.

Ley, R. E., Hamady, M., Lozupone, C., Turnbaugh, P. J., Ramey, R. R., Bircher, J. S., Schlegel, M. L., Tucker, T. A., Schrenzel, M. D., Knight, R., & Gordon, J. I. (2008). Evolution of mammals and their gut microbes. *Science, 320*, 1647–1651.

Liang, C., Tseng, H. C., Chen, H. M., Wang, W. C., Chiu, C. M., Chang, J. Y., Lu, K. Y., Weng, S. L., Chang, T. H., Chang, C. H., et al. (2017). Diversity and enterotype in gut bacterial community of adults in Taiwan. *BMC Genomics, 18*, 932.

Lim, M. Y., Rho, M., Song, Y. M., Lee, K., Sung, J., & Ko, G. (2014). Stability of gut enterotypes in Korean monozygotic twins and their association with biomarkers and diet. *Scientific Reports, 4*, 7348.

Lin, H. V., Frassetto, A., Kowalik, E. J., Jr., Nawrocki, A. R., Lu, M. M., Kosinski, J. R., Hubert, J. A., Szeto, D., Yao, X., Forrest, G., & Marsh, D. J. (2012). Butyrate and propionate protect against diet-induced obesity and regulate gut hormones via free fatty acid receptor 3-independent mechanisms. *PLoS One, 7*, e35240.

Liu, C., Finegold, S. M., Song, Y., & Lawson, P. A. (2008). Reclassification of *Clostridium coccoides, Ruminococcus hansenii, Ruminococcus hydrogenotrophicus, Ruminococcus luti, Ruminococcus productus* and *Ruminococcus schinkii* as *Blautia coccoides* gen. nov., comb. nov., *Blautia hansenii* comb. nov., *Blautia hydrogenotrophica* comb. nov., *Blautia luti* comb. nov., *Blautia producta* comb. nov., *Blautia schinkii* comb. nov. and description of *Blautia wexlerae* sp. nov., isolated from human faeces. *International Journal of Systematic and Evolutionary Microbiology, 58*, 1896–1902.

Lopez, C. A., Winter, S. E., Rivera-Chavez, F., Xavier, M. N., Poon, V., Nuccio, S. P., Tsolis, R. M., & Baumler, A. J. (2012). Phage-mediated acquisition of a type III secreted effector protein boosts growth of salmonella by nitrate respiration. *MBio, 3*, e00143–e00112.

Lopez-Siles, M., Khan, T. M., Duncan, S. H., Harmsen, H. J., Garcia-Gil, L. J., & Flint, H. J. (2012). Cultured representatives of two major phylogroups of human colonic *Faecalibacterium prausnitzii* can utilize pectin, uronic acids, and host-derived substrates for growth. *Applied and Environmental Microbiology, 78*, 420–428.

Macfarlane, G. T., Cummings, J. H., & Allison, C. (1986). Protein degradation by human intestinal bacteria. *Journal of General Microbiology, 132*, 1647–1656.

Macfarlane, G. T., Gibson, G. R., & Cummings, J. H. (1992). Comparison of fermentation reactions in different regions of the human colon. *The Journal of Applied Bacteriology, 72*, 57–64.

Magee, E. A., Richardson, C. J., Hughes, R., & Cummings, J. H. (2000). Contribution of dietary protein to sulfide production in the large intestine: An in vitro and a controlled feeding study in humans. *The American Journal of Clinical Nutrition, 72*, 1488–1494.

Martens, E. C., Chiang, H. C., & Gordon, J. I. (2008). Mucosal glycan foraging enhances fitness and transmission of a saccharolytic human gut bacterial symbiont. *Cell Host & Microbe, 4*, 447–457.

Maruo, T., Sakamoto, M., Ito, C., Toda, T., & Benno, Y. (2008). *Adlercreutzia equolifaciens* gen. nov., sp. nov., an equol-producing bacterium isolated from human faeces, and emended description of the genus *Eggerthella*. *International Journal of Systematic and Evolutionary Microbiology, 58*, 1221–1227.

Matthies, A., Blaut, M., & Braune, A. (2009). Isolation of a human intestinal bacterium capable of daidzein and genistein conversion. *Applied and Environmental Microbiology, 75*, 1740–1744.

McNulty, N. P., Wu, M., Erickson, A. R., Pan, C., Erickson, B. K., Martens, E. C., Pudlo, N. A., Muegge, B. D., Henrissat, B., Hettich, R. L., & Gordon, J. I. (2013). Effects of diet on resource utilization by a model human gut microbiota containing Bacteroides cellulosilyticus WH2, a symbiont with an extensive glycobiome. *PLoS Biology, 11*, e1001637.

Metcalf, A. M., Phillips, S. F., Zinsmeister, A. R., MacCarty, R. L., Beart, R. W., & Wolff, B. G. (1987). Simplified assessment of segmental colonic transit. *Gastroenterology, 92*, 40–47.

Minot, S., Sinha, R., Chen, J., Li, H., Keilbaugh, S. A., Wu, G. D., Lewis, J. D., & Bushman, F. D. (2011). The human gut virome: Inter-individual variation and dynamic response to diet. *Genome Research, 21*, 1616–1625.

Nava, G. M., Carbonero, F., Croix, J. A., Greenberg, E., & Gaskins, H. R. (2012). Abundance and diversity of mucosa-associated hydrogenotrophic microbes in the healthy human colon. *The ISME Journal, 6*, 57–70.

Qin, J., Li, R., Raes, J., Arumugam, M., Burgdorf, K. S., Manichanh, C., Nielsen, T., Pons, N., Levenez, F., Yamada, T., et al. (2010). A human gut microbial gene catalogue established by metagenomic sequencing. *Nature, 464*, 59–65.

Ragsdale, S. W. (2006). Metals and their scaffolds to promote difficult enzymatic reactions. *Chemical Reviews, 106*, 3317–3337.

Ramsay, A. G., Scott, K. P., Martin, J. C., Rincon, M. T., & Flint, H. J. (2006). Cell-associated alpha-amylases of butyrate-producing Firmicute bacteria from the human colon. *Microbiology, 152*, 3281–3290.

Reeves, A. R., Wang, G. R., & Salyers, A. A. (1997). Characterization of four outer membrane proteins that play a role in utilization of starch by Bacteroides thetaiotaomicron. *Journal of Bacteriology, 179*, 643–649.

Reichardt, N., Duncan, S. H., Young, P., Belenguer, A., McWilliam Leitch, C., Scott, K. P., Flint, H. J., & Louis, P. (2014). Phylogenetic distribution of three pathways for propionate production within the human gut microbiota. *The ISME Journal, 8*, 1323–1335.

Ridlon, J. M., Kang, D. J., & Hylemon, P. B. (2006). Bile salt biotransformations by human intestinal bacteria. *Journal of Lipid Research, 47*, 241–259.

Roediger, W. E. (1980). Role of anaerobic bacteria in the metabolic welfare of the colonic mucosa in man. *Gut, 21*, 793–798.

Salyers, A. A., Palmer, J. K., & Wilkins, T. D. (1977a). Laminarinase (beta-glucanase) activity in *Bacteroides* from the human colon. *Applied and Environmental Microbiology, 33*, 1118–1124.

Salyers, A. A., West, S. E., Vercellotti, J. R., & Wilkins, T. D. (1977b). Fermentation of mucins and plant polysaccharides by anaerobic bacteria from the human colon. *Applied and Environmental Microbiology, 34*, 529–533.

Savage, D. C. (1977). Microbial ecology of the gastrointestinal tract. *Annual Review of Microbiology, 31*, 107–133.

Schauer, K., Rodionov, D. A., & de Reuse, H. (2008). New substrates for TonB-dependent transport: Do we only see the 'tip of the iceberg'? *Trends in Biochemical Sciences, 33*, 330–338.

Schoefer, L., Mohan, R., Braune, A., Birringer, M., & Blaut, M. (2002). Anaerobic C-ring cleavage of genistein and daidzein by *Eubacterium ramulus*. *FEMS Mircrobiology Letters, 208*, 197–202.

Schroder, C., Matthies, A., Engst, W., Blaut, M., & Braune, A. (2013). Identification and expression of genes involved in the conversion of daidzein and genistein by the equol-forming bacterium Slackia isoflavoniconvertens. *Applied and Environmental Microbiology, 79*, 3494–3502.

Scott, K. P., Martin, J. C., Chassard, C., Clerget, M., Potrykus, J., Campbell, G., Mayer, C. D., Young, P., Rucklidge, G., Ramsay, A. G., & Flint, H. J. (2011). Substrate-driven gene expression in *Roseburia inulinivorans*: Importance of inducible enzymes in the utilization of inulin and starch. *Proceedings of the National Academy of Sciences of the United States of America, 108(Suppl 1)*, 4672–4679.

Sender, R., Fuchs, S., & Milo, R. (2016a). Are we really vastly outnumbered? Revisiting the ratio of bacterial to host cells in humans. *Cell, 164*, 337–340.

Sender, R., Fuchs, S., & Milo, R. (2016b). Revised estimates for the number of human and bacteria cells in the body. *PLoS Biology, 14*, e1002533.

Setchell, K. D., & Clerici, C. (2010). Equol: History, chemistry, and formation. *The Journal of Nutrition, 140*, 1355S–1362S.

Shipman, J. A., Cho, K. H., Siegel, H. A., & Salyers, A. A. (1999). Physiological characterization of SusG, an outer membrane protein essential for starch utilization by Bacteroides thetaiotaomicron. *Journal of Bacteriology, 181*, 7206–7211.

Shipman, J. A., Berleman, J. E., & Salyers, A. A. (2000). Characterization of four outer membrane proteins involved in binding starch to the cell surface of *Bacteroides thetaiotaomicron*. *Journal of Bacteriology, 182*, 5365–5372.

Sonnenburg, J. L., Xu, J., Leip, D. D., Chen, C. H., Westover, B. P., Weatherford, J., Buhler, J. D., & Gordon, J. I. (2005). Glycan foraging in vivo by an intestine-adapted bacterial symbiont. *Science, 307*, 1955–1959.

Sonnenburg, E. D., Zheng, H., Joglekar, P., Higginbottom, S. K., Firbank, S. J., Bolam, D. N., & Sonnenburg, J. L. (2010). Specificity of polysaccharide use in intestinal bacteroides species determines diet-induced microbiota alterations. *Cell, 141*, 1241–1252.

Sonnenburg, E. D., Smits, S. A., Tikhonov, M., Higginbottom, S. K., Wingreen, N. S., & Sonnenburg, J. L. (2016). Diet-induced extinctions in the gut microbiota compound over generations. *Nature, 529*, 212–215.

Suau, A., Bonnet, R., Sutren, M., Godon, J. J., Gibson, G. R., Collins, M. D., & Doré, J. (1999). Direct analysis of genes encoding 16S rRNA from complex communities reveals many novel molecular species within the human gut. *Applied and Environmental Microbiology, 65*, 4799–4807.

Sugihara, P. T., Sutter, V. L., Attebery, H. R., Bricknell, K. S., & Finegold, S. M. (1974). Isolation of *Acidaminococcus fermentans* and *Megasphaera elsdenii* from normal human feces. *Applied Microbiology, 27*, 274–275.

Suhr, M. J., & Hallen-Adams, H. E. (2015). The human gut mycobiome: Pitfalls and potentials – a mycologist's perspective. *Mycologia, 107*, 1057–1073.

Suhr, M. J., Banjara, N., & Hallen-Adams, H. E. (2016). Sequence-based methods for detecting and evaluating the human gut mycobiome. *Letters in Applied Microbiology, 62*, 209–215.

Tanaka, H., Hashiba, H., Kok, J., & Mierau, I. (2000). Bile salt hydrolase of *Bifidobacterium longum*-biochemical and genetic characterization. *Applied and Environmental Microbiology, 66*, 2502–2512.

Tang, C., Ahmed, K., Gille, A., Lu, S., Grone, H. J., Tunaru, S., & Offermanns, S. (2015). Loss of FFA2 and FFA3 increases insulin secretion and improves glucose tolerance in type 2 diabetes. *Nature Medicine, 21*, 173–177.

Tasse, L., Bercovici, J., Pizzut-Serin, S., Robe, P., Tap, J., Klopp, C., Cantarel, B. L., Coutinho, P. M., Henrissat, B., Leclerc, M., et al. (2010). Functional metagenomics to mine the human gut microbiome for dietary fiber catabolic enzymes. *Genome Research, 20*, 1605–1612.

Tazoe, H., Otomo, Y., Kaji, I., Tanaka, R., Karaki, S. I., & Kuwahara, A. (2008). Roles of short-chain fatty acids receptors, GPR41 and GPR43 on colonic functions. *Journal of Physiology and Pharmacology, 59(Suppl 2)*, 251–262.

Tolhurst, G., Heffron, H., Lam, Y. S., Parker, H. E., Habib, A. M., Diakogiannaki, E., Cameron, J., Grosse, J., Reimann, F., & Gribble, F. M. (2012). Short-chain fatty acids stimulate glucagon-like peptide-1 secretion via the G-protein-coupled receptor FFAR2. *Diabetes, 61*, 364–371.

Topping, D. L., & Clifton, P. M. (2001). Short-chain fatty acids and human colonic function: Roles of resistant starch and nonstarch polysaccharides. *Physiological Reviews, 81*, 1031–1064.

Turnbaugh, P. J., Hamady, M., Yatsunenko, T., Cantarel, B. L., Duncan, A., Ley, R. E., Sogin, M. L., Jones, W. J., Roe, B. A., Affourtit, J. P., et al. (2009). A core gut microbiome in obese and lean twins. *Nature, 457*, 480–484.

Wang, J., Linnenbrink, M., Kunzel, S., Fernandes, R., Nadeau, M. J., Rosenstiel, P., & Baines, J. F. (2014). Dietary history contributes to enterotype-like clustering and functional metagenomic content in the intestinal microbiome of wild mice. *Proceedings of the National Academy of Sciences of the United States of America, 111*, E2703–E2710.

Winter, S. E., Winter, M. G., Xavier, M. N., Thiennimitr, P., Poon, V., Keestra, A. M., Laughlin, R. C., Gomez, G., Wu, J., Lawhon, S. D., et al. (2013). Host-derived nitrate boosts growth of *E. coli* in the inflamed gut. *Science, 339*, 708–711.

Wolin, M. J., & Miller, T. L. (1983). Carbohydrate fermentation. In D. J. Hentges (Ed.), *Human intestinal microflora in health and disease* (p. 19). New York, London: Academic Press.

Worsoe, J., Fynne, L., Gregersen, T., Schlageter, V., Christensen, L. A., Dahlerup, J. F., Rijkhoff, N. J., Laurberg, S., & Krogh, K. (2011). Gastric transit and small intestinal transit time and motility assessed by a magnet tracking system. *BMC Gastroenterology, 11*, 145.

Woting, A., Clavel, T., Loh, G., & Blaut, M. (2010). Bacterial transformation of dietary lignans in gnotobiotic rats. *FEMS Microbiology Ecology, 72*, 507–514.

Wren, A. M., & Bloom, S. R. (2007). Gut hormones and appetite control. *Gastroenterology, 132*, 2116–2130.

Wu, G. D., Chen, J., Hoffmann, C., Bittinger, K., Chen, Y. Y., Keilbaugh, S. A., Bewtra, M., Knights, D., Walters, W. A., Knight, R., et al. (2011). Linking long-term dietary patterns with gut microbial enterotypes. *Science, 334*, 105–108.

Xiong, Y., Miyamoto, N., Shibata, K., Valasek, M. A., Motoike, T., Kedzierski, R. M., & Yanagisawa, M. (2004). Short-chain fatty acids stimulate leptin production in adipocytes through the G protein-coupled receptor GPR41. *Proceedings of the National Academy of Sciences of the United States of America, 101*, 1045–1050.

Xu, J., Mahowald, M. A., Ley, R. E., Lozupone, C. A., Hamady, M., Martens, E. C., Henrissat, B., Coutinho, P. M., Minx, P., Latreille, P., et al. (2007). Evolution of symbiotic bacteria in the distal human intestine. *PLoS Biology, 5*, e156.

Yao, C. K., Muir, J. G., & Gibson, P. R. (2016). Review article: Insights into colonic protein fermentation, its modulation and potential health implications. *Alimentary Pharmacology & Therapeutics, 43*, 181–196.

Ze, X., Duncan, S. H., Louis, P., & Flint, H. J. (2012). *Ruminococcus bromii* is a keystone species for the degradation of resistant starch in the human colon. *The ISME Journal, 6*, 1535–1543.

Zoetendal, E. G., Raes, J., van den Bogert, B., Arumugam, M., Booijink, C. C., Troost, F. J., Bork, P., Wels, M., de Vos, W. M., & Kleerebezem, M. (2012). The human small intestinal microbiota is driven by rapid uptake and conversion of simple carbohydrates. *The ISME Journal, 6*, 1415–1426.

3 Microbiome and Early Life

Mathias Hornef

Abstract

Under physiological conditions, the fetus is protected from exposure to viable microorganisms. With rupture of membranes and passage through the birth canal, the neonate becomes exposed to bacteria that colonize maternal body surfaces and the environment. These bacteria start to establish the enteric microbiota initially characterized by low bacterial diversity and high interindividual variation. This makes the neonatal and early infant microbiota particularly vulnerable to exogenous interference. On the other hand, the low colonization resistance allows the interventional modification of the early microbiota by oral administration of beneficial bacteria. With time, additional bacterial species colonize the intestine and increase the diversity of the microbiota composition. In combination with the influence of genetic determinants and environmental factors, this ultimately leads to the generation of a mature and highly diverse enteric microbiota that remains relatively stable throughout life. In this chapter we will discuss the establishment of the enteric microbiota after birth and the current understanding of its influence on disease susceptibility. We will also address interventional strategies that particularly during early life might be able to modify the microbiota and improve long-term health.

M. Hornef (✉)
Institute of Medical Microbiology, RWTH Aachen University, Aachen, Germany
e-mail: mhornef@ukaachen.de

3.1 Establishment of the Enteric Microbiota After Birth

Prior to birth, the fetal body is enclosed by the amniotic membranes that physically separate it from the uterus cavity and preserve a sterile environment. The first exposure to bacteria starts with rupture of membranes that in most cases happens few hours or even less prior to birth. It releases the amniotic fluid that surrounds the fetus and facilitates access of bacteria that ascend from the vaginal mucosa into the uterus cavity. Although this process is well known, the number and kind of bacteria that reach the embryo at this early time point have not been specifically studied. With increasing uterine contraction and opening of the cervix, the fetal body enters the birth canal. The majority of neonates pass the birth canal with the occiput ahead (so-called dorsoanterior occiput presentation) which means that the nasal and oral openings move closely along the densely colonized posterior mucosal surfaces of the birth canal. This intense physical contact facilitates efficient transfer of bacteria from the mother to the newborn. Consistently, the early enteric microbiota of vaginally delivered neonates resembles the maternal vaginal microbiota (Dominguez-Bello et al. 2010).

D. Haller (ed.), *The Gut Microbiome in Health and Disease*,
https://doi.org/10.1007/978-3-319-90545-7_3

Following delivery, the close contact between mother and child facilitates contact to bacteria colonizing the maternal skin. Soon the neonate starts suckling at the mother's breasts, which intensifies the contact to bacteria colonizing the maternal skin. Regular breast-feeding maintains the transfer and ingestion of resident bacteria. Clearly, breast milk is not sterile but contains significant numbers of bacteria (Martin et al. 2003; Toscano et al. 2017). Therefore, some groups have proposed the existence of an entero-mammary pathway, i.e., the direct transfer of viable bacteria from the mother's intestine to the breast gland that would reinforce transfer of gut bacteria to the neonate (Rodriguez 2014). However, compelling evidence or any mechanistic description of such a transfer of viable bacteria through the systemic circulation has not yet been provided. Alternatively, the opening of the milk ducts that drain the breast milk to the skin surface and an ascending colonization could explain the presence of bacteria in breast milk samples. Consistently, bacteria found in breast milk such as *Streptococci* and *Staphylococci* represent known constituents of the skin microbiota. Breast-feeding undoubtedly exerts a major influence on the establishment of the enteric microbiota, but this may not primarily be due to its role as source for new bacterial species but rather its important prebiotic and nutritional influence. Breast milk constituents exert a major influence on bacterial growth within the intestinal lumen. For example, bifidobacteria, prominent member of the neonate microbiota, metabolize milk oligosaccharides and increase in abundance in breast-fed neonates (Klaassens et al. 2009; Penders et al. 2006; Ward et al. 2007; Yu et al. 2013).

Subsequently, also bacteria from other sources such as siblings, pets, and livestock or the direct environment may become part of the developing microbiota (Azad et al. 2013; Hill et al. 2017; van Best et al. 2015). At this early period during the development of the microbiota, the bacterial diversity is low, and thus newly incoming bacteria do not have to compete for nutrients and space. Thus, in the neonate, exposure to most types of bacteria may result in successful colonization. The term used to describe this situation is low "colonization resistance" (Brugiroux et al. 2016; Olsan et al. 2017; Stecher et al. 2013). We will see later that this characteristic of a microbiota differs dramatically between neonates and adults. On the one hand, low colonization resistance allows rapid establishment of an increasingly complex microbiota that fulfills the metabolic and immunological requirements of the developing individual. On the other hand, it also renders the neonate particularly susceptible to infection.

Thus, the early neonatal enteric microbiota is of low diversity and characterized by low colonization resistance. In addition, it is highly individual. Microbiota analysis of 14 neonates in 1 study revealed major differences between all individuals but two—a pair of twin babies (Palmer et al. 2007). These results again stress the importance of the mother as source of bacteria for the initial microbiota but also indicate that the early process of colonization mostly depends on genetic and exogenous influences (Hill et al. 2017). Bacteria that first colonize the neonate intestine have also been named pioneer bacteria and are thought to play a particularly important role (Karlsson et al. 2011). However, this role has not been defined. Another aspect of the microbiota besides its composition is its density, i.e., the number of bacteria at a given anatomical site. The ability of many bacteria to multiply rapidly is well known. Fast-proliferating bacteria can divide every 20 min under nutrient-rich conditions theoretically resulting in more than 5×10^{10} bacteria from just one bacterium within 24 h. The high lactose and fat concentration in milk promote the growth of many bacteria. It might therefore not be surprising that although the bacterial diversity of the enteric microbiota requires years to develop (see below), the density reaches its threshold values already within few days (Palmer et al. 2007; Wesemann et al. 2013).

Over the time following birth, additional bacterial species become members of the infant enteric microbiota. This process is supported by changes in the environmental conditions within the intestinal lumen. For example, oxygen is present in the postnatal intestine limiting early colonization to oxygen-resistant so-called facultatively anaerobic bacteria. Their presence and metabolism soon deplete the oxygen allowing subsequent colonization by strictly anaerobic bacteria such as *Bifidobacterium*, *Clostridia*, and other *Firmicutes* (Reinhardt et al. 2009; van Best et al. 2015). Also, changes in the diet such as the cessation of breast-feeding and the introduction of solid food offer new nutritional niches. This leads to a shift toward *Bacteroidetes* and *Firmicutes* (Backhed et al. 2015). The succession of bacteria that colonize and expand as part of the enteric microbiota thus follows a certain order, although most regulatory mechanisms and circuits are probably still unknown. This process step-by-step increases the diversity of the enteric microbiota reaching values similar to adult individuals approximately at the age of 2–3 years (Backhed et al. 2015; Yatsunenko et al. 2012). At the same time, the individual variation has decreased generating something like a typical "mature" adult composition of the enteric microbiota (Yatsunenko et al. 2012). The underlying mechanisms are not entirely clear, but probably this development results from the overall more similar metabolic and environmental conditions in the adult intestine (Figs. 3.1 and 3.2).

3.2 Factors Influencing the Infant Microbiota

A number of factors influence the establishment and maturation of the enteric microbiota after birth. Most of them alter the timing or quality of important exogenous determinants. While determinants present throughout human evolution such as vaginal delivery, breast-feeding, and low personal and food hygiene might appear beneficial in respect to their effect on the establishment of a diverse and competitive enteric microbiota, we must be aware that they are associated with a significant risk for infection with pathogenic microorganisms (Table 3.1).

As described above, the newborn under physiological conditions first encounters bacteria from the maternal vaginal mucosa during passage of the birth canal. As expected, this exposure markedly changes when the neonate is born by cesarean section (Dominguez-Bello et al. 2010). Cesarean section represents a surgical intervention introduced to rescue the child's and mother's life in the event of acute bleeding, infection, impaired fetal blood circulation, or anatomical obstruction of the birth canal. It requires anesthesia of the mother and antibiotic prophylaxis to prevent infection. Several studies have shown that the early enteric microbiota of neonates delivered by cesarean section differs significantly during the first months after birth (Backhed et al. 2015; Dogra et al. 2015; Hesla et al. 2014; Jakobsson et al. 2014; Penders et al. 2013). It exhibits a reduced

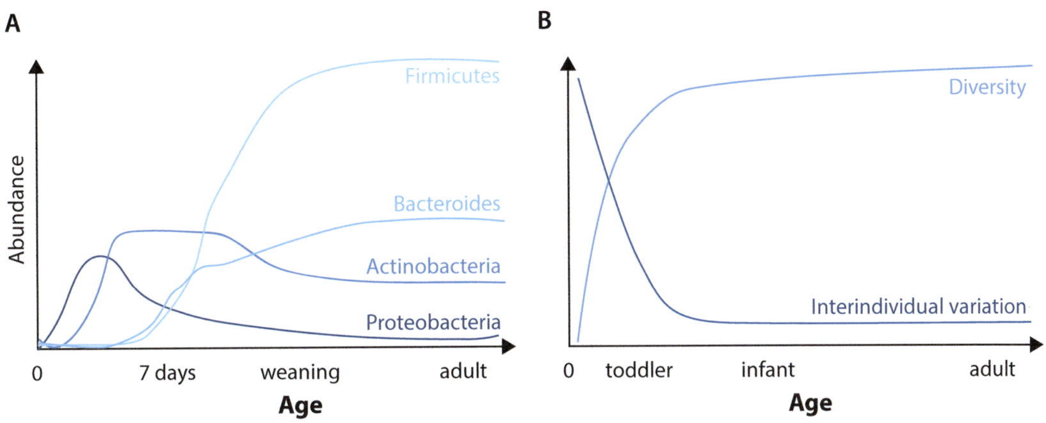

Fig. 3.1 Establishment of the enteric microbiota. (**a**) Bacterial composition. The dominant bacterial species within the microbiota change over time depending on, for example, the susceptibility of certain obligate anaerobic bacteria to luminal oxygen or the dietary substrates before or after weaning (i.e., breast milk versus solid food). (**b**) Overall microbial diversity. Shortly after birth, the neonate's microbiome exhibits a high individuality (i.e., all babies are different) but low diversity (i.e., the microbiome consists of few bacterial species). This changes rapidly with age, and approximately with 2–3 years of age in humans, a mature, highly diverse microbiome is established. Modified after (**a**) Reinhardt et al. (2009), (**b**) Yatsunenko et al. (2012)

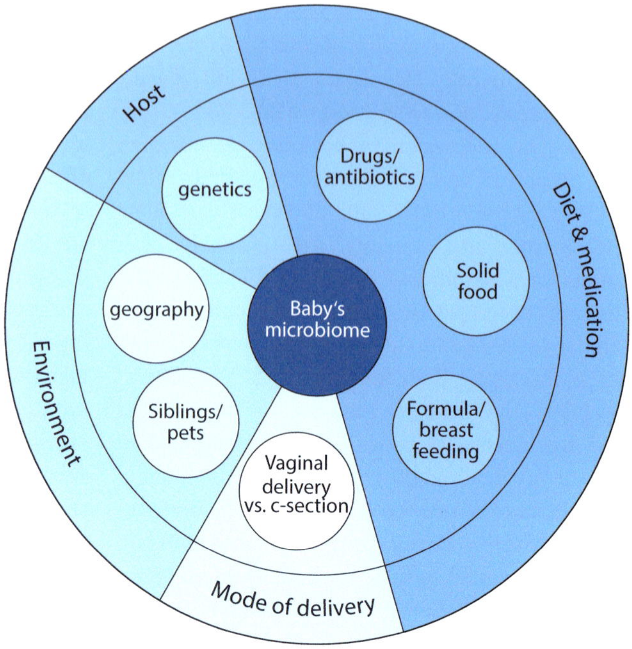

Fig. 3.2 Factors influencing the neonate's enteric microbiota. The model of delivery (vaginal delivery versus c-section), diet (breast milk versus formula), medication (antibiotics), host genetics, and environmental exposure (siblings, pets, geography) impact on the neonate's enteric microbiota

Table 3.1 Factors identified to influence the composition of the enteric microbiota and their impact on the abundance of specific commensal bacteria within the infant's enteric microbiome

Factor	Observed change	References
Geography	Differences in the microbiota composition have been noted, but the overall postnatal development is similar.	Yatsunenko et al. (2012), Vatanen et al. (2016)
Mode of delivery	The microbiota of neonates born by vaginal delivery resembles the mother's vaginal microbiota whereas cesarean section results in a more skin microbiota-like bacterial composition.	Dominguez-Bello et al. (2010), Penders et al. (2013), Hesla et al. (2014), Jakobsson et al. (2014), Backhed et al. (2015), Dogra et al. (2015), Hill et al. (2017)
Preterm birth	Enhanced abundance of Staphylococci, *Enterobacteriaceae*, and lower colonization by *Bifidobacteria*.	Arboleya et al. (2016), Roze et al. (2017), Hill et al. (2017)
Breast or formula feeding, solid food	Breast milk enhances the abundance of Bifidobacteria and Lactobacilli. Formula feeding increases microbial diversity, lowers the abundance of Bifidobacteria, and increases Bacteroidetes. The introduction of solid food and plant oligosaccharides promotes expansion of the phyla Bacteroidetes and Firmicutes.	Fallani et al. (2010), Bezirtzoglou et al. (2011), Penders et al. (2013), Backhed et al. (2015), Koenig et al. (2011), Kashtanova et al. (2016)
Early life antibiotic treatment	Reduced diversity, loss of certain species, increased abundance of bacteria with intrinsic antibiotic resistance like Enterococci.	Penders et al. (2006), Russell et al. (2012), Stefka et al. (2014), Rutten et al. (2015), Arboleya et al. (2015), Thiemann et al. (2016)
Pets and siblings	Increased diversity by exposure to pets.	Koplin et al. (2012), Azad et al. (2013)

abundance of *Bifidobacterium* spp. and *Bacteroides* spp., a reduced diversity of members of the *Bacteroidetes* phylum and enhanced numbers of *Enterobacteriaceae*. Not surprisingly, the early microbiota after cesarean section resembles the maternal skin microbiota rather than the maternal vaginal microbiota with enhanced numbers of *Staphylococcus, Corynebacterium* spp., and *Propionibacteria* (Dominguez-Bello et al. 2010). The observed differences might not only result from the alteration in the early exposure to bacteria. Cesarean section is also associated with an altered perinatal cytokine milieu (Liao et al. 2017; Lotz et al. 2006; Ulas et al. 2017), more medical interventions, reduced early mother child contact, delayed start of breast-feeding, and antibiotic prophylaxis (Chalmers et al. 2009). Interestingly, cesarean section was shown to enhance the risk of a number of noncommunicable diseases such as asthma (Huang et al. 2015), allergic disease (Magnus et al. 2011), obesity (Kuhle et al. 2015), diabetes (Cardwell et al. 2008), and coeliac disease (Decker et al. 2010). The enhanced disease susceptibility might result from an altered microbiota or other independent cesarean section-associated factors. Noteworthy, despite the recommendation of the WHO to perform cesarean delivery for no more than 15% of deliveries (Karlstrom et al. 2013) (WHO 1985), the rate in most industrialized countries rises continuously and now reaches 30%.

The group of Maria Dominguez-Bello has recently proposed a procedure named "vaginal seeding" to compensate for the lack of contact with bacteria of the maternal vaginal mucosa in neonates born by cesarean section (Dominguez-Bello et al. 2016). In their first study, they exposed cesarean section-born neonates with vaginal fluids directly after birth and could demonstrate partial restoration of the neonatal enteric microbiota. It remains to be seen whether this approach allows to also decrease the incidence of some of the associated diseases. A major concern represents the possible transmission of pathogenic microorganisms such as herpes simplex virus (HSV), GBS, or human immunodeficiency virus (HIV).

The best-studied and most powerful factor influencing the enteric microbiota composition in both infants and adults is diet. In neonates, this particularly addresses the issue of breast-feeding *versus* formula feeding. Whereas past centuries employed wet nurses to replace mothers that were unable or unwilling to breast-feed their babies, companies today offer formula diets that mimic the composition of the human breast milk. Breast-feeding is associated with enhanced abundance of bifidobacteria and lactobacilli (Kashtanova et al. 2016). In contrast, formula feeding results in higher microbial diversity, lower abundance of bifidobacteria, and higher numbers of bacteria of the phylum *Bacteroidetes* (Bezirtzoglou et al. 2011; Fallani et al. 2010). Although major efforts have been made to make formulas as similar to human breast milk as possible, significant differences remain. For example, human breast milk and in particular the breast milk produced during the first days after birth contains enormous concentrations of secretory (S) immunoglobulin A (SIgA). This SIgA is at least in part directed against commensal bacteria and contributes to the establishment of a beneficial microbiota composition (Cullender et al. 2013; Lindner et al. 2015; Palm et al. 2014). Also it contains specific human milk oligosaccharides (HMO), unconjugated complex carbohydrates with nutritional effects on bacteria (Yu et al. 2013). HMOs increase colonization with *Bifidobacterium* spp. and impede colonization by pathogenic bacteria (Asakuma et al. 2011). Efforts are made to identify specific carbohydrate molecules that mimic HMOs and confer a beneficial effect (Matsuki et al. 2016). Following the initial period of solely breast-feeding (recommended by the WHO to last for 4–6 months after birth) or formula feeding, solid food is introduced. This introduces a major shift in the enteric microbiota composition. The introduction of plant oligosaccharides promotes expansion of the phyla *Bacteroidetes* and *Firmicutes* (Koenig et al. 2011).

Another potent and well-established factor is the administration of antibiotics either perinatally to the mother or after birth to the neonate (Rutten et al. 2015; Thiemann et al. 2016). As described above, administration of a single dose of a prophylactic antibiotic (in most cases a second-generation cephalosporin) is recommended for all mothers undergoing cesarean section to prevent infection of the skin at the site of the surgical incision. Also, many countries recommend intrapartum antibiotic prophylaxis (IAP) with penicillin to all mothers with a positive vaginal swab for group B streptococci (GBS, also named *Streptococcus agalactiae*). GBS colonizes the urogenital tract of approximately one third of women and represent the most frequent causative agent of neonatal sepsis and meningitis (Ahmadzia and Heine 2014). Transmission occurs during birth by direct contact with colonized body surfaces. Third, some pregnant females develop fever with signs of ascending bacterial infection of the amniotic sac and placental tissue (chorioamnionitis) (Goldenberg et al. 2000). This condition is associated with preterm birth and neonatal sepsis and requires antibiotic treatment. In all three cases, both the mother and the neonate are exposed to the antibiotic. Additionally, the neonate after birth may exhibit clinical or laboratory signs of infection warranting antibiotic therapy. Notably, neonates and in particular preterm-born neonates are highly vulnerable to infection, and thus a delay in the initiation of an adequate antibiotic treatment may risk major complications or even death of the newborn. The alteration of the neonate's microbiota following antibiotic therapy depends on the substance used (i.e., its antibacterial spectrum) as well as the dose and regimen. Generally, antibiotic treatment increases the abundance of bacterial species with intrinsic antibiotic resistance such as *Enterococcus* spp., *Staphylococcus* spp., and members of the *Enterobacteriaceae*. Importantly, exposure to antibiotics after birth has been associated with an enhanced susceptibility to food allergy and asthma in preclinical models (Russell et al. 2012; Stefka et al. 2014). The mechanisms have not been resolved, but the altered microbiota may enhance mucosal antigen uptake and immune stimulation or dampen the efficacy of regulatory effector cells and mediators and thereby promote immune activation.

Additional factors exert a major influence on the microbiota composition. For example, antibiotic administration to both, the mother and the child, significantly alters the enteric microbiota (Arboleya et al. 2015; Penders et al. 2006). Preterm birth enhances the abundance of *Staphylococcus* and *Enterobacteriaceae* and lowers the colonization by *Bifidobacterium* (Arboleya et al. 2016; Hill et al. 2017; Roze et al. 2017). However, preterm birth is simultaneously associated with birth by cesarean section, prolonged hospitalization, risk of infection, and antibiotic administration, and it remains unclear whether any of these factors or the premature intestinal tissue per se is responsible for the observed change in the microbiota composition.

3.3 Postnatal Establishment of Host-Microbial Immune Homeostasis

Bacterial colonization of the till-then sterile mucosal surface after birth challenges the regulatory mechanisms of the innate immune system to prevent inappropriate immune receptor stimulation by microbial pattern molecules. A number of studies have identified mechanisms that contribute to maintain immune homeostasis and restrict the secretion of proinflammatory mediators during this transitional period (Hornef and Fulde 2014). For example, the presence of immunomodulatory mediators such as the epidermal growth factor (EGF), intestinal alkaline phosphatase (IAP), and the secretory leukocyte protease inhibitor (SLPI) in amniotic fluid and breast milk dampens postnatal mucosal immune stimulation. Also, enhanced expression of regulatory molecules of innate immune signalling prevents an exaggerated innate immune response during the postnatal time period (Good et al. 2012; Vereecke et al. 2010). Immunomodulatory cytokines such as interleukin (IL)-10 also contribute to maintain immune homeostasis during early microbial colonization (Glocker et al. 2009). In addition, inhibitory mechanisms activated upon first microbial encounter such as the postnatal induction of innate immune tolerance have been described. Postnatal exposure to bacterial endotoxin induces a state of innate immune non-responsiveness of the intestinal epithelium (Chassin et al. 2010; Lotz et al. 2006).

Interestingly, postnatal acquisition of innate immune tolerance is also observed at systemic body sites (Ulas et al. 2017).

Also, the neonate's mucosal adaptive immune system differs significantly from that of adult individuals (Torow et al. 2017). The development of lymphoid tissues and stromal support structures and the homing of thymus- and bone marrow-derived cells continue during the first days and weeks of life (Lindner et al. 2012; South et al. 1967; Thome et al. 2016; Torow et al. 2015). Modulatory mechanisms such as T regulatory cells and maternal SIgA delay early postnatal cell maturation and acquisition of effector function in response to microbial colonization under physiological conditions (Torow et al. 2015). Yet, the neonatal immune system is able to react and mount a protective immune response to pathogen challenge (Forsthuber et al. 1996; Sarzotti et al. 1996). It may therefore be more appropriate to speak about a "distinct neonatal" as opposed to "immature" immune system of the newborn host. Not unexpectedly, both systems, the establishment of the enteric microbiota and the maturation of the adaptive immune system, closely interfere and influence each other although the whole complexity of this interaction only starts to emerge (Gensollen et al. 2016).

For example, the microbiota was shown to influence B cell development and the B cell receptor repertoire in the intestinal lamina propria during a postnatal time window (Wesemann et al. 2013). A significant portion of B cells within the lamina propria exhibit Rag2-dependent pre-immune receptor diversification increasing until approximately at weaning followed by disappearance in the adult animal. Notably, this intestinal mucosal B cell receptor diversification was microbiota dependent and induced in germ-free mice upon bacterial colonization. This finding has changed our view that B cell development is restricted to the bone marrow and highlights the particular role of the postnatal period for and the influence of signals from the enteric microbiota on pre-immune B cell maturation.

Also B cell maturation and antibody production are stimulated by early microbial exposure. A study analyzed the influence of the early enteric microbiota on B cell effector function, namely, the production and secretion of IgE (Cahenzli et al. 2013). The authors found that mucosal B cells in the absence of an enteric microbiota (in - germ-free mice) switch to IgE resulting in elevated serum IgE levels. Interestingly, the presence of a diverse enteric microbiota during a postnatal time window was required to inhibit IgE induction and prevent oral-induced systemic anaphylaxis.

The third example addresses yet another aspect of immune maturation, namely, the postnatal homing of immune cells to peripheral tissue sites. In the absence of enteric bacteria, Olszak et al. observed increased homing of invariant natural killer cells (iNKT) to colonic and lung tissue associated with enhanced susceptibility to colitis and allergic asthma (Olszak et al. 2012). Early postnatal microbial colonization was required to regulate chemokine expression and restrict mucosal iNKT cell homing associated with lower disease susceptibility. Notably, bacterial colonization at later time points was unable to compensate the phenotype.

Thus, the enteric microbiota influences development and maturation of the mucosal innate and adaptive immune system. The described examples highlight the particular and non-redundant role of the postnatal period and the early microbiota establishment to reach long-term immune homeostasis and health. The existence of a critical time period has been described as "neonatal window of opportunity." It might explain the studies that identify the strong effect of microbiota-influencing factors such as postnatal antibiotic use, mode of delivery, or livestock exposure on a farm for the incidence of allergic diseases (Cox et al. 2014; Depner et al. 2013; Huang et al. 2015). Notably, we here only discuss the effect of viable bacteria as part of the enteric microbiota. Microbial constituents and metabolites might reach the embryonic and fetal organisms via the placental circulation and influence its development (Gomez de Agüero et al. 2016). Also, the effects of the microbiota are not restricted to the mucosal immune system but also include aspects of the systemic immune system (Hergott et al. 2016).

3.4 Dysbiosis and Disease

Alterations in the microbiota composition and diversity have been observed in various disease models, and comparative analyses have revealed significant differences between healthy individuals and patients suffering from inflammatory, metabolic, and immune-mediated diseases. Notably, many factors can explain these microbiota differences, and a direct causative contribution has been demonstrated for only few conditions.

One of the conditions for which the enteric microbiota is thought to contribute to the disease etiology is necrotizing enterocolitis of the newborn (NEC) (Berman and Moss 2011). This condition is almost exclusively found in preterm-born neonates (i.e., neonates that are born below the 37th week of gestation) and characterized by a severe inflammation and necrosis of colonic tissue. Despite surgical intervention it is associated with high mortality and morbidity. Interestingly, its incidence and severity increase with low gestational age of the preterm neonate, and it usually starts with a delay of 1–2 weeks after birth, i.e., after bacterial colonization of the intestine. An inappropriate immune stimulation of the still "immature" intestinal mucosa by members of the microbiota is thought to drive the tissue inflammation. Consistently, a number of immunoregulatory mechanisms that are thought to protect the mature neonate mucosa from inappropriate microbiota-induced immune stimulation were shown to be absent or functionally impaired in the preterm intestine (Hackam et al. 2013). Also, a significantly altered microbiota was noted in preterm neonates, and some changes were associated with the development of NEC (Roze et al. 2017). For example, a high abundance of *Enterobacteriaceae* in the human preterm intestine was demonstrated, and the outer cell membrane constituent of Gram-negative bacteria, lipopolysaccharide (LPS), and its receptor, Toll-like receptor (TLR)4, was identified to drive mucosal inflammation in animal models of NEC (Egan et al. 2016).

Infections represent the leading cause of death in the age group of neonates and young infants below the age of 1 year worldwide (GBD 2015 HIV Collaborators 2016). Among them, gastrointestinal and pulmonary infections account for the majority of cases (Liu et al. 2012). The still maturing innate and adaptive immune system and the absence of immune memory from previous exposures might contribute to this enhanced susceptibility. In addition, the previously described low colonization resistance of the early microbiota that results from low bacterial diversity might play a significant role. The low colonization resistance might also contribute to the remarkably different spectrums of pathogenic microorganisms that affect neonates. The most common causative agents of neonatal sepsis, group B streptococci (GBS, also called *S. agalactiae*), *Listeria monocytogenes*, and *E. coli* K1 are rarely observed in adults. The major causes of infant gastroenteritis are rotavirus, enterohemorrhagic, and enteropathogenic *E. coli*, but these pathogens rarely cause disease in adults (Kotloff et al. 2013).

The incidence of autoimmune diseases and allergies in industrialized countries has steadily increased during the last 30 years. Given the strong influence of our modern Western lifestyle on the microbiota composition, a possible causative role of the microbiota has been suggested. A recent study observed a significantly higher abundance of *Bacteroides* species in the microbiota of Finnish and Estonian as compared to Russian children (Vatanen et al. 2016). The low immunostimulatory potential of *Bacteroides* lipopolysaccharide (LPS) was suggested to preclude immune education and explain the enhanced incidence of autoimmune diseases in Western countries. Another study suggested that the presence of siblings and household pets lowers the incidence of food allergies by a microbiota-dependent effect (Koplin et al. 2012). Also the administration of antibiotics during early life through its effect on the enteric microbiota was proposed to contribute to the increase in immune-mediated diseases. A higher incidence of early signs of allergy was observed in both animal studies as well as human patient cohorts after repeated antibiotic treatment prior to or shortly after birth (Lapin et al. 2015; Russell et al. 2012; Stefka et al. 2014). Changes in the microbiota composition induced by antibiotic exposure are thought to contribute to enhanced susceptibility to allergic diseases (Russell et al. 2012; Stefka et al. 2014). The mechanisms involved, however, have not been identified, and other studies did not confirm this effect.

3.5 Probiotics and Prebiotics

The oral administration of specific bacterial species might occupy open niches and thus enhance colonization resistance preventing infection with pathogenic microorganisms. Also, inflammatory diseases induced by an altered microbiota (then called dysbiosis) might be targeted by correcting the microbiota composition through oral administration of beneficial bacteria. These beneficial bacteria are called probiotic bacteria, and the WHO defines probiotic bacteria as "viable microorganisms that exert a health-promoting effect on the host organism." It is important to keep in mind that probiotics by and large represent commensal bacteria or yeast. The most commonly used probiotic agents are *Saccharomyces cerevisiae* (the brewer's yeast), *Lactobacillus*, and *Bifidobacterium* species as well as *E. coli*. In addition to the administration of probiotic bacteria, the nutritional supplementation with substances that foster growth of beneficial bacteria (so-called prebiotics) has become a therapeutic strategy. As described, breast milk and its constituents strongly promote the growth of beneficial neonatal bacteria such as bifidobacteria and can be considered a natural prebiotic.

The prophylactic value of this strategy has been tested for all three examples for potentially dysbiosis-related diseases of the neonate discussed in the previous chapter. In several studies, premature neonates received Gram-positive bacteria in order to prevent the overgrowth of Gram-negative bacteria and the stimulation of the immature intestinal mucosa leading to NEC. A recent meta-analysis concluded that postnatal oral administration of probiotic bacteria to preterm human neonates provides a significant benefit lowering the incidence and severity of NEC (Ganguli and Walker 2011; Olsen et al. 2016). However, not all probiotic bacteria confer a beneficial effect, and further studies are needed to better define the optimal prophylactic strategy (Costeloe et al. 2016; Embleton et al. 2016). Besides probiotics, also the administration of breast milk as a physiological form of prebiotic (see above) has been shown to protect from NEC (Maayan-Metzger et al. 2012; McGuire and Anthony 2003). Also, the low colonization resistance in neonates and young children was addressed by oral administration of viable beneficial bacteria. Indeed a recent prospective controlled study in rural India revealed a strong protective effect of daily oral *Lactobacillus plantarum* plus fructooligosaccharide administration on the incidence of bacterial sepsis and lower respiratory tract infection (Panigrahi et al. 2017). The underlying mechanisms that allow outcompetition of enteropathogenic bacteria have been analyzed in preclinical studies (Kim et al. 2017). Additionally, the incidence of allergies in children born by cesarean section was reduced following oral administration of probiotic bacteria (Kuitunen et al. 2009). It is interesting that neonates and young infants appear to exhibit a greater responsiveness to the administration of probiotic bacteria as compared to adult individuals. In fact, the vast majority of studies on adult patients with various diseases failed to show a significant benefit of oral treatment with probiotic bacteria. Thus, the lower colonization resistance in neonates might make neonates more vulnerable to disturbance of the microbiota composition but at the same time render them more susceptible for the beneficial effects by oral administration of probiotic bacteria.

The enhanced susceptibility of the neonate and infant host should also raise concerns on possible safety aspects and secondary long-term effects of the administration of probiotic bacteria. Invasive infections caused by orally administered probiotic bacteria have been rarely reported. Also, orally administered bacteria disappeared from the infant microbiota shortly after cessation of oral administration (Bazanella et al. 2017). Nevertheless, comparative studies are needed to define the beneficial effect for each individual probiotic regimen. Also, robust quality control data are required before we treat this vulnerable patient population.

3.6 Future Perspectives

Although much has been learned during the last years and we now have an approximate idea of the development of the early enteric microbiota and influencing factors, much remains to be investigated. In particular, we only marginally understand how and to what extent the host directs and shapes the establishment of the enteric microbiota. Since essential physiological processes critically depend on signals from the enteric microbiota, it is likely that the host ensures the presence of the right bacteria providing the right signals at the right time. A better understanding of these mechanisms might help us to identify the overall beneficial aspects of the microbiota. And this is not restricted to bacteria but may also involve phages, fungi, and archaea (Schei et al. 2017).

A large number of studies performed during the last years demonstrated differences in the microbiota composition between healthy individuals and patients and discussed a possible causative role in the disease etiology or progression. However, few investigations were able to proof a functional link and provide insight in the mechanism. In many cases, the microbiota alterations might result from disease-associated behavioral changes or treatment-related factors. One recent example illustrates the problems but also the surprises that might emerge from this type of analysis. First a large study identified differences in the microbiota between healthy and diabetic (type 2) patients suggesting a role of the microbiota in the etiology of the disease (Qin et al. 2012). Then, the commonly administered antidiabetic drug, metformin, was identified to have caused much of the observed microbiota alteration (Forslund et al. 2015). Finally, the metformin-induced microbiota changes were shown to at least in part contribute to the therapeutic benefit of the drug (Wu et al. 2017). Thus, we have to overcome the descriptive analysis of pure differences and move forward to identify the responsible microbial species and underlying molecular mechanisms. Microbiota alterations might confer improved health or disease by different ways. The increased abundance of one commensal species might occupy the niche of a pathogenic one and thus prevent colonization (Brugiroux et al. 2016). Beneficial bacteria might contribute to nutrient metabolism and provide beneficial metabolic substrates or produce immunostimulatory molecules that promote immune homeostasis (Round et al. 2011). Even the bacterial production of human mediator homologues was described that might act at very low concentrations and significantly influence physiological function (Cohen et al. 2017).

Important in the context of the discussion of the neonatal, developing enteric microbiota is the fact that it displays a much higher individual variation, dynamic in its composition, and vulnerability to exogenous factors such as antibiotics (Cox et al. 2014). We therefore have to exert caution when administering antibiotics—if not strictly medically indicated—and other microbiota-modifying agents to neonates and young children. Few studies have started to address possible side effects both during treatment and in the long term. At the same time, this situation opens the possibility of (possibly lasting) interventional modification to prevent metabolic diseases and infections (Panigrahi et al. 2017). Thus, although many questions still need to be addressed, the microbiota represents a promising future target for prophylactic and interventional strategies, and neonates and young infants represent the most promising age group.

▶ **Controversy**

It is generally believed that the fetus in utero develops in the absence of viable microorganisms. Several groups recently challenged this view and proposed the existence of a placental microbiome, i.e., the transfer of a heterogeneous group of viable commensal bacteria from the mother to the fetus with colonization of the fetal gastrointestinal tract. For example, one group found low numbers of culturable bacteria in human cord blood samples after elective cesarean section (Jimenez et al. 2005) and meconium samples of healthy neonates born by either vaginal delivery or cesarean section (Jimenez et al. 2008). Others used PCR-based methods and detected bacterial DNA in placental tissue (Aagaard et al. 2014; Bassols et al. 2016; Rautava et al. 2012). Based on these results, the authors suggested the presence of a low-richness and low-diversity fetal microbiota (Collado et al. 2016). However, do the reported findings really support the existence of a "fetal microbiota" and of "microbial transfer at the feto-maternal interface"? PCR-based methods detect DNA and are by definition unable to differentiate between viable bacteria and dead bacteria or bacterial DNA. Modern high-throughput PCR methods are extremely sensitive and able to detect minute amounts of DNA. These low DNA concentrations were shown to be present in commercial enzyme preparations, column material, and even buffer solutions. In addition, many mucosal body surfaces are densely colonized, and we know that transient bacteremia occurs following minor manipulation of mucosal tissues, for example, after brushing the teeth or during labor. Thus, the detection of bacterial DNA and the culture of low bacterial numbers in clinical specimen do not proof the existence of a bona fide microbiota. Also, bacteria can proliferate very rapidly, and thus meconium released hours after birth is not expected to be sterile. Finally, low numbers of microorganisms are ubiquitously found in our environment, and tight precautions have to be put in place to avoid contamination of the samples during the analytical process (Lauder et al. 2016). The most compelling argument, however, comes from animal studies: the sterile recovery of mature mouse embryos can be used to generate germ-free animals. Thus, the current findings do not provide sufficient evidence to proof the existence of a fetal microbiota (Hornef and Penders 2017; Perez-Munoz et al. 2017).

History

In the mid-nineteenth century, microbiologists like Robert Koch and Louis Pasteur identified the first pathogenic bacteria and started to gather knowledge on the etiology of infectious disease and the epidemiology and transmission of microorganisms. This led to the development of methods and behavioral habits to avoid contact with pathogens and lower the incidence of infections. Technical inventions made during the last 100 years such as water toilets and sewage treatment, the production and distribution of drinking water, personal hygiene, and improved housing conditions (e.g., separate rooms for livestock and humans) and food storage (e.g., the introduction of refrigerators that reduce microbial growth and allow prolonged food storage) contributed to this. They significantly lowered the incidence of life-threatening infections in industrialized countries accompanied by an increase in life expectancy and a reduction in morbidity and mortality. Public and personal hygiene as a strategy to avoid contact with pathogenic microorganisms cumulated in habits such as sterilizing the pacifier or milk bottle from babies and infants prior to use. Today, we learn that these measures also deprive us from exposure to nonpathogenic microorganisms and microbial constituents that contribute to maturation of our immune system and host-microbial homeostasis. We start to understand that this deprivation from microbial exposure in particular during the early childhood might be associated with secondary effects and contribute to the increasing incidence of noncommunicable inflammatory and immune-mediated diseases such as allergic disease and inflammatory bowel disease (Riedler et al. 2001). This concept was initially proposed in the so-called hygiene hypothesis that particularly focused on the influence of infections during the postnatal and infant period for immune-mediated disease susceptibility in later life (Strachan 1989). It was later extended to also include changes in our exposure to commensal bacteria (Rook 2010). Today, we try to characterize the altered microbial exposure and identify the mechanisms that contribute to enhanced disease susceptibility. We also aim at reestablishing microbial exposure during early childhood to prevent or treat immune-mediated diseases, of course, without increasing the risk of exposure to pathogenic microorganisms.

Highlights

- First exposure to viable bacteria starts during birth with contact to the maternal vaginal mucosa.
- The early postnatal enteric microbiota is characterized by low diversity and high interindividual variation.
- Microbial diversity increases by incorporation of new bacterial taxa reaching an adult-like microbiota in humans at approximately 2–3 years of age.
- Important exogenous factors are the mode of delivery and diet (breast-feeding versus formula feeding).
- The neonatal/infant microbiota is significantly more susceptible than the adult microbiota to exogenous alteration, e.g., by antibiotic treatment or administration of probiotic bacteria.

References

Aagaard, K., Ma, J., Antony, K. M., Ganu, R., Petrosino, J., & Versalovic, J. (2014). The placenta harbors a unique microbiome. *Science Translational Medicine, 6*, 237ra265.

Ahmadzia, H. K., & Heine, R. P. (2014). Diagnosis and management of group B streptococcus in pregnancy. *Obstetrics and Gynecology Clinics of North America, 41*, 629–647.

Arboleya, S., Sanchez, B., Milani, C., Duranti, S., Solis, G., Fernandez, N., de los Reyes-Gavilan, C. G., Ventura, M., Margolles, A., & Gueimonde, M. (2015). Intestinal microbiota development in preterm neonates and effect of perinatal antibiotics. *The Journal of Pediatrics, 166*, 538–544.

Arboleya, S., Sanchez, B., Solis, G., Fernandez, N., Suarez, M., Hernandez-Barranco, A. M., Milani, C., Margolles, A., de Los Reyes-Gavilan, C. G., Ventura, M., & Gueimonde, M. (2016). Impact of prematurity and perinatal antibiotics on the developing intestinal microbiota: A functional inference study. *International Journal of Molecular Sciences, 17*, 649.

Asakuma, S., Hatakeyama, E., Urashima, T., Yoshida, E., Katayama, T., Yamamoto, K., Kumagai, H., Ashida, H., Hirose, J., & Kitaoka, M. (2011). Physiology of consumption of human milk oligosaccharides by infant gut-associated bifidobacteria. *The Journal of Biological Chemistry, 286*, 34583–34592.

Azad, M. B., Konya, T., Maughan, H., Guttman, D. S., Field, C. J., Sears, M. R., Becker, A. B., Scott, J. A., & Kozyrskyj, A. L. (2013). Infant gut microbiota and the hygiene hypothesis of allergic disease: Impact of household pets and siblings on microbiota composition and diversity. *Allergy, Asthma and Clinical Immunology, 9*, 15.

Backhed, F., Roswall, J., Peng, Y., Feng, Q., Jia, H., Kovatcheva-Datchary, P., Li, Y., Xia, Y., Xie, H., Zhong, H., et al. (2015). Dynamics and stabilization of the human gut microbiome during the first year of life. *Cell Host & Microbe, 17*, 852.

Bassols, J., Serino, M., Carreras-Badosa, G., Burcelin, R., Blasco-Baque, V., Lopez-Bermejo, A., & Fernandez-Real, J. M. (2016). Gestational diabetes is associated with changes in placental microbiota and microbiome. *Pediatric Research, 80*, 777–784.

Bazanella, M., Maier, T. V., Clavel, T., Lagkouvardos, I., Lucio, M., Maldonado-Gomez, M. X., Autran, C., Walter, J., Bode, L., Schmitt-Kopplin, P., & Haller, D. (2017). Randomized controlled trial on the impact of early-life intervention with bifidobacteria on the healthy infant fecal microbiota and metabolome. *The American Journal of Clinical Nutrition, 106*(5), 1274–1286.

Berman, L., & Moss, R. L. (2011). Necrotizing enterocolitis: An update. *Seminars in Fetal & Neonatal Medicine, 16*(3), 145–150.

Bezirtzoglou, E., Tsiotsias, A., & Welling, G. W. (2011). Microbiota profile in feces of breast- and formula-fed newborns by using fluorescence in situ hybridization (FISH). *Anaerobe, 17*, 478–482.

Brugiroux, S., Beutler, M., Pfann, C., Garzetti, D., Ruscheweyh, H. J., Ring, D., Diehl, M., Herp, S., Lotscher, Y., Hussain, S., et al. (2016). Genome-guided design of a defined mouse microbiota that confers colonization resistance against *Salmonella enterica* serovar Typhimurium. *Nature Microbiology, 2*, 16215.

Cahenzli, J., Koller, Y., Wyss, M., Geuking, M. B., & McCoy, K. D. (2013). Intestinal microbial diversity during early-life colonization shapes long-term IgE levels. *Cell Host & Microbe, 14*, 559–570.

Cardwell, C. R., Stene, L. C., Joner, G., Cinek, O., Svensson, J., Goldacre, M. J., Parslow, R. C., Pozzilli, P., Brigis, G., Stoyanov, D., et al. (2008). Caesarean section is associated with an increased risk of childhood-onset type 1 diabetes mellitus: A meta-analysis of observational studies. *Diabetologia, 51*, 726–735.

Chalmers, B., Kaczorowski, J., Levitt, C., Dzakpasu, S., O'Brien, B., Lee, L., Boscoe, M., Young, D., & Maternity Experiences Study Group of the Canadian Perinatal Surveillance, S., and Public Health Agency of, C. (2009). Use of routine interventions in vaginal labor and birth: Findings from the maternity experiences survey. *Birth, 36*, 13–25.

Chassin, C., Kocur, M., Pott, J., Duerr, C. U., Gutle, D., Lotz, M., & Hornef, M. W. (2010). miR-146a mediates protective innate immune tolerance in the neonate intestine. *Cell Host & Microbe, 8*, 358–368.

Cohen, L. J., Esterhazy, D., Kim, S. H., Lemetre, C., Aguilar, R. R., Gordon, E. A., Pickard, A. J., Cross, J. R., Emiliano, A. B., Han, S. M., et al. (2017). Commensal bacteria make GPCR ligands that mimic human signalling molecules. *Nature, 549*, 48–53.

Collado, M. C., Rautava, S., Aakko, J., Isolauri, E., & Salminen, S. (2016). Human gut colonisation may be initiated in utero by distinct microbial communities in the placenta and amniotic fluid. *Scientific Reports, 6*, 23129.

Costeloe, K., Hardy, P., Juszczak, E., Wilks, M., Millar, M. R., & and Probiotics in Preterm Infants Study Collaborative, G. (2016). Bifidobacterium breve BBG-001 in very preterm infants: A randomised controlled phase 3 trial. *Lancet, 387*, 649–660.

Cox, L. M., Yamanishi, S., Sohn, J., Alekseyenko, A. V., Leung, J. M., Cho, I., Kim, S. G., Li, H., Gao, Z., Mahana, D., et al. (2014). Altering the intestinal microbiota during a critical developmental window has lasting metabolic consequences. *Cell, 158*, 705–721.

Cullender, T. C., Chassaing, B., Janzon, A., Kumar, K., Muller, C. E., Werner, J. J., Angenent, L. T., Bell, M. E., Hay, A. G., Peterson, D. A., et al. (2013). Innate and adaptive immunity interact to quench microbiome flagellar motility in the gut. *Cell Host & Microbe, 14*, 571–581.

Decker, E., Engelmann, G., Findeisen, A., Gerner, P., Laass, M., Ney, D., Posovszky, C., Hoy, L., & Hornef, M. W. (2010). Cesarean delivery is associated with celiac disease but not inflammatory bowel disease in children. *Pediatrics, 125*, e1433–e1440.

Depner, M., Ege, M. J., Genuneit, J., Pekkanen, J., Roponen, M., Hirvonen, M. R., Dalphin, J. C., Kaulek, V., Krauss-Etschmann, S., Riedler, J., et al. (2013). Atopic sensitization in the first year of life. *The Journal of Allergy and Clinical Immunology, 131*, 781–788.

Dogra, S., Sakwinska, O., Soh, S. E., Ngom-Bru, C., Bruck, W. M., Berger, B., Brussow, H., Lee, Y. S., Yap, F., Chong, Y. S., et al. (2015). Dynamics of infant gut microbiota are influenced by delivery mode and gestational duration and are associated with subsequent adiposity. *MBio, 6*, e02419–e02414.

Dominguez-Bello, M. G., Costello, E. K., Contreras, M., Magris, M., Hidalgo, G., Fierer, N., & Knight, R. (2010). Delivery mode shapes the acquisition and structure of the initial microbiota across multiple body habitats in newborns. *Proceedings of the National Academy of Sciences of the United States of America, 107*, 11971–11975.

Dominguez-Bello, M. G., De Jesus-Laboy, K. M., Shen, N., Cox, L. M., Amir, A., Gonzalez, A., Bokulich, N. A., Song, S. J., Hoashi, M., Rivera-Vinas, J. I., et al. (2016). Partial restoration of the microbiota of cesarean-born infants via vaginal microbial transfer. *Nature Medicine, 22*, 250–253.

Egan, C. E., Sodhi, C. P., Good, M., Lin, J., Jia, H., Yamaguchi, Y., Lu, P., Ma, C., Branca, M. F., Weyandt, S., et al. (2016). Toll-like receptor 4-mediated lymphocyte influx induces neonatal necrotizing enterocolitis. *The Journal of Clinical Investigation, 126*, 495–508.

Embleton, N. D., Zalewski, S., & Berrington, J. E. (2016). Probiotics for prevention of necrotizing enterocolitis and sepsis in preterm infants. *Current Opinion in Infectious Diseases, 29*, 256–261.

Fallani, M., Young, D., Scott, J., Norin, E., Amarri, S., Adam, R., Aguilera, M., Khanna, S., Gil, A., Edwards, C. A., et al. (2010). Intestinal microbiota of 6-week-old infants across Europe: Geographic influence beyond delivery mode, breast-feeding, and antibiotics. *Journal of Pediatric Gastroenterology and Nutrition, 51*, 77–84.

Forslund, K., Hildebrand, F., Nielsen, T., Falony, G., Le Chatelier, E., Sunagawa, S., Prifti, E., Vieira-Silva, S., Gudmundsdottir, V., Pedersen, H. K., et al. (2015). Disentangling type 2 diabetes and metformin treatment signatures in the human gut microbiota. *Nature, 528*, 262–266.

Forsthuber, T., Yip, H. C., & Lehmann, P. V. (1996). Induction of TH1 and TH2 immunity in neonatal mice. *Science, 271*, 1728–1730.

Ganguli, K., & Walker, W. A. (2011). Probiotics in the prevention of necrotizing enterocolitis. *Journal of Clinical Gastroenterology, 45(Suppl)*, S133–S138.

GBD 2015 HIV Collaborators. (2016). Estimates of global, regional, and national incidence, prevalence, and mortality of HIV, 1980–2015: The Global Burden of Disease Study 2015. *Lancet HIV, 3*, e361–e387.

Gensollen, T., Iyer, S. S., Kasper, D. L., & Blumberg, R. S. (2016). How colonization by microbiota in early life shapes the immune system. *Science, 352*, 539–544.

Glocker, E. O., Kotlarz, D., Boztug, K., Gertz, E. M., Schaffer, A. A., Noyan, F., Perro, M., Diestelhorst, J., Allroth, A., Murugan, D., et al. (2009). Inflammatory bowel disease and mutations affecting the interleukin-10 receptor. *The New England Journal of Medicine, 361*, 2033–2045.

Goldenberg, R. L., Hauth, J. C., & Andrews, W. W. (2000). Intrauterine infection and preterm delivery. *The New England Journal of Medicine, 342*, 1500–1507.

Gomez de Agüero, M., Ganal-Vonarburg, S. C., Fuhrer, T., Rupp, S., Uchimura, Y., Li, H., Steinert, A., Heikenwalder, M., Hapfelmeier, S., Sauer, U., McCoy, K. D., & Macpherson, A. J. (2016). The maternal microbiota drives early postnatal innate immune development. *Science, 351*, 1296–1302.

Good, M., Siggers, R. H., Sodhi, C. P., Afrazi, A., Alkhudari, F., Egan, C. E., Neal, M. D., Yazji, I., Jia,

H., Lin, J., et al. (2012). Amniotic fluid inhibits Toll-like receptor 4 signaling in the fetal and neonatal intestinal epithelium. *Proceedings of the National Academy of Sciences of the United States of America, 109*, 11330–11335.

Hackam, D. J., Afrazi, A., Good, M., & Sodhi, C. P. (2013). Innate immune signaling in the pathogenesis of necrotizing enterocolitis. *Clinical & Developmental Immunology, 2013*, 475415.

Hergott, C. B., Roche, A. M., Tamashiro, E., Clarke, T. B., Bailey, A. G., Laughlin, A., Bushman, F. D., & Weiser, J. N. (2016). Peptidoglycan from the gut microbiota governs the lifespan of circulating phagocytes at homeostasis. *Blood, 127*, 2460–2471.

Hesla, H. M., Stenius, F., Jaderlund, L., Nelson, R., Engstrand, L., Alm, J., & Dicksved, J. (2014). Impact of lifestyle on the gut microbiota of healthy infants and their mothers-the ALADDIN birth cohort. *FEMS Microbiology Ecology, 90*, 791–801.

Hill, C. J., Lynch, D. B., Murphy, K., Ulaszewska, M., Jeffery, I. B., O'Shea, C. A., Watkins, C., Dempsey, E., Mattivi, F., Tuohy, K., et al. (2017). Evolution of gut microbiota composition from birth to 24 weeks in the INFANTMET Cohort. *Microbiome, 5*, 4.

Hornef, M. W., & Fulde, M. (2014). Ontogeny of intestinal epithelial innate immune responses. *Frontiers in Immunology, 5*, 474.

Hornef, M., & Penders, J. (2017). Does a prenatal bacterial microbiota exist? *Mucosal Immunology, 10*, 598–601.

Huang, L., Chen, Q., Zhao, Y., Wang, W., Fang, F., & Bao, Y. (2015). Is elective cesarean section associated with a higher risk of asthma? A meta-analysis. *The Journal of Asthma, 52*, 16–25.

Jakobsson, H. E., Abrahamsson, T. R., Jenmalm, M. C., Harris, K., Quince, C., Jernberg, C., Bjorksten, B., Engstrand, L., & Andersson, A. F. (2014). Decreased gut microbiota diversity, delayed Bacteroidetes colonisation and reduced Th1 responses in infants delivered by caesarean section. *Gut, 63*, 559–566.

Jimenez, E., Fernandez, L., Marin, M. L., Martin, R., Odriozola, J. M., Nueno-Palop, C., Narbad, A., Olivares, M., Xaus, J., & Rodriguez, J. M. (2005). Isolation of commensal bacteria from umbilical cord blood of healthy neonates born by cesarean section. *Current Microbiology, 51*, 270–274.

Jimenez, E., Marin, M. L., Martin, R., Odriozola, J. M., Olivares, M., Xaus, J., Fernandez, L., & Rodriguez, J. M. (2008). Is meconium from healthy newborns actually sterile? *Research in Microbiology, 159*, 187–193.

Karlsson, C. L., Molin, G., Cilio, C. M., & Ahrne, S. (2011). The pioneer gut microbiota in human neonates vaginally born at term-a pilot study. *Pediatric Research, 70*, 282–286.

Karlstrom, A., Lindgren, H., & Hildingsson, I. (2013). Maternal and infant outcome after caesarean section without recorded medical indication: Findings from a Swedish case-control study. *BJOG, 120*, 479–486, discussion 486.

Kashtanova, D. A., Popenko, A. S., Tkacheva, O. N., Tyakht, A. B., Alexeev, D. G., & Boytsov, S. A. (2016). Association between the gut microbiota and diet: Fetal life, early childhood, and further life. *Nutrition, 32*, 620–627.

Kim, Y. G., Sakamoto, K., Seo, S. U., Pickard, J. M., Gillilland, M. G., 3rd, Pudlo, N. A., Hoostal, M., Li, X., Wang, T. D., Feehley, T., et al. (2017). Neonatal acquisition of Clostridia species protects against colonization by bacterial pathogens. *Science, 356*, 315–319.

Klaassens, E. S., Boesten, R. J., Haarman, M., Knol, J., Schuren, F. H., Vaughan, E. E., & de Vos, W. M. (2009). Mixed-species genomic microarray analysis of fecal samples reveals differential transcriptional responses of bifidobacteria in breast- and formula-fed infants. *Applied and Environmental Microbiology, 75*, 2668–2676.

Koenig, J. E., Spor, A., Scalfone, N., Fricker, A. D., Stombaugh, J., Knight, R., Angenent, L. T., & Ley, R. E. (2011). Succession of microbial consortia in the developing infant gut microbiome. *Proceedings of the National Academy of Sciences of the United States of America, 108(Suppl 1)*, 4578–4585.

Koplin, J. J., Dharmage, S. C., Ponsonby, A. L., Tang, M. L., Lowe, A. J., Gurrin, L. C., Osborne, N. J., Martin, P. E., Robinson, M. N., Wake, M., et al. (2012). Environmental and demographic risk factors for egg allergy in a population-based study of infants. *Allergy, 67*, 1415–1422.

Kotloff, K. L., Nataro, J. P., Blackwelder, W. C., Nasrin, D., Farag, T. H., Panchalingam, S., Wu, Y., Sow, S. O., Sur, D., Breiman, R. F., et al. (2013). Burden and aetiology of diarrhoeal disease in infants and young children in developing countries (the Global Enteric Multicenter Study, GEMS): A prospective, case-control study. *Lancet, 382*, 209–222.

Kuhle, S., Tong, O. S., & Woolcott, C. G. (2015). Association between caesarean section and childhood obesity: A systematic review and meta-analysis. *Obesity Reviews, 16*, 295–303.

Kuitunen, M., Kukkonen, K., Juntunen-Backman, K., Korpela, R., Poussa, T., Tuure, T., Haahtela, T., & Savilahti, E. (2009). Probiotics prevent IgE-associated allergy until age 5 years in cesarean-delivered children but not in the total cohort. *The Journal of Allergy and Clinical Immunology, 123*, 335–341.

Lapin, B., Piorkowski, J., Ownby, D., Freels, S., Chavez, N., Hernandez, E., Wagner-Cassanova, C., Pelzel, D., Vergara, C., & Persky, V. (2015). Relationship between prenatal antibiotic use and asthma in at-risk children. *Annals of Allergy, Asthma & Immunology, 114*, 203–207.

Lauder, A. P., Roche, A. M., Sherrill-Mix, S., Bailey, A., Laughlin, A. L., Bittinger, K., Leite, R., Elovitz, M. A., Parry, S., & Bushman, F. D. (2016). Comparison of placenta samples with contamination controls does not provide evidence for a distinct placenta microbiota. *Microbiome, 4*, 29.

Liao, S. L., Tsai, M. H., Yao, T. C., Hua, M. C., Yeh, K. W., Chiu, C. Y., Su, K. W., Huang, S. Y., Kao, C. C., Lai, S. H., & Huang, J. L. (2017). Caesarean section is associated with reduced perinatal cytokine response, increased risk of bacterial colonization in the airway, and infantile wheezing. *Scientific Reports, 7*, 9053.

Lindner, C., Wahl, B., Fohse, L., Suerbaum, S., Macpherson, A. J., Prinz, I., & Pabst, O. (2012). Age, microbiota, and T cells shape diverse individual IgA repertoires in the intestine. *The Journal of Experimental Medicine, 209*, 365–377.

Lindner, C., Thomsen, I., Wahl, B., Ugur, M., Sethi, M. K., Friedrichsen, M., Smoczek, A., Ott, S., Baumann, U., Suerbaum, S., et al. (2015). Diversification of memory B cells drives the continuous adaptation of secretory antibodies to gut microbiota. *Nature Immunology, 16*, 880–888.

Liu, L., Johnson, H. L., Cousens, S., Perin, J., Scott, S., Lawn, J. E., Rudan, I., Campbell, H., Cibulskis, R., Li, M., et al. (2012). Global, regional, and national causes of child mortality: An updated systematic analysis for 2010 with time trends since 2000. *Lancet, 379*, 2151–2161.

Lotz, M., Gutle, D., Walther, S., Menard, S., Bogdan, C., & Hornef, M. W. (2006). Postnatal acquisition of endotoxin tolerance in intestinal epithelial cells. *The Journal of Experimental Medicine, 203*, 973–984.

Maayan-Metzger, A., Avivi, S., Schushan-Eisen, I., & Kuint, J. (2012). Human milk versus formula feeding among preterm infants: Short-term outcomes. *American Journal of Perinatology, 29*, 121–126.

Magnus, M. C., Haberg, S. E., Stigum, H., Nafstad, P., London, S. J., Vangen, S., & Nystad, W. (2011). Delivery by Cesarean section and early childhood respiratory symptoms and disorders: The Norwegian mother and child cohort study. *American Journal of Epidemiology, 174*, 1275–1285.

Martin, R., Langa, S., Reviriego, C., Jiminez, E., Marin, M. L., Xaus, J., Fernandez, L., & Rodriguez, J. M. (2003). Human milk is a source of lactic acid bacteria for the infant gut. *The Journal of Pediatrics, 143*, 754–758.

Matsuki, T., Tajima, S., Hara, T., Yahagi, K., Ogawa, E., & Kodama, H. (2016). Infant formula with galacto-oligosaccharides (OM55N) stimulates the growth of indigenous bifidobacteria in healthy term infants. *Beneficial Microbes, 7*, 453–461.

McGuire, W., & Anthony, M. Y. (2003). Donor human milk versus formula for preventing necrotising enterocolitis in preterm infants: Systematic review. *Archives of Disease in Childhood – Fetal and Neonatal, 88*, F11–F14.

Olsan, E. E., Byndloss, M. X., Faber, F., Rivera-Chavez, F., Tsolis, R. M., & Baumler, A. J. (2017). Colonization resistance: The deconvolution of a complex trait. *The Journal of Biological Chemistry, 292*, 8577–8581.

Olsen, R., Greisen, G., Schroder, M., & Brok, J. (2016). Prophylactic probiotics for preterm infants: A systematic review and meta-analysis of observational studies. *Neonatology, 109*, 105–112.

Olszak, T., An, D., Zeissig, S., Vera, M. P., Richter, J., Franke, A., Glickman, J. N., Siebert, R., Baron, R. M., Kasper, D. L., & Blumberg, R. S. (2012). Microbial exposure during early life has persistent effects on natural killer T cell function. *Science, 336*, 489–493.

Palm, N. W., de Zoete, M. R., Cullen, T. W., Barry, N. A., Stefanowski, J., Hao, L., Degnan, P. H., Hu, J., Peter, I., Zhang, W., et al. (2014). Immunoglobulin A coating identifies colitogenic bacteria in inflammatory bowel disease. *Cell, 158*, 1000–1010.

Palmer, C., Bik, E. M., DiGiulio, D. B., Relman, D. A., & Brown, P. O. (2007). Development of the human infant intestinal microbiota. *PLoS Biology, 5*, e177.

Panigrahi, P., Parida, S., Nanda, N. C., Satpathy, R., Pradhan, L., Chandel, D. S., Baccaglini, L., Mohapatra, A., Mohapatra, S. S., Misra, P. R., et al. (2017). A randomized synbiotic trial to prevent sepsis among infants in rural India. *Nature, 548*, 407–412.

Penders, J., Thijs, C., Vink, C., Stelma, F. F., Snijders, B., Kummeling, I., van den Brandt, P. A., & Stobberingh, E. E. (2006). Factors influencing the composition of the intestinal microbiota in early infancy. *Pediatrics, 118*, 511–521.

Penders, J., Gerhold, K., Stobberingh, E. E., Thijs, C., Zimmermann, K., Lau, S., & Hamelmann, E. (2013). Establishment of the intestinal microbiota and its role for atopic dermatitis in early childhood. *The Journal of Allergy and Clinical Immunology, 132*, 601–607.

Perez-Munoz, M. E., Arrieta, M. C., Ramer-Tait, A. E., & Walter, J. (2017). A critical assessment of the "sterile womb" and "in utero colonization" hypotheses: Implications for research on the pioneer infant microbiome. *Microbiome, 5*, 48.

Qin, J., Li, Y., Cai, Z., Li, S., Zhu, J., Zhang, F., Liang, S., Zhang, W., Guan, Y., Shen, D., et al. (2012). A metagenome-wide association study of gut microbiota in type 2 diabetes. *Nature, 490*, 55–60.

Rautava, S., Kainonen, E., Salminen, S., & Isolauri, E. (2012). Maternal probiotic supplementation during pregnancy and breast-feeding reduces the risk of eczema in the infant. *The Journal of Allergy and Clinical Immunology, 130*, 1355–1360.

Reinhardt, C., Reigstad, C. S., & Backhed, F. (2009). Intestinal microbiota during infancy and its implications for obesity. *Journal of Pediatric Gastroenterology and Nutrition, 48*, 249–256.

Riedler, J., Braun-Fahrlander, C., Eder, W., Schreuer, M., Waser, M., Maisch, S., Carr, D., Schierl, R., Nowak, D., von Mutius, E., & Team, A. S. (2001). Exposure to farming in early life and development of asthma and allergy: A cross-sectional survey. *Lancet, 358*, 1129–1133.

Rodriguez, J. M. (2014). The origin of human milk bacteria: Is there a bacterial entero-mammary pathway during late pregnancy and lactation? *Advances in Nutrition, 5*, 779–784.

Rook, G. A. (2010). 99th Dahlem conference on infection, inflammation and chronic inflammatory disorders: Darwinian medicine and the 'hygiene' or 'old friends' hypothesis. *Clinical and Experimental Immunology, 160*, 70–79.

Round, J. L., Lee, S. M., Li, J., Tran, G., Jabri, B., Chatila, T. A., & Mazmanian, S. K. (2011). The Toll-like receptor 2 pathway establishes colonization by a commensal of the human microbiota. *Science, 332*, 974–977.

Roze, J. C., Ancel, P. Y., Lepage, P., Martin-Marchand, L., Al Nabhani, Z., Delannoy, J., Picaud, J. C., Lapillonne, A., Aires, J., Durox, M., et al. (2017). Nutritional strategies and gut microbiota composition as risk factors for necrotizing enterocolitis in very-preterm infants. *The American Journal of Clinical Nutrition, 106*, 821–830.

Russell, S. L., Gold, M. J., Hartmann, M., Willing, B. P., Thorson, L., Wlodarska, M., Gill, N., Blanchet, M. R., Mohn, W. W., McNagny, K. M., & Finlay, B. B. (2012). Early life antibiotic-driven changes in microbiota enhance susceptibility to allergic asthma. *EMBO Reports, 13*, 440–447.

Rutten, N. B., Rijkers, G. T., Meijssen, C. B., Crijns, C. E., Oudshoorn, J. H., van der Ent, C. K., & Vlieger, A. M. (2015). Intestinal microbiota composition after antibiotic treatment in early life: The INCA study. *BMC Pediatrics, 15*, 204.

Sarzotti, M., Robbins, D. S., & Hoffman, P. M. (1996). Induction of protective CTL responses in newborn mice by a murine retrovirus. *Science, 271*, 1726–1728.

Schei, K., Avershina, E., Oien, T., Rudi, K., Follestad, T., Salamati, S., & Odegard, R. A. (2017). Early gut mycobiota and mother-offspring transfer. *Microbiome, 5*, 107.

South, M. A., Warwick, W. J., Wolheim, F. A., & Good, R. A. (1967). The IgA system. 3. IgA levels in the serum and saliva of pediatric patients – Evidence for a local immunological system. *The Journal of Pediatrics, 71*, 645–653.

Stecher, B., Berry, D., & Loy, A. (2013). Colonization resistance and microbial ecophysiology: Using gnotobiotic mouse models and single-cell technology to explore the intestinal jungle. *FEMS Microbiology Reviews, 37*, 793–829.

Stefka, A. T., Feehley, T., Tripathi, P., Qiu, J., McCoy, K., Mazmanian, S. K., Tjota, M. Y., Seo, G. Y., Cao, S., Theriault, B. R., et al. (2014). Commensal bacteria protect against food allergen sensitization. *Proceedings of the National Academy of Sciences of the United States of America, 111*, 13145–13150.

Strachan, D. P. (1989). Hay fever, hygiene, and household size. *BMJ, 299*, 1259–1260.

Thiemann, S., Smit, N., & Strowig, T. (2016). Antibiotics and the intestinal microbiome: Individual responses, resilience of the ecosystem, and the susceptibility to infections. *Current Topics in Microbiology and Immunology, 398*, 123–146.

Thome, J. J., Bickham, K. L., Ohmura, Y., Kubota, M., Matsuoka, N., Gordon, C., Granot, T., Griesemer, A., Lerner, H., Kato, T., & Farber, D. L. (2016). Early-life compartmentalization of human T cell differentiation and regulatory function in mucosal and lymphoid tissues. *Nature Medicine, 22*, 72–77.

Torow, N., Yu, K., Hassani, K., Freitag, J., Schulz, O., Basic, M., Brennecke, A., Sparwasser, T., Wagner, N., Bleich, A., et al. (2015). Active suppression of intestinal CD4(+)TCRalphabeta(+) T-lymphocyte maturation during the postnatal period. *Nature Communications, 6*, 7725.

Torow, N., Marsland, B. J., Hornef, M. W., & Gollwitzer, E. S. (2017). Neonatal mucosal immunology. *Mucosal Immunology, 10*, 5–17.

Toscano, M., De Grandi, R., Peroni, D. G., Grossi, E., Facchin, V., Comberiati, P., & Drago, L. (2017). Impact of delivery mode on the colostrum microbiota composition. *BMC Microbiology, 17*, 205.

Ulas, T., Pirr, S., Fehlhaber, B., Bickes, M. S., Loof, T. G., Vogl, T., Mellinger, L., Heinemann, A. S., Burgmann, J., Schoning, J., et al. (2017). S100-alarmin-induced innate immune programming protects newborn infants from sepsis. *Nature Immunology, 18*, 622–632.

van Best, N., Hornef, M. W., Savelkoul, P. H., & Penders, J. (2015). On the origin of species: Factors shaping the establishment of infant's gut microbiota. *Birth Defects Research. Part C, Embryo Today, 105*, 240–251.

Vatanen, T., Kostic, A. D., d'Hennezel, E., Siljander, H., Franzosa, E. A., Yassour, M., Kolde, R., Vlamakis, H., Arthur, T. D., Hamalainen, A. M., et al. (2016). Variation in microbiome LPS immunogenicity contributes to autoimmunity in humans. *Cell, 165*, 1551.

Vereecke, L., Sze, M., Mc Guire, C., Rogiers, B., Chu, Y., Schmidt-Supprian, M., Pasparakis, M., Beyaert, R., & van Loo, G. (2010). Enterocyte-specific A20 deficiency sensitizes to tumor necrosis factor-induced toxicity and experimental colitis. *The Journal of Experimental Medicine, 207*, 1513–1523.

Ward, R. E., Ninonuevo, M., Mills, D. A., Lebrilla, C. B., & German, J. B. (2007). In vitro fermentability of human milk oligosaccharides by several strains of bifidobacteria. *Molecular Nutrition & Food Research, 51*, 1398–1405.

Wesemann, D. R., Portuguese, A. J., Meyers, R. M., Gallagher, M. P., Cluff-Jones, K., Magee, J. M., Panchakshari, R. A., Rodig, S. J., Kepler, T. B., & Alt, F. W. (2013). Microbial colonization influences early B-lineage development in the gut lamina propria. *Nature, 501*, 112–115.

Wu, H., Esteve, E., Tremaroli, V., Khan, M. T., Caesar, R., Manneras-Holm, L., Stahlman, M., Olsson, L. M., Serino, M., Planas-Felix, M., et al. (2017). Metformin alters the gut microbiome of individuals with treatment-naive type 2 diabetes, contributing to the therapeutic effects of the drug. *Nature Medicine, 23*, 850–858.

Yatsunenko, T., Rey, F. E., Manary, M. J., Trehan, I., Dominguez-Bello, M. G., Contreras, M., Magris, M., Hidalgo, G., Baldassano, R. N., Anokhin, A. P., et al. (2012). Human gut microbiome viewed across age and geography. *Nature, 486*, 222–227.

Yu, Z. T., Chen, C., Kling, D. E., Liu, B., McCoy, J. M., Merighi, M., Heidtman, M., & Newburg, D. S. (2013). The principal fucosylated oligosaccharides of human milk exhibit prebiotic properties on cultured infant microbiota. *Glycobiology, 23*, 169–177.

4 Molecular Microbiome Analysis

Amélia Camarinha-Silva and W. Florian Fricke

Abstract

Traditional microbiological research largely depends on the cultivation and characterization of microorganisms under laboratory conditions. However, with the establishment of new sequencing-based methods over the last two decades that expanded the accessible fraction of the microbiota to non-cultivable members, microbiome research has gained significant momentum and popularity. Today, next-generation sequencing allows even smaller research groups to carry out massively parallel sequencing at affordable costs. Selective amplification and sequencing of universal phylogenetic marker genes such as those of the small subunit ribosomal RNA still represent a cornerstone of the taxonomic composition analysis that is typically used to describe and compare microbiome samples. At the same time, shotgun sequencing of metagenomes that represent all members of a microbial community is becoming increasingly popular as a more expensive but also more comprehensive alternative to amplicon sequencing. Both approaches generate large amounts of sequence data, which require bioinformatic support for processing, analysis, and visualization. The following chapter provides an overview of the typical steps involved in microbiome projects, starting from sample collection and storage, over 16S rRNA and other marker gene amplification, amplicon, and metagenome sequencing to bioinformatic sequence analysis.

4.1 Capturing the Complexity of the Microbiome by Cultivation

In the seventeenth century, Anton van Leeuwenhoek, the "Father of Microbiology," was the first to discover the vast complexity of microbial communities associated with the human body by studying feces through his newly developed microscope (Egerton 2006). Since then scientists have been trying to characterize microorganisms by cultivating them. However, while the true proportion of uncultivable or not-yet-cultivable microorganisms remains controversial (Lagkouvardos et al. 2017), only a minor fraction of microorganisms identified in microbiome samples can typically be grown and kept under laboratory conditions (Gutleben et al. 2018; Nichols 2007). This discrepancy between the number of microbial cells in a sample and the number of colonies growing on a plate has been coined "the great plate count anomaly" (Staley and Konopka 1985). Different approaches have been used to increase the number of cultivable

A. Camarinha-Silva · W. F. Fricke (✉)
University of Hohenheim, Stuttgart, Germany
e-mail: amelia.silva@uni-hohenheim.de; W.Florian.Fricke@uni-hohenheim.de

D. Haller (ed.), *The Gut Microbiome in Health and Disease*,
https://doi.org/10.1007/978-3-319-90545-7_4

microorganisms such as cultivation chips, which integrate cell adhesion and nutrient diffusion in a matrix (Hesselman et al. 2012), microfluidics for single-cell cultivation (Boitard et al. 2015) and "culturomics," and high-throughput approaches to incubate microbiome samples under numerous culture conditions and to further analyze cultures with a combination of mass spectrometry and sequencing (Lagier et al. 2012). These techniques have expanded our microbial culture collections with novel species, which now include over 11,000 bacterial and archaeal representatives (Parte 2014). Still, a large fraction of the human microbiota remains elusive to cultivation today.

The List of Prokaryotic Names with Standing in Nomenclature (LPSN), which includes all validly classified prokaryotic microorganisms, only comprises 6% of all species that are present in the current version of the SILVA database of bacterial and archaeal ribosomal RNA genes (Parte 2014). If the majority of microorganisms have yet to be cultured, research must depend on the direct analysis of environmental samples, e.g., by next-generation sequencing. However, using a direct comparison of cultivation and metagenomic sequencing of human fecal samples, Browne et al. showed that for a particular dataset representative of the human gut microbiota, 73.5% of the 741 metagenomic species that were bioinformatically identified were also present in a culture collection from 6 stool donors (Browne et al. 2016). While this might suggest that a larger fraction of the (gastrointestinal) human microbiota could be accessible to cultivation than previously thought, it is clear that microbiome research will continue to depend on both direct and indirect analytical methods, such as cultivation and sequencing, respectively.

The gastrointestinal tract, the body site that harbors the largest and most diverse microbial community of the human body, is generally estimated to comprise between 100 and 1000 species per individual (Browne et al. 2016), although historically estimates for these figures have varied substantially (Clavel et al. 2016). While the often cited estimate of a 10:1 ratio of microbial to human cells has recently been revisited and corrected to 1:1 (~3×10^{13} cells with a total of 0.2 kg for a 70 kg reference man) (Sender et al. 2016), it is unlikely that cultivation-based methods alone will be sufficient to comprehensively study the human microbiome. In the last decades, culture-independent, sequencing-based approaches have had tremendous success in transforming our knowledge of microbial communities. The field of microbial ecology has advanced rapidly, and microbial communities from various environmental niches, such as water, soil, plants, animals, and humans, have been characterized in depth. Yarza and colleagues estimated that by the end of 2017, a total of 400,000 species of bacteria and archaea will be discovered using high-throughput sequencing technologies (Yarza et al. 2014). This advancement has been tightly linked to the technical progress of next-generation sequencing and the decline in sequencing costs. Today, even mid-sized laboratories can afford a next-generation sequencing platform and generate up to 50 Mio. sequence reads of 300 bp length in less than 60 h.

Genomics and other high-throughput omics technologies have improved to an extent that allows us to dissect microbial communities, their functions, and metabolic processes at unprecedented resolution. In order to collect such high-resolution data in a systematic manner and enable significant data integration and analysis, standard operating procedures (SOPs) for sample collection; storage and preservation; isolation; sequencing and other processing of nucleic acids, proteins, and metabolites; data processing; storage; and sharing have become crucial. Recently, consortia like the Earth Microbiome Project (EMP) (Gilbert et al. 2011), the Human Microbiome Project (HMP) (Human Microbiome Project Consortium 2012), and the human gut metagenome initiative MetaHIT (Qin et al. 2010) have established methodologies in order to enable the comparison between diverse studies across different countries. Method standardization is important to ensure complete representation of bacterial communities and to make studies comparable.

4.2 Collection and Storage of Microbiome Samples?

The best conditions for microbiome sample collection and storage are dependent on the intended application. Samples intended purely for analysis should be immediately preserved at the time of collection, including nucleic acids (metagenome and metatranscriptome), proteins (metaproteome), and metabolites (metabonome; see Box 4.1). It is important to note, however, that most standard protocols for microbiome sample preservation are focused on the stabilization of nucleic acids and in the process kill large fractions of the microbiota making them unsuitable to preserve live, cultivable microorganisms.

DNA, RNA, protein, and metabolite profiles of microbiome samples can rapidly change at room temperature due to degradation, protein modification, and gene expression, as the microbiota can maintain metabolic activity. Snap freezing by immersion in liquid nitrogen is a common method to stabilize samples for RNA analysis, which is limited by the availability of the necessary materials at the sampling site and the expenses of cold shipping and storage (Mutter et al. 2004). Storage of microbiome samples for nucleic acid extraction at −70 to −80 °C is the most commonly used method and considered the best approach to preserve biological samples for further analysis (Rissanen et al. 2010). However, if microbiome sampling takes place at remote locations or nonclinical sites—for example, in studies where participants have to self-sample at home—immediate and continuous freezing is not always possible (Choo et al. 2015). For these conditions, chemical preservation agents are often used to store microbiome samples and protect DNA and RNA from degradation between sample collection, storage, and extraction.

There is an ongoing debate in the scientific community about the best storage conditions for microbiome samples, with some studies suggesting relatively minor effects of even extended storage at room temperature (Tedjo et al. 2015). This controversy might have to do with uncertainty about the best parameter to describe and functionally characterize a microbiota and thus the problem to properly evaluate the effects of different storage conditions.

It should be pointed out that the only clinically relevant application of the microbiota today is the use of gut microbiome samples for fecal microbiota transplantation (FMT), which is assumed to depend on the presence of live, cultivable microbes, although this view has recently been challenged (Ott et al. 2017). Stock cultures of individual bacterial isolates are typically stored in glycerol at −80 °C, and similar long-term storage conditions are typically applied to preserve microbiome samples in stool banks intended for fecal microbiota transplantation (Satokari et al. 2015).

4.3 Extraction of Nucleic Acids from Microbiome Samples

Depending on the sample type and composition, microbiome samples can be difficult to process with the goal to extract nucleic acids. Microorganisms have evolved strategies to cope with damaging environmental conditions, such as UV radiation, oxidative stress, desiccation, low pH, etc., which include complex cell wall structures or the formation of spores, i.e., metabolically inactive, durable forms. These same strategies make it difficult to break open microbial cells to extract nucleic acids. They can also lead to biases in the representation of microbial genomes or transcriptomes in DNA or RNA extracted from microbiome samples, if extraction protocols favor easier-to-lyse over harder-to-lyse microorganisms.

For example, bacteria can be divided into Gram-negative and Gram-positive bacteria, based on their cell wall structure. Gram-negative bacteria have a thin layer of peptidoglycan between the inner and outer lipid membranes, while Gram-positive bacteria have a single lipid membrane enclosed by a thick layer of peptidoglycan and lipoteichoic acid (Brown et al. 2015). The layer of peptidoglycan is responsible for the strength of the wall and in case of Gram-positive bacteria is hard to disrupt, making the cell lysis process more challenging.

Nucleic acid extraction from microbiome samples typically involves chemical, enzymatic, and mechanical lysis of viral, bacterial, archaeal, and eukaryotic microorganisms, including efficient disruption of Gram-positive bacteria and inactive bacterial spores. Most commercial nucleic acid extraction kits incorporate a lysis step, but protocols differ in the efficacy for even microbiome lysis and thus represent one important factor to explain study biases and inconsistencies in the results of microbiota analyses between studies. As a consequence, inefficient cell lysis can result in the underrepresentation of Gram-positive bacteria in microbiome sample composition data.

The best-known chemical agents for cell lysis are surfactants or detergents, which help dissolve membrane proteins and lipids by disrupting the boundary between hydrophobic and hydrophilic systems (e.g., Triton X-100 or sodium dodecyl sulfate (SDS)). Chelating agents (e.g., ethylenediaminetetraacetic acid (EDTA)) bind metal ions such as magnesium ions and make them inaccessible for other chemical reactions. Chelating agents can lyse Gram-negative cells by binding cations and creating holes in their cell walls and, at the same time, prevent degradation of nucleic acids by DNases, which depend on the availability of Mg^{2+} to complex DNA. Organic solvents (e.g., ether or chloroform) permeate the cell membranes and walls and lyse the cells. Chaotropic agents (e.g., urea or guanidine) are also used for cell lysis as they disrupt hydrophobic interactions between solute molecules.

Enzymatic treatment of microbiome samples typically involves lysozyme, which degrades Gram-positive bacterial cell walls; trypsin, which is used to release cells from tissues; and Proteinase K, which inactivates nucleases and prevents nucleic acid degradation and is active at elevated temperatures and resistant to SDS (Salazar and Asenjo 2007). Similar to lysozyme, other naturally found enzymes have been made commercially available to increase the efficiency of lysing-specific microbiota components, e.g., glucanases, to disrupt fungal cell walls and others.

Mechanical disruption is used to break open thick-walled microorganisms and microbial spores by physical force. This is achieved by mixing microbiome samples with glass, zirconium, silica, ceramic, or stainless steel beads and oscillating the mixture at high frequency. Optimization of this bead-beating process is important, as the physical stress can heat up the sample and longtime homogenization might shear nucleic acid molecules and compromise the quality and yield of the isolate (de Boer et al. 2010).

4.4 Simultaneous Extraction of DNA, RNA, and Proteins for Multiple Omics Analyses

The classical method for DNA extraction uses a phenol-chloroform combination to extract mixtures of molecules based on their solubility in two immiscible solutions (Kirby 1956). With the inclusion of guanidinium thiocyanate, Chomczynski and Sacchi expanded this protocol into a single-step method for the simultaneous isolation of RNA, DNA, and proteins (Chomczynski and Sacchi 2006). Several companies enhanced the method, e.g., by limiting RNase activity during cell disruption, and sell it worldwide under different designations, the most common being the TRIzol® reagent from Invitrogen.

Besides the proprietary TRIzol agent, this protocol requires chloroform, ethanol, isopropyl alcohol, sodium citrate, and sodium hydroxide. In essence, after thorough mixing, incubation, and centrifugation, the protocol generates a separation of the sample into three phases: an upper colorless aqueous phase that contains the RNA, an interphase, and a lower red phenol-chloroform phase that contains DNA and proteins. DNA, RNA, and proteins are then individually precipitated, washed, and resuspended from the two separate phases (Fig. 4.1). Total RNA, DNA, and proteins can be quantified with a spectrophotometer or fluorometer (DNA and RNA) or Bradford assay (proteins), used in downstream applications or stored at −70 °C (RNA) or −20 °C (DNA and proteins).

Simultaneous extraction of nucleic acids and proteins can be advantageous in projects with small amounts of starting material, where it is critical to acquire sufficient material for downstream applications in a single round of extraction. Another great advantage of this method is that all studies can be performed on the same sample, allowing direct comparisons between the results obtained from metagenomics, metaproteomics, and metatranscriptomics (Hummon et al. 2007).

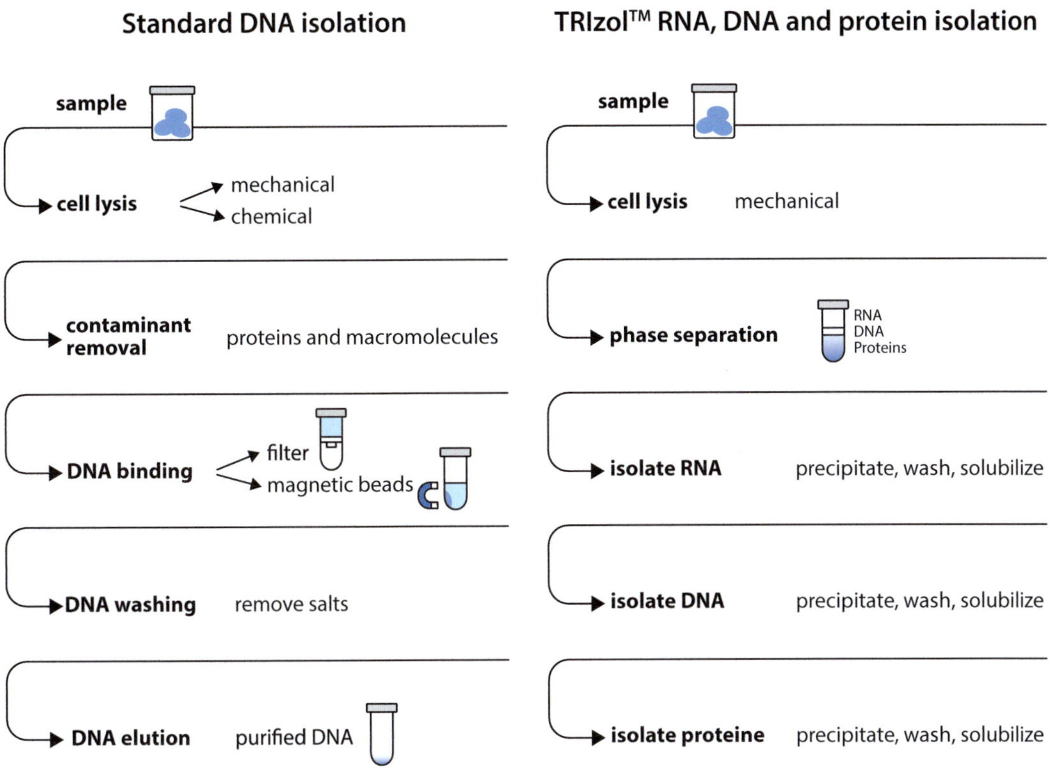

Fig. 4.1 Protocol overview for the processing of microbiome samples, using standard commercial kits for the isolation of metagenomic DNA or the TRIzolTM-based protocol for the simultaneous isolation of RNA, DNA and protein

4.5 Isolation of Metagenomic DNA from the Microbiome

Various DNA extraction kits have been made commercially available with user-friendly protocols that can handle a dozen or more samples in less than 2 h. A further increase in sample throughput can be achieved by using automated robotic solutions. Specific kits exist to isolate DNA from bacteria, algae, fungi, feces, food, hairs, plants, roots, seeds, soil, tissue, skin, wastewater, and soil among others. Despite the diversity of applications and small differences in the protocol, these kits follow similar principles and share a general overall protocol: cell lysis, contaminant removal, DNA binding, washing, and elution (Fig. 4.1). Most of the commercial kits are suitable for processing 200–400 mg of sample material. Often, reducing the quantity of starting material improves lysis efficiency and DNA purity.

A distinguishing feature between commercial kits for DNA isolation, especially for human microbiome studies, can be the removal of contaminants and inhibitors of downstream applications. For example, co-extraction of humic acids with metagenomic DNA is a frequent problem in the analysis of human fecal samples, which can inhibit DNA polymerases used for taxonomic microbiota composition analysis by targeted nucleic acid amplification (see below). Some kits include a contaminant removal step to remove humic acids, cell debris, and proteins, which can diminish DNA purity and also inhibit downstream applications. Under conditions of high salt concentrations, DNA is either bound to a silica membrane in a filter or, often in combination with automated systems, to silica-cladded magnetic beads. DNA is then washed with ethanol-based solutions that keep the DNA precipitated but solve and remove residual salts, humic acids, and other contaminants. A proper removal by centrifugation of the ethanol-based solution is crucial to improve downstream applications. At last, DNA is eluted with water or low-salt buffers and can be directly checked in an agarose gel, quantified, used in downstream applications, or frozen (−20 °C to −80 °C). A successful DNA extraction protocol should generate high yields of pure DNA from all the different species present in the sample.

In practice, the success of all nucleic acid extraction protocols is also largely dependent on the removal of host DNA and RNA from the sample. The presence of nucleic acids from the host can affect the success of downstream applications such as metagenomics, metaproteomics, and metatranscriptomics, either directly by inhibiting amplification reactions or indirectly by reducing the fraction of microbial data in the resulting data output, effectively increasing the cost of sequencing (Oyola et al. 2013).

4.6 Sequencing the Microbiome

With the advent of next-generation sequencing, which dramatically reduced the effort, footprint, and cost for the installation and use of high-throughput sequencing infrastructures, the microbiome research field has significantly expanded. Instead of depending on the isolation and characterization of cultivable microorganisms, it allowed for a standardized high-throughput, large-scale comparative characterization of complex microbial communities from hundreds of samples, including known and unknown, cultivable and non-cultivable microorganisms. As a consequence, microbiome analysis has become a widely used molecular tool with increasing applications in various areas of research, including genomics, immunology, and physiology.

Two main sequencing-based approaches are mostly commonly applied to microbiome research: targeted amplicon sequencing (Metataxonomics) and metagenomic shotgun sequencing (Fig. 4.2). The first reduces the cost of microbiome analysis while maintaining a comparatively high resolution by focusing on specific phylogenetic marker sequences that are extracted, amplified, and sequenced. These target loci are typically amplified by polymerase chain reaction (PCR) and sequenced in parallel, effectively reducing the required sequencing effort to (at least theoretically) one sequence per microorganism and generating a sequence-based phylogenetic profile of the original sample. As a downside, no further information from the studied microbiome is available beside the taxonomic profile of the targeted microbiota component.

Metagenomic shotgun sequencing on the other hand omits the target selection and amplification steps and directly proceeds to sequencing the complex metagenomic DNA mixture as isolated from the entire microbial community in the sample. As substantially greater sequencing efforts are required to characterize a microbiome sample in depth, this second approach is much more expensive than targeted amplicon sequencing. However, because of the lack of specific marker sequence amplifications, it also provides a more comprehensive and less biased insight into the entire microbiota, including members of all three domains of life, i.e., phages and viruses, bacteria and archaea, fungi, protozoa, and other eukaryotes. In addition, not only insights into phylogenetic but also functional microbiome profiles can be gained from metagenomic shotgun sequencing.

The famous questions "Who is in there?", "What are they doing?", and "How are they doing it?" can be even better answered by complementing genomic tools with further "omics" technologies linking microbial community characterization to ecological processes. Targeted amplicon sequencing and shotgun metagenomics fail to reveal the metabolic activity of the microbiome because they are not capable to distinguish between expressed and non-expressed genes. Metatranscriptomics gives information on regulatory and expression profiles and networks of a microbiome sample, by sequencing, quantifying, and comparing relative transcript levels from all microorganisms. Metabonomics generates metabolite profiles of the sample and can be used to study the actual consequences of all metabolic activities in a sample, irrespective of their microbial or host origin.

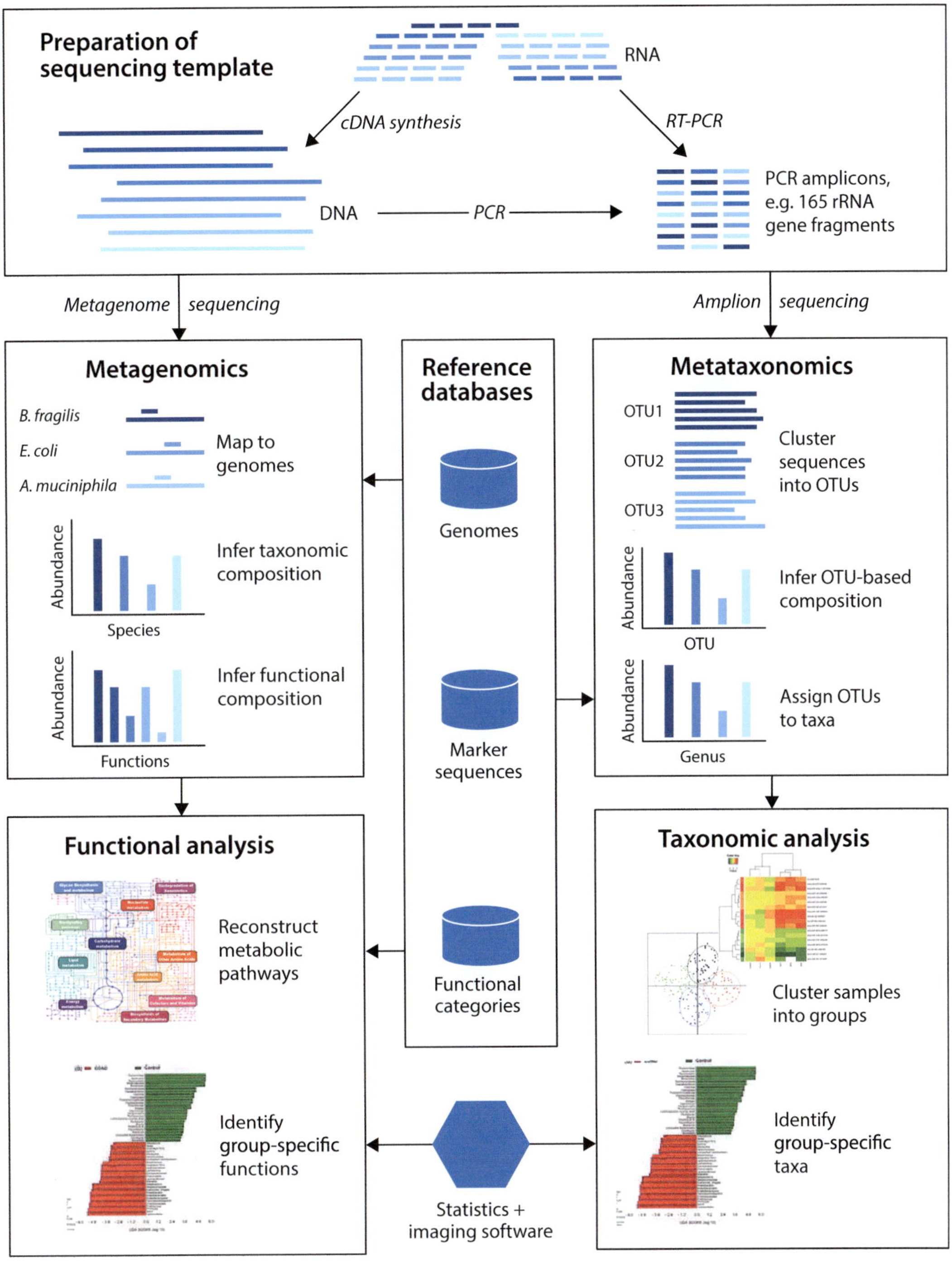

Fig. 4.2 Workflow overview for amplicon or metagenome sequencing-based microbiome projects

4.7 Targeted Amplicon Sequencing: Generating Phylogenetic Profiles of Specific Microbiome Members

Most recent insights into host-microbiome interactions have resulted from next-generation sequencing of gene fragments of the small subunit of the bacterial ribosome, the 16S rRNA gene. Similarly, less widely applied methods exist for the small rRNA subunit of Eukarya, i.e., the 18S rRNA gene, and of rRNA gene fragments and intergenic regions of fungi, i.e., internal transcribed spacers (ITSs). These genetic marker loci have been selected based on two crucial properties: (1) conserved regions with little to no sequence variation between the targeted organisms, which serve as binding sites for universal primers during the PCR amplification step, and (2) hypervariable regions that vary even between closely related organisms and thus carry the phylogenetic signal of the sequenced PCR amplicons for the taxonomic assignment back to its source organism. These two features have most extensively been studied for the bacterial 16S rRNA gene, and degenerate primer pairs have been described with relatively even binding capacities across the entire bacterial kingdom (Claesson et al. 2010). In eukaryotes, due to greater sequence variations, even between closely related organisms, fewer primer pairs have been described for microbiota analysis, which typically only focus on smaller taxonomic groups, such as fungi or protozoa.

Besides biological considerations, the selection of target loci for amplicon sequencing is currently also limited by sequencing technologies. The read length generated by the sequencing platform dictates the length of the target locus that is amplified for sequencing. For example, most bacterial microbiome studies have been using targeted amplicon sequencing of only a fragment instead of the entire bacterial 16S rRNA gene (>1500 bp) or the even larger 23S rRNA gene (>2900 bp), which would allow for a higher phylogenetic resolution of the analysis but cannot be covered in a single sequence read or two overlapping paired-end reads.

The bacterial 16S ribosomal RNA gene has been used for many years for the taxonomic typing of bacteria and in culture-independent microbiome analyses. Applications include quantitative PCR, terminal restriction fragment length polymorphism (T-RFLP), and amplicon sequencing. Originally, 16S rRNA gene PCR products were cloned in plasmid libraries and used for Sanger sequencing, today high-throughput applications rely on next-generation sequencing. The drawback of current methods is that due to the limited read length of the most cost-efficient sequencing technologies, only two or three consecutive regions of the 16S rRNA gene are covered by the amplification (Kircher and Kelso 2010). As these regions do not always carry large variations between closely related bacteria, this method loses power in the classification at genus and species levels (Claesson et al. 2010). In addition, not all hypervariable regions of the 16S rRNA gene carry the exact same phylogenetic signal in all bacteria, leading to different recommendations for primer pairs, depending on the intended application. While previous studies mostly favored the hypervariable regions V2–V4, others preferred V3–V5, V2–V3, V1–V2, or in some cases even just a single hypervariable region V4 or V6 (Claesson et al. 2010). An important conclusion is that microbiome analyses will generate variable community compositions, i.e., show different bacteria as present or absent or vary in their relative abundance profiles, depending on the targeted hypervariable region used (Claesson et al. 2010). As a consequence, besides other factors, primer biases are responsible for the incompatibility of microbiome data from different studies or study batch effects, as data generated in one study should not be compared with those from another study, even if similar types of samples were analyzed.

As the eukaryotic homologue of the 16S rRNA gene in prokaryotes, mitochondria, and chloroplasts, the 18S rRNA gene is used to depict eukaryotic diversity (Meyer et al. 2008). However, widespread usage of this gene is still limited because targeted regions are typically longer than the sequence length that is available from current sequencing platforms (Amaral-Zettler et al. 2009) and because eukaryotic 18S rRNA genes show greater variations in copy numbers between species than bacterial 16S rRNA genes (Prokopowich et al. 2003). In silico analyses of available reference sequences suggest that the best coverage and taxonomic resolution of eukaryotic taxonomic variation can be achieved with regions comprising more than 500 nucleotides (Bradley et al. 2016).

Intergenic transcribed spacers between the 18S and 5.8S rRNA genes (ITS1) and between the 5.8S and 26S rRNA genes (ITS2) of the eukaryotic ribosomal RNA operon have become key molecular targets for amplicon sequencing-based taxonomy characterizations of fungal communities. ITS sequence variations provide the broadest classification of fungi today and have been proposed by the Fungal Barcoding Consortium to be used as the primary fungal marker (Schoch et al. 2012). In 2014 the National Center for Biotechnology Information (NCBI) created a curated public database of ITS reference sequences, the RefSeq Targeted Loci (RTL) database (Schoch et al. 2014).

Ultimately, it is the parallel analysis of multiple microbiome samples in a single sequencing reaction, which leads to the drastic cost-effectiveness of targeted amplicon sequencing compared to metagenomic shotgun sequencing methods. This is typically achieved by using sample-specific sequence barcodes, which are attached to the 5′ ends of the PCR primer primers and thus added to each amplicon during the amplification reactions. Barcoding systems that employ hundreds of unique sequence tags of 8–12 nucleotide length have been published, allowing for the pooling and parallel sequencing of hundreds of samples in a single sequencing reaction (Hamady et al. 2008; Sogin et al. 2006).

4.8 Metagenomic Shotgun Sequencing: Taxonomic and Functional Profiling of the Entire Microbiome

Metagenomics refers to the analysis of DNA mixtures from microbial communities, i.e., instead of focusing on individual microbial genomes or selected target regions from multiple genomes, the entire "meta"genome of all microorganisms combined is studied (Eisen 2007). Because of the required sequencing depth to provide sufficient resolution for hundreds of genomes simultaneously, metagenomics usually relies on short-read next-generation sequencing platforms that provide the best cost-per-base pair ratio. In order to generate an even distribution of all microbial genomes in the metagenomic sequence dataset, DNA isolates need to be sheared into fragments of equal length. The term "shotgun sequencing" refers to the generation of sequencing libraries by randomly fragmenting DNA templates, irrespective of differences between microbial genomes, such as specific sequence motifs, G+C contents, secondary structures, etc. As such, the generation of metagenomic sequencing libraries is not fundamentally different from those generated for single microbial or eukaryotic genomes, i.e., template DNA is mechanically or enzymatically sheared and resulting fragments are ligated to adaptors used for sequencing, amplified, and sequenced (Fleischmann et al. 1995).

4.9 Other Omics Technologies: Insights into Gene Expression and Metabolic Activities

Metatranscriptomics refers to the analysis of transcriptional activities of microbial communities by whole-transcriptome shotgun sequencing or "RNA-Seq." RNA isolated from microbiome samples is reverse-transcribed into cDNA, typically using random hexamer primers. cDNA fragments can be sequenced using the same methods and platforms as metagenomic DNA. This metatranscriptomic approach has its own specific challenges, as it can be difficult to extract sufficient amounts of pure RNA devoid of DNA and to evenly reverse-transcribe this RNA to high-quality cDNA that can be sequenced (Simon and Daniel 2011).

Metaproteomics identifies and characterizes the proteins of all microorganisms that are present in a microbiome sample at a particular time point. Peptide identification is achieved using liquid chromatography combined with mass spectrometry. Some considerations have to be taken into consideration during sample preparation and to choose the protein extraction protocol to improve the total number and even representation of proteins from the samples (Wilmes et al. 2015).

Traditionally, metabolomics refers to the identification and quantification of the metabolites released by a clonal culture of a single microorganism or (homogenous) tissue sample, which provides information about all its metabolic activities. To expand this analysis to complex, heterogeneous microbial communities, the term metabonomics has been proposed (Marchesi and Ravel 2015), to avoid using the clumsy term "meta-metabolomics." The metabonome is a good indicator of the health of an environment and can be characterized with nuclear magnetic resonance (NMR) spectroscopy and mass spectrometry (MS) linked to a liquid chromatography separation system (Patejko et al. 2017).

4.10 Bioinformatic Principles of Sequence-Based Microbiome Analysis

Sequence-based microbiome analysis relies on the identification and quantification of similarities between sequences or sequence fragments to infer phylogenetic relationships and shared functionalities. In general, the taxonomic compositions of microbiome samples are determined by 16S rRNA gene amplicon sequencing using phylogenetic relationships determined by multiple sequence alignments (Lane et al. 1985). A more comprehensive profile of both taxonomic and functional microbiota characteristics can be determined by metagenomic shotgun sequencing based on the alignment of sequences to taxonomic or functional reference databases (Morgan and Huttenhower 2014). Several widely used bioinformatic platforms support 16S rRNA gene amplicon and metagenomic shotgun sequence analysis, including open-source and proprietary software that is available for download and local installation or web-based execution on online servers (Table 4.1).

In case of targeted, amplicon sequencing-based microbiota analysis, phylogenetic relationships are determined on the basis of sequence similarities and used for taxonomic assignments, e.g., if 16S rRNA genes with at least 97% sequence similarity are assigned to the same bacterial species. The thresholds of minimal sequence similarity used to define taxonomic units such as the bacterial species are empirically determined based on the in silico comparison of available sequence data from the genome databases.

Groups of sequences sharing at least the similarity defined in the taxon threshold are combined into *operational taxonomic units* or *OTUs*, which are equivalent of biological taxa but defined statistically based on sequence similarity. For efficiency,

Table 4.1 Bioinformatic support platforms

16S rRNA gene amplicon sequence analysis		
Mothur	Open-source software package for amplicon sequencing-based microbiota analysis	Schloss et al. (2009)
QIIME 2	Open-source software package for amplicon sequencing-based microbiota analysis	Caporaso et al. (2010)
RDP	Open-source bacterial and archaeal 16S and fungal 28S rRNA sequence database and amplicon sequence analysis tools	Cole et al. (2014)
Silva	Reference database of small and large subunit rRNA sequences for all three domains of life	Quast et al. (2013)
Metagenomic shotgun sequence analysis		
MG-RAST	Automated, web-based, open-source pipeline for metagenomic sequence analysis	Keegan et al. (2016)
MetaPhlAn	Open-source software for taxonomic profiling of metagenomic sequence data	Segata et al. (2012)
HUMANn	Open-source software for metabolic pathway prediction from metagenomic or metatranscriptomic sequence data	Abubucker et al. (2012)
KEGG	Reference database of metabolic pathways	Kanehisa et al. (2017)

the clustering approach to generate OTUs is typically performed as a two-step process, which is referred to as "open-reference OTU picking" (Rideout et al. 2014): new sequences are first compared against a reference database ("closed-reference OTU picking"), and then the remaining sequences, i.e., those without close relatives in the reference database, are compared against each other ("de novo OTU picking"). If a known, classified reference sequence falls within the similarity threshold of a specific OTU, its "name" or taxonomic lineage is adopted. The use of the "artificial," sequence-based OTU unit has the advantage to allow for the identification, classification, and comparison of previously unknown taxa, i.e., taxa without representatives in the reference databases. For example, the comparison of fecal samples from two patient cohorts by 16S rRNA gene fragment amplicon sequence analysis could identify a group of previously unknown bacteria (i.e., an OTU of sequences with $>=$ 97% sequence similarity), which is overrepresented in one compared to the other cohort. While the specific bacterial species might be unknown, it could potentially still be classified at a higher taxonomic level, e.g., as a species within the family *Enterobacteriaceae*, and thus serve as an important biomarker for the specific patient cohort and as a target for further analysis.

It is important to note that the relationship between sequence similarity and phylogenetic relationships can differ between taxa and target amplicons, due to varying selective pressures, mutation rates, or simply the length of the amplicon. For example, different selective pressures shape the evolution of the bacterial 16S rRNA gene, the hypervariable regions V1–9 within the 16S rRNA gene, or the internal transcribed spacer (ITS) amplicons, which include both rRNA gene fragments and intergenic regions. In addition, in case of multi-copy target amplicons, such as the bacterial 16S rRNA gene, individual copies have been shown to vary (Větrovský and Baldrian 2013).

Metagenomic shotgun sequencing has the potential to characterize the entire genetic potential of a microbiome sample. As genomic DNA isolates from the sample are directly submitted to sequencing, there is no restriction imposed by the conditions of the PCR amplification of specific target loci that is required for amplicon sequence analysis. As a result, at least theoretically, the complete genome sequences, including all protein-coding and noncoding sequences from all organisms in the sample, can be obtained, including viral, bacterial, archaeal, fungal, and other eukaryotic microorganisms. However, lack of a targeted amplification step also means less control over the sequencing template. Human DNA, which is typically not the target of microbiome projects, can represent a considerable problem for metagenomic shotgun sequencing of microbiome samples. The "contamination" of microbiome samples with human DNA is less of a problem for microbiota analyses of fecal samples with relatively few human cells or DNA, which is one reason why fecal microbiota analysis has become so popular. However, as human cells contain roughly 1000× more DNA than bacterial cells, human DNA contamination is more problematic for other types of microbiome samples, such as epidermal or mucosal samples, vaginal or skin swabs, and biopsies or tissue samples, among other reasons resulting in fewer available metagenomic sequence data from these project types. Removing human DNA from metagenomes using selective hybridization before sequencing or increasing the sequencing effort for problematic samples have been proposed to address this problem; bioinformatically, the removal of human reads from sequence data poses less of a problem (Ferretti et al. 2017).

Metagenomic shotgun sequence analysis is generally based on the comparison of individual sequence reads to a reference database, with the analysis output type depending on the focus and functionality of the database. Reference databases for taxonomic composition analysis include curated datasets of complete genome sequences (e.g., NCBI's RefSeq) or datasets of biomarkers that have been bioinformatically identified based on comparative genome analysis as specific for individual microbial taxa [e.g., as available through the MetaPhlAn2 program (Truong et al. 2015)]. Similarly, mapping of metagenomic shotgun reads against orthologs of functionally related genes, such as the KEGG database (http://www.genome.jp/kegg/), can be used to determine functional abundance composition profiles of a microbiome sample. As on the one hand phylogenetically distinct bacteria have been shown to perform similar ecological functions in a microbiome and, on the other hand, phylogenetically similar bacteria can phenotypically differ substantially, functional profiles have been suggested to better characterize microbiomes than taxonomic profiles.

Bioinformatically, the alignment of metagenomic and reference sequences is performed by mapping individual reads against reference markers. Mapping refers to an alignment, which only allows for minor differences between query and reference, typically less than a few single nucleotide variants per 100 bp. This limitation makes mapping bioinformatically efficient enough to process even large metagenomic datasets within less than 1 day.

4.11 What Is the Best Feature to Describe the Microbiome?

The vast majority of publications in microbiome research have been characterizing the microbiota based on taxonomic compositions determined by 16S rRNA gene amplicon sequencing, with the goal to identify specific taxa that would be over- or underrepresented in samples of interest, such as patients, infants, geographic locations, ethnicities, etc. This approach has led to several widely cited observations, such as the association of obesity with altered ratios of the two dominant bacterial phyla in the gut, *Firmicutes* and *Bacteroidetes* (Turnbaugh et al. 2006), or the classification and assignment of healthy individuals to so-called enterotype (Arumugam et al. 2011). However, these findings have also been discussed controversially (Knights et al. 2014). Moreover, many more examples exist where a general association of health features with the microbiome has been established—typically based on antibiotic interventions or studies in germ-free mice—but where no direct link to a specific microbial taxon could be established. This is obvious from unclear or even contradictory findings that have been published in the context of health problems such as obesity and inflammatory bowel disease.

One potential reason for inconsistent findings from different studies could be functional redundancy. It is now well-established that different microbial taxa can perform similar tasks within the human microbiome. This becomes clear when comparing microbiome samples based on taxonomic and functional profiles: despite substantial taxonomic differences, microbiomes contain similar relative abundance profiles of functional categories.

From the human perspective, the contribution and relevance of those microbial features that are typically the subject of microbiome projects are not always clear. For example, a specific microbial metabolic pathway could be involved in the adaptation of a microorganism to a specific environment, without directly affecting the human host. A number of promising recent studies have therefore attempted to characterize the microbiota based on functional parameters with direct relevance to the human host, such as their recognition by the immune system. These methods have, for example, identified bacterial taxa from different phylogenetic branches that stimulate regulatory T cells (Atarashi et al. 2011) or the production of IL-17 (Ivanov et al. 2008).

In summary, ample evidence supports the importance of the human microbiome for health and disease. Improving our understanding of the specific factors responsible for the microbiome role for human health will depend on expanding and standardizing available research tools, including genomic and bioinformatic protocols and methods.

Box 4.1: The Terminology of Microbiome Research

- OTU (operational taxonomic unit): group of sequences that are separated from others by using clustering approaches under a specific sequence identity threshold. Sometimes referred as phylotypes.
- Microbiome: the entire habitat including the microorganisms, their genes, genomes, proteomes, and metabolomes.

- Microbiota: the group of microorganism (bacteria, archaea, or lower eukaryotes) inhabiting a specific niche. The microbiota is defined based on molecular approaches that amplify 16S rRNA genes, 18S rRNA genes, ITS regions, or specific marker genes.
- Metagenome: the collection of genes and genomes of the microbiota members and its encoded functions.
- Metatranscriptomics: the study of the expressed RNAs with information on the regulation and expression profiles.
- Metaproteomics: protein identification and characterization of a specific sample at a given point in time.
- Metabolomics: quantitative and qualitative determination of the metabolites of a microorganism, a tissue, or a single cell.
- Metabonomics: metabolomic analysis of multiple organisms, equivalent to "meta-metabolomics."

Box 4.2: History of Microbiome Analysis

1681 Leeuwenhoek examined feces under a microscope and described different microorganisms.

1977 Development of dideoxy chain terminator sequencing by Fred Sanger.

1995 Completion of the first complete genome sequences: *H. influenzae* and *M. genitalium*.

2005 Introduction of the first next-generation, high-throughput sequencing platform (Roche/454 GS20).

2007 Launch of the currently mostly widely used sequencing platform with the Illumina Genome Analyzer II

2008 Completion of the Human Microbiome Project (USA), microbial characterization of different sites of the human body, and MetaHIT (Europe), metagenomics of the human intestinal tract.

2010 Earth Microbiome Project: microbial communities across the globe.

Box 4.3: Microbiome Research: Hype and Hubris of an Emerging Field

Human microbiome research continues to rapidly expand with currently well beyond 100 new publications being released on the subject per week. This rapid growth and the enthusiasm that accompanies it come at the cost of several important weaknesses, limitations, and inaccuracies that will be outlined in the following:

- Terminology: A number of poorly or inconsistently defined expressions are widely found in the literature on microbiome research. Following recommendations made by Marchesi and Ravel (2015), Box 4.3 provides an overview of the essential microbiome vocabulary.
- Interpretation of findings: The vast majority of all existing microbiome studies, particularly those from 16S rRNA gene amplicon sequencing project, continue to report proportional data, although protocols for quantitative microbiome profiling have been proposed (Vandeputte et al. 2017). This means that microbial taxa or functions are listed as fractions of 100%, irrespective of absolute quantities, such as the number of microbial cells in a sample. Still, when comparing two groups of compared samples, e.g., patients with a disease and healthy controls, differences are often interpreted as showing shifts or increases in one member of the microbiota and decreases in another. In reality, the total number of microorganisms could increase or decrease, leading to misleading interpretations of dependent relative abundance values. It is therefore likely that quantitative microbiome analyses will become more widely used in the future.
- Controls: Due to the sensitivity of the bacterial 16S rRNA gene PCR amplification and the omnipresence of bacteria in the environment, laboratory, and even consumables, there is a huge risk of reporting false-positive microbiome results, particularly when studying low-biomass samples. It is possible that a number of the enthusiastic, early microbiome studies have been misinterpreting contamination. For example, the existence of a placenta microbiome remains controversial (Perez-Muñoz et al. 2017). As a consequence, extreme care should be applied when processing microbiome samples, and negative controls to assess potential contamination during nucleic acid extraction and PCR amplification should be included, processed, sequenced, and analyzed side by side with samples (Kim et al. 2017).
- Transparency and reproducibility: As datasets grow larger and bioinformatic and biostatistic protocols become more complex, full and open access to microbiome sequence data, source data and background sample information (so-called metadata), as well as to scripts and programs used for data

processing, should be mandatory. Full transparency is a critical prerequisite to afford reproducibility and maintain integrity of the field.

Box 4.4: Summary and Highlights

- The future and long-term success of microbiome research will depend to a large extent on standard operating procedures (SOPs), i.e., community-validated, standardized, and efficient protocols, for sample collection, storage and preservation, isolation, and sequencing.
- Successful nucleic acid extraction protocols involve steps to eliminate inhibitory substances, unwanted DNA or RNA from microbiome samples, as well nucleic acids from the host to improve the outcome of downstream analysis.
- Targeted amplicon sequencing can be used to generate sequence-based phylogenetic and taxonomic profiles of specific microbiome members and components, while metagenomic shotgun sequencing provides insights into both phylogenetic and functional profiles of the entire microbiota.
- Bioinformatic sequence analysis for microbiome analysis relies on the identification and quantification of sequence similarities to infer phylogenetic relationships and shared functionalities.
- Microbiome research is an active, growing field in the process of developing the best set of experimental and bioinformatic tools, protocols, and parameters to study and determine the ecological, physiological, immunological, and clinical relevance of the microbiome.

References

Abubucker, S., Segata, N., Goll, J., Schubert, A. M., Izard, J., Cantarel, B. L., Rodriguez-Mueller, B., Zucker, J., Thiagarajan, M., Henrissat, B., et al. (2012). Metabolic reconstruction for metagenomic data and its application to the human microbiome. *PLoS Computational Biology, 8*, e1002358.

Amaral-Zettler, L. A., McCliment, E. A., Ducklow, H. W., & Huse, S. M. (2009). A method for studying protistan diversity using massively parallel sequencing of V9 hypervariable regions of small-subunit ribosomal RNA genes. *PLoS One, 4*, e6372.

Arumugam, M., Raes, J., Pelletier, E., Le Paslier, D., Yamada, T., Mende, D. R., Fernandes, G. R., Tap, J., Bruls, T., Batto, J.-M., et al. (2011). Enterotypes of the human gut microbiome. *Nature, 473*, 174–180.

Atarashi, K., Tanoue, T., Shima, T., Imaoka, A., Kuwahara, T., Momose, Y., Cheng, G., Yamasaki, S., Saito, T., Ohba, Y., et al. (2011). Induction of colonic regulatory T cells by indigenous clostridium species. *Science (New York, NY), 331*, 337–341.

Boitard, L., Cottinet, D., Bremond, N., Baudry, J., & Bibette, J. (2015). Growing microbes in millifluidic droplets. *Engineering in Life Sciences, 15*, 318–326.

Bradley, I. M., Pinto, A. J., & Guest, J. S. (2016). Design and evaluation of illumina MiSeq-compatible, 18S rRNA gene-specific primers for improved characterization of mixed phototrophic communities. *Applied and Environmental Microbiology, 82*, 5878–5891.

Brown, L., Wolf, J. M., Prados-Rosales, R., & Casadevall, A. (2015). Through the wall: Extracellular vesicles in Gram-positive bacteria, mycobacteria and fungi. *Nature Reviews. Microbiology, 13*, 620–630.

Browne, H. P., Forster, S. C., Anonye, B. O., Kumar, N., Neville, B. A., Stares, M. D., Goulding, D., & Lawley, T. D. (2016). Culturing of 'unculturable' human microbiota reveals novel taxa and extensive sporulation. *Nature, 533*, 543–546.

Caporaso, J. G., Kuczynski, J., Stombaugh, J., Bittinger, K., Bushman, F. D., Costello, E. K., Fierer, N., Peña, A. G., Goodrich, J. K., Gordon, J. I., et al. (2010). QIIME allows analysis of high-throughput community sequencing data. *Nature Methods, 7*, 335–336.

Chomczynski, P., & Sacchi, N. (2006). The single-step method of RNA isolation by acid guanidinium thiocyanate-phenol-chloroform extraction: Twenty-something years on. *Nature Protocols, 1*, 581–585.

Choo, J. M., Leong, L. E. X., & Rogers, G. B. (2015). Sample storage conditions significantly influence faecal microbiome profiles. *Scientific Reports, 5*, 16350.

Claesson, M. J., Wang, Q., O'Sullivan, O., Greene-Diniz, R., Cole, J. R., Ross, R. P., & O'Toole, P. W. (2010). Comparison of two next-generation sequencing technologies for resolving highly complex microbiota composition using tandem variable 16S rRNA gene regions. *Nucleic Acids Research, 38*, e200.

Clavel, T., Lagkouvardos, I., & Hiergeist, A. (2016). Microbiome sequencing: Challenges and opportunities for molecular medicine. *Expert Review of Molecular Diagnostics, 16*, 795–805.

Cole, J. R., Wang, Q., Fish, J. A., Chai, B., McGarrell, D. M., Sun, Y., Brown, C. T., Porras-Alfaro, A., Kuske, C. R., & Tiedje, J. M. (2014). Ribosomal Database Project: Data and tools for high throughput rRNA analysis. *Nucleic Acids Research, 42*, D633–D642.

de Boer, R., Peters, R., Gierveld, S., Schuurman, T., Kooistra-Smid, M., & Savelkoul, P. (2010). Improved detection of microbial DNA after bead-beating before DNA isolation. *Journal of Microbiological Methods, 80*, 209–211.

Egerton, F. N. (2006). A history of the ecological sciences, Part 19: Leeuwenhoek's microscopic natural history. *The Bulletin of the Ecological Society of America, 87*, 47–58.

Eisen, J. A. (2007). Environmental shotgun sequencing: Its potential and challenges for studying the hidden world of microbes. *PLoS Biology, 5*, e82.

Ferretti, P., Farina, S., Cristofolini, M., Girolomoni, G., Tett, A., & Segata, N. (2017). Experimental metagenomics and ribosomal profiling of the human skin microbiome. *Experimental Dermatology, 26*, 211–219.

Fleischmann, R., Adams, M., White, O., Clayton, R., Kirkness, E., Kerlavage, A., Bult, C., Tomb, J., Dougherty, B., Merrick, J., et al. (1995). Whole-genome random sequencing and assembly of Haemophilus influenzae Rd. *Science, 269*, 496–512.

Gilbert, J., O'Dor, R., King, N., & Vogel, T. (2011). The importance of metagenomic surveys to microbial ecology: Or why Darwin would have been a metagenomic scientist. *Microbial Informatics and Experimentation, 1*, 5.

Gutleben, J., Chaib De Mares, M., van Elsas, J. D., Smidt, H., Overmann, J., & Sipkema, D. (2018). The multi-omics promise in context: From sequence to microbial isolate. *Critical Reviews in Microbiology, 44*(2), 212–229.

Hamady, M., Walker, J. J., Harris, J. K., Gold, N. J., & Knight, R. (2008). Error-correcting barcoded primers for pyrosequencing hundreds of samples in multiplex. *Nature Methods, 5*, 235–237.

Hesselman, M. C., Odoni, D. I., Ryback, B. M., de Groot, S., van Heck, R. G., Keijsers, J., Kolkman, P., Nieuwenhuijse, D., van Nuland, Y. M., Sebus, E., et al. (2012). A multi-platform flow device for microbial (co-)cultivation and microscopic analysis. *PLoS One, 7*, e36982.

Human Microbiome Project Consortium. (2012). A framework for human microbiome research. *Nature, 486*, 215–221.

Hummon, A. B., Lim, S. R., Difilippantonio, M. J., & Ried, T. (2007). Isolation and solubilization of proteins after TRIzol(®) extraction of RNA and DNA from patient material following prolonged storage. *Biotechniques, 42*, 467–472.

Ivanov, I. I., de Llanos Frutos, R., Manel, N., Yoshinaga, K., Rifkin, D. B., Sartor, R. B., Finlay, B. B., & Littman, D. R. (2008). Specific microbiota direct the differentiation of Th17 cells in the mucosa of the small intestine. *Cell Host & Microbe, 4*, 337–349.

Kanehisa, M., Furumichi, M., Tanabe, M., Sato, Y., & Morishima, K. (2017). KEGG: New perspectives on genomes, pathways, diseases and drugs. *Nucleic Acids Research, 45*, D353–D361.

Keegan, K. P., Glass, E. M., & Meyer, F. (2016). MG-RAST, a metagenomics service for analysis of microbial community structure and function. *Methods in Molecular Biology, 16*, 207–233.

Kim, D., Hofstaedter, C. E., Zhao, C., Mattei, L., Tanes, C., Clarke, E., Lauder, A., Sherrill-Mix, S., Chehoud, C., Kelsen, J., et al. (2017). Optimizing methods and dodging pitfalls in microbiome research. *Microbiome, 5*, 52.

Kirby, K. S. (1956). A new method for the isolation of ribonucleic acids from mammalian tissues. *Biochemical Journal, 64*, 405–408.

Kircher, M., & Kelso, J. (2010). High-throughput DNA sequencing – concepts and limitations. *BioEssays, 32*, 524–536.

Knights, D., Ward, T. L., McKinlay, C. E., Miller, H., Gonzalez, A., McDonald, D., & Knight, R. (2014). Rethinking "Enterotypes". *Cell Host & Microbe, 16*, 433–437.

Lagier, J. C., Armougom, F., Million, M., Hugon, P., Pagnier, I., Robert, C., Bittar, F., Fournous, G., Gimenez, G., Maraninchi, M., et al. (2012). Microbial culturomics: Paradigm shift in the human gut microbiome study. *Clinical Microbiology and Infection, 18*, 1185–1193.

Lagkouvardos, I., Overmann, J., & Clavel, T. (2017). Cultured microbes represent a substantial fraction of the human and mouse gut microbiota. *Gut Microbes, 8*, 493–503.

Lane, D. J., Pace, B., Olsen, G. J., Stahl, D. A., Sogin, M. L., & Pace, N. R. (1985). Rapid determination of 16S ribosomal RNA sequences for phylogenetic analyses. *Proceedings of the National Academy of Sciences of the United States of America, 82*, 6955–6959.

Marchesi, J. R., & Ravel, J. (2015). The vocabulary of microbiome research: A proposal. *Microbiome, 3*, 31.

Meyer, F., Paarmann, D., D'Souza, M., Olson, R., Glass, E. M., Kubal, M., Paczian, T., Rodriguez, A., Stevens, R., Wilke, A., Wilkening, J., & Edwards, R. A. (2008). The metagenomics RAST server – A public resource for the automatic phylogenetic and functional analysis of metagenomes. *BMC Bioinformatics, 9*, 386. https://doi.org/10.1186/1471-2105-9-386.

Morgan, X. C., & Huttenhower, C. (2014). Meta'omic analytic techniques for studying the intestinal microbiome. *Gastroenterology, 146*, 1437–1448. e1431.

Mutter, G. L., Zahrieh, D., Liu, C., Neuberg, D., Finkelstein, D., Baker, H. E., & Warrington, J. A. (2004). Comparison of frozen and RNALater solid tissue storage methods for use in RNA expression microarrays. *BMC Genomics, 5*, 88.

Nichols, D. (2007). Cultivation gives context to the microbial ecologist. *FEMS Microbiology Ecology, 60*, 351–357.

Ott, S. J., Waetzig, G. H., Rehman, A., Moltzau-Anderson, J., Bharti, R., Grasis, J. A., Cassidy, L., Tholey, A., Fickenscher, H., Seegert, D., et al. (2017). Efficacy of sterile fecal filtrate transfer for treating patients with clostridium difficile infection. *Gastroenterology, 152*, 799–811.e797.

Oyola, S. O., Gu, Y., Manske, M., Otto, T. D., O'Brien, J., Alcock, D., MacInnis, B., Berriman, M., Newbold, C. I., Kwiatkowski, D. P., et al. (2013). Efficient depletion of host DNA contamination in malaria clinical sequencing. *Journal of Clinical Microbiology, 51*, 745–751.

Parte, A. C. (2014). LPSN—list of prokaryotic names with standing in nomenclature. *Nucleic Acids Research, 42*, D613–D616.

Patejko, M., Jacyna, J., & Markuszewski, M. J. (2017). Sample preparation procedures utilized in microbial metabolomics: An overview. *Journal of Chromatography B, 1043*, 150–157.

Perez-Muñoz, M. E., Arrieta, M.-C., Ramer-Tait, A. E., & Walter, J. (2017). A critical assessment of the "sterile womb" and "in utero colonization" hypotheses: Implications for research on the pioneer infant microbiome. *Microbiome, 5*, 48.

Prokopowich, C. D., Gregory, T. R., & Crease, T. J. (2003). The correlation between rDNA copy number and genome size in eukaryotes. *Genome, 46*, 48–50.

Qin, J., Li, R., Raes, J., Arumugam, M., Burgdorf, K. S., Manichanh, C., Nielsen, T., Pons, N., Levenez, F., Yamada, T., et al. (2010). A human gut microbial gene catalogue established by metagenomic sequencing. *Nature, 464*, 59–65.

Quast, C., Pruesse, E., Yilmaz, P., Gerken, J., Schweer, T., Yarza, P., Peplies, J., & Glöckner, F. O. (2013). The SILVA ribosomal RNA gene database project: Improved data processing and web-based tools. *Nucleic Acids Research, 41*, D590–D596.

Rideout, J. R., He, Y., Navas-Molina, J. A., Walters, W. A., Ursell, L. K., Gibbons, S. M., Chase, J., McDonald, D., Gonzalez, A., Robbins-Pianka, A., Clemente, J. C., Gilbert, J. A., Huse, S. M., Zhou, H. W., Knight, R., & Caporaso, J. G. (2014). Subsampled open-reference clustering creates consistent, comprehensive OTU definitions and scales to billions of sequences. *PeerJ, 2*, e545.

Rissanen, A. J., Kurhela, E., Aho, T., Oittinen, T., & Tiirola, M. (2010). Storage of environmental samples for guaranteeing nucleic acid yields for molecular microbiological studies. *Applied Microbiology and Biotechnology, 88*, 977–984.

Salazar, O., & Asenjo, J. A. (2007). Enzymatic lysis of microbial cells. *Biotechnology Letters, 29*, 985–994.

Satokari, R., Mattila, E., Kainulainen, V., & Arkkila, P. E. T. (2015). Simple faecal preparation and efficacy of frozen inoculum in faecal microbiota transplantation for recurrent Clostridium difficile infection – an observational cohort study. *Alimentary Pharmacology & Therapeutics, 41*, 46–53.

Schloss, P. D., Westcott, S. L., Ryabin, T., Hall, J. R., Hartmann, M., Hollister, E. B., Lesniewski, R. A., Oakley, B. B., Parks, D. H., Robinson, C. J., et al. (2009). Introducing mothur: Open-source, platform-independent, community-supported software for describing and comparing microbial communities. *Applied and Environmental Microbiology, 75*, 7537–7341.

Schoch, C. L., Seifert, K. A., Huhndorf, S., Robert, V., Spouge, J. L., Levesque, C. A., Chen, W., & Consortium, F. B. (2012). Nuclear ribosomal internal transcribed spacer (ITS) region as a universal DNA barcode marker for Fungi. *Proceedings of the National Academy of Sciences, 109*, 6241–6246.

Schoch, C. L., Robbertse, B., Robert, V., Vu, D., Cardinali, G., Irinyi, L., Meyer, W., Nilsson, R. H., Hughes, K., Miller, A. N., et al. (2014). Finding needles in haystacks: Linking scientific names, reference specimens and molecular data for Fungi. *Database (Oxford), 2014*, bau061.

Segata, N., Waldron, L., Ballarini, A., Narasimhan, V., Jousson, O., & Huttenhower, C. (2012). Metagenomic microbial community profiling using unique clade-specific marker genes. *Nature Methods, 10*, 811–814.

Sender, R., Fuchs, S., & Milo, R. (2016). Revised estimates for the number of human and bacteria cells in the body. *PLoS Biology, 14*, e1002533.

Simon, C., & Daniel, R. (2011). Metagenomic analyses: Past and future trends. *Applied and Environmental Microbiology, 77*, 1153–1161.

Sogin, M. L., Morrison, H. G., Huber, J. A., Mark Welch, D., Huse, S. M., Neal, P. R., Arrieta, J. M., & Herndl, G. J. (2006). Microbial diversity in the deep sea and the underexplored "rare biosphere". *Proceedings of the National Academy of Sciences of the United States of America, 103*, 12115–12120.

Staley, J. T., & Konopka, A. (1985). Measurement of in situ activities of nonphotosynthetic microorganisms in aquatic and terrestrial habitats. *Annual Review of Microbiology, 39*, 321–346.

Tedjo, D. I., Jonkers, D. M. A. E., Savelkoul, P. H., Masclee, A. A., van Best, N., Pierik, M. J., & Penders, J. (2015). The effect of sampling and storage on the fecal microbiota composition in healthy and diseased subjects. *PLoS One, 10*, e0126685.

Truong, D. T., Franzosa, E. A., Tickle, T. L., Scholz, M., Weingart, G., Pasolli, E., Tett, A., Huttenhower, C., & Segata, N. (2015). MetaPhlAn2 for enhanced metagenomic taxonomic profiling. *Nature Methods, 12*, 902–903.

Turnbaugh, P. J., Ley, R. E., Mahowald, M. A., Magrini, V., Mardis, E. R., & Gordon, J. I. (2006). An obesity-associated gut microbiome with increased capacity for energy harvest. *Nature, 444*, 1027–1131.

Vandeputte, D., Kathagen, G., D'hoe, K., Vieira-Silva, S., Valles-Colomer, M., Sabino, J., Wang, J., Tito, R. Y., De Commer, L., Darzi, Y., Vermeire, S., Falony, G., & Raes, J. (2017). Quantitative microbiome profiling links gut community variation to microbial load. *Nature, 551*, 507–511.

Větrovský, T., & Baldrian, P. (2013). The variability of the 16S rRNA gene in bacterial genomes and its consequences for bacterial community analyses. *PLoS One, 8*, e57923.

Wilmes, P., Heintz-Buschart, A., & Bond, P. L. (2015). A decade of metaproteomics: Where we stand and what the future holds. *Proteomics, 15*, 3409–3417.

Yarza, P., Yilmaz, P., Pruesse, E., Glöckner, F. O., Ludwig, W., Schleifer, K. H., Whitman, W. B., Euzéby, J., Amann, R., & Rosselló-Móra, R. (2014). Uniting the classification of cultured and uncultured bacteria and archaea using 16S rRNA gene sequences. *Nature Reviews Microbiology, 12*, 635–645.

Evolutionary Perspectives on the Human Gut Microbiome

5

Shauni Doms, Britt-Marie Hermes, and John F. Baines

Abstract

The renewed interest in human gut microbiome research spawned by modern developments in metagenomics resulted in many fascinating new results, but confusion and seeming contradictions are still common in this nascent field. As for other subdisciplines of biology, evolutionary biology serves as a unifying principle in studying host-microbe interactions. However, the range of perspectives offered by evolution is often not considered or fully appreciated in human gut microbiome research. In this chapter we provide a broad overview of evolutionary perspectives on the human gut microbiome, which range from the origin of holobionts to strain-level microbial variation within a host's lifetime.

The realization of the human gut microbiota as an important determinant for health and disease, together with the relative ease of generating metagenomic sequencing-based profiles of microbial communities, resulted in tremendous renewed interest in microbiome research from many different areas within biology. This attention resulted in many fascinating new results but also leads to some degree of "microbiomania" (Eisen 2014) and confusion among researchers coming from diverse backgrounds, in part due to misunderstandings and the inherent complexity of describing and interpreting metagenomic data.

One frequent source of confusion is also one of the most fundamental questions that concerns human gut microbiome researchers, namely, "What are the origins and consequences of variability in the gut microbiome?" Part of the challenge in answering this basic question is that a second, equally important question is often not carefully considered: "At what level are you referring to?" Indeed, the gut microbiota is a complex ecological community, whose differences can range from subtle to dramatic, independently in terms of both (1) composition and/or structure and (2) their potential impact on the host. Thus, many considerations must be taken when designing and interpreting the results of comparative studies. Excellent review material is available on a wide range of topics including diet, development, genetics, disease, and ecology and evolution (Goodrich et al. 2017; Hall et al. 2017; Ley et al. 2006; Singh et al. 2017; Spor et al. 2011). Still, the full scope of evolutionary considerations for gut microbiome research, which range from the origins of "holobionts" (see definitions in Box 5.1) to strain-level microbial evolution within a host's lifetime, is often not sufficiently appreciated. Thus, in this

S. Doms · B.-M. Hermes · J. F. Baines (✉)
Section for Evolutionary Medicine, Institute for Experimental Medicine, Christian-Albrechts-University of Kiel, Kiel, Germany

Max-Planck-Institute for Evolutionary Biology, Plön, Germany
e-mail: doms@evolbio.mpg.de; hermes@evolbio.mpg.de; baines@evolbio.mpg.de

D. Haller (ed.), *The Gut Microbiome in Health and Disease*,
https://doi.org/10.1007/978-3-319-90545-7_5

chapter, we provide a broad overview of evolutionary perspectives on the gut microbiome and, following the lead of Dobzhansky's oft-quoted essay (Dobzhansky 1973) and other renditions (Varki 2006), here boldly assert that "nothing in gut microbiome research makes sense, except in the light of evolution".

5.1 On the Origin of Holobionts and Their Hologenomes

An appropriate place to begin our considerations is at the origin of eukaryotes and multicellular organisms, as these entities arose in a world already dominated by microbes and were likely shaped by interactions with them throughout their evolution (Margulis 1993; McFall-Ngai et al. 2013; Zilber-Rosenberg and Rosenberg 2008) (Fig. 5.1). The existence of these ancient interactions is evidenced by, e.g., the existence of mitochondria and chloroplasts and the conservation of microbial pattern recognition receptor pathways across animals (Franzenburg et al. 2012). Contemporary observations, on the other hand, find every plant and animal examined to be associated with diverse microbes, ranging from strict to loose associations. A largely agreed-upon term to describe these entities, comprised of macrobes and their associated microbes, is the "holobiont" (Gordon et al. 2013; Margulis and Fester 1991). In turn, the collective genomic content of the host genome, the organelles, and the associated microbiome is referred to as the "hologenome" (Zilber-Rosenberg and Rosenberg 2008).

The utility, definition, and implications of the hologenome concept have been a matter of discussion (Moran and Sloan 2015; Theis et al.

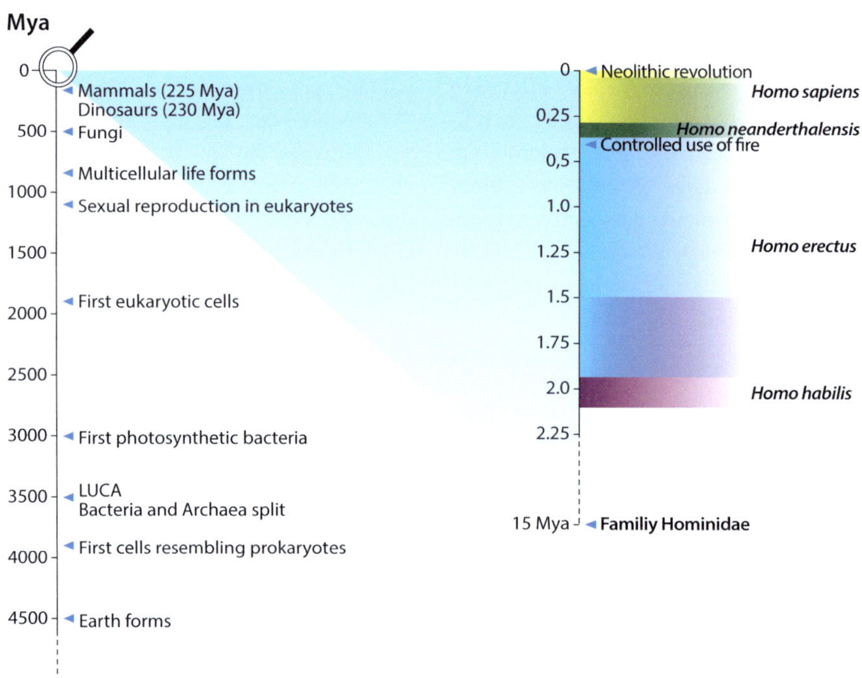

Fig. 5.1 Timeline of history of life on earth highlighting key events in the evolution of the human holobiont. *Homo habilis* lived between 2.1 and 1.5 Mya. Fossil evidence dates *Homo erectus* from about 1.9 Mya to about 150,000 BC. Neanderthals, *Homo neanderthalensis*, lived from about 350,000 to 40,000 BC. Our species, *Homo sapiens*, first appeared about 300,000 BC. We conservatively dated the first controlled use of fire at 400 Mya based on unequivocal archaeological evidence (Karkanas et al. 2007). Less conclusive archaeological evidence suggests that fire use by *Homo erectus* may have occurred as early as one million years ago (Berna et al. 2012). The Neolithic Revolution began about 10,000 BC

2016) (Box 5.2). However, after careful consideration and definition of the term, viewing evolutionary processes through the lens of holobionts and their hologenomic content offers a number of valuable principles that aid in conceptual understanding of host-microbiome interactions and, importantly, help guide research (Bordenstein and Theis 2015). Among these principles is that holobionts and hologenomes represent a critical unit of biological organization, which includes the nuclear genome, organelles, and microbiome. Beneficial, neutral, and deleterious changes can occur at each of the above three sublevels and contribute to variation in holobiont traits, which is in turn the raw material upon which selection can act. Accordingly, variation in the microbiome can also contribute to adaptation and speciation (Bordenstein and Theis 2015).

5.2 The Bacterial 16S rRNA Gene as a Phylogenetic Marker

An important consideration for this chapter is that the vast majority of data collected for the microbial component of holobionts to date is based on bacterial 16S rRNA gene profiles, which itself is a favored marker gene due to its evolutionary properties (Woese and Fox 1977). Specifically, it contains both conserved, slowly evolving regions that enable "universal" primers (i.e., those that can amplify widely divergent bacterial taxa) to be designed and also fast-evolving, hypervariable regions that provide more fine-scale resolution to identify individual taxa. After classifying 16S rRNA gene sequences using widely available public databases such as the Ribosomal Database Project, or "RDP" (Wang et al. 2007), researchers can obtain abundance profiles of groups of bacteria, ranging from the phylum to genus level of classification. An alternative and complementary approach to classifying sequences to a known taxonomy is to cluster sequences into operational taxonomic units (OTUs) (Box 5.1) based on a similarity threshold, whereby 97% sequence similarity is typically used as a cutoff to denote species-level taxa.

Accordingly, unless otherwise noted, the comparative studies discussed throughout the chapter in large part rely on overall bacterial community similarity/dissimilarity between individual samples (holobionts), as determined by shared presence and/or abundance of bacterial taxa determined by the 16S rRNA gene. Differences in bacterial community composition between samples are typically quantified by various measures of beta diversity (Box 5.1). Thus, we will mostly be discussing "evolution" of the microbiome in terms of inferred bacterial community-level changes over time, based on comparing hosts at varying degrees of relatedness, ranging from among mammals to differences between individual humans.

5.3 The Gut Microbiome Across Mammalian Evolution

Phylogenetic comparison is one of the most common means of inference used to test evolutionary hypotheses, whereby similarities and differences between species are interpreted in terms of their evolutionary history. To understand the composition of the human gut microbiome in the broader context of mammalian evolution, researchers have applied this method and analyzed the relative roles of evolutionary history and changes in diet across host species, which revealed many important insights (Groussin et al. 2017; Ley et al. 2008a). A first and important point to consider is that ancestral mammals were carnivores (Stevens and Hume 2004), raising the question of how herbivory evolved in their descendants. On the one hand, different lineages evolved either foregut or hindgut enlargement of their existing gastrointestinal (GI) tracts, which provided the increased gut retention times necessary for bacterial fermentation. In contrast, multiple analyses indicate that the bacterial lineages required for the digestion of a plant-based diet in herbivores did not evolve from the bacterial lineages already present in carnivores but were rather independently recruited from the environment (Groussin et al. 2017; Ley et al. 2008a). This is likely one reason for disparities between studies in terms of the relative impact of host phylogeny (Box 5.1) and the environment/diet. In this regard, Groussin et al. provide extremely important insight. Using their novel method termed "beta diversity through time" (BDTT), which allows bacterial communities to be analyzed over a range of phylogenetic scales, they demonstrate that major dietary shifts were associated with the horizontal acquisition of ancient bacterial lineages, whereas associations with host phylogeny are more apparent among recent bacterial lineages (Groussin et al. 2017). As such, individual bacterial genera that display signatures of cospeciation (Box 5.1) could be identified, whose more intimate evolutionary relationship with the host also intriguingly appears to make them more likely to be associated with immune disorders such as inflammatory bowel disease (Groussin et al. 2017).

5.4 Evolutionary Patterns Among Hominids

As highlighted above, it is important to consider the phylogenetic scale at which evolutionary changes among components of the hologenome take place. In this section we will discuss the intriguing patterns that have emerged from focusing on microbiome comparisons among humans and our closest living relatives. In the first study examining fecal samples from free-living representatives of the great apes, i.e., humans (*Homo sapiens)*, three chimpanzee subspecies (*Pan troglodytes troglodytes*, *P. t. schweinfurthii*, and *P. t. ellioti)*, bonobos (*Pan paniscus)*, and eastern and western lowland gorillas (*Gorilla beringei* and *G. gorilla* resp.), Ochman et al. observed that the relationship of fecal microbial communities among these five species parallels the phylogeny of the host species (Ochman et al. 2010). Together with similar observations in numerous other unrelated animal hosts, these results helped inspire the term "phylosymbiosis" (Box 5.1), which refers to the phenomenon whereby the ecological relatedness of host-associated microbial communities parallels the phylogeny of their hosts (Brooks et al. 2016). The existence of phylosymbiosis implies that factors such as vertical transmission and the accumulation of host genetic differences over time significantly impact gut bacterial community composition. A follow-up study of great apes by Moeller et al. provides an additional level of resolution by performing taxonomic profiling using a faster-evolving molecular barcode, the variable region of DNA gyrase subunit B (gyrB) (Moeller et al. 2016). This reveals numerous individual taxa belonging to the families *Bacteroidaceae* and *Bifidobacteriaceae* to have phylogenies congruent with those of their host, which is consistent with cospeciation. Lineages belonging to the *Lachnospiraceae*, on the other hand, display more evidence of host switching. Interestingly, this may be related to *Lachnospiraceae* containing spore-forming members that can survive outside the gut, which is not the case for *Bacteroidaceae* and *Bifidobacteriaceae.* These observations demonstrate how evolutionary analyses help interpret contrasting patterns among individual members of the microbiome.

In another study, Moeller et al. provide an additional critical observation, namely, that in terms of community composition, the human gut microbiome has experienced accelerated evolution relative to that of other wild hominids (Moeller et al. 2014). Specifically, the authors found that changes in bacterial abundances steadily accumulated among African apes, but humans in particular have experienced an accelerated loss of ancestral microbes over the course of hominization. Interesting possible explanations for accelerated evolution of the human gut microbiome are recent culturally defined practices, e.g., meat consumption, the consumption of processed starches, and a hygienic living environment. For example, the "expensive tissue hypothesis" argues that a dietary increase in meat by *Homo* species throughout the Pleistocene period (2.5 MYA to 12 KYA) fostered the development of the large brain size by providing essential nutrition necessary for this evolutionary development, at the expense of larger guts (Aiello and Wheeler 1995; Ley et al. 2008b). Consistent with a change toward animal-based diets, the human gut microbiome consists of a higher relative abundance of the genus *Bacteroides*, a taxon associated with a diet rich in animal products

(Moeller et al. 2014; Wu et al. 2011), whereas that of wild apes, who generally eat a varied diet rich in plants, is enriched with taxa known to promote the degradation of plant materials. A further cultural factor to consider is the habitual use of fire for food preparation beginning approximately 350,000 years ago (Shimelmitz et al. 2014) (Fig. 5.1), which may have reduced the selective pressure to retain certain microbes, such as those responsible for digesting more resistant components of plant-based nutrition (Gillings et al. 2015). The use of fire could also presumably disrupt previous means of transmission/acquisition of ancestral microbes.

5.5 Variation in the Gut Microbiome Among Human Populations

Following dietary changes in our more distant human ancestors, the Neolithic Revolution (about 10K BC) marks the transition from hunter-gatherer civilizations to settled communities with the widespread adoption of agricultural practices and animal domestication (Richards 2002). Evidence of adaptation to this transition has been left in the human genome, such as amylase (*AMY1*) copy number. This enzyme hydrolyzes starch and correlates positively with human salivary protein levels. Individuals living in communities consuming a diet rich in starches, namely, agricultural societies, have higher copy numbers of *AMY1* than individuals from communities that traditionally eat diets low in starch, like rain forest and circum-arctic hunter-gatherer groups (Ley et al. 2008b; Perry et al. 2007). The selective pressure for an increase in amylase activity is also hypothesized to be in response to conflict with members of the gut microbiota, whereby greater energy gains can be achieved by faster uptake of starch monomers in the upper GI tract compared to the lower GI tract, where fermentation of starch by microbes occurs (Walter and Ley 2011).

Studying extant hunter-gatherer groups also provides important insight, as biogeographic differences in their diet influence the composition of their gut microbiome. For example, hunter-gatherer societies in arid environments rely on starchy plants like tubers to supplement their diet (Perry et al. 2007). Analysis of the Hadza hunter-gatherer tribe residing in the dry lands of Tanzania found that their gut microbiome contains a higher abundance of *Prevotella* and a lower abundance of *Bacteroides* compared to the gut microbiome of Italian subjects (Schnorr et al. 2014). As in wild apes, a higher abundance of *Prevotella* is associated with human diets rich in starchy carbohydrates (Moeller et al. 2012; Wu et al. 2011). Further, self-reporting vegetarians with a diet rich in carbohydrates and simple sugars also tend to have relatively high amounts of *Prevotella* in their gut microbiome, whereas individuals eating a standard Western diet heavy in meat products display higher *Bacteroides* (Wu et al. 2011).

In Fig. 5.2, we display human microbiome data originally published by Yatsunenko et al., highlighting how diet and lifestyle impact community structure and diversity across populations (Yatsunenko et al. 2012). The data are derived from sampling three populations: (1) Amerindians in the Amazonas of the State of Venezuela, (2) individuals living in rural communities in Malawi, and (3) Americans residing in metropolitan areas of the USA. The gut microbiomes of individuals living in a Western society (USA) are notably different from those living in rural environments (Malawi and Venezuela) with respect to the most abundant genera present in feces (Fig. 5.2a, c). US individuals had high amounts of *Faecalibacterium*, whereas Venezuelan and Malawi samples were dominated by *Prevotella*, which is consistent with the rural Hadza samples. There were also significant differences in the bacterial diversity within the fecal samples (Fig. 5.2b). Venezuelan samples were the most diverse, whereas US American samples had the fewest number of unique genera. These findings parallel the results by Schnorr et al. who found Western populations to have reduced bacterial diversity compared to hunter-gatherer populations (Schnorr et al. 2014). Taken together, cultural differences appear to greatly impact gut microbiome composition. However, Moeller et al. found that regardless of whether humans were living in urban or rural/agrarian settings, they harbored fewer unique bacterial

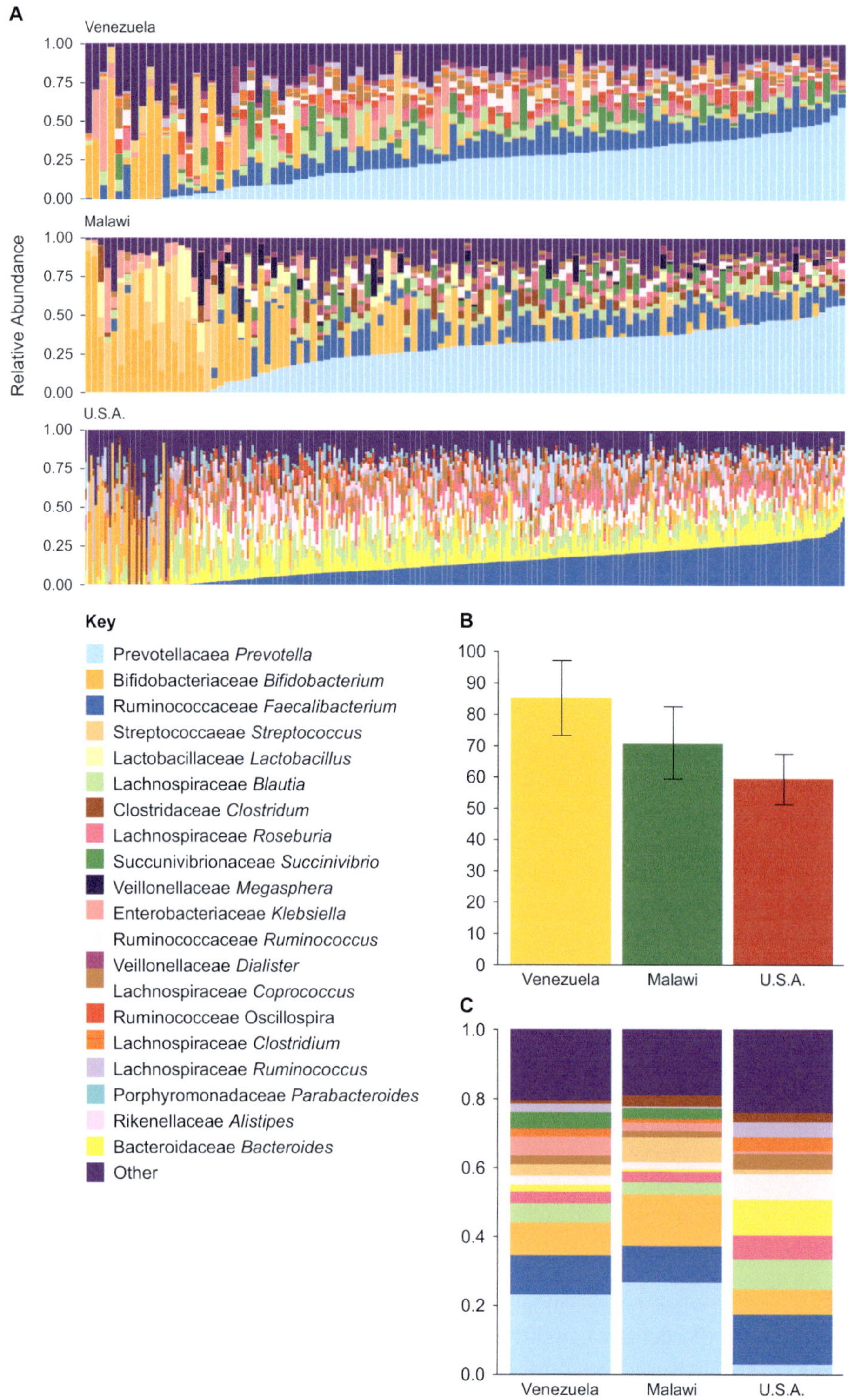

Fig. 5.2 Bacterial diversity and composition within the gut microbiome of rural and urban populations. (**a**) Genus-level composition of individuals from an American population compared to a rural Malawi community and in the Amazonas of Venezuela. The 14 most abundant genera are shown. The samples are ordered in ascending

genera in their gut microbial communities relative to nonhuman primates (Moeller et al. 2014). Thus, it seems that hominization universally impacted our microbial inhabitants but that modernization, especially in Western society, is further accelerating a decrease in diversity of our gut microbiome (Moeller 2017).

5.6 Interindividual Variation

From an evolutionary perspective, understanding the origins of variation among individuals within a population is of fundamental importance, as it is these differences that are ultimately converted to variation between populations and species. This is relevant to each level of the hologenome, although we here focus on the origins of microbial community variation that can be attributed to variation in the host genome. In particular, the extent to which microbial traits display heritability, i.e., a genetic basis, determines the degree to which natural selection can act upon host-microbial traits. Heritability is defined as the proportion of phenotypic variance in a trait that is due to genotypic differences within a population. One way to estimate heritability for microbial traits is to use a twin-based study, whereby within-pair similarity of microbial taxon abundances in monozygotic twins is compared to dizygotic twins. Two twin studies performed in this manner in 2009 and 2012 showed little to no evidence of heritability in the gut microbiome (Turnbaugh et al. 2009; Yatsunenko et al. 2012). However, a recent study consisting of a larger sample cohort (1126 twin pairs) showed that 8.8% of gut microbial taxa are heritable (Goodrich et al. 2016). The authors further demonstrate that these heritable taxa are temporally stable over long periods, suggesting a smaller effect of environmental factors on their relative abundances.

An important alternative approach to study the role of host genetic variation in shaping patterns of interindividual variation is to perform a genome-wide association study (GWAS), which offers the further advantage of being able to identify the responsible host genes. These studies are challenging to accomplish due to the sample sizes required to obtain sufficient statistical power, but a handful have been performed to date with promising results (Goodrich et al. 2017). Among these initial studies is that by Wang et al., who examined 1812 individuals from Northern Germany (Wang et al. 2016). Similar to other human cohorts examined, these individuals display a large degree of interindividual variability (Fig. 5.3). Because it is well established that nongenetic influences including diet, lifestyle, demographic, and environmental factors influence the gut microbiome, the authors accounted for these variables when scanning >6 million SNPs throughout the human genome for genetic effects. Importantly, this enabled the proportion of overall variation in the gut microbial community due to genetic and other factors to be estimated, whereby genetic effects explain 10% of the overall variation. Evidence for the robustness of these results is provided by an overrepresentation of genes expressed in the digestive tract and overlap between syntenic regions of the mouse genome that were identified by similar genetic mapping approaches. Interestingly, further functional examination of candidate genes, including the vitamin D receptor (VDR), reveals bile acid metabolism as a key pathway, which was supported by correlations with bile acid measurements (Wang et al. 2016).

At this point, further clarification of the term "role of host genetics" is however warranted. As pointed out by Goodrich et al., it must be emphasized that estimates of genetic effects such as those provided by Wang et al. are based on the subset of human genes that actually contain functionally related variation (Goodrich et al.

Fig. 5.2 (continued) relative abundance of the most abundant genus in a given population. (**b**) Number of unique genera per individual. The sequencing depth was rarefied to the minimum number of reads present in an individual. (**c**) Average relative abundance of the 15 most abundant genera across populations. Data are from Yatsunenko et al. (2012)

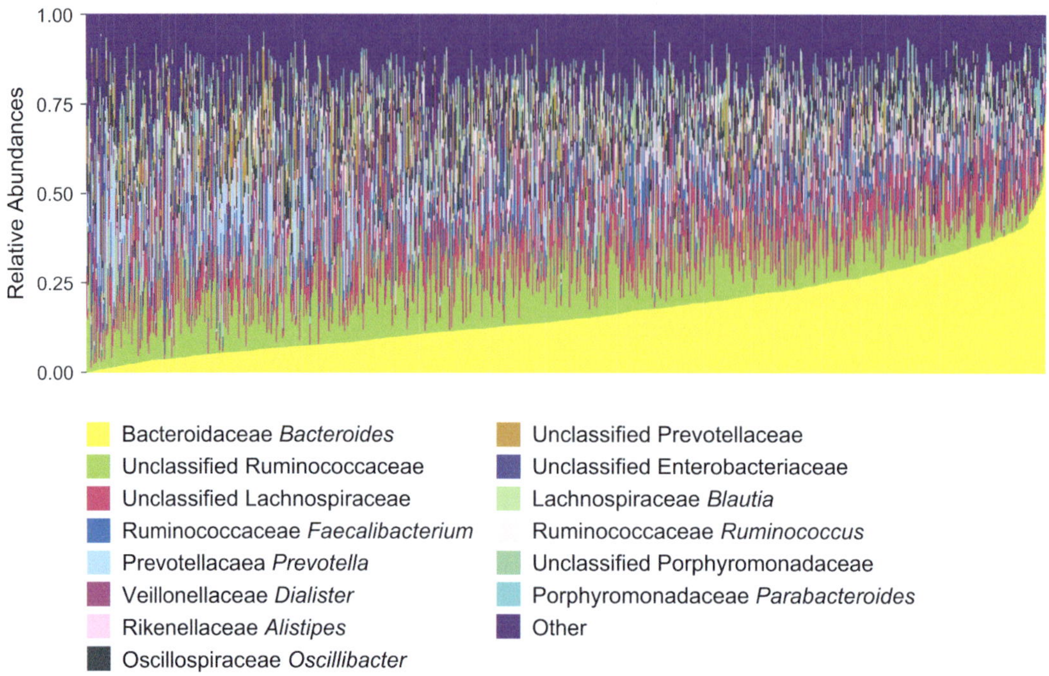

Fig. 5.3 Interindividual variation in gut microbiome composition. Genus-level composition of individuals belonging to the Northern German "PopGen" cohort (Wang et al. 2016). The 14 most abundant genera are shown. The samples are ordered in ascending relative abundance of the most abundant genus. $N = 914$

2017; Wang et al. 2016). Because of the important functions that gut microbes provide to the host and the challenge to contain and regulate them, many if not most genes involved in host-microbe interactions may experience predominantly negative or "purifying" selection (Box 5.1) against mutations disrupting their function. As a consequence, these genes would not be expected to display appreciable variation within a host species. Thus, 10% of the standing variation in the gut microbiome being attributed to genetic variation is not synonymous with concluding that "only 10% of the gut microbiome is determined by host genes." Furthermore, feasibility and technical aspects also limit the identification of host-microbial traits. Although it was previously hypothesized that bacteria inhabiting the intestinal mucosa may be more influenced by host genetics due to more intimate contact with the host (Spor et al. 2011), the majority of studies are based on feces due to limited access to biopsy material. Indeed, subsequent studies found evidence of stronger genetic effects in the intestinal mucosa of free-living mice (Linnenbrink et al. 2013), and candidate genes observed to influence mucosal-attached communities (Rausch et al. 2011; Tong et al. 2014) fail to be replicated in studies based on fecal material (Davenport et al. 2016). Thus, it is likely that a portion of functionally related genetic variation influencing the gut microbiome may be currently overlooked due to differences in habitat within the intestine.

5.7 Conclusion and Outlook

In this chapter we have outlined the range of perspectives from which evolutionary considerations help understand the composition, function, and changes in the human gut microbiome. It is also very likely that evolutionary biology will play key roles in the future of microbiome research. One important question to address, for example, is why does host genetic variation for gut microbiome traits exist? A key example is the association of the human *Lactase* (*LCT*) gene and *Bifidobacterium* abundance (Blekhman et al. 2015), where variation exists as a by-product of adaptive evolution in the human genome (lactase persistence mutations evolved multiple times independently in response to the advent of dairy farming) (Goodrich et al. 2017). A similar possibility pointed out by Blekhman et al. are so-called bystander effects (Benson 2015; Blekhman et al. 2015), where alleles providing resistance against pathogens also happen to influence the commensal microbiome. It is also conceivable that such effects may not always represent mere "bystanders" but may also ensue "collateral damage" when the changes are undesirable. Given the prevalence of pathogen resistance-associated variation in the human genome (Andrés et al. 2009; Fumagalli et al. 2009), systematically evaluating the overlap between pathogen- and gut microbiome-associated loci may represent a promising strategy to identify variation relevant to personalized medicine.

Finally, although very little human data exist to date, it appears that individual members of the gut microbiome undergo adaptive evolution within individual people (Zhao et al. 2017). Given previous demonstrations of adaptive radiation after colonizing germ-free mice with a single *Escherichia coli* isolate (De Paepe et al. 2011), this may not be surprising. However, based on these initial observations and that estimates of novel mutations within the gut microbiome are on the order of billions per day (Zhao et al. 2017), it is now clear that evolutionary considerations must be extended to the finest scale possible, i.e., to the evolution of our own personal microbiomes. This insight offers particular promise for personalized medicine, whereby not only "dysbiosis" but also *evolutionary* processes that occur in the context of diseases involving the microbiome may ultimately be targeted for therapeutic purposes.

▶ Box 5.1: Definitions

- **Beta diversity:** in ecology this term refers to the ratio between regional and local species diversity. In microbiome research, the term typically refers to differences in community composition between samples.
- **Cospeciation**: a form of coevolution in which the speciation of one species dictates speciation of another species.
- **Holobiont**: a host organism plus its associated microorganisms.
- **Hologenome**: the genomic content of the host genome, the organelles, and the associated microbiome.
- **Operational taxonomic unit (OTU)**: a cluster of organisms grouped by DNA sequence similarity of a specific taxonomic marker gene, usually the 16S rRNA gene. If similarity at the 16S rRNA gene is greater than or equal to 97%, the term OTU serves as a convenient proxy for "species."
- **Phylogeny**: inferred evolutionary relationships among various biological species or other entities.
- **Phylosymbiosis**: phenomenon whereby the ecological relatedness of host-associated microbial communities parallels the phylogeny of their hosts.
- **Purifying selection**: also termed negative selection. This refers to the phenomenon whereby deleterious genetic variation, i.e., that which reduces an organism's fitness, is removed from populations through natural selection.

▶ Box 5.2: Controversy

A large part of the controversy surrounding the hologenome concept relates to the term coevolution. An almost ubiquitous statement appearing in the introductions of host-microbiome studies is that hosts and their associated microbes represent the outcome of millions of years of coevolution. However, as Moran and Sloan point out, the definition of coevolution is more restrictive, requiring that respective partners undergo reciprocal evolutionary change due to selective forces imposed on each other, and many other alternative phenomena can produce patterns of host-microbial association appearing consistent with coevolution (Moran and Sloan 2015). Although coevolution within mammalian holobionts has most certainly occurred, its likelihood is dependent on the mode of transmission of microbes, and the extent and nature of coevolution are still largely to be determined.

A second contention pertains to whether the holobiont is the primary unit upon which natural selection acts or, rather, whether the original conception of the hologenome *requires* that it be the primary unit of selection (Moran and Sloan 2015; Theis et al. 2016). A problem with assuming that the holobiont is the primary unit of selection is that akin to problems with group selection, it requires selection at a higher level (i.e., the holobiont) to outweigh conflicts between its individual members. Given the diversity of the microbiota in terms of its composition and modes of transmission/acquisition, clear opportunities for conflicts of interest exist, such as an individual microbe increasing its own transmission at the expense of the host. However, as noted by Theis et al., the hologenome concept does not require it to be the primary unit of selection but rather embraces multilevel selection, including those traits for which the holobiont could be a primary unit of selection (Theis et al. 2016).

Box 5.3: Highlights

- Evolutionary considerations for gut microbiome research range from the origins of holobionts to strain-level microbial evolution within a host's lifetime.
- The concept of holobionts and their hologenomes provides a number of valuable principles that aid in the conceptual understanding of host-microbiome interactions and, importantly, help guide research.
- Humans have experienced accelerated evolution of their gut microbiome, most likely in response to dietary changes.
- Future research needs to account for bacterial strain-level evolution within the gut microbiome of individuals over their lifetime.

References

Aiello, L. C., & Wheeler, P. (1995). The expensive-tissue hypothesis: The brain and the digestive system in human and primate evolution. *Current Anthropology, 36*, 199–221.

Andrés, A. M., Hubisz, M. J., Indap, A., Torgerson, D. G., Degenhardt, J. D., Boyko, A. R., Gutenkunst, R. N., White, T. J., Green, E. D., Bustamante, C. D., et al. (2009). Targets of balancing selection in the human genome. *Molecular Biology and Evolution, 26*, 2755–2764.

Benson, A. K. (2015). Host genetic architecture and the landscape of microbiome composition: Humans weigh in. *Genome Biology, 16*, 203.

Berna, F., Goldberg, P., Horwitz, L. K., Brink, J., Holt, S., Bamford, M., & Chazan, M. (2012). Microstratigraphic evidence of in situ fire in the Acheulean strata of Wonderwerk Cave, Northern Cape province, South Africa. *Proceedings of the National Academy of Sciences, 109*, E1215–E1220.

Blekhman, R., Goodrich, J. K., Huang, K., Sun, Q., Bukowski, R., Bell, J. T., Spector, T. D., Keinan, A., Ley, R. E., Gevers, D., et al. (2015). Host genetic variation impacts microbiome composition across human body sites. *Genome Biology, 16*, 191.

Bordenstein, S. R., & Theis, K. R. (2015). Host biology in light of the microbiome: Ten principles of holobionts and hologenomes. *PLoS Biology, 13*, e1002226.

Brooks, A. W., Kohl, K. D., Brucker, R. M., van Opstal, E. J., & Bordenstein, S. R. (2016). Phylosymbiosis: Relationships and functional effects of microbial communities across host evolutionary history. *PLoS Biology, 14*, e2000225.

Davenport, E. R., Goodrich, J. K., Bell, J. T., Spector, T. D., Ley, R. E., & Clark, A. G. (2016). ABO antigen and secretor statuses are not associated with gut microbiota composition in 1,500 twins. *BMC Genomics, 17*, 941.

De Paepe, M., Gaboriau-Routhiau, V., Rainteau, D., Rakotobe, S., Taddei, F., & Cerf-Bensussan, N. (2011). Trade-off between bile resistance and nutritional competence drives *Escherichia coli* diversification in the mouse gut. *PLoS Genetics, 7*, e1002107.

Dobzhansky, T. (1973). Nothing in biology makes sense except in the light of evolution. *The American Biology Teacher, 35*(3), 125–129.

Eisen, J. A. (2014, June). *The tree of life: Microbiomania and "overselling the microbiome"* [Blog post]. Retrieved from https://phylogenomics.blogspot.de/p/blog-page.html

Franzenburg, S., Fraune, S., Künzel, S., Baines, J. F., Domazet-Loso, T., & Bosch, T. C. G. (2012). MyD88-deficient Hydra reveal an ancient function of TLR signaling in sensing bacterial colonizers. *Proceedings of the National Academy of Sciences of the United States of America, 109*, 19374–19379.

Fumagalli, M., Cagliani, R., Pozzoli, U., Riva, S., Comi, G. P., Menozzi, G., Bresolin, N., & Sironi, M. (2009).

Widespread balancing selection and pathogen-driven selection at blood group antigen genes. *Genome Research, 19*, 199–212.

Gillings, M. R., Paulsen, I. T., & Tetu, S. G. (2015). Ecology and evolution of the human microbiota: Fire, farming and antibiotics. *Genes, 6*, 841–857.

Goodrich, J. K., Davenport, E. R., Beaumont, M., Jackson, M. A., Knight, R., Ober, C., Spector, T. D., Bell, J. T., Clark, A. G., & Ley, R. E. (2016). Genetic determinants of the gut microbiome in UK twins. *Cell Host & Microbe, 19*, 731–743.

Goodrich, J. K., Davenport, E. R., Clark, A. G., & Ley, R. E. (2017). The relationship between the human genome and microbiome comes into view. *Annual Review of Genetics, 51*, 413–433.

Gordon, J., Knowlton, N., Relman, D. A., Rohwer, F., & Youle, M. (2013). Superorganisms and holobionts. *Microbe Magazine, 8*, 152–153.

Groussin, M., Mazel, F., Sanders, J. G., Smillie, C. S., Lavergne, S., Thuiller, W., & Alm, E. J. (2017). Unraveling the processes shaping mammalian gut microbiomes over evolutionary time. *Nature Communications, 8*, 14319.

Hall, A. B., Tolonen, A. C., & Xavier, R. J. (2017). Human genetic variation and the gut microbiome in disease. *Nature Reviews Genetics, 18*, 690.

Karkanas, P., Shahackgross, R., Ayalon, A., Barmatthews, M., Barkai, R., Frumkin, A., Gopher, A., & Stiner, M. (2007). Evidence for habitual use of fire at the end of the Lower Paleolithic: Site-formation processes at Qesem Cave, Israel. *Journal of Human Evolution, 53*, 197–212.

Ley, R. E., Peterson, D. A., & Gordon, J. I. (2006). Ecological and evolutionary forces shaping microbial diversity in the human intestine. *Cell, 124*, 837–848.

Ley, R. E., Hamady, M., Lozupone, C., Turnbaugh, P. J., Ramey, R. R., Bircher, J. S., Schlegel, M. L., Tucker, T. A., Schrenzel, M. D., Knight, R., et al. (2008a). Evolution of mammals and their gut microbes. *Science, 320*, 1647–1651.

Ley, R. E., Lozupone, C. A., Hamady, M., Knight, R., & Gordon, J. I. (2008b). Worlds within worlds: Evolution of the vertebrate gut microbiota. *Nature Reviews Microbiology, 6*, 776–788.

Linnenbrink, M., Wang, J., Hardouin, E. A., Künzel, S., Metzler, D., & Baines, J. F. (2013). The role of biogeography in shaping diversity of the intestinal microbiota in house mice. *Molecular Ecology, 22*, 1904–1916.

Margulis, L. (1993). *Symbiosis in cell evolution: Microbial communities in the Archean and Proterozoic eons*. New York: Freeman.

Margulis, L., & Fester, R. (Eds.). (1991). *Symbiosis as a source of evolutionary innovation: Speciation and morphogenesis*. Cambridge, MA: MIT Press.

McFall-Ngai, M., Hadfield, M. G., Bosch, T. C. G., Carey, H. V., Domazet-Lošo, T., Douglas, A. E., Dubilier, N., Eberl, G., Fukami, T., Gilbert, S. F., et al. (2013). Animals in a bacterial world, a new imperative for the life sciences. *Proceedings of the National Academy of Sciences of the United States of America, 110*, 3229–3236.

Moeller, A. H. (2017). The shrinking human gut microbiome. *Current Opinion in Microbiology, 38*, 30–35.

Moeller, A. H., Degnan, P. H., Pusey, A. E., Wilson, M. L., Hahn, B. H., & Ochman, H. (2012). Chimpanzees and humans harbour compositionally similar gut enterotypes. *Nature Communications, 3*, 1179.

Moeller, A. H., Li, Y., Mpoudi Ngole, E., Ahuka-Mundeke, S., Lonsdorf, E. V., Pusey, A. E., Peeters, M., Hahn, B. H., & Ochman, H. (2014). Rapid changes in the gut microbiome during human evolution. *Proceedings of the National Academy of Sciences, 111*, 16431–16435.

Moeller, A. H., Caro-Quintero, A., Mjungu, D., Georgiev, A. V., Lonsdorf, E. V., Muller, M. N., Pusey, A. E., Peeters, M., Hahn, B. H., & Ochman, H. (2016). Cospeciation of gut microbiota with hominids. *Science, 353*, 380–382.

Moran, N. A., & Sloan, D. B. (2015). The hologenome concept: Helpful or hollow? *PLoS Biology, 13*, e1002311.

Ochman, H., Worobey, M., Kuo, C.-H., Ndjango, J.-B. N., Peeters, M., Hahn, B. H., & Hugenholtz, P. (2010). Evolutionary relationships of wild hominids recapitulated by gut microbial communities. *PLoS Biology, 8*, e1000546.

Perry, G. H., Dominy, N. J., Claw, K. G., Lee, A. S., Fiegler, H., Redon, R., Werner, J., Villanea, F. A., Mountain, J. L., Misra, R., et al. (2007). Diet and the evolution of human amylase gene copy number variation. *Nature Genetics, 39*, 1256–1260.

Rausch, P., Rehman, A., Kunzel, S., Hasler, R., Ott, S. J., Schreiber, S., Rosenstiel, P., Franke, A., & Baines, J. F. (2011). Colonic mucosa-associated microbiota is influenced by an interaction of Crohn disease and FUT2 (Secretor) genotype. *Proceedings of the National Academy of Sciences, 108*, 19030–19035.

Richards, M. P. (2002). A brief review of the archaeological evidence for Palaeolithic and Neolithic subsistence. *European Journal of Clinical Nutrition, 56*, 16 p following 1262.

Schnorr, S. L., Candela, M., Rampelli, S., Centanni, M., Consolandi, C., Basaglia, G., Turroni, S., Biagi, E., Peano, C., Severgnini, M., et al. (2014). Gut microbiome of the Hadza hunter-gatherers. *Nature Communications, 5*, 3654.

Shimelmitz, R., Kuhn, S. L., Jelinek, A. J., Ronen, A., Clark, A. E., & Weinstein-Evron, M. (2014). "Fire at will": The emergence of habitual fire use 350,000 years ago. *Journal of Human Evolution, 77*, 196–203.

Singh, R. K., Chang, H.-W., Yan, D., Lee, K. M., Ucmak, D., Wong, K., Abrouk, M., Farahnik, B., Nakamura, M., Zhu, T. H., et al. (2017). Influence of diet on the gut microbiome and implications for human health. *Journal of Translational Medicine, 15*, 73.

Spor, A., Koren, O., & Ley, R. (2011). Unravelling the effects of the environment and host genotype on the gut microbiome. *Nature Reviews Microbiology, 9*, 279–290.

Stevens, C. E., & Hume, I. D. (2004). *Comparative physiology of the vertebrate digestive system*. New York: Cambridge University Press.

Theis, K. R., Dheilly, N. M., Klassen, J. L., Brucker, R. M., Baines, J. F., Bosch, T. C. G., Cryan, J. F., Gilbert, S. F., Goodnight, C. J., Lloyd, E. A., et al. (2016). Getting the hologenome concept right: An eco-evolutionary framework for hosts and their microbiomes. *MSystems, 1*, e00028–16.

Tong, M., McHardy, I., Ruegger, P., Goudarzi, M., Kashyap, P. C., Haritunians, T., Li, X., Graeber, T. G., Schwager, E., Huttenhower, C., et al. (2014). Reprograming of gut microbiome energy metabolism by the FUT2 Crohn's disease risk polymorphism. *The ISME Journal, 8*, 2193–2206.

Turnbaugh, P. J., Hamady, M., Yatsunenko, T., Cantarel, B. L., Duncan, A., Ley, R. E., Sogin, M. L., Jones, W. J., Roe, B. A., Affourtit, J. P., et al. (2009). A core gut microbiome in obese and lean twins. *Nature, 457*, 480–484.

Varki, A. (2006). Nothing in glycobiology makes sense, except in the light of evolution. *Cell, 126*, 841–845.

Walter, J., & Ley, R. (2011). The human gut microbiome: Ecology and recent evolutionary changes. *Annual Review of Microbiology, 65*, 411–429.

Wang, Q., Garrity, G. M., Tiedje, J. M., & Cole, J. R. (2007). Naïve Bayesian classifier for rapid assignment of rRNA sequences into the new bacterial taxonomy. *Applied and Environmental Microbiology, 73*, 5261–5267.

Wang, J., Thingholm, L. B., Skiecevičienė, J., Rausch, P., Kummen, M., Hov, J. R., Degenhardt, F., Heinsen, F.-A., Rühlemann, M. C., Szymczak, S., et al. (2016). Genome-wide association analysis identifies variation in vitamin D receptor and other host factors influencing the gut microbiota. *Nature Genetics, 48*, 1396–1406.

Woese, C. R., & Fox, G. E. (1977). Phylogenetic structure of the prokaryotic domain: The primary kingdoms. *Proceedings of the National Academy of Sciences of the United States of America, 74*, 5088–5090.

Wu, G. D., Chen, J., Hoffmann, C., Bittinger, K., Chen, Y.-Y., Keilbaugh, S. A., Bewtra, M., Knights, D., Walters, W. A., Knight, R., et al. (2011). Linking long-term dietary patterns with gut microbial enterotypes. *Science, 334*, 105–108.

Yatsunenko, T., Rey, F. E., Manary, M. J., Trehan, I., Dominguez-Bello, M. G., Contreras, M., Magris, M., Hidalgo, G., Baldassano, R. N., Anokhin, A. P., et al. (2012). Human gut microbiome viewed across age and geography. *Nature, 486*, 222–227.

Zhao, S., Lieberman, T. D., Poyet, M., Groussin, M., Gibbons, S. M., Xavier, R. J., & Alm, E. J. (2017). Adaptive evolution within the gut microbiome of individual people. *BioRxiv*. https://doi.org/10.1101/208009 [Preprint] March 9, 2018.

Zilber-Rosenberg, I., & Rosenberg, E. (2008). Role of microorganisms in the evolution of animals and plants: The hologenome theory of evolution. *FEMS Microbiology Reviews, 32*, 723–735.

6 Microbiome and Diet

Francesca De Filippis and Danilo Ercolini

Abstract

The importance of gut microbiome in influencing human health has been widely assessed. The gut microbiome may vary according to several extrinsic factors, among which diet can be considered one of the most important. Substrates provided through diet are metabolized by the gut microbiome, with the possible production of beneficial or harmful metabolites. In the past decades, dietary habits in the Western world have strongly changed, with an increase in the consumption of foods of animal origin and a decrease in the intake of fiber and complex polysaccharides. These changes in the diet impacted our microbial symbionts, possibly playing a role in the development of several diseases. The understanding of these relationships will allow, in a next future, a targeted modulation of the gut microbiome through ad hoc dietary interventions for therapeutic or preventive purposes. In this chapter, recent findings about the existing interconnections between gut microbiome, diet, and human health are discussed, highlighting possible future perspectives.

6.1 The Human Gut Microbiome

Gut microbiota of healthy adults is commonly dominated by two bacterial *phyla*, *Firmicutes* and *Bacteroidetes*, with interindividual variability in their proportions. *Proteobacteria*, *Actinobacteria*, and *Verrucomicrobia* are present at lower levels (Lozupone et al. 2012). In spite of the great inter- and intraindividual variability, in the past years, a classification of subjects based on the most abundant genera in their gut microbiome was proposed. Some years ago, it was observed that all the subjects may be classified in three discrete clusters, named "enterotypes," based on the prevalence of *Prevotella*, *Bacteroides*, or *Ruminococcus* in their gut microbiome (Arumungam et al. 2011). However, the "enterotype" concept was lately criticized, since a rigorous categorization may lead to an oversimplified vision of the gut microbiome (Knights et al. 2014; Jeffery et al. 2012). On the contrary, although this classification may be attractive for understanding microbial variation in health and disease, the existence of a smooth gradient of the dominant *taxa* is more plausible, where the abundance of dominant genera varies continuously in the human population going from an enterotype to another. The important role of gut microbiota in

F. De Filippis · D. Ercolini (✉)
Division of Microbiology, Department of Agricultural Sciences, University of Naples Federico II, Portici (NA), Italy

Task Force on Microbiome Studies, University of Naples Federico II, Naples, Italy
e-mail: francesca.defilippis@unina.it; ercolini@unina.it

D. Haller (ed.), *The Gut Microbiome in Health and Disease*,
https://doi.org/10.1007/978-3-319-90545-7_6

influencing human well-being is widely recognized. It plays a primary function in host health by shaping the development of the immune system, metabolizing dietary nutrients and drugs, and synthesizing vitamins, bioactive molecules, and other beneficial or detrimental metabolites.

In the past decades, gut dysbiosis has been linked to the development of several kinds of diseases, including obesity (Turnbaugh et al. 2006; Le Chatelier et al. 2013), diabetes (Qin et al. 2012), inflammatory bowel disease (Marchesi et al. 2016), and cardiovascular diseases (Koeth et al. 2013).

Moreover, the gut microbiome may influence human behavior by the bidirectional communication path between the gastrointestinal (GI) and the central nervous system, namely, the gut-brain axis. This happens by a microbiome-mediated production of molecules that have neuroactive effects, such as serotonin and γ-aminobutyric acid (GABA). Nutrients and microbial metabolites interact with the enteroendocrine cells (EECs) located along the GI tract and containing most of the nutrient receptors. Interactions with EEC receptors are crucial in mechanisms such as the regulation of appetite and insulin secretion (Furness et al. 2013). Indeed, recent studies suggest that bacterial proteins may influence the appetite-controlling pathways, acting locally in the gut with a short-term effect on satiation (Breton et al. 2016). Moreover, plasmatic levels of specific bacterial proteins may activate host anorexigenic circuitries with a long-term regulation of the feeding pattern (Breton et al. 2016).

6.2 Diet-Induced Signatures in Gut Microbiota Composition and Functions in Rural Populations Around the World

Diet can be considered as the primary factor influencing gut microbiota composition and functionality. The first studies highlighting the key role played by diet found an association of *Bacteroides* with a high-fat/low-fiber diet, while a high abundance of *Prevotella* was linked to a diet rich in fiber and low in foods of animal origin (David et al. 2014; Wu et al. 2011). Indeed, a short-term dietary switch from high-fat/low-fiber to high-fiber/low-fat diet causes reproducible changes in the gut microbiome that are, however, not long-lasting. On the contrary, long-term, habitual diet is the primary factor shaping our gut microbiome (David et al. 2014; Wu et al. 2011). Over recent decades, modern dietary patterns in Western countries have undergone major compositional changes, with an increase in the consumption of red meat, high-fat foods, and refined sugars. This "Westernization" of the diet, together with changes in the lifestyle, has surely partially contributed to the higher incidence of inflammatory disorders, such as obesity, diabetes, cardiovascular diseases, and allergies. The study of agrarian populations living in South America or Africa helped to understand the complex relationships between habitual diet and the gut microbiome (Table 6.1). Comparison of the gut microbiomes of agrarian and Western populations may help in understanding how dietary changes have affected the gut symbionts. All these studies consistently found that the composition of the gut microbiota dramatically differs between urbanized, Western people eating a high-fat and protein diet and rural populations still consuming a subsistent, agrarian diet, with low intake of products of animal origin and high consumption of fruit, vegetables, fibrous tubers, and roots. Most of these studies highlighted the loss of microbial diversity in Westernized populations (Clemente et al. 2015; Martínez et al. 2015; Obregon-Tito et al. 2015; Schnorr et al. 2014; De Filippo et al. 2010) and overlapping signatures in gut microbiota composition of the traditional population studied (Table 6.1). *Prevotella*, *Xylanibacter*, *Treponema*, *Succinivibrio*, *Lachnospira*, and other fiber-degrading bacteria were reported as enriched in traditional populations consuming an agrarian diet (Fig. 6.1). *Treponema* is a genus belonging to the *Spirochaetes*. It was only reported to be present in the gut microbiota of nonhuman primates (Gomez et al. 2016a; Ley et al. 2008; Ochman et al. 2010) and of human traditional populations (Schnorr et al. 2014; De Filippo et al. 2010; Obregon-Tito et al. 2015; Gomez et al. 2016b; Ou et al. 2013) suggesting that these symbionts were lost in urban, Westernized people. The genomes of two distinct

Table 6.1 Principal gut microbiome signatures reported in traditional populations across the world

Traditional population	Location	Dietary habits	Gut microbiota signatures[a]	Reference
Burkinabe	Africa	Agricultural products	↑ Bacteroidetes, *Prevotella, Xylanibacter, Butyrivibrio, Treponema* ↓ Firmicutes, *Bacteroides, Enterobacteriaceae*	De Filippo et al. (2010)
Malawian	Africa	Agricultural products	↑ *Prevotella*	Yatsunenko et al. (2012)
Guahibo Amerindians	South America	Agricultural products	↑ *Prevotella*	Yatsunenko et al. (2012)
Hazda	Africa	Hunting and gathering	↑ *Prevotella, Succinivibrio, Treponema, Eubacterium* ↓ *Bacteroides, Blautia, Dorea*	Schnorr et al. (2014)
BaAka	Africa	Hunting and gathering	↑ Bacteroidetes, *Prevotella, Treponema*	Gomez et al. (2016b)
Bantu	Africa	Agricultural products	↑ Firmicutes	Gomez et al. (2016b)
Tunapuco	South America	Agricultural products	↑ *Treponema, Prevotella*	Obregon-Tito et al. (2015)
Matses	South America	Hunting and gathering	↑ *Eubacterium, Lachnospira, Catenibacterium, Treponema, Clostridium*	Obregon-Tito et al. (2015)
Rural South Africans	Africa	Agricultural products	↑ *Prevotella, Oscillospira, Succinivibrio, Treponema*	Ou et al. (2013)
Arctic Inuit	Canada	Animal fats and proteins rich diet	↓ *Prevotella*	Girard et al. (2017)
Yanomami Amerindian	South America	Hunting and gathering	↑ *Prevotella, Spirochaeta, Desulfovibrio, Helicobacter*	Clemente et al. (2015)
Asaro and Sausi	Papua New Guineas	Agricultural products	↑ *Prevotella, Streptococcus Bacteroides, Odoribacter* ↓ *Parabacteroides, Alistipes, Bilophila*	Martínez et al. (2015)

[a]The increase or decrease refers to a comparison with Western subjects

strains of *Treponema* from gut metagenomes of South Americans were reconstructed (Obregon-Tito et al. 2015). These strains turned out to be clearly different from known pathogenic *Treponema* strains with their genomic potential suggesting a selective adaptation to the gut environment including the capability to use complex polysaccharides, abundant in the diet of non-Westernized subjects.

Gomez et al. (2016b) characterized the gut microbiome of two traditional African populations (BaAka and Bantu) with different diet and lifestyle: BaAka are hunter-gatherers living in the rainforest and consuming a diet rich in fibrous tubers, such as manioc roots and wild yams, while Bantu have switched to a subsistence agriculture in the past century. Interestingly, when comparing their gut microbiomes with those of American control subjects, they found that Bantu showed a gut microbiota composition intermediate between BaAka and Americans. Indeed, a decrease in the abundance of fiber-degrading bacteria was observed from BaAka to Americans, with intermediate levels in Bantu, reflecting the different subsistence patterns and highlighting the effect of a recent transition to a more modern lifestyle in Bantu (Gomez et al. 2016b).

On the contrary, the gut microbiota of Western subjects is often enriched in bile-tolerant microorganisms, such as *Bacteroides*, *Alistipes*, and *Bilophila*, associated with the consumption of a diet rich in fat and protein and poor in indigestible carbohydrates (Fig. 6.1).

Such a difference in microbiota composition between rural and Western populations results in a different functionality. Indeed, metagenomes of

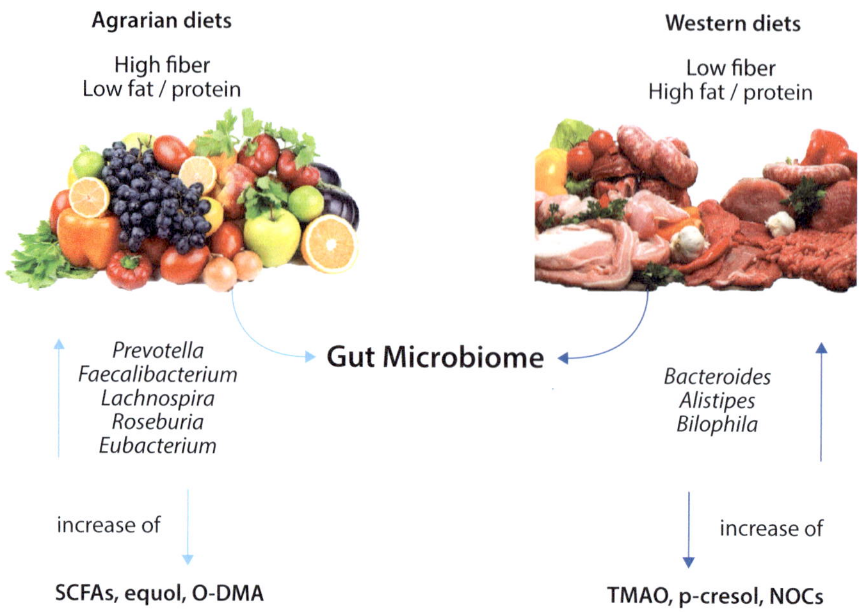

Fig. 6.1 Effect of diet on gut microbiota and metabolome

Hadza hunter-gatherer and urbanized Italians showed a different potential activity (Rampelli et al. 2015). The Hadza microbiome contains a more diverse pool of genes that encodes for enzymes able to break down a broader set of polysaccharides compared to Italians. On the contrary, metabolic pathways related to the degradation of several xenobiotics of human origin (e.g., naphthalene, xylene, or benzoate), ubiquitous in industrialized environments, were enriched in the Italian gut microbiome (Rampelli et al. 2015). Moreover, Western diet, richer in fat and proteins, led to the selection of a gut microbiome with higher capacity to degrade amino acids and convert primary bile acids in secondary bile acids, as well as simple sugars (Yatsunenko et al. 2012).

6.3 Diet-Mediated Production of Beneficial or Detrimental Metabolites by the Gut Microbiome

Different dietary habits and gut microbiomes promote the production of a distinctive fecal metabolome in traditional and Western populations. Undigested dietary components reach the large intestine, where they are fermented by the microbial community to produce a wide pattern of metabolites, reflecting both the chemical diversity of the available substrates and the different metabolic potential of the gut microbiota. Most studies focused on fecal levels of short-chain fatty acids (SCFAs). The three most abundant SCFAs, acetic, propionic, and butyric acids, are produced by bacterial fermentation of indigestible polysaccharides and are normally present in a molar ratio of about 3:1:1 (Louis et al. 2014). Nondigestible carbohydrates include the structural polysaccharides of plant cell walls (non-starch polysaccharides, such as celluloses, xylans, pectins, and glucans), resistant starch, and soluble oligosaccharides (e.g., fructo-oligosaccharides) (Flint et al. 2012). SCFAs are an important energy source for human colonocytes and have been often associated with several health-promoting effects, such as anti-inflammatory and anticarcinogenic (Louis et al. 2014; O'Keefe 2016). Indeed, the abundance of fecal SCFAs was significantly higher in the agrarian and rural populations studied around the world (De Filippo et al. 2010; Ou et al. 2013; Schnorr et al. 2014), and this is undoubtedly correlated with the substantial enrichment of fiber in their diet, which selects for a fiber-degrading and SCFA-producing gut microbiota. Besides SCFAs, gut microbiota can produce other metabolites, with a potentially beneficial effect on the host health (Fig. 6.1).

Plant cell walls englobe a variety of complex micronutrients, collectively named phytochemicals, which are unabsorbed in the upper gastrointestinal tract. In the colon, microbial fermentation releases these compounds, and some bacteria can metabolize and convert them to a wide range on bioactive molecules (Cardona et al. 2013). As an example, the soy isoflavone, daidzein, can be converted to equol by some gut bacteria, such as species belonging to *Faecalibacterium*, *Eubacterium*, *Bifidobacterium*, and *Clostridium*. According to some scientific evidence, equol may possess anticarcinogenic properties (Cardona et al. 2013). However, due to the wide variety of chemical molecules commonly included in the phytochemical class, it can be expected that different gut microbiota members may be differently involved in their degradation, with the consequent production of a wide range of bioactive metabolites.

Gut microbiota can also produce detrimental metabolites (Fig. 6.1). High-protein intake results in an increase in degradation of proteins in the colon and consequently higher levels of fecal amino acid-derived products, such as branched-chain fatty acids and phenylacetic acid (Ou et al. 2013; Russell et al. 2011). High levels of branched-chain amino acids in the blood have been associated with insulin resistance and development of type 2 diabetes (Pedersen et al. 2016). In addition, some bacteria, such as *Bacteroides* spp., can ferment aromatic amino acids to produce phenylacetic acid, phenols, indoles, and *p*-cresol, associated with a pro-inflammatory and carcinogenic effect (Louis et al. 2014). Sulfate-reducing bacteria (e.g., *Desulfovibrio* spp.) produce sulfides through the catabolism of sulfur amino acids and taurine. Sulfides are pro-inflammatory and toxic for colonocytes (O'Keefe 2016). Moreover, amine derived from microbial fermentation of proteins in the colon can be nitrosated to produce *N*-nitroso compounds (NOCs), which exert a carcinogenic and mutagenic effect and are correlated to the incidence of colorectal cancer (Loh et al. 2011). Indeed, increased fecal NOCs and phenylacetic acid and decreased SCFA levels can be found after administration of a high-protein/low-carbohydrate diet (Russell et al. 2011). In addition, high-fat diet leads to an increase in bile secretion and consequent higher quantity of bile acids in the colon. Bile salt hydrolases can cleave glycine and taurine residues from the primary bile acids, converting them into several secondary bile acids, mainly deoxycholic and lithocholic acids. Ou et al. (2013) observed that microbial genes encoding for secondary bile acid production and fecal secondary bile acid concentration were more abundant in African Americans compared to rural native Africans, also showing lower colorectal cancer risk. Accordingly, when African Americans and rural Africans switched their diets for 2 weeks, a reduction in fecal secondary bile acids and in colonic mucosal inflammation markers was observed in African Americans, associated with the decrease in the abundance of *Bilophila wadsworthia*. On the contrary, the high-fat intervention in rural Africans was linked to an increase in *Fusobacterium nucleatum*, previously found in human colon cancer tissues (O'Keefe et al. 2015).

Another important microbial metabolite associated with a detrimental effect for host health is trimethylamine-*N*-oxide (TMAO; Fig. 6.1). Intestinal microbiota catabolism of choline, phosphatidylcholine, and L-carnitine produces trimethylamine (TMA), which is further oxidized in the liver resulting in TMAO. The latter considered a risk factor in the development of cardiovascular diseases (CVDs) and atherosclerosis (Tang et al. 2013; Wang et al. 2011).

6.4 Gut Microbiome and Dietary Habits in the Western World

Western diets are not necessarily detrimental for the gut microbiome. In Westernized countries, healthier dietary patterns such as vegetarian and vegan diets are becoming increasingly popular. According to an American poll in 2016, approximately 3.3% of American adults are vegetarians or vegans. This percentage increases to 6% when considering only young adults (18–34 years), while only 2% of people 65 years or older are vegetarians. Accordingly, sales of alternative meat products reached $553 million in 2012, an 8% increase in 2 years (Melina et al. 2016). A well-planned vegetarian diet contains vegetables, fruits, whole grains, legumes, nuts, and seeds, besides some products of animal origin, such as eggs, milk, and derivates thereof. Both vegetarian and vegan diets are devoid of flesh foods (such as meat, poultry, wild game, seafood, and their products). In addition, vegans do not consume any food of animal origin. The adoption of a vegetarian diet may cause a reduced intake of certain nutrients; however, deficiencies can be readily avoided by appropriate dietary planning and, if necessary, the consumption of supplements (Melina et al. 2016). Although the spread of these dietary patterns is growing, only few studies addressed the question if these diets select for distinctive traits in the gut microbiome. Recently, gut microbiome and metabolome were studied in a cohort of 153 vegetarian, vegan, and omnivore Italians (De Filippis et al. 2016). Vegans and vegetarians showed higher abundance of plant-degrading bacteria, such as *Lachnospira* and *Prevotella*. Moreover, these genera were positively correlated to the fecal levels of the three main SCFAs and negatively correlated with the urinary concentration of TMAO, with the former being higher and the latter being lower in vegetarians/vegans compared to omnivores (De Filippis et al. 2016). Moreover, increased plasma levels of metabolites likely derived from microbial catabolism of plant polyphenolic compounds as well as increased abundance of equol were found in a small cohort of vegans (Wu et al. 2016). However, differences in equol concentration were not linked to the consumption of foods rich in equol precursors (such as soy-based products). Indeed, only about 30–40% of Western adults are likely able to convert isoflavones to equol, regardless the dietary intake, against about the 70% of Asians (Magee 2011). Thus, these data emphasize that both the consumption of the right substrates and the presence of specific microbial metabolic capacity jointly determine the potential development of a health-promoting metabolome.

Beyond the strict vegetarian or vegan regimes, the consumption of a healthy and diverse dietary pattern, such as that based on the Mediterranean-style, can help to develop a health-associated microbiome and metabolome, without completely banning meat and other products of animal origin. The Mediterranean diet has been recognized by UNESCO as intangible cultural heritage (http://www.unesco.org/culture/ich/RL/00884). The Mediterranean dietary pattern is characterized by high intake of fruit, vegetables, legumes, nuts, and whole grains, moderate consumption of fish, and low intake of saturated fat, meat, and dairy products (Trichopoulou et al. 1995). It has been demonstrated to be beneficial for the treatment of obesity, type 2 diabetes, and inflammatory and cardiovascular diseases (Santoro et al. 2014; Estruch et al. 2013; Salas-Salvadó et al. 2011). Studying the gut microbiota and metabolome in an Italian cohort of adults with different diary habits, De Filippis et al. (2016) also evaluated the adherence level to the Mediterranean diet, calculating the Mediterranean dietary score (Agnoli et al. 2011). They observed a progressive increase in the concentration of fecal SCFA going from low-adherence to high-adherence subjects (Fig. 6.2a). Intriguingly, even when considering the omnivore group alone, the higher the adherence to the Mediterranean diet, the higher the concentration of fecal SCFAs (Fig. 6.2b). Therefore, a healthy dietary pattern, such as the Mediterranean-style diet, can provide appropriate substrates and shape the gut microbiome to promote the production of health-related metabolites even in an overall omnivore-type diet.

6.5 Modulation of Gut Microbiome Through Diet: Possible Implications for Human Health

Several studies tried to address the question whether we can modulate the microbiome through diet. Indeed, the presence/abundance of specific gut microbiota members was shown to be adjustable through a specific dietary intervention. However, the consumption of Westernized diets may induce changes in the gut microbiota composition that are not reversible. Indeed, experiments on humanized mice showed that the consumption of a diet low in microbiota-accessible carbohydrates (low-MACs)

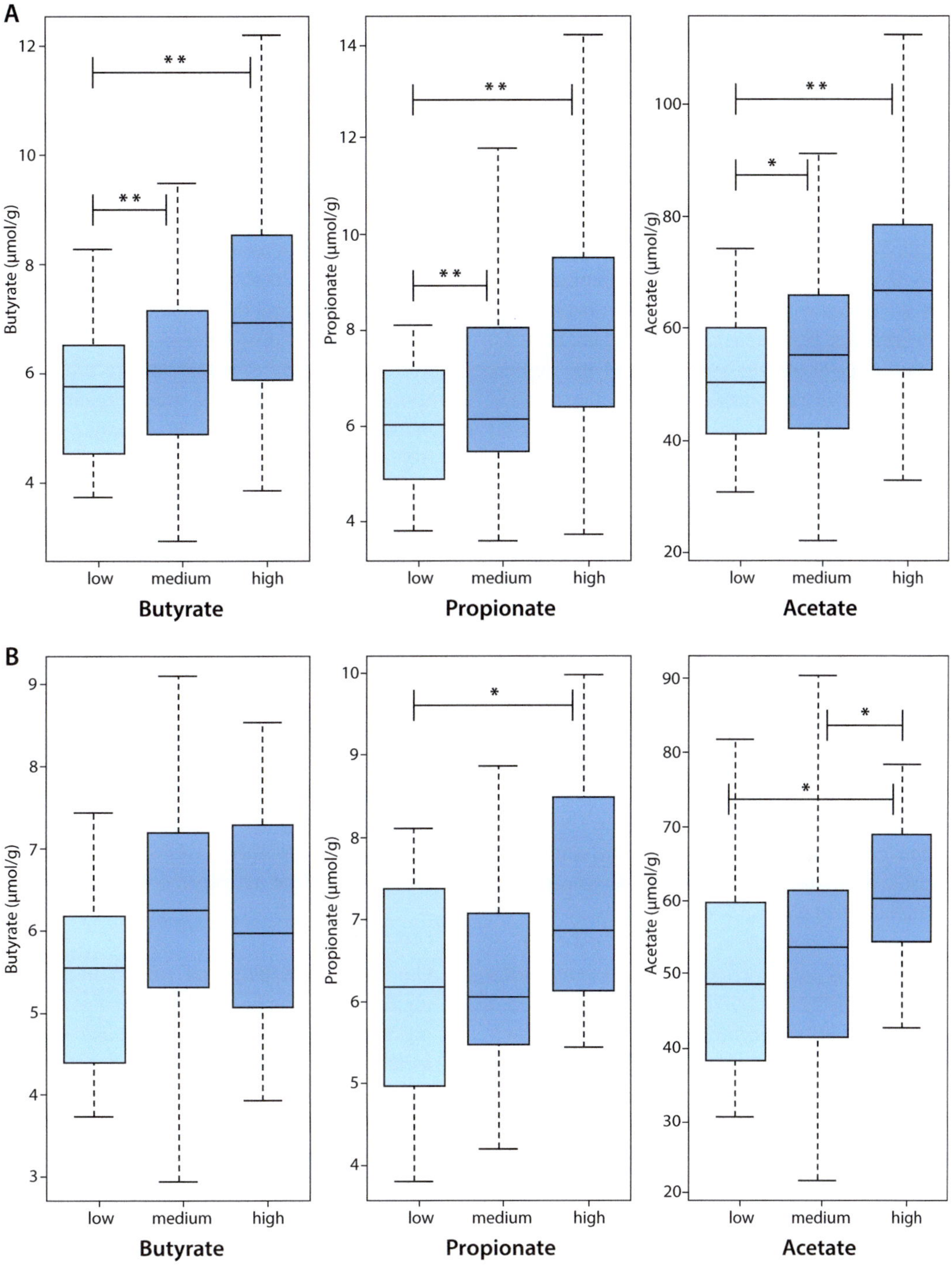

Fig. 6.2 Box plots showing the fecal concentrations of butyrate, propionate, and acetate in Italian omnivores, vegetarians, and vegans analyzed in De Filippis et al. (2016). Panel **a** shows the data for the subjects grouped according to adherence level to the Mediterranean diet, while panel **b** shows the data for the omnivore subjects. ($^{*}p < 0.05$ and $^{**}p < 0.01$). Reproduced from "High-level adherence to a Mediterranean diet beneficially impacts the gut microbiota and associated metabolome.", De Filippis F. et al. 2016. *Gut*, 65:1812–1821 with permission from BMJ Publishing Group Ltd

over several generations results in a progressive loss of microbial diversity in the offspring, which is not recoverable after the reintroduction of dietary MACs (Sonnenburg et al. 2016). However, response to dietary intervention is person-specific and has been shown to be strongly linked to the gut microbiota composition. Indeed, Kovatcheva-Datchary and co-workers (2015) demonstrated that a dietary intervention with barley kernel fiber may induce changes in the gut microbiome which are associated with improved glucose metabolism. Nevertheless, the supplementation did not have the same effect on all the subjects, and the authors suggested that responders have peculiar traits in their gut microbiota, such as higher *Prevotella* abundance, and that this microbe may be responsible for increased glycogen storage in the liver. On the contrary, subjects showing lower abundance of *P. copri* did not show a metabolic response to the fiber supplementation. Accordingly, people eating identical meals showed different metabolic responses, consistently associated with gut microbiota features. Integration of blood parameters, dietary habits, and anthropometric and gut microbiota features in a complex algorithm enables an accurate prediction of postprandial glycemic response, linked to type 2 diabetes development (Zeevi et al. 2015). Moreover, such algorithm may be successfully used in setting up personalized dietary intervention to lower postprandial blood glucose.

Since the application of generalized medical practices did not always ensure consistent results among subjects, the possibility of a personalized nutrition strategy recently emerged. The knowledge acquired in the past decades about gut microbiome composition and functionality and the presence of diet-responsive members will be used in the near future in personalized therapeutic approaches based on a targeted modulation of intestinal microbial communities.

▶ Controversy

Recent studies highlighted the possibility of boosting changes in the gut microbiome through specific and appropriate dietary interventions. Although this opportunity is tantalizing, there are several issues to consider. All intervention studies must face the presence of many confounding factors, e.g., different physical activity or lifestyle and consumption of alcohol, drugs, or medications in the cohort studied, all of which may obscure the effect of the treatment. Thus, controlling for external factors in microbiome studies is extremely important, disentangling the effect of the treatment from the effects of confounders. Moreover, high intersubject variability in the gut microbiome composition exists. Interindividual differences in the gut microbiome at baseline may lead to subject-specific responses to the same treatment. Indeed, results reported in literature are often contrasting, and even in the same study, different subjects in the cohort may show distinct outcomes in response to the same dietary intervention. Integration of blood parameters, dietary habits, and anthropometric and gut microbiota features in complex algorithms can help to accurately predict metabolic response to different meals and to formulate subject-specific dietary regimes. Personalized nutrition strategies based on individual gut microbiome features are recently emerging and in a next future will allow developing new therapeutic or disease-preventive approaches based on a targeted modulation of gut microbiome through diet.

History

Gut microbiota plays an important role in regulating human health/disease status, affecting the development of the immune system, metabolizing dietary nutrients and drugs, and synthesizing vitamins, bioactive molecules, or other beneficial or detrimental metabolites. Several factors may influence gut microbiome, leading to gut dysbiosis, which was linked to the development of several diseases. Diet may be considered as the primary factor influencing gut microbiota composition and functionality. In a recent past, dietary habits in the Western world have strongly changed, with increased consumption of animal origin fat/proteins and reduced intake of complex polysaccharides. Comparison of the gut microbiome of urbanized, Western subjects with traditional, rural populations living in Africa and South America highlighted several differences, suggesting that our microbiome coevolved with humans, shaped by diet and lifestyle. Westernized subjects show lower microbial diversity and lose specific fiber-degrading bacteria in their gut microbiome. Differences in gut microbiota composition reflect its functionality. Indeed, gut microbiome produces several health-promoting (e.g., short-chain fatty acids, equol) or detrimental (e.g., sulfides, N-nitroso compounds, trimethylamine) metabolites, depending on the substrates that are made available through the diet. Specific dietary interventions may induce change in the gut microbiome. However, recent studies highlighted subject-specific responses to the same intervention, possibly due to specific features of the gut microbiome.

Highlights

- Gut microbiome may influence host health through the production of beneficial or detrimental metabolites, depending on the type of substrates provided by diet.
- What is food for us is also food for our microbial symbionts; thus, gut microbiome is strongly influenced by the habitual diet.
- Changes in dietary habits and lifestyles in Western populations may have affected their gut microbiome.
- Short dietary intervention may induce changes in the gut microbiome.
- Recent research provides the bases for the development of new therapeutic strategies based on a targeted modulation of the gut microbiome through diet.

References

Agnoli, C., Krogh, V., Grioni, S., et al. (2011). A priori–defined dietary patterns are associated with reduced risk of stroke in a large Italian cohort. *The Journal of Nutrition, 141*, 1552–1558.

Arumungam, M., Raes, J., Pelletier, E., et al. (2011). Enterotypes of the human gut microbiome. *Nature, 473*(7346), 174–180.

Breton, J., Tennoune, N., Lucas, N., et al. (2016). Gut commensal *E. coli* proteins activate host satiety pathways following nutrient-induced bacterial growth. *Cell Metabolism, 23*(2), 324–334.

Cardona, F., Andrés-Lacueva, C., Tulipani, S., et al. (2013). Benefits of polyphenols on gut microbiota and implications in human health. *The Journal of Nutritional Biochemistry, 24*, 1415–1422.

Clemente, J. C., Pehrsson, E. C., Blaser, M. J., et al. (2015). The microbiome of uncontacted Amerindians. *Science Advances, 1*(3), e1500183. https://doi.org/10.1126/sciadv.1500183.

David, L. A., Maurice, C. F., Carmody, R. N., et al. (2014). Diet rapidly and reproducibly alters the human gut microbiome. *Nature, 505*(7484), 559–563.

De Filippis, F., Pellegrini, N., Vannini, L., et al. (2016). High-level adherence to a Mediterranean diet beneficially impacts the gut microbiota and associated metabolome. *Gut, 65*(11), 1812–1821.

De Filippo, C., Cavalieri, D., Di Paola, M., et al. (2010). Impact of diet in shaping gut microbiota revealed by a comparative study in children from Europe and rural Africa. *Proceedings of the National Academy of Sciences of the United States of America, 107*(33), 14691–14696.

Estruch, R., Ros, E., Salas-Salvadó, J., et al. (2013). Primary prevention of cardiovascular disease with a Mediterranean diet. *The New England Journal of Medicine, 368*(14), 1279–1290.

Flint, H. J., Scott, K. P., Duncan, S. H., et al. (2012). Microbial degradation of complex carbohydrates in the gut. *Gut Microbes, 3*(4), 289–306.

Furness, J. B., Rivera, L. R., Cho, H.-J., et al. (2013). The gut as a sensory organ. *Nature Reviews Gastroenterology & Hepatology, 10*, 729–740.

Girard, C., Tromas, N., Amyot, M., et al. (2017). Gut microbiome of the Canadian arctic Inuit. *mSphere, 2*(1), e00297-16.

Gomez, A., Petrzelkova, K. J., Burns, M. B., et al. (2016a). Gut microbiome of coexisting BaAka pygmies and Bantu reflects gradients of traditional subsistence patterns. *Cell Reports, 14*(9), 2142–2153.

Gomez, A., Rothman, J. M., Petrzelkova, K., et al. (2016b). Temporal variation selects for diet–microbe co-metabolic traits in the gut of Gorilla spp. *The ISME Journal, 10*, 514–526.

Jeffery, I. B., Claesson, M. J., O'Toole, P. W., & Shanahan, F. (2012). Categorization of the gut microbiota: Enterotypes or gradients? *Nature Reviews Microbiology, 10*(9), 591–592.

Knights, D., Ward, T. L., McKinlay, C. E., et al. (2014). Rethinking "enterotypes". *Cell Host & Microbe, 16*(4), 433–437.

Koeth, R. A., Wang, Z., Levison, B. S., et al. (2013). Intestinal microbiota metabolism of L-carnitine, a nutrient in red meat, promotes atherosclerosis. *Nature Medicine, 19*(5), 576–585.

Kovatcheva-Datchary, P., Nilsson, A., Akrami, R., et al. (2015). Dietary fiber-induced improvement in glucose metabolism is associated with increased abundance of *Prevotella*. *Cell Metabolism, 22*(6), 971–982.

Le Chatelier, E., Nielsen, T., Qin, J., et al. (2013). Richness of human gut microbiome correlates with metabolic markers. *Nature, 500*(7464), 541–546.

Ley, R. E., Hamady, M., Lozupone, C., et al. (2008). Evolution of mammals and their gut microbes. *Science, 320*, 1647–1651.

Loh, Y. H., Jakszyn, P., Luben, R. N., et al. (2011). N-nitroso compounds and cancer incidence: The European Prospective Investigation into Cancer and Nutrition (EPIC)-Norfolk Study. *The American Journal of Clinical Nutrition, 93*(5), 1053–1061.

Louis, P., Hold, G. L., & Flint, H. J. (2014). The gut microbiota, bacterial metabolites and colorectal cancer. *Nature Reviews Microbiology, 12*(10), 661–672.

Lozupone, C. A., Stombaugh, J. I., Gordon, J. I., et al. (2012). Diversity, stability and resilience of the human gut microbiota. *Nature, 489*(7415), 220–230.

Magee, P. J. (2011). Is equol production beneficial to health? *The Proceedings of the Nutrition Society, 70*(1), 10–18.

Marchesi, J. R., Adams, D. H., Fava, F., et al. (2016). The gut microbiota and host health: A new clinical frontier. *Gut, 65*, 330–339.

Martínez, I., Stegen, J. C., Maldonado-Gómez, M. X., et al. (2015). The gut microbiota of rural Papua New Guineans: Composition, diversity patterns, and ecological processes. *Cell Reports, 11*(4), 527–538.

Melina, V., Craig, W., & Levin, S. (2016). Position of the academy of nutrition and dietetics: Vegetarian diets. *Journal of the Academy of Nutrition and Dietetics, 116*(12), 1970–1980.

O'Keefe, S. J. (2016). Diet, microorganisms and their metabolites, and colon cancer. *Nature Reviews Gastroenterology & Hepatology, 13*(12), 691–706.

O'Keefe, S. J., Li, J. V., Lahti, L., et al. (2015). Fat, fibre and cancer risk in African Americans and rural Africans. *Nature Communications, 6*, 6342. https://doi.org/10.1038/ncomms7342.

Obregon-Tito, A. J., Tito, R. Y., Metcalf, J., et al. (2015). Subsistence strategies in traditional societies distinguish gut microbiomes. *Nature Communications, 6*, 6505. https://doi.org/10.1038/ncomms7505.

Ochman, H., Worobey, M., Kuo, C. H., et al. (2010). Evolutionary relationships of wild hominids recapitulated by gut microbial communities. *PLoS Biology, 8*(11), e10000546. https://doi.org/10.1371/journal.pbio.1000546.

Ou, J., Carbonero, F., Zoetendal, E. G., et al. (2013). Diet, microbiota, and microbial metabolites in colon cancer risk in rural Africans and African Americans. *The American Journal of Clinical Nutrition, 98*(1), 111–120.

Pedersen, H. K., Gudmundsdottir, V., Nielsen, H. B., et al. (2016). Human gut microbes impact host serum metabolome and insulin sensitivity. *Nature, 535*(7612), 376–381.

Qin, J., Li, Y., Cai, Z., et al. (2012). A metagenome-wide association study of gut microbiota in type 2 diabetes. *Nature, 490*(7418), 55–60.

Rampelli, S., Schnorr, S. L., Consolandi, C., et al. (2015). Metagenome sequencing of the Hadza hunter-gatherer gut microbiota. *Current Biology, 25*(13), 1682–1693.

Russell, W. R., Gratz, S. W., Duncan, S. H., et al. (2011). High-protein, reduced-carbohydrate weight-loss diets promote metabolite profiles likely to be detrimental to colonic health. *The American Journal of Clinical Nutrition, 93*(5), 1062–1072.

Salas-Salvadó, J., Bulló, M., Babio, N., et al. (2011). Reduction in the incidence of type 2 diabetes with the Mediterranean diet results of the PREDIMED-Reus nutrition intervention randomized trial. *Diabetes Care, 34*(1), 14–19.

Santoro, A., Pini, E., Scurti, M., et al. (2014). Combating inflammaging through a Mediterranean whole diet approach: The NU-AGE project's conceptual framework and design. *Mechanisms of Ageing and Development, 136-137*, 3–13.

Schnorr, S. L., Candela, M., Rampelli, S., et al. (2014). Gut microbiome of the Hadza hunter-gatherers. *Nature Communications, 5*, 3654. https://doi.org/10.1038/ncomms4654.

Sonnenburg, E. D., Smits, S. A., Tikhonov, M., et al. (2016). Diet-induced extinctions in the gut microbiota compound over generations. *Nature, 529*(7585), 212–215.

Tang, W. H. W., Wang, Z., Levison, B. S., et al. (2013). Intestinal microbial metabolism of phosphatidylcholine and cardiovascular risk. *The New England Journal of Medicine, 368*, 1575–1584.

Trichopoulou, A., Kouris-Blazos, A., Wahlqvist, M. L., et al. (1995). Diet and overall survival in elderly people. *BMJ, 311*, 1457–1460.

Turnbaugh, P. J., Ley, R. E., Mahowald, M. A., et al. (2006). An obesity-associated gut microbiome with increased capacity for energy harvest. *Nature, 444*(7122), 1027–1031.

Wang, Z., Klipfell, E., Bennett, B. J., et al. (2011). Gut flora metabolism of phosphatidylcholine promotes cardiovascular disease. *Nature, 472*(7341), 57–63.

Wu, G. D., Chen, J., Hoffmann, C., et al. (2011). Linking long term dietary patterns with gut microbial enterotypes. *Science, 334*(6052), 105–108.

Wu, G. D., Compher, C., Chen, E. Z., et al. (2016). Comparative metabolomics in vegans and omnivores reveal constraints on diet-dependent gut microbiota metabolite production. *Gut, 65*, 63–72.

Yatsunenko, T., Rey, F. E., Manary, M. J., et al. (2012). Human gut microbiome viewed across age and geography. *Nature, 486*(7402), 222–227.

Zeevi, D., Korem, T., Zmora, N., et al. (2015). Personalized nutrition by prediction of glycemic responses. *Cell, 163*(5), 1079–1094.

Microbiome and Gut Immunity: The Epithelium

7

Claudia Günther

Abstract

The intestinal epithelium not only plays a fundamental role in negotiating a homeostatic host-microbial relation but also represents the first line of defense against pathogenic microbes and microbial agents. As a consequence intestinal epithelial cells have developed a variety of mechanisms to respond to commensal and non-commensal microbes. Accordingly intestinal epithelial cells can physically restrict the translocation of potentially harmful microorganisms from the intestinal tract into the surrounding tissue by providing a physical barrier but also release antimicrobial peptides and mucus that control microbial composition and location. Despite its barrier function, the intestinal epithelium has an important function in translating luminal signals from the barrier surface to the underlying mucosal immune system. Defects in one of these functions can have tremendous effects on intestinal homeostasis and have been identified as key factors in the pathogenesis of intestinal inflammation. In this chapter we will discuss the role of the intestinal epithelium during host-microbe interactions.

C. Günther (✉)
Kussmaul Campus für Medizinische Forschung, Universitätsklinikum Erlangen – Medizinische Klinik I, Erlangen, Germany
e-mail: c.guenther@uk-erlangen.de

7.1 Introduction

The primary function of the gastrointestinal (GI) tract is the digestion and absorption of nutrients present in the intestinal lumen. Beside these food-derived nutrients, exogenous microorganisms such as bacteria, fungi, and viruses can also enter the gut. The abundance of innate and adaptive immune cells that reside together with trillions of beneficial commensal microorganisms in the gastrointestinal tract requires an effective barrier in order to define host-microbial interactions and to conserve tissue homeostasis. The intestinal tract is lined by the intestinal epithelium, which forms a physical and biochemical barrier at the interface between the host and its microbial-lined environment (Fig. 7.1). The intestinal epithelium consists of a monolayer of specialized intestinal epithelial cells that can exert diverse functions (van der Flier and Clevers 2009). This cell layer has an enormous self-renewing capacity (Heath 1996; van der Flier and Clevers 2009). Accordingly, the adult intestinal epithelium is renewed every 4–5 days to maintain optimal function (Crosnier et al. 2006). The life cycle of intestinal epithelial cells is determined by the time span in which these cells migrate from their place of origin to the villus tip, where these cells are exposed to the luminal environment. Cells that reach this status are expelled from the epithelial layer through a complex cytoskeletal remodeling process. The organization of the epithelium is adapted to the specific

D. Haller (ed.), *The Gut Microbiome in Health and Disease*,
https://doi.org/10.1007/978-3-319-90545-7_7

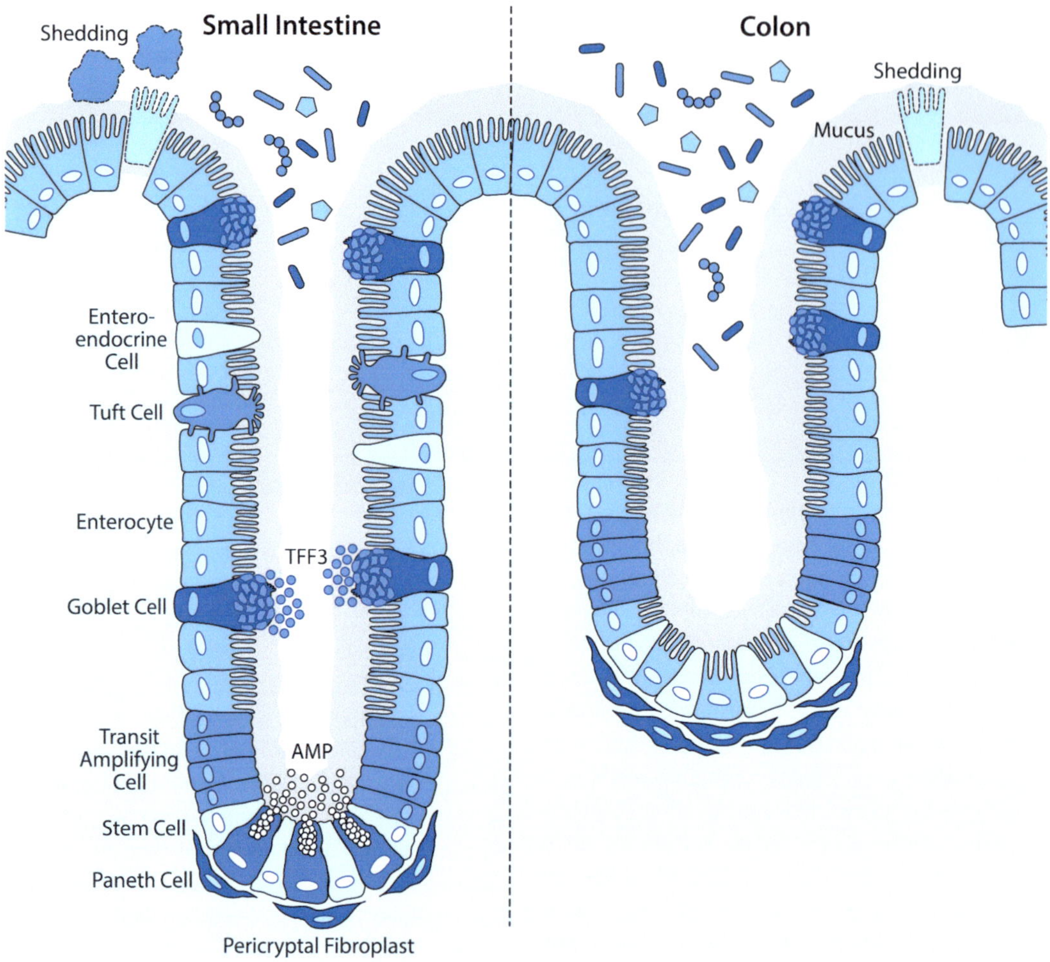

Fig. 7.1 Structure of the intestinal epithelium. In the small intestine (left panel), stem cells are located between Paneth cells at the crypt base. Stem cells continuously generate a pool of proliferating transit-amplifying cells (TA cells). These TA cells give rise to differentiated IECs with specialized functions. Enterocytes, goblet cells, enteroendocrine cells, and tuft cells replace cells that are lost by shedding at the villus tip. Paneth cells are located in the crypt next to stem cells. This epithelial monolayer is covered by a thick mucus film layer, which protects the intestinal epithelium. Both Paneth cells and pericryptal fibroblasts supply essential factors (including WNT, the Notch ligand delta-like 4 (DLL4), epidermal growth factor (EGF), and Noggin) to regulate the survival and function of the intestinal stem cells. In the colon, crypts lack Paneth cells (right panel)

functional requirements of the different regions of the intestinal tract. To allow efficient nutrient absorption, the small intestine is folded to form a large number of tubular invaginations, denoted as crypts, and fingerlike villus structures (Marshman et al. 2002; Ziv and Bendayan 2000). This enables the host to dramatically enlarge the surface area. Particularly in the lower small intestine (ileum), crypt regions are interrupted by aggregated lymphoid follicles, the Peyer's patches (PP) (Mowat 2003). Peyer's patches contain groups of lymphoid aggregates located in the submucosa of the small intestine that contain many immune cells, including B cells, T cells, and dendritic cells (Jung et al. 2010). Above the epithelium these lymphoid follicles consist of specialized epithelial cells, called microfold (M) cells (Corr et al. 2008; Mabbott et al. 2013; Miller et al. 2007). M cells are described as IECs that allow luminal contents to pass through and encounter antigen-presenting cells (APCs) below. Absorption of water and compacting of stool for rapid excretion take place in the large intestine (Kvietys and Granger 2010). Since villi would hamper this process, the colonic epithelium is arranged into multiple extended crypts associated with a flat luminal surface (Barker 2014). The colon also contains gut-associated lymphoid tissue (isolated lymphoid follicles) resembling Peyer's patches of the small intestine (Gibbons and Spencer 2011). In addition to the complex anatomical features of the lower GI tract (small intestine and colon), established by the three-dimensional structure of the gut surface, the intestinal epithelium represents a highly heterogeneous cell population containing various different cell types with diverse functions (Barker 2014; Gunther et al. 2013; Sancho et al. 2003; van der Flier and Clevers 2009). The life cycle of an intestinal epithelial cells starts at the crypt base, where small populations of adult stem cells regularly divide to produce highly proliferative progenitors known as transit-amplifying (TA) cells. TA cells spend approximately 2 days in the crypt before they undergo cell cycle arrest and differentiate into specialized intestinal epithelial cells (Barker 2014; Crosnier et al. 2006). Differentiation of these precursor cells occurs along the crypt-villus axis. Intestinal epithelial cells can be subdivided into two major groups: cells of the absorptive lineage and cells imprinted for the secretory lineage (Barker 2014; van der Flier and Clevers 2009). Senescent epithelial cells reach the tip of the villus (or surface epithelium in the colon) where they are shed into the intestinal lumen which allows a continual epithelial turnover (Bullen et al. 2006; Marchiando et al. 2011). The balance of cell shedding at the villus tip and renewal of stem cells in the crypts are indispensable to maintain small intestinal morphology and functionality. The current model of cell shedding predicts that overcrowding due to proliferation and migration induces live cell extrusion at the villus tip to maintain homeostatic cell numbers in the epithelium (Eisenhoffer et al. 2012).

7.2 Intestinal Epithelial Cells: Regulators of Barrier Function and Immune Homeostasis

As mentioned above intestinal epithelial stem cells can give rise to cells of either the secretory or absorptive lineage. Beside these main cell types, researchers have described some lesser known cell types, such as tuft cells, cup cells, and Peyer's patch-associated M cells (Gerbe et al. 2012; Madara 1982; Man et al. 2004; Ramirez and Gebert 2003). These specialized epithelial cells provide a tight barrier and achieve frontline defense by an incessant production of an array of secreted factors that limit direct contact between the epithelium and infectious agents. These major functions of intestinal epithelial cells are underlined by the fact that enteric pathogens have evolved strategies to target these defense mechanisms (Fig. 7.2).

7.2.1 Enterocytes

The majority (around 80%) of cells bordering the intestinal lumen are absorptive enterocytes (alternatively termed columnar cells) (de Santa Barbara et al. 2003; Snoeck et al. 2005). Enterocytes provide the physical barrier of the intestinal epithelium by forming close contacts via tight junction molecules. Since efficient absorption and transport of nutrients from the luminal side across the epithelium is their main function, enterocytes have further enlarged their luminal surface by carrying apically located microvilli (microscopic cellular membrane protrusions). These microvilli can be nicely visualized by electron microscopy, but not light microscopy; that is why they are mostly referred as brush border (due to their nappy appearance by light microscopy) (Ziv and Bendayan 2000). However, enterocytes, apart from their participation in digestive processes, exert more than a passive barrier function and are often directly involved in immune processes. Enterocytes express class I and II MHC molecules which allow them to deliver and present foreign antigens (AG) to the underlying

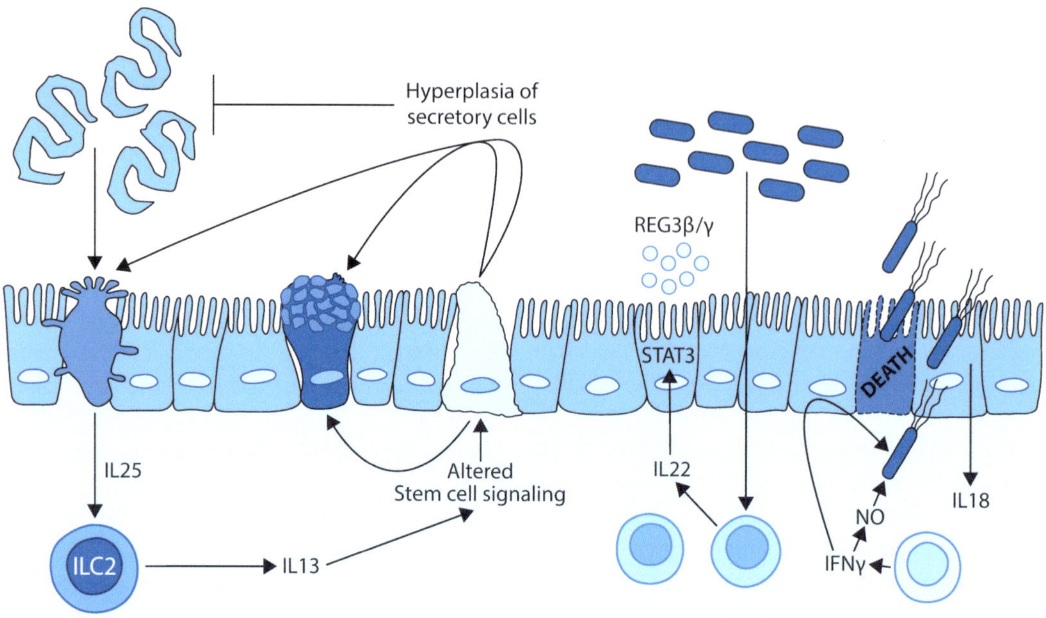

Fig. 7.2 Intestinal epithelial barrier function is essential for maintaining intestinal homeostasis. The intestinal epithelium can sense and respond to luminal contents by various direct and indirect mechanisms. IECs can directly respond to pathogens by the production of antimicrobial peptides or mucus. Elimination of infected cells by various cell death mechanisms can be initiated by IEC intrinsic signaling pathways or by factors produced by the underlying immune cell compartment. IECs are capable to produce and release immunogenic molecules that can shape the mucosal immune system. Tuft cells, ILC2s, and pericryptal fibroblasts can control differentiation of IEC progenitors into cells of the secretory lineage upon infection to efficiently clear pathogens. NO, nitic oxide (antimicrobial activity)

gut-associated lymphoid tissue to respond appropriately to luminal antigens (immunity versus tolerance) (Campbell et al. 1999; Hershberg et al. 1997; Kambayashi and Laufer 2014; Shao et al. 2005). In order to act as antigen-presenting cells (APCs), enterocytes must be able to internalize and process antigens and contact lymphocytes within the epithelium (mainly intraepithelial lymphocytes) and T cells in the underlying lamina propria via basolateral projection (Dahan et al. 2007; Vitale et al. 2016). However the role of MHCII expression (expression induced under inflammatory conditions) by IECs is not fully (if at all) understood, and it remains debated whether and how IECs can affect mucosal immunity. During homeostasis, enterocytes do not express costimulatory molecules, resulting in the induction of energy and thus tolerance. Under pathologic conditions, such as inflammation or infections, enterocytes function as professional APCs and stimulate immune responses. In addition enterocytes are capable to produce a variety of proteins with antimicrobial activity (such as C-type lectin regenerating islet-derived protein III, Reg3γ) and antimicrobial peptides (e.g., β-defensins that are expressed by colonic enterocytes) and thus contribute to the antimicrobial defense (Gallo and Hooper 2012; Wittkopf et al. 2015).

In addition to absorptive enterocytes, intestinal epithelial stem cells also give rise to other terminally differentiated cell types, namely, goblet cells, Paneth cells, and enteroendocrine cells. They all belong to the pool of secretory intestinal epithelial cells.

7.2.2 Enteroendocrine Cells

Over 30 different hormones are produced in the gastrointestinal tract. Therefore the gut has been described as "the largest endocrine organ" in the body (Gunawardene et al. 2011). Enteroendocrine cells are characterized by the presence of secretory vesicles, which are either large dense-core vesicles (LDCVs) or the smaller synaptic-like microvesicles (SLMVs) similar to those found in neurons. Thus these cells (also known as neuroendocrine cells) represent the link between the central and enteric neuroendocrine system and coordinate gut function by the secretion of specific gut hormones (Buffa et al. 1978; Furness et al. 2013; Sternini et al. 2008). These hormones coordinate diverse functions of the gut such as secretion (5-HT), peristalsis (somatostatin), and enterocyte proliferation (GLP-2) (Gunawardene et al. 2011). Enteroendocrine cells are a highly heterogeneous population of intestinal epithelial cells. There are up to 15 different subtypes defined by their morphology and expression of specific peptide hormones or marker genes (Furness et al. 2013). They are scattered as individual cells and can be found throughout the mucosa, but especially in the upper part of the small intestine. With approximately 1% of the cells lining the intestinal lumen, they represent a minor population of IECs (Barker et al. 2008; Gunawardene et al. 2011).

7.2.3 Goblet Cells

A key element of the intestinal strategy for maintaining a homeostatic host-microbial relation is to limit the contact between the dense luminal microbial community and the intestinal epithelial cell surface (Belkaid and Hand 2014). This is accomplished by goblet cells which represent the most abundant cell type of the secretory lineage (Kim and Ho 2010; Specian and Oliver 1991). The proportion of goblet cells among all other intestinal epithelial cells increases from the proximal small intestine (around 4%) to the distal colon (around 16%) (van der Flier and Clevers 2009). They provide a protective function against physical and chemical injury by the secretion of high molecular weight glycoproteins called mucins (Kim and Ho 2010; Pelaseyed et al. 2014; Specian and Oliver 1991). These mucins are composed of a polymeric protein backbone structure, linked to numerous hygroscopic and hydrophilic oligosaccharide side chains that assemble into a viscous gel-like matrix, covering the intestinal epithelium (Johansson et al. 2013; Lamblin et al. 1992). The composition of the mucus layer varies between the small and the large intestine (Donaldson et al. 2016). In the small intestine, the mucus layer delineates a protected zone which hampers access and survival of bacteria that are directly adjacent to the epithelium. In the colon the mucus layer is separated in an inner and outer layer. The inner mucus layer separates the commensal bacteria from the host epithelium, is resistant to bacterial penetration, and thus limits the direct bacterial contact with intestinal epithelial cells. The outer colonic mucus layer is the natural habitat for commensal bacteria (Pelaseyed et al. 2014). Mice with a targeted inactivation of the most abundant secreted gastrointestinal mucin (Muc2) lack this bacterial-free zone and over time suffer from spontaneous inflammation and frequently develop adenomas (Van der Sluis et al. 2006; Velcich et al. 2002). These experimental studies demonstrate the importance of the mucus layer in maintaining a symbiotic relationship with the microbiota.

7.2.4 Paneth Cells

A second immune mechanism that limits bacteria-epithelial cell contact is the secretion of antimicrobial peptides (AMPs) by gut epithelial cells. Paneth cells are the main source of endogenous AMPs like α-defensins, lysozyme, or phospholipase A (Ayabe et al. 2004; Salzman 2010). Morphologically, they contain an extensive endoplasmic reticulum and Golgi network structures that already point toward an intensive secretory activity (Bevins and Salzman 2011; Porter et al. 2002). Within Paneth cells, AMPs are stored in large apical secretory granules, from which they can be released by exocytosis into the gut lumen and thereby contribute to host defense against a broad-spectrum bacteria, fungi, and some viruses (Elphick and Mahida 2005; Wehkamp and Stange 2006). Some of them are constitutively released into the mucus layer, whereas secretion of others is induced in response to specific external stimuli (Bevins and Salzman 2011). Some AMPs, such as the α-defensins (a Paneth cell-specific AMP), are secreted as precursors and have to be processed (cleaved by matrix metalloproteinase-7) in order to acquire full antimicrobial activity (Wilson et al. 2009). In addition to these AMPs, Paneth cell granules contain IgA (immunoglobulin A) and cytokines (e.g., TNF-α) (Ouellette 2010). Paneth cells are not evenly distributed in the gut epithelium. They are restricted to the small intestine, where they are located at the crypt base together with intestinal stem cells (Barker et al. 2008). In contrast to all other differentiated epithelial cell types, Paneth cells escape from the migration toward the villus tip and instead settle down at the crypt base, where they can survive for more than 3 weeks (Bjerknes and Cheng 1981, 2005; Ireland et al. 2005). The relocation of Paneth cells at the crypt base has an important function, since Paneth cells constitute the niche for stem cells in intestinal crypts (Sato et al. 2011). Paneth cells express EGF (epidermal growth factor), TGFα (transforming growth factor alpha), Wnt3, and the Notch ligand DLL4 (delta-like canonical notch ligand 4), all essential signals for stem cell maintenance. Dysfunction and death of Paneth cells has been associated with intestinal inflammation both in mice and men (Adolph et al. 2013; Bevins and Salzman 2011; Cadwell et al. 2008; Gunther et al. 2011; Kaser and Blumberg 2008; Kobayashi et al. 2005; Wehkamp et al. 2007).

7.2.5 Tuft Cells

Tuft cells (sometimes also referred to as brush cells) have been considered as a mysterious component of the intestinal epithelium for a long time. They only represent 0,4–1% of epithelial cells but have important functions in sensing and responding to intestinal protozoa and helminth parasites (Gerbe and Jay 2016; Gerbe et al. 2012, 2016). It is currently believed that tuft cells, Paneth cells, and goblet cells have a unique precursor, which would identify them as members of the secretory lineage (Gerbe et al. 2012). Indeed tuft cells are known to produce and secrete endogenous opioids and represent the major source of secreted interleukin-25 (IL-25) in the gut (Gronke and Diefenbach 2016; Kokrashvili et al. 2009; von Moltke et al. 2016). This prominent IL-25 production by tuft cells indicates a key role for these cells for the initiation of type 2 immune responses in the small intestine (Gronke and Diefenbach 2016).

7.3 Interaction Between the Epithelium and Immune Cells

Despite its barrier function, the intestinal epithelium has an important function in translating luminal signals from the barrier surface to the underlying mucosal immune system (Artis 2008; Gunther et al. 2016; Okumura and Takeda 2017; Peterson and Artis 2014; Ramanan and Cadwell 2016). IECs are capable of releasing immunogenic molecules that can shape the mucosal immune system and antimicrobial effectors that strongly influence the luminal microbiota both in the steady state and during inflammation or infection (Fig. 7.2). Moreover, IECs not only produce and release various factors affecting immune cells; they also respond to factors produced by the subjacent lymphocytes. This communication between the intestinal epithelium and resident immune cells might be particularly crucial to exert an appropriate immune or epithelial response during microbial challenge.

During intestinal infection, lamina propria cells and/or IECs express a wide range of inflammatory cytokines, thereby providing constant signals about the activation status of both compartments (Wittkopf et al. 2014). This axis plays a major role for the defense against intestinal pathogens. Pathogen colonization resistance can be regulated by either direct inhibition by the host microbiota or indirect by microbiota-mediated host immune responses (Buffie and Pamer 2013; Kamada et al. 2013). For example, commensal microbes contribute to colonization resistance against invading pathogens such as *Salmonella typhimurium* by priming the immune system (Buffie and Pamer 2013; Thiemann et al.

2017). This is achieved by a distinct microbiota that enhances the antibacterial IFN-γ production by innate and adaptive immune cells (Thiemann et al. 2017). IFN-γ can induce defense mechanisms in the intestinal epithelium such as secretion of antimicrobial peptides (Farin et al. 2014) and generation of mucus-filled vacuoles by goblet cells and mucus release into the gut lumen (Raetz et al. 2013; Songhet et al. 2011). Similar IL-22, released by lymphocytes (including mainly CD4+ T cells but also γδ T cells and subsets of innate lymphoid cells (ILCs)) in the early phase of infections, plays an important role for the communication between IECs and lamina propria cells. One of the main functions of IL-22 is to support and maintain the gastrointestinal epithelial barrier as well as to facilitate barrier defense mechanisms against bacterial pathogens such as *Clostridium difficile* (Hasegawa et al. 2014), *Citrobacter rodentium* (Munoz et al. 2015; Zheng et al. 2008), and *Toxoplasma gondii* (Munoz et al. 2015). Defense mechanisms supported by IL-22 include epithelial cell survival, proliferation of IECs, and release of proteins with antimicrobial activity (REG3γ) (Pickert et al. 2009; Wittkopf et al. 2015). IL-22 acts directly on intestinal epithelial cells and activates STAT3 (Hall 1997). Epithelial STAT3 in turn activates proliferation and wound healing but also promotes the secretion of REG3γ, which further controls *C. rodentium* infection (Wittkopf et al. 2015). IECs can also directly sense intestinal pathogens and respond with various mechanisms such as death or cytokine release. For this approach IECs express inflammasome components, which form cytosolic multi-protein complexes that initiate innate immune responses against invasive pathogens by activating inflammatory pathways (Blazejewski et al. 2017; Sellin et al. 2015; Strowig et al. 2012). This includes activation of cell death and rapid expulsion of infected IECs as well as caspase-1-dependent processing and release of pro-inflammatory interleukin-1β and interleukin-18 (Rauch et al. 2017). IL-18 then in turn synergizes with IL-23 to promote IL-22 expression by innate lymphoid cells. As mentioned above, elevated levels of IL-22 act on the intestinal epithelium to enhance antimicrobial defense. While these examples illustrate direct functions of IECs induced by pathogens or immune cell-derived cytokines, factors released by immune cells in response to infections can also shift epithelial cell differentiation toward the secretory lineage to improve innate immune responses (Gerbe et al. 2016; Grencis and Worthington 2016). Interleukin-33 (IL-33) is an IL-1-like cytokine with host-protective functions. It can be expressed by both hematopoietic and non-hematopoietic cells such as IECs and pericryptal fibroblasts, and it is particularly expressed at barrier tissues (Mahapatro et al. 2016; Oboki et al. 2010). In the gut, IL-33 is involved in helminth expulsion and defense against pathogenic bacteria such as *Salmonella typhimurium* (Alves-Filho et al. 2010; Rostan et al. 2015). IL-33 reprograms intestinal epithelial progenitor cell differentiation toward a specific expansion of cells of the secretory lineage such as goblet and Paneth cells. This is achieved by the modulation of cell intrinsic Notch signaling in the stem cell compartment (Mahapatro et al. 2016). Whether IL-33 directly acts on IECs or indirectly via the release of IL-13 by innate lymphoid cells (ILC2s) is still not clear. IL-33 also promotes the expansion of tuft cells. Interestingly, after helminth infection, tuft cell-derived IL-25 activates ILC2 to secrete IL-13, which acts either directly or indirectly via IL-33 on epithelial crypt progenitors to promote differentiation of tuft and goblet cells, leading to increased frequencies of both (von Moltke et al. 2016). This nicely illustrates how the intestine controls the communication between the host and its environment to maintain intestinal homeostasis.

7.4 Epithelial Dysfunctions in Inflammatory Diseases

Intestinal epithelial cells perform several cell intrinsic functions that can directly influence IEC homeostasis and function. Defects in such pathways can have tremendous effects on intestinal barrier function. This includes as mentioned above cell death and differentiation but also antimicrobial peptide expression and secretion. It has been shown

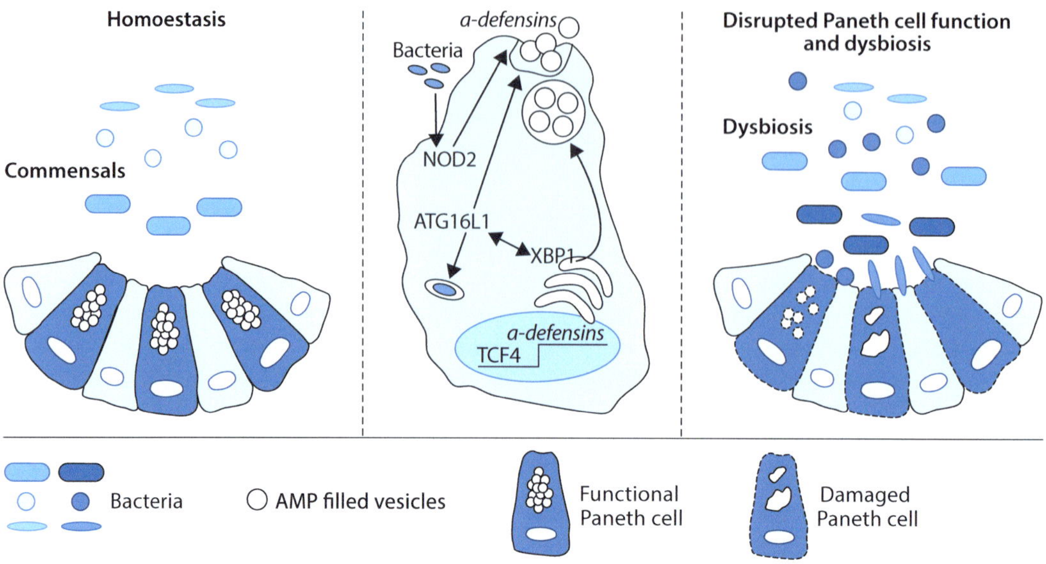

Fig. 7.3 Paneth cell dysfunction. Paneth cells play an essential role in maintaining a homeostatic host-microbial relation by secreting host endogenous antimicrobial peptides (left panel). Paneth cells are subject to damage via diverse environmental stressors and rendered even more vulnerable by genetic predisposition. NOD2 (Kobayashi et al. 2005), TCF4 (Wehkamp et al. 2007), XBP1 (Adolph et al. 2013; Kaser et al. 2008), and ATG16L1 each promote antimicrobial protein expression and secretion by Paneth cells. ATG16L1 (Cadwell et al. 2008) is critical for antibacterial autophagy. Polymorphisms in the corresponding genes are associated with Paneth cell dysfunction promoting microbial dysbiosis and increasing the risk for inflammatory bowel disease (right panel)

that several genes that increase susceptibility to Crohn's disease directly implicate Paneth cells and their antimicrobial peptides (Fig. 7.3) (Wehkamp and Stange 2010). One of the first genes that was described in that context was NOD2 (*nucleotide-binding oligomerization domain-containing 2*), a gene that is commonly mutated in individuals with Crohn's disease (around one third of adult CD patients carry at least one heterozygous *NOD2* mutation; detailed information can be found in [22346247]) (Hugot et al. 2001; Ogura et al. 2001). Individuals who carry either homozygous or heterozygous germ line variations of NOD2 have a 40-fold increased risk to develop ileal Crohn's disease (Cuthbert et al. 2002). In general, NOD2 is a protein that is expressed in macrophages, dendritic cells, as well as intestinal epithelial cells, including Paneth cells (Ogura et al. 2003). NOD2 is an intracellular receptor for muramyl dipeptide, which can be found in bacterial peptidoglycans (Girardin et al. 2003a, b; Inohara et al. 2003). Activation of this receptor therefore provokes immune responses to intracellular bacteria. Animal studies uncovered that NOD2 signaling is linked to expression of antimicrobial peptides (defensins) and important for host immunity to *Yersinia pseudotuberculosis*, *Listeria monocytogenes*, and *Citrobacter rodentium* (Kobayashi et al. 2005; Sorbara and Philpott 2011). Moreover challenge of NOD2-deficient animals with *Helicobacter hepaticus*, a pathogen that is associated with IBD, results in the development of pathological lesions, called granulomas that are characteristically associated with Crohn's disease (Biswas et al. 2010). Notably, NOD2 is well known for its potential to interact with the autophagy protein ATG16L1 (Travassos et al. 2010). NOD1 and NOD2 recruit ATG16L1 to the plasma membrane and can target cyto-invasive bacteria at the point of entry. Autophagy, in general, is a pathway in which cellular material, including organelles, is sequestered in double-membrane-coated vesicles and transported to lysosomes for degradation and recycling. Thus it is not surprising that packing and secretion of granules within Paneth cells is strongly influenced by autophagy.

Patients carrying homozygous ATG16L1 risk alleles and mice with hypomorphic expression of ATG16L1 are both associated with Paneth cell and goblet cell abnormalities (Adolph et al. 2013; Cadwell et al. 2008, 2010; Lassen et al. 2014). Additionally, deletion of the unfolded protein response (UPR) transcription factor X-box binding protein-1 (*Xbp1*) in intestinal epithelial cells results in endoplasmic reticulum (ER) stress, Paneth cell impairment, and spontaneous enteritis (Kaser et al. 2008). Importantly, variants of XBP1 have also been associated with an increased risk of Crohn's disease (Kaser et al. 2008). XBP1 is a transcription factor involved in maintaining ER function and required for expansion of the endoplasmic reticulum, which is essential for the functionality of highly secretory cells such as Paneth cells. It is believed that XBP1 deficiency induces ER stress in Paneth cells, leading to an epithelium that is overly reactive to bacterial products (such as flagellin) and inflammatory mediators (such TNF-α), which is associated with Paneth cell dysfunction.

As described above, impairment in either UPR (*Xbp1*) or autophagy function (*Atg16l*) in intestinal epithelial cells in mouse models both impairs Paneth cell function and promotes severe spontaneous Crohn's disease-like transmural ileitis. It is currently believed that both factors are probably co-dependent and amplify the development of spontaneous inflammation in the small intestine (Adolph et al. 2013). In this scenario, the UPR probably sets the threshold for susceptibility of the host with hypofunctional autophagy to develop inflammation. In summary, Paneth cells with their complex functionality contribute to the maintenance of a beneficial bacterial composition at the mucosal surface and are thus poised to broadly influence host physiology. Thus it is not surprising that disruption of Paneth cell function is strongly linked to Crohn's disease. Nevertheless, XBP1, NOD2, and ATG16L1 are no specific products of Paneth cells. Thus their function in other cells may be also relevant for the pathogenesis of Crohn's disease.

▶ **Controversy**

To allow sufficient nutrient absorption to sustain life, the small intestine is folded to form a large number of tubular invaginations, denoted as crypts, and fingerlike villus structures. This formation massively expands the intestinal epithelial surface area. Villus morphogenesis is a process where the flat pseudostratified intestine begins to remodel and give rise to villus structures. Murine studies have increased the understanding about the regulation of villus morphogenesis and have highlighted significant species-specific differences in this process. In humans, villus formation starts around 51 and 54 days of gestation correlating with the beginning of villus morphogenesis at E14.5 in the mouse embryo. Importantly, while previously it was thought that the human and mouse intestine initially formed micro-lumens in the flat embryonic epithelium, which then fuse and give rise to villi, recent studies demonstrated that micro-lumens may be present only in the developing human intestine, whereas in mice the lumen is continuous during villus formation.

History

It has been extremely difficult to establish in vitro propagation of primary adult intestinal epithelium without inducing genetic transformations. However recent advances in stem cell biology have enabled the in vitro generation of complex 3D structures resembling whole organs. These organ-like structures denoted as organoids have been isolated from various vertebrates, including human, mouse, cow, chicken, and pig tissues. Organoids have been successfully generated from many regions of the gastrointestinal tract, e.g., the stomach, small intestine, and colon. They can recapitulate the in vivo architecture, functionality, and genetic signature of intestinal tissues. Thus, organoid technology can be applied to reveal novel insights into stem cell biology, organogenesis, host-microbial interaction, and many pathologies (including inflammation, infection, and cancer).

Highlights

- The intestine is covered by a monolayer of epithelial cells. It is organized into crypts (invaginations containing stem cells and transit-amplifying cells) and villi (protrusions that are covered with terminally differentiated cells).
- The intestinal epithelium develops from the embryonic endoderm, which is one of the three primary germ layers derived during gastrulation. In the mouse, the intestinal epithelium originates as a single-layered structure lining the inner surface of the primitive gut (around embryonic day 9.5). The regenerative capacity of the adult epithelium is established between embryonic day 16.5 and postnatal day 7, when mature crypts harboring adult stem cells and progenitor cells develop from a pool of cells located at the base of the embryonic villi (Noah et al. 2011).
- In the adult intestinal epithelium, pluripotent stem cells in the crypt region give rise to a pool of transit-amplifying cells. These cells will (after 2–3 days) terminally differentiate into one of the epithelial cell lineages. This includes absorptive cells called enterocytes, which form the majority, as well as secretory cells including Paneth cells, enteroendocrine cells, goblet cells, M (microfold) cells, tuft cells, and cup cells.
- The intestinal epithelium is the most vigorously self-renewing tissue. The *life cycle* of intestinal epithelial cells (4–5 days) is determined by the time span in which these *cells* migrate from their place of origin at the crypt base to the villus tip, from where they are expelled from the epithelial layer through a complex cytoskeletal remodeling process.
- IECs are crucial mediators of intestinal homeostasis. They provide physical barrier functions and enable the establishment of an immunological environment permissive to colonization by commensal bacteria.

References

Adolph, T. E., Tomczak, M. F., Niederreiter, L., Ko, H. J., Bock, J., Martinez-Naves, E., Glickman, J. N., Tschurtschenthaler, M., Hartwig, J., Hosomi, S., et al. (2013). Paneth cells as a site of origin for intestinal inflammation. *Nature, 503*, 272–276.

Alves-Filho, J. C., Sonego, F., Souto, F. O., Freitas, A., Verri, W. A., Jr., Auxiliadora-Martins, M., Basile-Filho, A., McKenzie, A. N., Xu, D., Cunha, F. Q., & Liew, F. Y. (2010). Interleukin-33 attenuates sepsis by enhancing neutrophil influx to the site of infection. *Nature Medicine, 16*, 708–712.

Artis, D. (2008). Epithelial-cell recognition of commensal bacteria and maintenance of immune homeostasis in the gut. *Nature Reviews Immunology, 8*, 411–420.

Ayabe, T., Ashida, T., Kohgo, Y., & Kono, T. (2004). The role of Paneth cells and their antimicrobial peptides in innate host defense. *Trends in Microbiology, 12*, 394–398.

Barker, N. (2014). Adult intestinal stem cells: Critical drivers of epithelial homeostasis and regeneration. *Nature Reviews Molecular Cell Biology, 15*, 19–33.

Barker, N., van de Wetering, M., & Clevers, H. (2008). The intestinal stem cell. *Genes & Development, 22*, 1856–1864.

Belkaid, Y., & Hand, T. W. (2014). Role of the microbiota in immunity and inflammation. *Cell, 157*, 121–141.

Bevins, C. L., & Salzman, N. H. (2011). Paneth cells, antimicrobial peptides and maintenance of intestinal homeostasis. *Nature Reviews Microbiology, 9*, 356–368.

Biswas, A., Liu, Y. J., Hao, L., Mizoguchi, A., Salzman, N. H., Bevins, C. L., & Kobayashi, K. S. (2010). Induction and rescue of Nod2-dependent Th1-driven granulomatous inflammation of the ileum. *Proceedings of the National Academy of Sciences of the United States of America, 107*, 14739–14744.

Bjerknes, M., & Cheng, H. (1981). The stem-cell zone of the small intestinal epithelium. IV. Effects of resecting 30% of the small intestine. *The American Journal of Anatomy, 160*, 93–103.

Bjerknes, M., & Cheng, H. (2005). Gastrointestinal stem cells. II. Intestinal stem cells. *American Journal of Physiology. Gastrointestinal and Liver Physiology, 289*, G381–G387.

Blazejewski, A. J., Thiemann, S., Schenk, A., Pils, M. C., Galvez, E. J. C., Roy, U., Heise, U., de Zoete, M. R., Flavell, R. A., & Strowig, T. (2017). Microbiota normalization reveals that canonical caspase-1 activation exacerbates chemically induced intestinal inflammation. *Cell Reports, 19*, 2319–2330.

Buffa, R., Capella, C., Fontana, P., Usellini, L., & Solcia, E. (1978). Types of endocrine cells in the human colon and rectum. *Cell and Tissue Research, 192*, 227–240.

Buffie, C. G., & Pamer, E. G. (2013). Microbiota-mediated colonization resistance against intestinal pathogens. *Nature Reviews Immunology, 13*, 790–801.

Bullen, T. F., Forrest, S., Campbell, F., Dodson, A. R., Hershman, M. J., Pritchard, D. M., Turner, J. R., Montrose, M. H., & Watson, A. J. (2006). Characterization of epithelial cell shedding from human small intestine. *Laboratory Investigation, 86*, 1052–1063.

Cadwell, K., Liu, J. Y., Brown, S. L., Miyoshi, H., Loh, J., Lennerz, J. K., Kishi, C., Kc, W., Carrero, J. A., Hunt, S., et al. (2008). A key role for autophagy and the autophagy gene Atg16l1 in mouse and human intestinal Paneth cells. *Nature, 456*, 259–263.

Cadwell, K., Patel, K. K., Maloney, N. S., Liu, T. C., Ng, A. C., Storer, C. E., Head, R. D., Xavier, R., Stappenbeck, T. S., & Virgin, H. W. (2010). Virus-plus-susceptibility gene interaction determines Crohn's disease gene Atg16L1 phenotypes in intestine. *Cell, 141*, 1135–1145.

Campbell, N., Yio, X. Y., So, L. P., Li, Y., & Mayer, L. (1999). The intestinal epithelial cell: Processing and presentation of antigen to the mucosal immune system. *Immunological Reviews, 172*, 315–324.

Corr, S. C., Gahan, C. C., & Hill, C. (2008). M-cells: Origin, morphology and role in mucosal immunity and microbial pathogenesis. *FEMS Immunology and Medical Microbiology, 52*, 2–12.

Crosnier, C., Stamataki, D., & Lewis, J. (2006). Organizing cell renewal in the intestine: Stem cells, signals and combinatorial control. *Nature Reviews Genetics, 7*, 349–359.

Cuthbert, A. P., Fisher, S. A., Mirza, M. M., King, K., Hampe, J., Croucher, P. J., Mascheretti, S., Sanderson, J., Forbes, A., Mansfield, J., et al. (2002). The contribution of NOD2 gene mutations to the risk and site of disease in inflammatory bowel disease. *Gastroenterology, 122*, 867–874.

Dahan, S., Roth-Walter, F., Arnaboldi, P., Agarwal, S., & Mayer, L. (2007). Epithelia: Lymphocyte interactions in the gut. *Immunological Reviews, 215*, 243–253.

de Santa Barbara, P., van den Brink, G. R., & Roberts, D. J. (2003). Development and differentiation of the intestinal epithelium. *Cellular and Molecular Life Sciences, 60*, 1322–1332.

Donaldson, G. P., Lee, S. M., & Mazmanian, S. K. (2016). Gut biogeography of the bacterial microbiota. *Nature Reviews Microbiology, 14*, 20–32.

Eisenhoffer, G. T., Loftus, P. D., Yoshigi, M., Otsuna, H., Chien, C. B., Morcos, P. A., & Rosenblatt, J. (2012). Crowding induces live cell extrusion to maintain homeostatic cell numbers in epithelia. *Nature, 484*, 546–549.

Elphick, D. A., & Mahida, Y. R. (2005). Paneth cells: Their role in innate immunity and inflammatory disease. *Gut, 54*, 1802–1809.

Farin, H. F., Karthaus, W. R., Kujala, P., Rakhshandehroo, M., Schwank, G., Vries, R. G., Kalkhoven, E., Nieuwenhuis, E. E., & Clevers, H. (2014). Paneth cell extrusion and release of antimicrobial products is directly controlled by immune cell-derived IFN-gamma. *The Journal of Experimental Medicine, 211*, 1393–1405.

Furness, J. B., Rivera, L. R., Cho, H. J., Bravo, D. M., & Callaghan, B. (2013). The gut as a sensory organ. *Nature Reviews Gastroenterology & Hepatology, 10*, 729–740.

Gallo, R. L., & Hooper, L. V. (2012). Epithelial antimicrobial defence of the skin and intestine. *Nature Reviews Immunology, 12*, 503–516.

Gerbe, F., & Jay, P. (2016). Intestinal tuft cells: Epithelial sentinels linking luminal cues to the immune system. *Mucosal Immunology, 9*, 1353–1359.

Gerbe, F., Legraverend, C., & Jay, P. (2012). The intestinal epithelium tuft cells: Specification and function. *Cellular and Molecular Life Sciences, 69*, 2907–2917.

Gerbe, F., Sidot, E., Smyth, D. J., Ohmoto, M., Matsumoto, I., Dardalhon, V., Cesses, P., Garnier, L., Pouzolles, M., Brulin, B., et al. (2016). Intestinal epithelial tuft cells initiate type 2 mucosal immunity to helminth parasites. *Nature, 529*, 226–230.

Gibbons, D. L., & Spencer, J. (2011). Mouse and human intestinal immunity: Same ballpark, different players; different rules, same score. *Mucosal Immunology, 4*, 148–157.

Girardin, S. E., Boneca, I. G., Viala, J., Chamaillard, M., Labigne, A., Thomas, G., Philpott, D. J., & Sansonetti, P. J. (2003a). Nod2 is a general sensor of peptidoglycan through muramyl dipeptide (MDP) detection. *The Journal of Biological Chemistry, 278*, 8869–8872.

Girardin, S. E., Travassos, L. H., Herve, M., Blanot, D., Boneca, I. G., Philpott, D. J., Sansonetti, P. J., & Mengin-Lecreulx, D. (2003b). Peptidoglycan molecular requirements allowing detection by Nod1 and Nod2. *The Journal of Biological Chemistry, 278*, 41702–41708.

Grencis, R. K., & Worthington, J. J. (2016). Tuft cells: A new flavor in innate epithelial immunity. *Trends in Parasitology, 32*, 583–585.

Gronke, K., & Diefenbach, A. (2016). Tuft cell-derived IL-25 activates and maintains ILC2. *Immunology and Cell Biology, 94*, 221–223.

Gunawardene, A. R., Corfe, B. M., & Staton, C. A. (2011). Classification and functions of enteroendocrine cells of the lower gastrointestinal tract. *International Journal of Experimental Pathology, 92*, 219–231.

Gunther, C., Martini, E., Wittkopf, N., Amann, K., Weigmann, B., Neumann, H., Waldner, M. J., Hedrick, S. M., Tenzer, S., Neurath, M. F., & Becker, C. (2011).

Caspase-8 regulates TNF-alpha-induced epithelial necroptosis and terminal ileitis. *Nature, 477*, 335–339.

Gunther, C., Neumann, H., Neurath, M. F., & Becker, C. (2013). Apoptosis, necrosis and necroptosis: Cell death regulation in the intestinal epithelium. *Gut, 62*, 1062–1071.

Gunther, C., Josenhans, C., & Wehkamp, J. (2016). Crosstalk between microbiota, pathogens and the innate immune responses. *International Journal of Medical Microbiology, 306*, 257–265.

Hall, C. A. (1997). Patient management in head injury care: A nursing perspective. *Intensive & Critical Care Nursing, 13*, 329–337.

Hasegawa, M., Yada, S., Liu, M. Z., Kamada, N., Munoz-Planillo, R., Do, N., Nunez, G., & Inohara, N. (2014). Interleukin-22 regulates the complement system to promote resistance against pathobionts after pathogen-induced intestinal damage. *Immunity, 41*, 620–632.

Heath, J. P. (1996). Epithelial cell migration in the intestine. *Cell Biology International, 20*, 139–146.

Hershberg, R. M., Framson, P. E., Cho, D. H., Lee, L. Y., Kovats, S., Beitz, J., Blum, J. S., & Nepom, G. T. (1997). Intestinal epithelial cells use two distinct pathways for HLA class II antigen processing. *The Journal of Clinical Investigation, 100*, 204–215.

Hugot, J. P., Chamaillard, M., Zouali, H., Lesage, S., Cezard, J. P., Belaiche, J., Almer, S., Tysk, C., O'Morain, C. A., Gassull, M., et al. (2001). Association of NOD2 leucine-rich repeat variants with susceptibility to Crohn's disease. *Nature, 411*, 599–603.

Inohara, N., Ogura, Y., Fontalba, A., Gutierrez, O., Pons, F., Crespo, J., Fukase, K., Inamura, S., Kusumoto, S., Hashimoto, M., et al. (2003). Host recognition of bacterial muramyl dipeptide mediated through NOD2. Implications for Crohn's disease. *Journal of Biological Chemistry, 278*, 5509–5512.

Ireland, H., Houghton, C., Howard, L., & Winton, D. J. (2005). Cellular inheritance of a cre-activated reporter gene to determine Paneth cell longevity in the murine small intestine. *Developmental Dynamics, 233*, 1332–1336.

Johansson, M. E., Sjovall, H., & Hansson, G. C. (2013). The gastrointestinal mucus system in health and disease. *Nature Reviews Gastroenterology & Hepatology, 10*, 352–361.

Jung, C., Hugot, J. P., & Barreau, F. (2010). Peyer's patches: The immune sensors of the intestine. *International Journal of Inflammation, 2010*, 823710.

Kamada, N., Chen, G. Y., Inohara, N., & Nunez, G. (2013). Control of pathogens and pathobionts by the gut microbiota. *Nature Immunology, 14*, 685–690.

Kambayashi, T., & Laufer, T. M. (2014). Atypical MHC class II-expressing antigen-presenting cells: Can anything replace a dendritic cell? *Nature Reviews Immunology, 14*, 719–730.

Kaser, A., & Blumberg, R. S. (2008). Paneth cells and inflammation dance together in Crohn's disease. *Cell Research, 18*, 1160–1162.

Kaser, A., Lee, A. H., Franke, A., Glickman, J. N., Zeissig, S., Tilg, H., Nieuwenhuis, E. E., Higgins, D. E., Schreiber, S., Glimcher, L. H., & Blumberg, R. S. (2008). XBP1 links ER stress to intestinal inflammation and confers genetic risk for human inflammatory bowel disease. *Cell, 134*, 743–756.

Kim, Y. S., & Ho, S. B. (2010). Intestinal goblet cells and mucins in health and disease: Recent insights and progress. *Current Gastroenterology Reports, 12*, 319–330.

Kobayashi, K. S., Chamaillard, M., Ogura, Y., Henegariu, O., Inohara, N., Nunez, G., & Flavell, R. A. (2005). Nod2-dependent regulation of innate and adaptive immunity in the intestinal tract. *Science, 307*, 731–734.

Kokrashvili, Z., Rodriguez, D., Yevshayeva, V., Zhou, H., Margolskee, R. F., & Mosinger, B. (2009). Release of endogenous opioids from duodenal enteroendocrine cells requires Trpm5. *Gastroenterology, 137*, 598–606, 606 e591–592.

Kvietys, P. R., & Granger, D. N. (2010). Role of intestinal lymphatics in interstitial volume regulation and transmucosal water transport. *Annals of the New York Academy of Sciences, 1207(Suppl 1)*, E29–E43.

Lamblin, G., Aubert, J. P., Perini, J. M., Klein, A., Porchet, N., Degand, P., & Roussel, P. (1992). Human respiratory mucins. *The European Respiratory Journal, 5*, 247–256.

Lassen, K. G., Kuballa, P., Conway, K. L., Patel, K. K., Becker, C. E., Peloquin, J. M., Villablanca, E. J., Norman, J. M., Liu, T. C., Heath, R. J., et al. (2014). Atg16L1 T300A variant decreases selective autophagy resulting in altered cytokine signaling and decreased antibacterial defense. *Proceedings of the National Academy of Sciences of the United States of America, 111*, 7741–7746.

Mabbott, N. A., Donaldson, D. S., Ohno, H., Williams, I. R., & Mahajan, A. (2013). Microfold (M) cells: Important immunosurveillance posts in the intestinal epithelium. *Mucosal Immunology, 6*, 666–677.

Madara, J. L. (1982). Cup cells: Structure and distribution of a unique class of epithelial cells in guinea pig, rabbit, and monkey small intestine. *Gastroenterology, 83*, 981–994.

Mahapatro, M., Foersch, S., Hefele, M., He, G. W., Giner-Ventura, E., McHedlidze, T., Kindermann, M., Vetrano, S., Danese, S., Gunther, C., et al. (2016). Programming of intestinal epithelial differentiation by IL-33 derived from pericryptal fibroblasts in response to systemic infection. *Cell Reports, 15*, 1743–1756.

Man, A. L., Prieto-Garcia, M. E., & Nicoletti, C. (2004). Improving M cell mediated transport across mucosal barriers: Do certain bacteria hold the keys? *Immunology, 113*, 15–22.

Marchiando, A. M., Shen, L., Graham, W. V., Edelblum, K. L., Duckworth, C. A., Guan, Y., Montrose, M. H., Turner, J. R., & Watson, A. J. (2011). The epithelial barrier is maintained by in vivo tight junction expansion during pathologic intestinal epithelial shedding. Gastroenterology *140*, 1208–1218 e1201–1202.

Marshman, E., Booth, C., & Potten, C. S. (2002). The intestinal epithelial stem cell. *BioEssays, 24*, 91–98.

Miller, H., Zhang, J., Kuolee, R., Patel, G. B., & Chen, W. (2007). Intestinal M cells: The fallible sentinels? *World Journal of Gastroenterology, 13*, 1477–1486.

Mowat, A. M. (2003). Anatomical basis of tolerance and immunity to intestinal antigens. *Nature Reviews Immunology, 3*, 331–341.

Munoz, M., Eidenschenk, C., Ota, N., Wong, K., Lohmann, U., Kuhl, A. A., Wang, X., Manzanillo, P., Li, Y., Rutz, S., et al. (2015). Interleukin-22 induces interleukin-18 expression from epithelial cells during intestinal infection. *Immunity, 42*, 321–331.

Noah, T. K., Donahue, B., & Shroyer, N. F. (2011). Intestinal development and differentiation. *Experimental Cell Research, 317*, 2702–2710.

Oboki, K., Ohno, T., Kajiwara, N., Saito, H., & Nakae, S. (2010). IL-33 and IL-33 receptors in host defense and diseases. *Allergology International, 59*, 143–160.

Ogura, Y., Bonen, D. K., Inohara, N., Nicolae, D. L., Chen, F. F., Ramos, R., Britton, H., Moran, T., Karaliuskas, R., Duerr, R. H., et al. (2001). A frameshift mutation in NOD2 associated with susceptibility to Crohn's disease. *Nature, 411*, 603–606.

Ogura, Y., Lala, S., Xin, W., Smith, E., Dowds, T. A., Chen, F. F., Zimmermann, E., Tretiakova, M., Cho, J. H., Hart, J., et al. (2003). Expression of NOD2 in Paneth cells: A possible link to Crohn's ileitis. *Gut, 52*, 1591–1597.

Okumura, R., & Takeda, K. (2017). Roles of intestinal epithelial cells in the maintenance of gut homeostasis. *Experimental & Molecular Medicine, 49*, e338.

Ouellette, A. J. (2010). Paneth cells and innate mucosal immunity. *Current Opinion in Gastroenterology, 26*, 547–553.

Pelaseyed, T., Bergstrom, J. H., Gustafsson, J. K., Ermund, A., Birchenough, G. M., Schutte, A., van der Post, S., Svensson, F., Rodriguez-Pineiro, A. M., Nystrom, E. E., et al. (2014). The mucus and mucins of the goblet cells and enterocytes provide the first defense line of the gastrointestinal tract and interact with the immune system. *Immunological Reviews, 260*, 8–20.

Peterson, L. W., & Artis, D. (2014). Intestinal epithelial cells: Regulators of barrier function and immune homeostasis. *Nature Reviews Immunology, 14*, 141–153.

Pickert, G., Neufert, C., Leppkes, M., Zheng, Y., Wittkopf, N., Warntjen, M., Lehr, H. A., Hirth, S., Weigmann, B., Wirtz, S., et al. (2009). STAT3 links IL-22 signaling in intestinal epithelial cells to mucosal wound healing. *The Journal of Experimental Medicine, 206*, 1465–1472.

Porter, E. M., Bevins, C. L., Ghosh, D., & Ganz, T. (2002). The multifaceted Paneth cell. *Cellular and Molecular Life Sciences, 59*, 156–170.

Raetz, M., Hwang, S. H., Wilhelm, C. L., Kirkland, D., Benson, A., Sturge, C. R., Mirpuri, J., Vaishnava, S., Hou, B., Defranco, A. L., et al. (2013). Parasite-induced TH1 cells and intestinal dysbiosis cooperate in IFN-gamma-dependent elimination of Paneth cells. *Nature Immunology, 14*, 136–142.

Ramanan, D., & Cadwell, K. (2016). Intrinsic defense mechanisms of the intestinal epithelium. *Cell Host & Microbe, 19*, 434–441.

Ramirez, C., & Gebert, A. (2003). Vimentin-positive cells in the epithelium of rabbit ileal villi represent cup cells but not M-cells. *The Journal of Histochemistry and Cytochemistry, 51*, 1533–1544.

Rauch, I., Deets, K. A., Ji, D. X., von Moltke, J., Tenthorey, J. L., Lee, A. Y., Philip, N. H., Ayres, J. S., Brodsky, I. E., Gronert, K., & Vance, R. E. (2017). NAIP-NLRC4 inflammasomes coordinate intestinal epithelial cell expulsion with eicosanoid and IL-18 release via activation of caspase-1 and -8. *Immunity, 46*, 649–659.

Rostan, O., Arshad, M. I., Piquet-Pellorce, C., Robert-Gangneux, F., Gangneux, J. P., & Samson, M. (2015). Crucial and diverse role of the interleukin-33/ST2 axis in infectious diseases. *Infection and Immunity, 83*, 1738–1748.

Salzman, N. H. (2010). Paneth cell defensins and the regulation of the microbiome: Detente at mucosal surfaces. *Gut Microbes, 1*, 401–406.

Sancho, E., Batlle, E., & Clevers, H. (2003). Live and let die in the intestinal epithelium. *Current Opinion in Cell Biology, 15*, 763–770.

Sato, T., van Es, J. H., Snippert, H. J., Stange, D. E., Vries, R. G., van den Born, M., Barker, N., Shroyer, N. F., van de Wetering, M., & Clevers, H. (2011). Paneth cells constitute the niche for Lgr5 stem cells in intestinal crypts. *Nature, 469*, 415–418.

Sellin, M. E., Maslowski, K. M., Maloy, K. J., & Hardt, W. D. (2015). Inflammasomes of the intestinal epithelium. *Trends in Immunology, 36*, 442–450.

Shao, L., Kamalu, O., & Mayer, L. (2005). Non-classical MHC class I molecules on intestinal epithelial cells: Mediators of mucosal crosstalk. *Immunological Reviews, 206*, 160–176.

Snoeck, V., Goddeeris, B., & Cox, E. (2005). The role of enterocytes in the intestinal barrier function and antigen uptake. *Microbes and Infection, 7*, 997–1004.

Songhet, P., Barthel, M., Stecher, B., Muller, A. J., Kremer, M., Hansson, G. C., & Hardt, W. D. (2011). Stromal IFN-gammaR-signaling modulates goblet cell function during Salmonella Typhimurium infection. *PLoS One, 6*, e22459.

Sorbara, M. T., & Philpott, D. J. (2011). Peptidoglycan: A critical activator of the mammalian immune system during infection and homeostasis. *Immunological Reviews, 243*, 40–60.

Specian, R. D., & Oliver, M. G. (1991). Functional biology of intestinal goblet cells. *The American Journal of Physiology, 260*, C183–C193.

Sternini, C., Anselmi, L., & Rozengurt, E. (2008). Enteroendocrine cells: A site of 'taste' in gastrointestinal chemosensing. *Current Opinion in Endocrinology, Diabetes, and Obesity, 15*, 73–78.

Strowig, T., Henao-Mejia, J., Elinav, E., & Flavell, R. (2012). Inflammasomes in health and disease. *Nature, 481*, 278–286.

Thiemann, S., Smit, N., Roy, U., Lesker, T. R., Galvez, E. J. C., Helmecke, J., Basic, M., Bleich, A., Goodman, A. L., Kalinke, U., et al. (2017). Enhancement of IFNgamma production by distinct commensals ameliorates salmonella-induced disease. *Cell Host & Microbe, 21*, 682–694 e685.

Travassos, L. H., Carneiro, L. A., Ramjeet, M., Hussey, S., Kim, Y. G., Magalhaes, J. G., Yuan, L., Soares, F., Chea, E., Le Bourhis, L., et al. (2010). Nod1 and Nod2 direct autophagy by recruiting ATG16L1 to the plasma membrane at the site of bacterial entry. *Nature Immunology, 11*, 55–62.

van der Flier, L. G., & Clevers, H. (2009). Stem cells, self-renewal, and differentiation in the intestinal epithelium. *Annual Review of Physiology, 71*, 241–260.

Van der Sluis, M., De Koning, B. A., De Bruijn, A. C., Velcich, A., Meijerink, J. P., Van Goudoever, J. B., Buller, H. A., Dekker, J., Van Seuningen, I., Renes, I. B., & Einerhand, A. W. (2006). Muc2-deficient mice spontaneously develop colitis, indicating that MUC2 is critical for colonic protection. *Gastroenterology, 131*, 117–129.

Velcich, A., Yang, W., Heyer, J., Fragale, A., Nicholas, C., Viani, S., Kucherlapati, R., Lipkin, M., Yang, K., & Augenlicht, L. (2002). Colorectal cancer in mice genetically deficient in the mucin Muc2. *Science, 295*, 1726–1729.

Vitale, S., Picascia, S., & Gianfrani, C. (2016). The cross-talk between enterocytes and intraepithelial lymphocytes. *Molecular and Cellular Pediatrics, 3*, 20.

von Moltke, J., Ji, M., Liang, H. E., & Locksley, R. M. (2016). Tuft-cell-derived IL-25 regulates an intestinal ILC2-epithelial response circuit. *Nature, 529*, 221–225.

Wehkamp, J., & Stange, E. F. (2006). Paneth cells and the innate immune response. *Current Opinion in Gastroenterology, 22*, 644–650.

Wehkamp, J., & Stange, E. F. (2010). Paneth's disease. *Journal of Crohn's & Colitis, 4*, 523–531.

Wehkamp, J., Wang, G., Kubler, I., Nuding, S., Gregorieff, A., Schnabel, A., Kays, R. J., Fellermann, K., Burk, O., Schwab, M., et al. (2007). The Paneth cell alpha-defensin deficiency of ileal Crohn's disease is linked to Wnt/Tcf-4. *Journal of Immunology, 179*, 3109–3118.

Wilson, C. L., Schmidt, A. P., Pirila, E., Valore, E. V., Ferri, N., Sorsa, T., Ganz, T., & Parks, W. C. (2009). Differential processing of {alpha}- and {beta}-defensin precursors by matrix metalloproteinase-7 (MMP-7). *The Journal of Biological Chemistry, 284*, 8301–8311.

Wittkopf, N., Neurath, M. F., & Becker, C. (2014). Immune-epithelial crosstalk at the intestinal surface. *Journal of Gastroenterology, 49*, 375–387.

Wittkopf, N., Pickert, G., Billmeier, U., Mahapatro, M., Wirtz, S., Martini, E., Leppkes, M., Neurath, M. F., & Becker, C. (2015). Activation of intestinal epithelial Stat3 orchestrates tissue defense during gastrointestinal infection. *PLoS One, 10*, e0118401.

Zheng, Y., Valdez, P. A., Danilenko, D. M., Hu, Y., Sa, S. M., Gong, Q., Abbas, A. R., Modrusan, Z., Ghilardi, N., de Sauvage, F. J., & Ouyang, W. (2008). Interleukin-22 mediates early host defense against attaching and effacing bacterial pathogens. *Nature Medicine, 14*, 282–289.

Ziv, E., & Bendayan, M. (2000). Intestinal absorption of peptides through the enterocytes. *Microscopy Research and Technique, 49*, 346–352.

8 Microbiome and Gut Immunity: Innate Immune Cells

Till Strowig, Sophie Thiemann, and Andreas Diefenbach

Abstract

The innate immune system not only serves as a first line of defense against infections with pathogenic microorganisms but also plays an important role in the balanced interplay with the intestinal microbiota. Distinct subsets of innate immune cells such as macrophages, dendritic cells, granulocytes, mast cells, and innate lymphoid cells are found spread throughout the intestinal tissue as well as organized in tissue-specific lymphoid structures. These cells constantly survey the intestinal tissue for the presence of live microbes to prevent the spread of invading microbes and fine-tune the intestinal barrier. Specifically, metabolites from the commensal microbiota such as short-chain fatty acids have been identified to maintain tolerogenic conditions in the intestine, e.g., promoting regulatory T cells and down-modulating pro-inflammatory signaling pathways. In turn, aberrant recognition and handling of commensal microbes by the innate immune system or the excessive immune activation after pathogen sensing have been demonstrated to promote inflammatory conditions such as inflammatory bowel diseases in the intestine. Hence, the detailed understanding of the interplay between the microbiota and innate immune system may enable novel therapeutic interventions to promote human health and, specifically, to prevent auto-inflammatory diseases.

T. Strowig (✉) · S. Thiemann
Research Group Microbial Immune Regulation, Helmholtz Centre for Infection Research, Braunschweig, Germany
e-mail: till.strowig@helmholtz-hzi.de

A. Diefenbach
Institute of Microbiology, Charité – Universitätsmedizin Berlin, Berlin, Germany

Berlin Institute of Health (BIH), Berlin, Germany

The intestine contributes to important physiological functions of the organism such as the resorption of nutrients but is also used by many pathogens as entry site to colonize the host. In order to limit the local infection and prevent systemic pathogen spread, the innate immune system serves as the first line of defense in the intestinal tract. Yet, besides infrequently invading pathogens, the intestinal lumen continuously harbors diverse and dense communities of microorganisms. As a consequence, the tissue-resident immune system is constantly exposed to vast amounts of different foreign molecules derived from infectious agents or harmless origins, i.e., food or commensal bacteria. Hence, the immune system has to avoid raising strong responses against a broad range of harmless bacteria, whereas it has to induce an effective protective response against invading pathogens (Thaiss et al. 2016b). This constantly requires the cells of

D. Haller (ed.), *The Gut Microbiome in Health and Disease*,
https://doi.org/10.1007/978-3-319-90545-7_8

the innate immune system to make the decision whether to favor tolerance toward foreign molecules and promote intestinal homeostasis or to trigger potent but potentially harmful and destructive immune effector mechanisms. In this chapter, the members of the innate immune system in the intestinal tract will be introduced, and their complex interplay with commensal microbes and pathogens will be explored. The specific focus lies on the decision-making process between immune homeostasis and inflammation, which is important to prevent aberrant responses to the indigenous intestinal microbiota.

8.1 Recognition of Microbiota and Pathogens by Innate Immune Receptors

In order to serve as first line of defense in the intestine, the innate immune system needs to recognize threats to the host, i.e., the presence of pathogenic microorganisms or other disturbances of the homeostatic conditions. In general, this is achieved by direct recognition of conserved structures, so-called microbe-associated molecular patterns (MAMPs), in viruses, bacteria, and fungi via pattern recognition receptors (PRRs) expressed in immune and nonimmune cells or indirectly by recognition of imbalances in the normal function of cells and tissues as a consequence of infection or tissue damage (Magalhaes et al. 2007; Palm and Medzhitov 2009). Germline-encoded PRRs include membrane-bound proteins from the Toll-like receptor (TLR) family and C-type lectin receptors (CLRs) as well as cytoplasmic proteins from the families of RIG-I-like receptors (RLRs), NOD-like receptors (NLRs), AIM2-like receptors (ALRs), and OAS-like receptors (OLRs) (Patten and Collett 2013; Parlato and Yeretssian 2014; Brubaker et al. 2015; Thaiss et al. 2016b). These receptors recognize very diverse molecular structures such as lipopolysaccharides (LPS), lipoproteins, small peptides, or specific types of nucleic acids, typically not present in the host. Subsequently, activation of PRRs by distinct cognate ligands leads to the activation of pro-inflammatory pathways through various transcriptional or posttranscriptional regulatory mechanisms, for instance, the induction of pro-inflammatory gene expression such as interleukin (IL)-6 after sensing of LPS by TLR4. Other immune sensors with important functions in the intestine are inflammasomes, which are multi-protein complexes that recruit and activate caspase-1 (Fig. 8.1) (Strowig et al. 2012; Rathinam and Fitzgerald 2016). They are formed upon detection of diverse activating ligands by multiple sensors from the NLR and ALR families, e.g., NLRP3 or AIM2, respectively, resulting in activation of caspase-1, which in turn activates pro-inflammatory cytokines by posttranslational cleavage. Strikingly, inflammasomes are an example for shared effector systems induced by many types of foreign- or host-derived molecules,

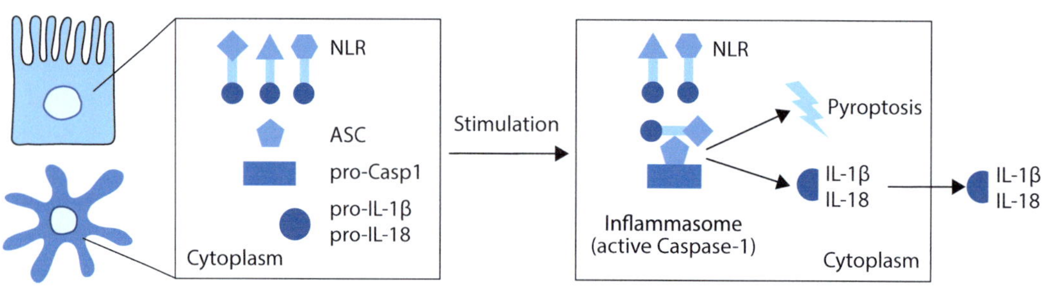

Fig. 8.1 Inflammasomes are modular platforms controlling the activity of Caspase-1. Microbial- and host-derived molecules stimulate the formation of inflammasomes resulting in the activation of the protease caspase-1, which subsequently cleaves pro-IL-1b and pro-IL-18, which then modulate intestinal immunity, e.g., mucus secretion and helper T cell activation. Different inflammasome sensor proteins are expressed in diverse cell subsets, which determines to which activating signals each cell type can respond

e.g., cytosolic viral dsDNA during viral infections or high extracellular concentrations of potassium ions as consequence of cell death, being recognized by various receptors. Of note, impairments in inflammasome function are linked to both increased susceptibility to enteric pathogens and impaired cross talk with the microbiota (Elinav et al. 2011; Robertson and Girardin 2013; Stewart and Cookson 2016). Specifically, caspase-1-deficient mice are more susceptible to infection with *Salmonella*, but at the same time, these mice are also characterized by an altered microbiota composition in comparison to wild-type (WT) mice, highlighting that specific immune receptors have important functions in both the hosts' cross talk with the microbiota and with pathogens.

Proper integration of the downstream signaling is important for the type and outcome of immune responses triggered by these receptors (Thaiss et al. 2016a). Importantly, prevention of overt inflammation involves the recognition of tolerogenic signals and subsequent blockade of pro-inflammatory pathways. In this regard, sensing of many microbial-derived metabolites produced by the commensal microbiota such as short-chain fatty acids (SCFAs) promotes intestinal homeostasis (Tan et al. 2014). SCFAs are produced by bacterial fermentation and are water-soluble, enabling their fast diffusion even across the mucus layer normally separating the luminal bacteria from the host. They are recognized by different G-protein-coupled cell surface receptors (GPCR) such as GPR41, GPR43, and GPR109 that are expressed by immune and nonimmune cells in the intestine and also throughout the organism (Tan et al. 2014). Signaling induced by binding of SCFAs to the cognate receptors, for instance, increases the threshold for the induction of pro-inflammatory signaling in macrophages and epithelial cells, thereby preventing overt immune activation (Park et al. 2007). Other important sensors that integrate environmental signals derived from the microbiota and diet into immune cells are the aryl hydrocarbon receptor and farnesoid X receptor (Kiss and Vonarbourg 2012; Jia et al. 2017), which both predominantly promote tissue homeostasis and regeneration. Together, sensing of commensal bacteria by innate immune pathways as well as their metabolic activity strongly influences an efficient functioning of the epithelial barrier and the differentiation and maturation of the mucosal immune system.

8.2 Interplay of Innate Immune Cells with the Microbiota

The largest proportion of the immune system, estimated up to 70% of total immune cells, reside in the intestine. These cells are located throughout the gastrointestinal tract in different compartments, either scattered in the epithelial cell layer (intraepithelial cells, IEL) and in the lamina propria (lamina propria leukocytes, LPL) or organized in tissue-specific lymphoid structures such as Peyer's patches (PPs), isolated lymphoid follicles (ILFs), and cryptopatches (CP) (Eberl and Lochner 2009). Together, these tissue-specific organized lymphoid structures in the intestine form the gut-associated lymphoid tissues (GALT) and are responsible for initiating and regulating immune responses in the gut. GALT development during ontogeny and its adaptation during adulthood are profoundly regulated by the presence of the microbiota (Eberl and Lochner 2009). Specifically, germ-free mice are characterized by an underdeveloped GALT with reduced numbers of IELs and LPLs as well as limited development of PPs and ILFs. Several bacteria, in particular segmented filamentous bacteria (SFB), have been studied intensely in mice for their ability to modulate GALT and, specifically, the expansion of PPs and ILFs in the small intestine resulting in the recruitment of immune cells to the intestine (Cebra et al. 1998).

Additionally, commensal bacteria influence the proper development and maturation of immune cell populations in the intestine through modulation of distinct phases of hematopoiesis, i.e., the development of mature immune cells from pluripotent hematopoietic stem cells (HSC) present in the bone marrow (Gorjifard and Goldszmid 2016). In a stepwise differentiation process, those HSC give rise to common myeloid progenitor cells (CMPs) and common lymphoid progenitor cells (CLPs). While CMPs give eventually rise via distinct progenitor cells to different cells of the innate immune cells, such as monocytes, macrophages, dendritic cells, granulocytes, and mast cells, CLPs give rise to cells of the lymphoid lineages including innate cells such as natural killer (NK) cells and innate lymphoid cells (ILCs). Commensal bacteria modulate this process already at early steps of hematopoiesis such as intestinal bacteria are required for myeloid cell development, i.e., fecal transplantation of a complex microbiota composition into germ-free mice restores defects in global myelopoiesis (Khosravi et al. 2014). Moreover, the microbiota influences the abundance and function of innate immune cells on multiple additional levels, and the effects on specific cell types are discussed in the following.

8.2.1 Granulocytes

Granulocytes belong to the group of myeloid cells and form polymorphic-shaped nuclei, typically looped into three segments (Leiding 2017). They are also referred to as polymorphonuclear leukocytes. They can be distinguished into distinct subsets (neutrophils, basophils, and eosinophils) and are commonly characterized by the presence of many granules containing enzymes and potent antimicrobial agents in their cytoplasm. Upon contact to microbes, these granules are released and are essential to fight pathogenic invaders such as extracellular bacteria.

Neutrophils are the most abundant leukocyte population in the human blood, from where they are recruited to inflamed tissues through the activity of chemokines secreted, e.g., by epithelial cells and by macrophages (Leiding 2017). In the intestine, neutrophils combat microbes that crossed from the lumen into the epithelium and the lamina propria. Moreover, neutrophils are also able to contribute to the recruitment of circulating monocytes from the blood by producing chemoattractants. Besides their ability to release pre-stored compounds with antimicrobial activity, neutrophils also release distinct cytokines such as IL-17 and IL-10 as well as matrix metalloproteases (MMPs), which cleave a number of different chemokines important for the recruitment of additional effector cells, thereby regulating mucosal wound healing of epithelial cells. The microbiota has a crucial role in the development of neutrophils, locally and systemically. In contrast to conventionally colonized animals, germ-free rats, which have never been exposed to microbes, display a neutropenic phenotype with generalized lower phagocytic function (Ohkubo et al. 1999). Peptidoglycans derived from the cell wall of intestinal bacteria are sufficient to restore systemic neutrophil function via the nucleotide oligomerization domain 1 (Nod1) receptor even in the absence of live bacteria (Clarke et al. 2010), demonstrating that microbiota-derived immune stimulation is required for local and systemic immunity.

Eosinophils develop from eosinophil progenitor cells, and in particular IL-3, IL-5, and granulocyte-macrophage colony-stimulating factor (GM-CSF) promote eosinophil development in the bone marrow (Mori et al. 2009). After differentiation, eosinophils circulate in the bloodstream and enter mucosal tissues (Weller and Spencer 2017). Eotaxin-1, which is released from macrophages in the mucosal tissues, strongly influences trafficking of eosinophils to the lamina propria of the intestine, but it is not known to which degree this process is regulated by the microbiota or differences in microbiota composition. Tissue homing via eotaxin-1 is highly regulated by the ligation with C–C chemokine-receptor (CCR) 3 (Humbles et al. 2002). Tissue-residing eosinophils only survive in the presence of survival-promoting cytokine signals. In addition to its function in eosinophil development, IL-5 also promotes activation and survival and is induced in the intestine upon colonization with some commensal protozoa or during helminth infection (Gieseck et al. 2018). Under homeostatic condition, eosinophils have been shown to be involved in the development and maintenance of immunoglobulin (Ig)A-producing cells, which are required for neutralizing intestinal bacteria (Chu et al. 2014). Moreover, eosinophils can release their cytotoxic granule proteins to target bacteria. However, eosinophils have also been demonstrated to play a critical role during pathogenesis, even promoting inflammatory bowel diseases (IBD). Specifically, increased numbers of tissue-residing eosinophils have been demonstrated in patients with ulcerative colitis (Raab et al. 1998; Saitoh et al. 1999), and depletion of eosinophils protects mice from spontaneous ileitis (Masterson et al. 2011). Currently, a monoclonal antibody binding to eotaxin-1, Bertilimumab, is under investigation for the indication of IBD (ClinicalTrials.gov identifier: NCT01671956). In contrast to this pathogenic function, microbiota-dependent stimulation of eosinophils by IL-25 has been shown to reduce mortality during *Clostridium difficile* infection (Buonomo et al. 2016).

Basophils are the least common granulocytes in the circulating blood, representing <1% of white blood cells, making it difficult to study their roles, especially during homeostatic conditions. Together with mast cells, basophils are innate effector cells of IgE-dependent allergic inflammation, as both cell types express the high-affinity receptor FcεR1. Basophils can be activated by IL-3 released from $CD4^+$ T cells or by binding to IgE. Upon activation, basophils can release histamine as well as proteases, diverse cytokines, heparin, and other proteoglycans. Release of large quantities of IL-4 and IL-6 by basophils initiates and supports allergic T helper type 2 (Th2) responses (Gomez et al. 2014). Commensal bacteria in the intestine regulate basophil development, therefore influencing circulating basophil populations. Specifically, it has been shown that disruption of the microbiota upon antibiotic use is associated with elevated IgE levels and augmented basophil population by influencing bone marrow-resident basophil precursors and in turn worsening of allergic inflammation (Hill et al. 2012; Russell et al. 2013). Moreover, higher amount of basophils has been detected in the inflamed intestine of IBD patients and has been hypothesized to contribute to the deterioration of the disease (Chapuy et al. 2014).

8.2.2 Mast Cells

Mast cells are granulated tissue-resident cells and are first responders encountering microbes colonizing the gut (Wouters et al. 2016). Mast cell progenitors circulate in the blood and migrate into tissues such as the intestine via the CXC chemokine receptor (CXCR) 2 (Kunii et al. 2011). Depending on the microenvironment, those progenitor cells differentiate further in the tissue. In the human gastrointestinal tract, 2–3% of all cells in the lamina propria represent mast cells as well as 1% of all cells in the submucosa (Bischoff 2009). Diverse inflammatory mediators are located in the granules of the mast cells, such as heparin, histamine, and proteases, which can be released via the IgE-mediated pathway driving allergic reactions but also as response to different substances including cytokines and chemokines. Mast cells contribute to initiate and maintain the inflammatory circuitry in the gut (St John and Abraham 2013). Intestinal commensal bacterial signals are important for migration of mast cells from the blood to the intestine. Specifically, germ-free mice are characterized by a reduced relative abundance of mast cells in the small intestine compared to conventionally housed mice but instead have elevated numbers of mast cells in the blood. This is linked to lower levels of CXCR2 ligands, which control homing of mast cells into the gut, within germ-free mice (Kunii et al. 2011)

8.2.3 Mononuclear Phagocytes

Mononuclear phagocytic cells (MNPs) serve as vital immunoregulatory switch in the intestine controlling the induction of immune responses and the maintenance of tolerogenic mechanisms (Joeris et al. 2017). MNPs in the intestine consist of both dendritic cells (DCs) and macrophages that fulfill both overlapping and distinct functions (Cerovic et al. 2014). To fulfill their respective functions, macrophages and DCs express highly specialized machineries including a diverse repertoire of PRR to detect microbes and subsequently release immunomodulatory cytokines with pro- or anti-inflammatory function. The specialization of these MNP subsets for specific tasks allows the innate immune system to fine-tune the immune response to the complex environment of the intestine and specific requirements of the host.

8.2.3.1 Macrophages

Macrophages are found scattered directly within and underneath the epithelial layer, throughout the lamina propria as well as in organized lymphoid follicles such as PPs (Fig. 8.2). They are largely resident cells, i.e., they do not leave the intestine upon terminal differentiation. As part of their functions, the macrophage population is characterized by continuous turnover in the intestine and is replaced by bone marrow-derived progenitor cells even in the steady state (Bain et al. 2014). They specifically develop from Ly6C^{high} monocytes circulating in the blood, which enter the intestinal mucosa depending on CCR2. Lineage-tracing experiments have revealed that immigrating monocytes downregulate markers of monocytes such as Ly6C and in parallel acquire the expression of markers of intestinal macrophage such as CXCR1, CD64, and F4/80. The continuous replenishment of macrophages in the lamina propria by Ly6C^{high} monocytes is influenced by commensal gut bacteria via the induction of ligands for CCR2 required for attracting monocytes to the intestine. Notably, this replacement from blood-derived progenitors is rather atypically as tissue-resident macrophages in most other tissues, e.g., the skin and liver, are replaced from tissue-resident progenitors that entered the respective tissue already during fetal development (Lavin et al. 2015).

In the intestinal tissue, mature macrophages serve as innate effector cells, i.e., they are known for their abilities to phagocytose microbes and apoptotic materials, but also promote intestinal

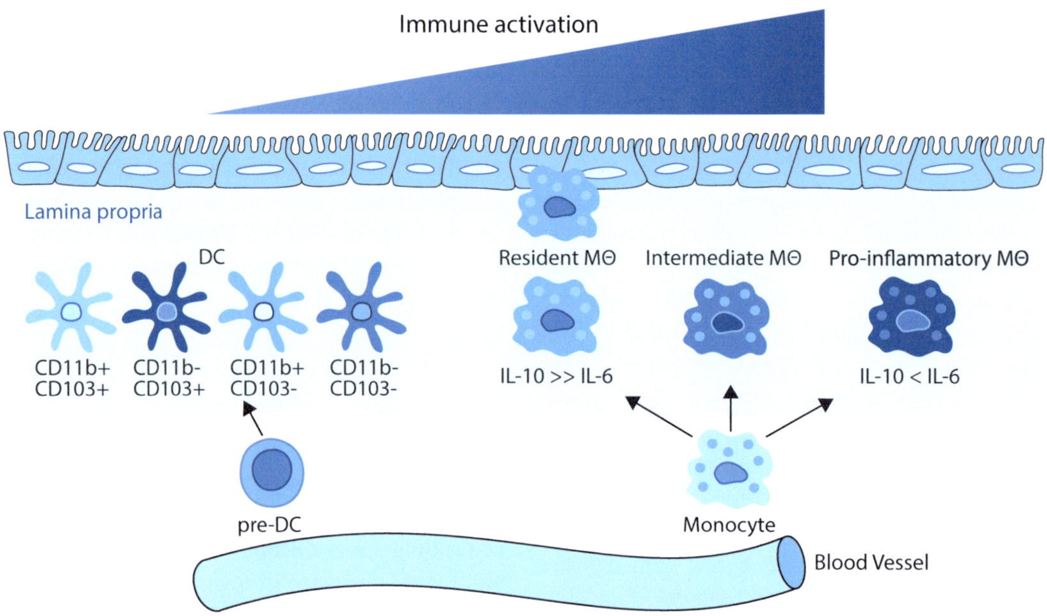

Fig. 8.2 Intestinal mononuclear phagocytic cells terminally differentiate in situ from circulating progenitor cells. Circulating monocytes are recruited via the CCL2-CCR2 axis to the intestinal tissue, where they differentiate into resident macrophages (MΦ). This recruitment is modulated by the microbiota by influencing local CCL2 production. The phenotype and functions of developing macrophages are strongly influenced by the local microenvironment and inflammatory signals. The development of pro-inflammatory macrophages during infections contributes to immune defense but also predisposes the host to intestinal inflammation. Circulating pre-cDCs develop into distinct intestinal dendritic cell (DC) subsets, which are distinguished based on the expression of CD11b and CD103. The abundance and functionality of DC subsets are determined by anatomical site and the local microenvironment, e.g., signals from the microbiota

homeostasis (Mowat et al. 2017). Specifically, the release of substantial amounts of anti-inflammatory mediators such as IL-10 and PGE2 is important to maintain epithelial barrier integrity and regulatory T cell (Treg) development upon microbial sensing in the steady state (Ueda et al. 2010). IL-10 production by macrophages is modulated by the microbiota, as macrophages from germ-free mice produce less IL-10 after TLR stimulation (Rivollier et al. 2012). Notably, the surprising absence of pro-inflammatory mediators commonly produced by macrophages after microbial stimulation has been attributed to different factors including the downregulation of signaling molecules such as myeloid differentiation primary response gene 88 (MyD88) necessary for PRR signal transduction (Smythies et al. 2010). Moreover, microbiota-derived SCFAs have been reported to modulate the function of macrophages in the lamina propria due to the inhibition of histone deacetylases, rendering macrophages hyporesponsive to intestinal bacteria by limiting the expression of pro-inflammatory cytokines (Chang et al. 2014). Notably, sensing of SCFAs by macrophages in peripheral organs such as the brain influences their maturation and function, demonstrating the long reach of commensal-derived metabolites on innate immune cells (Erny et al. 2015). Similar to DCs (see below) intestinal macrophages present antigens on MHC class II molecules to T cells, but since they perform this task only in the intestinal tissue rather than in lymphoid tissues, they largely interact with effector T cell subsets (Cerovic et al. 2014). Specifically, the microbiota-induced expression of IL-10 favors the expansion of Foxp3-expressing Treg subsets. However, not only the postnatal microbial colonization is crucial for immune cell development, but already the maternal microbiota influences macrophage development in the offspring. In this context, it has been recently shown that molecular metabolites of the maternal microbiota impact the development of F4/80$^+$ mononuclear cells already in utero (Gomez de Agüero et al. 2016).

However, if the strength of pro-inflammatory signals increases or those of anti-inflammatory signals decreases, newly immigrating monocytes acquire potent pro-inflammatory properties, e.g., the ability to secrete IL-6 and TNF-α, instead of becoming anti-inflammatory. Importantly, recruitment of monocytes/macrophages during infections with intestinal bacteria and protozoa is essential for pathogen clearance and prevention of excessive pathology. Their predominantly pro-inflammatory functions are further corroborated by the ability to exacerbate intestinal inflammation after chemically induced damage to the intestinal barrier during dextran sulfate sodium (DSS)-induced colitis (Platt et al. 2010). Notably, even during inflammation, resident macrophages largely maintain their anti-inflammatory properties, e.g., the ability to produce IL-10, which is thought to be important for resolution of inflammation and subsequent tissue repair (Weber et al. 2011). The dynamic switch between predominantly anti- or pro-inflammatory differentiation has been hypothesized to happen not only during enteric infections but potentially also in the absence of microbiota-derived SCFAs as consequence of consumption of diets low in fiber linking intestinal dysbiosis to impairments of tolerogenic macrophage functions and eventually the development of auto-inflammation in the intestine as observed during human IBD.

8.2.3.2 Dendritic Cells

Similar to macrophages DCs are scattered throughout the intestinal epithelium, mucosa, and lymphoid structures (Fig. 8.2). They share the phagocytic capabilities with macrophages but in addition possess the exquisite ability to carry antigens to lymphoid organs (Cerovic et al. 2014). There, DCs prime antigen-specific immune responses including antibody-producing B cells as well as tolerogenic and pro-inflammatory T cell subsets, thereby serving as relay between innate and adaptive immunity (Merad et al. 2013) (Fig. 8.3). The induction of opposing adaptive immune effector populations is achieved via highly coordinated mechanisms on the level of individual cells, e.g., the ability to integrate multiple signals from the environment, or on the level of the tissue, e.g., the functional specialization of DC subsets. Specifically, intestinal DC subsets are distinguished by the expression of the cell adhesion molecules CD11b and CD103 as well as of chemokine receptors such as CX3CR1 (Cerovic et al. 2014). The expression patterns of these

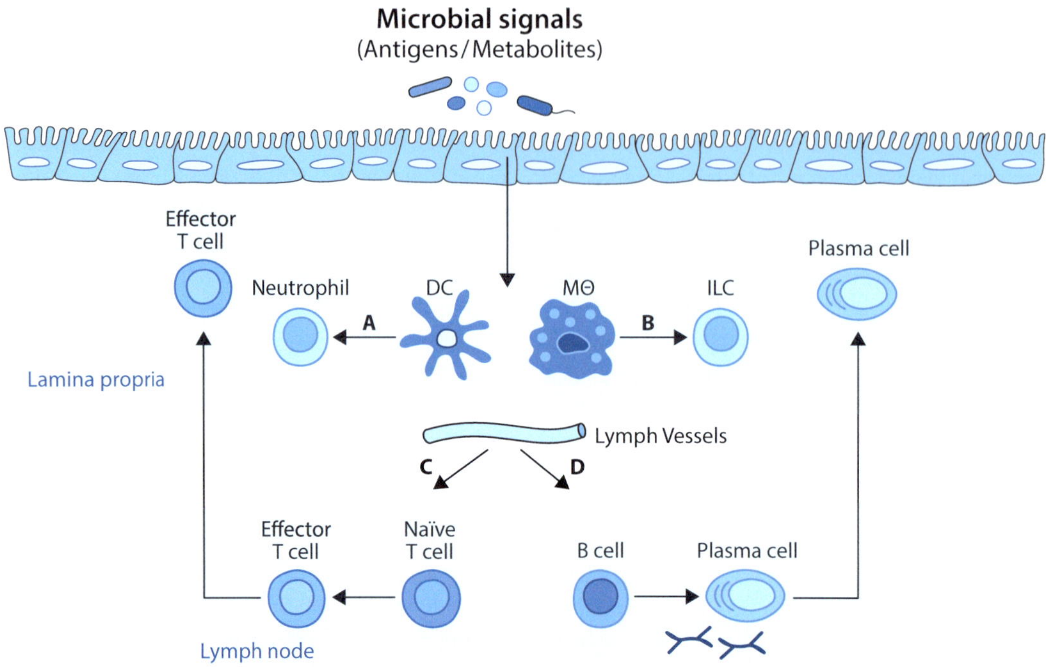

Fig. 8.3 Macrophages and dendritic cells orchestrate intestinal innate and adaptive immune responses. Exposure of intestinal dendritic cells (DC) and macrophages (MΦ) to microbiota-derived molecules results in the coordinated activation of distinct arms of the innate and adaptive immune system. (**a**) Sensing of pro-inflammatory stimuli may result in the recruitment of immune cells, e.g., neutrophils, that are normally largely absent from the intestine to sites of local inflammation. (**b**) Additionally, DCs and MΦ will also activate resident immune cells, e.g., ILCs, which then produce antimicrobial and immunomodulatory compounds. (**c** and **d**) Upon antigen uptake and immune stimulation, DCs also migrate via the lymph to gut-draining LNs, where they prime naive adaptive immune cells and direct their differentiation. Depending on the co-stimulatory signals, $CD4^+$ T cells may differentiate into regulatory or effector T cell subsets (**c**). Gut-draining DCs predominantly induce the differentiation of IgA-producing plasma cells, which play an important role in the defense against intestinal pathogens and the spatial segregation of host and microbiota (**d**)

markers allow the identification of four distinct DC subsets that differ in phenotype and presumably also in functionality. $CD103^+CD11b^+$ DCs have been demonstrated to induce pathogen- and microbiota-specific IgA-producing plasma cells as well as Tregs, Th1, and Th17 cells (Cerovic et al. 2014). They are the most frequent DC subset in the small intestine but are less abundant in the colon. $CD103^+CD11b^-$ DCs efficiently cross-present soluble antigens to $CD8^+$ T cells but also express the enzyme RALDH that promotes the induction of $Foxp3^+$ Tregs. $CD103^+CD11b^-$ DCs are found in the small intestine and colon as well as lymphoid organs such as PPs and mLNs. Together, $CD103^+$ DC subsets are considered to induce Tregs by releasing significant amounts of TGFβ and retinoic acid, yet the depletion of neither of the $CD103^+$ subsets alone is sufficient to break the induction of tolerance, which only occurs after ablation of all $CD103^+$ DCs, suggesting that a redundancy between these subsets exists (Welty et al. 2013). $CD103^-CD11b^+CX3CR1^{int}$ DCs resemble in many aspects intestinal macrophages and have been suggested to be able to take up luminal antigens by extending dendrites into the lumen via transcellular pores in M cells within PPs (Lelouard et al. 2012). Finally, the least abundant subset identified as $CD103^-CD11b^-$ DCs is thought to be enriched in ILFs and PPs as they are absent from mice lacking organized lymphoid structures in the intestine. Due to their low abundance, their functional specialization is less well defined, but it has been suggested that they are capable to induce Th17 cells (Cerovic et al. 2013).

Since the epithelial barrier separating the host and lumen has to be tight to prevent loss of water and nutrients, the nature of antigen uptake from the lumen by DCs has been an area of intense research. In the small intestine, PPs are an important entry site for both dietary and microbial-derived antigens, whose transfer from the lumen to the tissue is mediated through specialized M cells that pinocytose luminal content and then release it to underlying DCs and macrophages on the basolateral site (Ohno 2016). In addition, several other mechanisms have been suggested to contribute to the uptake of luminal antigens by DCs in the LP or in lymphoid structures such as paracellular leaks between epithelial cells, retrograde transport through goblet cells, and the formation of transepithelial dendrites (TED) (Niess et al. 2005; McDole et al. 2012). The relative contribution of each of those pathways for the uptake of soluble dietary antigens or particular microbial antigens from the lumen is still a matter of debate, but specifically, retrograde transport through goblet cells seems to have an important role to induce tolerance against microbial antigen postnatally (Knoop et al. 2017). Another source of microbial antigens originating from pathogens but also commensals is the colonization of host cells by these microbes in the PPs and potentially other lymphoid structures in the intestine, after which they are transferred intact to the mLN (Fung et al. 2016). Migration of DCs from the lamina propria to the mLNs occurs in a CCR7-dependent manner, where they interact with naïve T cells and B cells. Signals including those directly or indirectly derived from the microbiota play an important role in directing the type of the adaptive immune responses, e.g., the induction of Th17 cells or Tregs. For instance, the presence of the microbiota is required to acquire the ability to induce tolerogenic T cells in DCs located in the mLN (Cording et al. 2013). Yet, the microbiota also influences DC abundance and activity on multiple other levels, e.g., the absolute numbers of intestinal DCs are highly decreased in mLNs of germ-free mice (Walton et al. 2006). Moreover, bone marrow egress of monocytes and plasmacytoid DCs is modulated by the microbiota through induction of CCL2 for the aforementioned CCR2-dependent migration, demonstrating a systemic influence of the microbiota on trafficking of myeloid subsets in the host (Swiecki et al. 2017). Finally, microbiota-dependent pre-stimulation of DCs is required for their functional licensing in the spleen, i.e., their ability to efficiently prime immune responses even at systemic sites (Ganal et al. 2012). Whether this licensing depends on sensing of circulating bacteria, their metabolites or a microbiota-induced host factor is not known yet.

In summary, intestinal macrophages and DCs play an important part in the mutualism between host and microbiota. They induce and preserve tolerance to harmless microbes, while activating powerful antimicrobial responses when encountering potentially harmful pathogens. Signals derived from the microbiota including metabolites such as SCFA and structural bacterial components such as LPS are essential to fine-tune the balance between these opposing types of immunity. Imbalances in the microbiota have been associated with modulation of macrophage and DC function resulting in the development of host pathologies including IBD.

8.2.4 Innate Lymphoid Cells

Innate lymphoid cells (ILCs) are the lymphoid branch of the innate immune system. In analogy to the bifurcation in the T cell lineage ($CD4^+$ helper T cells and $CD8^+$ cytotoxic T cells), two major ILC lineages are recognized, cytotoxic ILC (i.e., conventional NK cells, cNK cells) and helper-like ILC (i.e., ILC1, ILC2, ILC3) (Fig. 8.4). The three helper-like ILC subsets roughly resemble the various CD4+ T helper cell effector states (i.e., Th1, Th2, Th17). Recently, a ILCreg population was recognized that seems developmentally distinct from the other ILC subsets (Wang et al. 2017).

8.2.4.1 Cytotoxic ILC: Conventional NK Cells

NK cells are cytotoxic lymphocytes belonging to the innate immune system. In contrast to the myeloid cell types described above, they derive

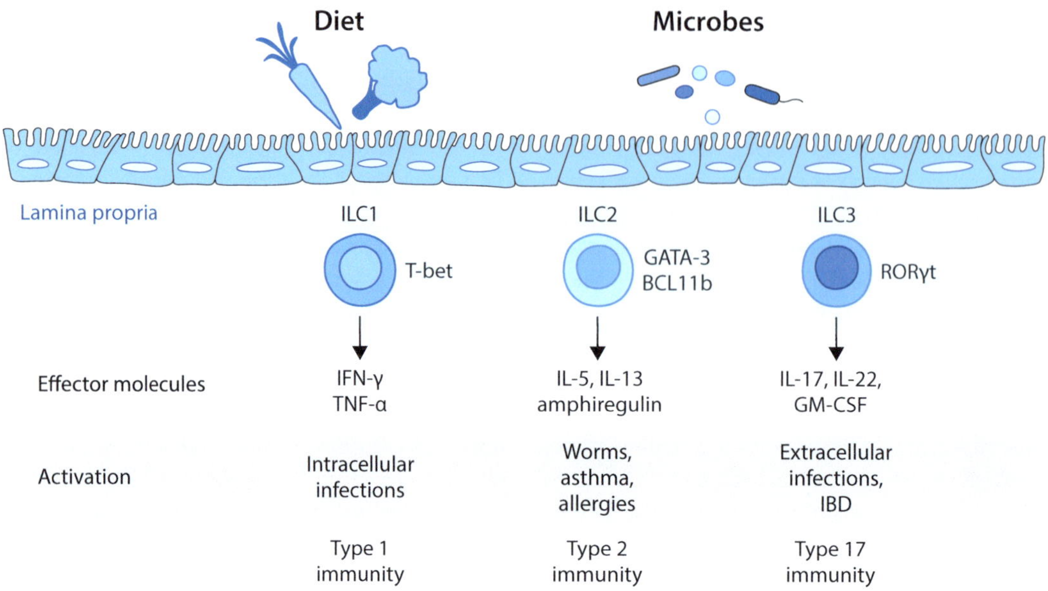

Fig. 8.4 Innate lymphoid cells can be phenotypically and functionally divided into three subsets. Dietary compounds and metabolites of the intestinal microbiota influence the recruitment, development, and functions of ILCs in the intestine. Subsets of ILCs are distinguished by the expression of specific transcription factors and effector molecules such as cytokines. These subsets are activated during different infections and contribute to properly shape the immune system

from CLPs in the bone marrow. The early ILC progenitor (EILP) marks the branching off of the T and B cell lineage and has the potential to develop into all ILC lineages (NK cells and helper-like ILCs but not B and T cells) (Yang et al. 2015). Downstream of the EILP is a precursor referred to as CHILP (Klose et al. 2014; Constantinides et al. 2014) (characterized by the expression of Id2 and Id2 plus PLZF, respectively). CHILP can differentiate into all helper-like ILCs subsets but not into NK cells (or any adaptive lymphoid lineage). Thus, while NK cells and helper-like ILCs share a common progenitor (the EILP), cNK cells and helper-like ILCs form separate ILC lineages (Diefenbach et al. 2014).

NK cells are systemically present in the bone marrow, blood, and secondary lymphoid organs but also in nonlymphoid tissues such as the liver, lung, and intestine, where NK cells acquire organ-specific functions (Björkström et al. 2016). In their mature state, NK cells have granules in their cytoplasm containing perforin, granzymes, and proteases, which allow NK cells to kill transformed or infected cells. Target cell recognition by NK cells is determined by various inhibitory (NKG2A in humans and mice, KIR molecules in humans, and Ly49 receptors in mice) and activating receptors (e.g., NKG2D, NKp30, NKp44, NKp46, DNAM1). Inhibitory receptors recognize either directly (Ly49 and KIR) or indirectly (NKG2A) MHC class I molecules (Raulet et al. 2001). As all healthy cells express MHC class I, NK cell inhibition by MHC class I is an important mechanism to keep NK cells tolerant against normal self cells. In the context of viral infection and tumor transformation, MHC class I is often downregulated allowing for NK cell activation ("missing self-recognition") (Kärre et al. 1986). While the "missing self-recognition" mode for NK cell activation has been recognized for a long time, the activating mode has only been investigated more recently. Initial work identified ligands for the activating receptor NKG2D, MICA/B in humans (Bauer et al. 1999) and the Rae1 family of ligands in humans and mice (Cerwenka et al. 2000; Diefenbach et al. 2000). Importantly, these MHC class I-related NKG2D ligands are not expressed on healthy cells but are upregulated after cellular insults such as tumor transformation, infection, DNA damage, and senescence (Diefenbach and Raulet 2003). Once cells acquire NKG2D ligand expression, they become targets for NK cells, and the NKG2D receptor/ligand system is a substantial barrier against tumor development (Guerra et al. 2008). It is believed that there is a certain hierarchy for NK cell activation with some receptors being very potent when triggered in isolation (such as CD16, a component of the Fc receptor), whereas others were not able to trigger resting NK cells in the absence of priming signals (see below) (Bryceson et al. 2006). The activation of NK cells through activating receptors or through type I interferons (IFN), IL-12, and IL-18 leads to production of other cytokines and chemokines such as tumor necrosis factor (TNF), IFNγ, and granulocyte-macrophage colony-stimulating factor (GM-CSF) important to act as immunoregulators.

It has recently become clear that there is considerable diversity between NK cells in the circulation and those located in tissues. The expression of CD69 as well as CD103 and CD49a are characteristic for NK cells residing in the tissue and distinguish them from circulating cNK cells (Björkström et al. 2016). It remains to be seen if the various tissue-resident NK cell subsets are indeed tissue-resident NK cells or belong to separate ILC lineages.

Although NK cells are considered "naturally active," there is a substantial cell extrinsic component for priming of NK cell function. Indeed, freshly isolated NK cells from SPF mice show very poor cytotoxic activity and do not produce cytokines. However, after injections of microbial compounds such as poly(I:C), LPS, and CpG, NK cells very effectively recognize target cells. This process is often referred to as "NK cell priming" and requires the presence of mononuclear phagocytes, in particular DCs. Mice ablated of DCs cannot prime NK cell function after injection of PAMP (Ganal et al. 2012). On a molecular level, various cytokines (such as IL-2, IL-12, or IL-15) produced by mononuclear phagocytes have been implicated in the priming process. More recently, it became clear that NK cell priming by components of the myeloid system requires the presence of the microbiota. It has been demonstrated that germ-free and antibiotic-treated mice show reduced NK cell activity, due to the inability of mononuclear phagocytes to release cytokines required for NK cell priming (Luu et al. 2016). On a mechanistic level, stimulation of mononuclear phagocytes resulted in normal initiation of the signal transduction cascade downstream of PRR. However, the activated transcription factors (i.e., the NF-κB subunit p65 and IRF3) could not bind to the promoter regions of cytokine genes because of an inappropriately resolved chromatin barrier. These data indicate that signals controlled by the microbiota calibrate mononuclear phagocytes involving epigenetic processes that prepare these cells to respond with a powerful cytokine response upon pathogen encounter (Ganal et al. 2012). It is unknown if this microbiota-controlled program is mediated by distinct microbes and distinct metabolites or by soluble factors produced by cells "sensing" commensal microbiota.

8.2.4.2 Helper-Like ILC

Helper-like ILCs are a recently identified group of innate lymphocytes, which are deposited in tissues during ontogeny and then are maintained in tissues either by self-renewal or by differentiation from tissue-resident progenitors without appreciable renewal from circulating cells as demonstrated by parabiosis experiments. This extreme sedentary lifestyle indicates close links of ILCs to organ function and organ homeostasis. In the intestine, direct or indirect stimulation through microbial induces distinct subsets of ILCs to release diverse sets of cytokines themselves, some of which directly act on IEC, leading, for example, to enhanced production of antimicrobial peptides and proteins, impacting the composition of the gut microbiota.

ILC1 express the T-box transcription factor T-bet but not Eomes (characteristic for NK cells), produce IFNγ and TNF, and are involved in the early immune response against intracellular pathogens (Daussy et al. 2014; Klose et al. 2014). ILC2 express high levels of Gata-3 and developmentally depend on the transcription factor

Bcl11b (Hoyler et al. 2012). They produce IL-5 and IL-13 and play important roles in type 2 immunity against worms. If inappropriately activated, ILC2 contribute to the pathogenesis of allergic diseases. Finally, ILC3 represent the "type 17 module" of ILCs. ILC3 are characterized by the expression of the transcription factor RORγt, and they are strictly required for the formation of secondary lymphoid organs such as lymph nodes, PPs, CPs, and ILFs. ILC3 produce IL-22, IL-17A, and IL-17F and are required for immunity to certain types of bacterial infections (i.e., "attaching-and-effacing" intestinal pathogens) and viral infections (e.g., rotavirus, norovirus). The complex cross talk between the microbiota and ILCs has been reviewed in detail (Sonnenberg and Artis 2012; Britanova and Diefenbach 2017).

8.3 Summary and Outlook

The cells of the innate immune system play an important role in the balanced interplay with the intestinal microbiota. During health, distinct subsets of innate immune cells derived from the myeloid and lymphoid lineages contribute to the sensing of the normal microbiota and integrate multiple signals such as MAMPs and metabolites to promote tolerogenic conditions. The disruption of this balance as consequence of infections or impaired barrier function may lead to excessive immune activation, which promotes local inflammatory conditions such as inflammatory bowel diseases, but has been also implicated in systemic diseases such as metabolic syndrome. In the future it will be of high interest to gain an advanced molecular understanding of the cross talk between innate immune system and microbiota, i.e., the identification of microbial metabolites and the corresponding host receptors and effectors, to enable novel therapeutic interventions to promote health and, specifically, to prevent auto-inflammatory diseases. Moreover, the importance of the imbalanced interplay of innate immunity and the microbiota for other unrelated complex diseases such as neurodegeneration remains to be investigated. Their potential involvement in these diseases with high socioeconomic costs is likely to further expand the recognition of the importance of this evolutionary ancient arm of the immune system for human health.

Controversy

Limitations of Laboratory Mice: A Bias Introduced by the Microbiota?

Most of our insights into the immune system have been gathered by analyzing laboratory mice. Yet, many mouse models of diseases have been criticized because they do not always reflect relevant aspects of the human immune system. One potential explanation for this limitation has been recently identified. While laboratory mice are housed under increasingly hygienic conditions (so-called SPF status), recent data has shown that the cellular and genetic signatures of the immune system in laboratory mouse strains are more reminiscent of human newborns but not of adult humans (Beura et al. 2016; Rosshart et al. 2017). For example, laboratory mice lack effector-differentiated and mucosally distributed T cells and ILCs. However, such populations were found in feral mice or in pet shop mice, whose microbiota composition and exposure to pathogens were profoundly different. Interestingly cohousing of laboratory mice with pet shop mice introduced wide-ranging changes so that immune cell signatures were now resembling adult humans, whereas gene expression signatures found in newborns were suppressed (Beura et al. 2016; Rosshart et al. 2017). Such changes had wide-ranging functional consequences. Laboratory mice exposed to a "dirty" microbiome were more resistant to bacterial and viral infections and developed less inflammation-induced cancer. These data underscore that the exposure to diverse microbial factors has a profound impact on the constitution of our immune system.

History

The Discovery of Dendritic Cells

The ability of specific subsets of host cells to phagocytose microbes and to thereby kill them was first described by Ilya Ilyich Metchnikov. He specifically observed the ability of distinct cells in starfish larvae to engulf and degrade bacteria and hypothesized that this instinctive response was an important part of the immune system. In 1908 he received the Nobel prize for his discoveries that were the foundation of what is known today as innate immune responses. Large advances were made in the following decades describing the cellular and biochemical processes including the discovery of phagosome acidification and the production of reactive oxygen species that explained the exquisite properties of macrophages to contribute to the innate immune response. The discovery of the link between innate and adaptive immunity, i.e., the ability of phagocytes to present microbial antigens to adaptive immune cells for the eventual formation of memory responses, was made by Ralph Steinman and colleagues. Before his groundbreaking studies, scientists were already aware of a need for accessory cells in the priming of adaptive immune responses, but their exact identity was not known, since these cells were relatively scarce. Ralph Steinman then developed new methods to isolate and visualize these critical cells in the 1970s and coined the term dendritic cells for them (Steinman and Cohn 1973). These cells differed from classical macrophages in their ability to phagocytose microbes and protein antigens for subsequent presentation of peptide antigens on major histocompatibility complex class I and II molecules. Many researchers initially disbelieved this discovery due to their difficulties to isolate pure populations of dendritic cells, which in turn also contaminated macrophage preparations. However, advances in the methodology to isolate dendritic cells and study their functions in vivo have proven Steinman's initial hypothesis of the distinct functionality of this cell population. The underlying model of antigen presentation and priming of adaptive immunity involving a central role for dendritic cells in this process has been widely accepted since and culminated in the Nobel Prize for the late Ralph Steinman in 2011. It is important to note that this model does not preclude a role of antigen presentation in macrophages, which occurs in both lymphoid organs and peripheral tissues such as the gastrointestinal tract, but to a much lower level and without the exquisite co-stimulatory abilities of dendritic cells.

Highlights

- The innate immune system is not only central to recognize and respond to pathogens but also to commensals in the intestine
- The innate immune system senses microbes via a variety of MAMPs such as bacterial cell wall components and microbe-specific metabolites.
- Resident mononuclear phagocytes, i.e., macrophages and dendritic cells, are essential for immunosurveillance in the tissue as well as for transmitting information to gut-draining lymphoid organs.
- Specialized subsets of innate lymphoid cells are an immediate source of immunoregulatory mediators contributing to the balanced interplay between microbiota and host.
- Impaired or dysregulated innate immunity promotes the development of imbalanced intestinal ecosystems and dysbiosis-associated diseases.

References

Bain, C. C., Bravo-Blas, A., Scott, C. L., et al. (2014). Constant replenishment from circulating monocytes maintains the macrophage pool in the intestine of adult mice. *Nature Immunology, 15*, 929–937.

Bauer, S., Groh, V., Wu, J., et al. (1999). Activation of NK cells and T cells by NKG2D, a receptor for stress-inducible MICA. *Science, 285*, 727–729.

Beura, L. K., Hamilton, S. E., Bi, K., et al. (2016). Normalizing the environment recapitulates adult human immune traits in laboratory mice. *Nature, 532*, 512–516.

Bischoff, S. C. (2009). Physiological and pathophysiological functions of intestinal mast cells. *Seminars in Immunopathology, 31*, 185–205.

Björkström, N. K., Ljunggren, H.-G., & Michaëlsson, J. (2016). Emerging insights into natural killer cells in human peripheral tissues. *Nature Reviews Immunology, 16*, 310–320.

Britanova, L., & Diefenbach, A. (2017). Interplay of innate lymphoid cells and the microbiota. *Immunological Reviews, 279*, 36–51.

Brubaker, S. W., Bonham, K. S., Zanoni, I., & Kagan, J. C. (2015). Innate immune pattern recognition: A cell biological perspective. *Annual Review of Immunology, 33*, 257–290.

Bryceson, Y. T., March, M. E., Ljunggren, H.-G., & Long, E. O. (2006). Synergy among receptors on resting NK cells for the activation of natural cytotoxicity and cytokine secretion. *Blood, 107*, 159–166.

Buonomo, E. L., Cowardin, C. A., Wilson, M. G., et al. (2016). Microbiota-regulated IL-25 increases eosinophil number to provide protection during *Clostridium difficile* infection. *Cell Reports, 16*, 432–443.

Cebra, J. J., Periwal, S. B., Lee, G., et al. (1998). Development and maintenance of the gut-associated lymphoid tissue (GALT): The roles of enteric bacteria and viruses. *Developmental Immunology, 6*, 13–18.

Cerovic, V., Houston, S. A., Scott, C. L., et al. (2013). Intestinal CD103(-) dendritic cells migrate in lymph and prime effector T cells. *Mucosal Immunology, 6*, 104–113.

Cerovic, V., Bain, C. C., Mowat, A. M., & Milling, S. W. F. (2014). Intestinal macrophages and dendritic cells: What's the difference? *Trends in Immunology, 35*, 270–277.

Cerwenka, A., Bakker, A. B., McClanahan, T., et al. (2000). Retinoic acid early inducible genes define a ligand family for the activating NKG2D receptor in mice. *Immunity, 12*, 721–727.

Chang, P. V., Hao, L., Offermanns, S., & Medzhitov, R. (2014). The microbial metabolite butyrate regulates intestinal macrophage function via histone deacetylase inhibition. *Proceedings of the National Academy of Sciences of the United States of America, 111*, 2247–2252.

Chapuy, L., Bsat, M., Mehta, H., et al. (2014). Basophils increase in Crohn disease and ulcerative colitis and favor mesenteric lymph node memory TH17/TH1 response. *Journal of Allergy and Clinical Immunology, 134*, 978–981.e1.

Chu, V. T., Beller, A., Rausch, S., et al. (2014). Eosinophils promote generation and maintenance of immunoglobulin-A-expressing plasma cells and contribute to gut immune homeostasis. *Immunity, 40*, 582–593 d.

Clarke, T. B., Davis, K. M., Lysenko, E. S., et al. (2010). Recognition of peptidoglycan from the microbiota by Nod1 enhances systemic innate immunity. *Nature Medicine, 16*, 228–231.

Constantinides, M. G., McDonald, B. D., Verhoef, P. A., & Bendelac, A. (2014). A committed precursor to innate lymphoid cells. *Nature, 508*, 397–401.

Cording, S., Fleissner, D., Heimesaat, M. M., et al. (2013). Commensal microbiota drive proliferation of conventional and Foxp3(+) regulatory CD4(+) T cells in mesenteric lymph nodes and Peyer's patches. *European Journal of Microbiology and Immunology, 3*, 1–10.

Daussy, C., Faure, F., Mayol, K., et al. (2014). T-bet and Eomes instruct the development of two distinct natural killer cell lineages in the liver and in the bone marrow. *The Journal of Experimental Medicine, 211*, 563–577.

Diefenbach, A., & Raulet, D. H. (2003). Innate immune recognition by stimulatory immunoreceptors. *Current Opinion in Immunology, 15*, 37–44.

Diefenbach, A., Jamieson, A. M., Liu, S. D., et al. (2000). Ligands for the murine NKG2D receptor: Expression by tumor cells and activation of NK cells and macrophages. *Nature Immunology, 1*, 119–126.

Diefenbach, A., Colonna, M., & Koyasu, S. (2014). Development, differentiation, and diversity of innate lymphoid cells. *Immunity, 41*, 354–365.

Eberl, G., & Lochner, M. (2009). The development of intestinal lymphoid tissues at the interface of self and microbiota. *Mucosal Immunology, 2*, 478–485.

Elinav, E., Strowig, T., Kau, A. L., et al. (2011). NLRP6 inflammasome regulates colonic microbial ecology and risk for colitis. *Cell, 145*, 745–757.

Erny, D., Hrabě de Angelis, A. L., Jaitin, D., et al. (2015). Host microbiota constantly control maturation and function of microglia in the CNS. *Nature Neuroscience, 18*, 965–977.

Fung, T. C., Bessman, N. J., Hepworth, M. R., et al. (2016). Lymphoid-tissue-resident commensal bacteria promote members of the IL-10 cytokine family to establish mutualism. *Immunity, 44*, 634–646.

Ganal, S. C., Sanos, S. L., Kallfass, C., et al. (2012). Priming of natural killer cells by nonmucosal mononuclear phagocytes requires instructive signals from commensal microbiota. *Immunity, 37*, 171–186.

Gieseck, R. L., Wilson, M. S., & Wynn, T. A. (2018). Type 2 immunity in tissue repair and fibrosis. *Nature Reviews Immunology, 18*, 62–76.

Gomez de Agüero, M., Ganal-Vonarburg, S. C., Fuhrer, T., et al. (2016). The maternal microbiota drives early postnatal innate immune development. *Science, 351*, 1296–1302.

Gomez, M. R., Talke, Y., Hofmann, C., et al. (2014). Basophils control T-cell responses and limit disease activity in experimental murine colitis. *Mucosal Immunology, 7*, 188–199.

Gorjifard, S., & Goldszmid, R. S. (2016). Microbiota-myeloid cell crosstalk beyond the gut. *Journal of Leukocyte Biology, 100*, 865–879.

Guerra, N., Tan, Y. X., Joncker, N. T., et al. (2008). NKG2D-deficient mice are defective in tumor surveillance in models of spontaneous malignancy. *Immunity, 28*, 571–580.

Hill, D. A., Siracusa, M. C., Abt, M. C., et al. (2012). Commensal bacteria-derived signals regulate basophil hematopoiesis and allergic inflammation. *Nature Medicine, 18*, 538–546.

Hoyler, T., Klose, C. S. N., Souabni, A., et al. (2012). The transcription factor GATA-3 controls cell fate and maintenance of type 2 innate lymphoid cells. *Immunity, 37*, 634–648.

Humbles, A. A., Lu, B., Friend, D. S., et al. (2002). The murine CCR3 receptor regulates both the role of eosinophils and mast cells in allergen-induced airway inflammation and hyperresponsiveness. *Proceedings of the National Academy of Sciences of the United States of America, 99*, 1479–1484.

Jia, W., Xie, G., & Jia, W. (2017). Bile acid-microbiota crosstalk in gastrointestinal inflammation and carcinogenesis. *Nature Reviews Gastroenterology & Hepatology, 5*, 172ra122.

Joeris, T., Müller-Luda, K., Agace, W. W., & Mowat, A. M. (2017). Diversity and functions of intestinal mononuclear phagocytes. *Mucosal Immunology, 10*, 845–864.

Kärre, K., Ljunggren, H. G., Piontek, G., & Kiessling, R. (1986). Selective rejection of H-2-deficient lymphoma variants suggests alternative immune defence strategy. *Nature, 319*, 675–678.

Khosravi, A., Yáñez, A., Price, J. G., et al. (2014). Gut microbiota promote hematopoiesis to control bacterial infection. *Cell Host & Microbe, 15*, 374–381.

Kiss, E. A., & Vonarbourg, C. (2012). Aryl hydrocarbon receptor: A molecular link between postnatal lymphoid follicle formation and diet. *Gut Microbes, 3*, 577–582.

Klose, C. S. N., Flach, M., Möhle, L., et al. (2014). Differentiation of type 1 ILCs from a common progenitor to all helper-like innate lymphoid cell lineages. *Cell, 157*, 340–356. https://doi.org/10.1016/j.cell.2014.03.030.

Knoop, K. A., Gustafsson, J. K., McDonald, K. G., et al. (2017). Microbial antigen encounter during a preweaning interval is critical for tolerance to gut bacteria. *Science Immunology, 2*, eaao1314.

Kunii, J., Takahashi, K., Kasakura, K., et al. (2011). Commensal bacteria promote migration of mast cells into the intestine. *Immunobiology, 216*, 692–697.

Lavin, Y., Mortha, A., Rahman, A., & Merad, M. (2015). Regulation of macrophage development and function in peripheral tissues. *Nature Reviews Immunology, 15*, 731–744.

Leiding, J. W. (2017). Neutrophil evolution and their diseases in humans. *Frontiers in Immunology, 8*, 1009.

Lelouard, H., Fallet, M., de Bovis, B., et al. (2012). Peyer's patch dendritic cells sample antigens by extending dendrites through M cell-specific transcellular pores. *Gastroenterology, 142*, 592–601.e3.

Luu, T. T., Ganesan, S., Wagner, A. K., et al. (2016). Independent control of natural killer cell responsiveness and homeostasis at steady-state by CD11c+ dendritic cells. *Scientific Reports, 6*, 37996.

Magalhaes, J. G., Tattoli, I., & Girardin, S. E. (2007). The intestinal epithelial barrier: How to distinguish between the microbial flora and pathogens. *Seminars in Immunology, 19*, 106–115.

Masterson, J. C., McNamee, E. N., Jedlicka, P., et al. (2011). CCR3 blockade attenuates eosinophilic ileitis and associated remodeling. *The American Journal of Pathology, 179*, 2302–2314.

McDole, J. R., Wheeler, L. W., McDonald, K. G., et al. (2012). Goblet cells deliver luminal antigen to CD103 + dendritic cells in the small intestine. *Nature, 483*, 345–349.

Merad, M., Sathe, P., Helft, J., et al. (2013). The dendritic cell lineage: Ontogeny and function of dendritic cells and their subsets in the steady state and the inflamed setting. *Annual Review of Immunology, 31*, 563–604.

Mori, Y., Iwasaki, H., Kohno, K., et al. (2009). Identification of the human eosinophil lineage-committed progenitor: Revision of phenotypic definition of the human common myeloid progenitor. *The Journal of Experimental Medicine, 206*, 183–193.

Mowat, A. M., Scott, C. L., & Bain, C. C. (2017). Barrier-tissue macrophages: Functional adaptation to environmental challenges. *Nature Medicine, 23*, 1258–1270.

Niess, J. H., Brand, S., Gu, X., et al. (2005). CX3CR1-mediated dendritic cell access to the intestinal lumen and bacterial clearance. *Science, 307*, 254–258.

Ohkubo, T., Tsuda, M., Suzuki, S., et al. (1999). Peripheral blood neutrophils of germ-free rats modified by in vivo granulocyte-colony-stimulating factor and exposure to natural environment. *Scandinavian Journal of Immunology, 49*, 73–77.

Ohno, H. (2016). Intestinal M cells. *Journal of Biochemistry, 159*, 151–160.

Palm, N. W., & Medzhitov, R. (2009). Pattern recognition receptors and control of adaptive immunity. *Immunological Reviews, 227*, 221–233.

Park, J.-S., Lee, E.-J., Lee, J.-C., et al. (2007). Anti-inflammatory effects of short chain fatty acids in IFN-gamma-stimulated RAW 264.7 murine macrophage cells: Involvement of NF-kappaB and ERK signaling pathways. *International Immunopharmacology, 7*, 70–77.

Parlato, M., & Yeretssian, G. (2014). NOD-like receptors in intestinal homeostasis and epithelial tissue repair. *International Journal of Molecular Sciences, 15*, 9594–9627.

Patten, D. A., & Collett, A. (2013). Exploring the immunomodulatory potential of microbial-associated molecular patterns derived from the enteric bacterial microbiota. *Microbiology (Reading, England), 159*, 1535–1544.

Platt, A. M., Bain, C. C., Bordon, Y., et al. (2010). An independent subset of TLR expressing CCR2-dependent macrophages promotes colonic inflammation. *Journal of Immunology, 184*, 6843–6854.

Raab, Y., Fredens, K., Gerdin, B., & Hallgren, R. (1998). Eosinophil activation in ulcerative colitis: Studies on mucosal release and localization of eosinophil granule constituents. *Digestive Diseases and Sciences, 43*, 1061–1070.

Rathinam, V. A. K., & Fitzgerald, K. A. (2016). Inflammasome complexes: Emerging mechanisms and effector functions. *Cell, 165*, 792–800.

Raulet, D. H., Vance, R. E., & McMahon, C. W. (2001). Regulation of the natural killer cell receptor repertoire. *Annual Review of Immunology, 19*, 291–330.

Rivollier, A., He, J., Kole, A., et al. (2012). Inflammation switches the differentiation program of Ly6Chi monocytes from antiinflammatory macrophages to inflammatory dendritic cells in the colon. *The Journal of Experimental Medicine, 209*, 139–155.

Robertson, S. J., & Girardin, S. E. (2013). Nod-like receptors in intestinal host defense: Controlling pathogens, the microbiota, or both? *Current Opinion in Gastroenterology, 29*, 15–22.

Rosshart, S. P., Vassallo, B. G., Angeletti, D., et al. (2017). Wild mouse gut microbiota promotes host fitness and improves disease resistance. *Cell, 171*, 1015–1028.e13.

Russell, S. L., Gold, M. J., Willing, B. P., et al. (2013). Perinatal antibiotic treatment affects murine microbiota, immune responses and allergic asthma. *Gut Microbes, 4*, 158–164.

Saitoh, O., Kojima, K., Sugi, K., et al. (1999). Fecal eosinophil granule-derived proteins reflect disease activity in inflammatory bowel disease. *The American Journal of Gastroenterology, 94*, 3513–3520.

Smythies, L. E., Shen, R., Bimczok, D., et al. (2010). Inflammation anergy in human intestinal macrophages is due to Smad-induced IkappaBalpha expression and NF-kappaB inactivation. *The Journal of Biological Chemistry, 285*, 19593–19604.

Sonnenberg, G. F., & Artis, D. (2012). Innate lymphoid cell interactions with microbiota: Implications for intestinal health and disease. *Immunity, 37*, 601–610.

St John, A. L., & Abraham, S. N. (2013). Innate immunity and its regulation by mast cells. *Journal of Immunology, 190*, 4458–4463.

Steinman, R. M., & Cohn, Z. A. (1973). Identification of a novel cell type in peripheral lymphoid organs of mice. I. Morphology, quantitation, tissue distribution. *The Journal of Experimental Medicine, 137*, 1142–1162.

Stewart, M. K., & Cookson, B. T. (2016). Evasion and interference: Intracellular pathogens modulate caspase-dependent inflammatory responses. *Nature Reviews Microbiology, 14*, 346–359.

Strowig, T., Henao-Mejia, J., Elinav, E., & Flavell, R. (2012). Inflammasomes in health and disease. *Nature, 481*, 278–286.

Swiecki, M., Miller, H. L., Sesti-Costa, R., et al. (2017). Microbiota induces tonic CCL2 systemic levels that control pDC trafficking in steady state. *Mucosal Immunology, 10*, 936–945.

Tan, J., McKenzie, C., Potamitis, M., et al. (2014). The role of short-chain fatty acids in health and disease. *Advances in Immunology, 121*, 91–119.

Thaiss, C. A., Levy, M., Itav, S., & Elinav, E. (2016a). Integration of innate immune signaling. *Trends in Immunology, 37*, 84–101.

Thaiss, C. A., Zmora, N., Levy, M., & Elinav, E. (2016b). The microbiome and innate immunity. *Nature, 535*, 65–74.

Ueda, Y., Kayama, H., Jeon, S. G., et al. (2010). Commensal microbiota induce LPS hyporesponsiveness in colonic macrophages via the production of IL-10. *International Immunology, 22*, 953–962.

Walton, K. L. W., He, J., Kelsall, B. L., et al. (2006). Dendritic cells in germ-free and specific pathogen-free mice have similar phenotypes and in vitro antigen presenting function. *Immunology Letters, 102*, 16–24.

Wang, S., Xia, P., Chen, Y., et al. (2017). Regulatory innate lymphoid cells control innate intestinal inflammation. *Cell, 171*, 201–216.e18.

Weber, B., Saurer, L., Schenk, M., et al. (2011). CX3CR1 defines functionally distinct intestinal mononuclear phagocyte subsets which maintain their respective functions during homeostatic and inflammatory conditions. *European Journal of Immunology, 41*, 773–779.

Weller, P. F., & Spencer, L. A. (2017). Functions of tissue-resident eosinophils. *Nature Reviews Immunology, 17*, 746–760.

Welty, N. E., Staley, C., Ghilardi, N., et al. (2013). Intestinal lamina propria dendritic cells maintain T cell homeostasis but do not affect commensalism. *The Journal of Experimental Medicine, 210*, 2011–2024.

Wouters, M. M., Vicario, M., & Santos, J. (2016). The role of mast cells in functional GI disorders. *Gut, 65*, 155–168.

Yang, Q., Li, F., Harly, C., et al. (2015). TCF-1 upregulation identifies early innate lymphoid progenitors in the bone marrow. *Nature Immunology, 16*, 1044–1050.

Microbiome and Gut Immunity: T Cells

9

Joern Pezoldt, Juhao Yang, Mangge Zou, and Jochen Huehn

Abstract

The gastrointestinal tract is colonized with a huge number of microbes, which are instrumental for the development, homeostasis, and fine-tuning of the immune system. Recent evidence suggests that microbiota very efficiently modulates conventional and regulatory T cell responses that are required for effective host defense against invading pathogens and avoidance of autoimmunity and other immunopathologic conditions, respectively. In this review, we discuss the interplay between the microbiota and T cells, with a particular focus on the de novo induction of regulatory T cells within gut-draining lymph nodes (LNs), the impact of microbiota-derived metabolites on T cell differentiation, and the functional role of unique regulatory T cell subsets within the intestinal immune system.

J. Pezoldt · J. Yang · M. Zou · J. Huehn (✉)
Department Experimental Immunology, Helmholtz Centre for Infection Research, Braunschweig, Germany
e-mail: Jochen.Huehn@helmholtz-hzi.de

9.1 Intestinal T Cell Immune Homeostasis

9.1.1 T Cell Development Within the Thymus

The backbone for the vast array of antigen-specific responses across different T cell populations is first established in the thymus, the primary lymphoid organ for T cell development. Lymphoid progenitors from the bone marrow seed the thymus (Cui et al. 2009; Rothenberg et al. 2008) and can develop into either γ:δ or α:β T cells after random rearrangement of their T cell receptor-encoding gene loci in a multistep process (Ciofani and Zuniga-Pflucker 2010; Vantourout and Hayday 2013). Subsequent to positive selection, α:β thymocytes further differentiate within the thymic medulla into either $CD4^+$ or $CD8^+$ cells, depending on the class of MHC molecule TCR specificity (Klein et al. 2009). Self-reactive T cells are negatively selected upon recognition of self-antigenic peptide being ectopically expressed within the thymus and undergo depletion (Hogquist et al. 2005; Klein et al. 2009). Mature $CD4^+$ and $CD8^+$ thymocytes egress from the thymus and migrate via the bloodstream to the peripheral lymphoid organs (the spleen and lymph nodes) to form the pool of naïve T cells (Fig. 9.1) (Weinreich and Hogquist 2008).

D. Haller (ed.), *The Gut Microbiome in Health and Disease*,
https://doi.org/10.1007/978-3-319-90545-7_9

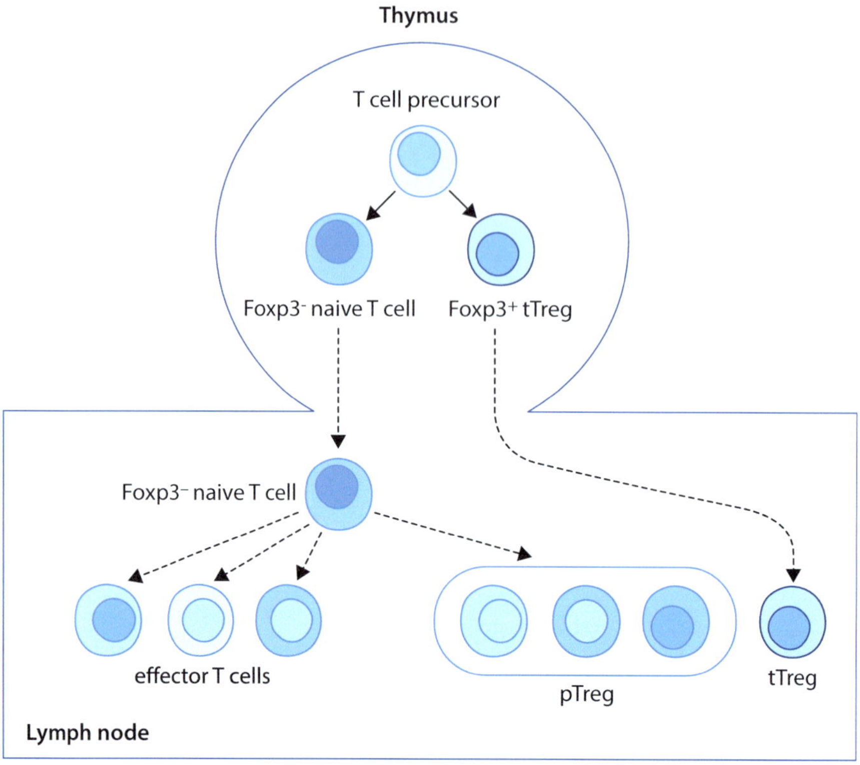

Fig. 9.1 T cell development. Both Foxp3^{-} naïve T cells and Foxp3^{+} Tregs develop from a common precursor in the thymus and egress via the blood to secondary lymphoid tissues, e.g., lymph nodes, where they can further differentiate into effector T cells or give rise to peripherally induced Tregs (pTregs)

Here, we will focus on adaptive immune responses of CD4^{+} T helper (Th) cells in the intestinal immune system. We will summarize mechanisms that are utilized to balance immune responses against gastrointestinal pathogens and at the same time allow for a symbiotic relationship with the intestinal microbiota.

9.1.2 Peripheral T Helper Cell Subset Differentiation

The pool of naïve T cells is constantly patrolling the secondary lymphoid organs positioned throughout the body to scan antigen-presenting cells (APCs) for their cognate antigens. Upon antigen recognition, the naïve CD4^{+} T cells get activated, clonally expand, and can further differentiate into various distinct Th cell subsets, dependent on the type of costimulatory as well as cytokine signals (Josefowicz et al. 2012a). Th1 cells, characterized by the expression of the lineage specification transcription factor T-bet, can secret interferon-γ (IFNγ) upon stimulation to confer protection against intracellular pathogens like *Listeria monocytogenes* (Amsen et al. 2009; Szabo et al. 2000). Th2 cells, defined by the expression of the transcription factor GATA-binding protein 3 (GATA3), produce interleukin (IL)-4, IL-5, and IL-13 upon stimulation (Abbas et al. 1996; Zheng and Flavell 1997), which promote humoral immunity and host defense against extracellular pathogens such as helminths (Paul and Zhu 2010). Th17 cells are characterized by the transcription factor retinoic acid receptor-related orphan receptor gamma t (RORγt) and can produce IL-17 and IL-22 upon stimulation, which allow efficient targeting of extracellular bacteria as well as fungi and support barrier function (He et al. 2017; Medzhitov 2007). Th9 cells, which preferentially express the transcription factor PU.1, can secret IL-9 upon stimulation and are involved in host defense against helminth infections as well as airway hypersensitivity reactions (Gerlach et al. 2014; Kaplan 2013). Finally, Th22 cells, an only incompletely defined Th population, produce IL-22 upon stimulation and are implicated in inflammatory skin diseases, such as psoriasis, and are strongly reduced in ulcerative colitis (UC) patients (Eyerich et al. 2009; Leung et al. 2014; Trifari et al. 2009). Viewed as a whole, various effector T cells are present within the body, showing highly specialized functions to support the eradication of specific pathogens and to maintain barrier function and tissue integrity.

9.2 Effector T Cell Subsets Are Tightly Balanced in Intestinal Immune Responses

9.2.1 Effector T Cells and Gastrointestinal Infections

The gastrointestinal tract constitutes the largest surface of the body and thus has developed a multitude of mechanisms to either prevent pathogen entry or to efficiently eliminate invading pathogens. During gastrointestinal infections, T cells undergo proliferation and differentiation upon cognate antigen recognition in the presence of certain cytokines produced by innate immune cells (Abbas et al. 1996; Ebbo et al. 2017; Mosmann and Coffman 1989; Sternberg 2006). Recognition of intracellular bacteria or virus infection elicits the production of IL-12 by dendritic cells (DC) and macrophages together with IFNγ by innate lymphoid cells (ILCs) 1s, which support Th1 differentiation (Ebbo et al. 2017; Littman and Rudensky 2010; Macatonia et al. 1995; Trinchieri 1994). In contrast, infections with parasitic worms induce production of IL-4 by cells of the innate immune system (Ebbo et al. 2017) and promote differentiation of Th2 cells that produce IL-4, as well as IL-5 and IL-13, cytokines involved in controlling expulsion of the helminths (Allen and Maizels 2011). Signals from transforming growth factor (TGF)β and IL-4 promote Th9 cells differentiation, which also support anti-helminth responses (Kaplan et al. 2015). The preferential production of IL-17 by T cells during infection with *Borrelia burgdorferi*, *Mycobacterium tuberculosis* (Infante-Duarte et al. 2000), *Klebsiella pneumonia* (Happel et al. 2005; Ye et al. 2001), and fungal species (LeibundGut-Landmann et al. 2007) suggests that Th17 cells are critical for host defense against a variety of pathogens at mucosal surfaces.

9.2.2 Imbalanced Intestinal T Cell Immunity Drives Inflammation

The intestinal immune system including antigen-specific T cell responses has to be tightly balanced to permit accurate and rapid protective responses against pathogens but also avoid deleterious immune responses by provoking overexuberant inflammatory processes (Littman and Rudensky 2010).

Several inflammatory bowel diseases (IBD) have been associated with dysregulated, skewed T cell responses. Crohn's disease (CD) is thought to be a Th1-mediated disease, underlined by higher amounts of IFNγ and IL-2 detected in mucosal T cells from CD patients (Breese et al. 1993). UC is in part mediated by Th2 responses (Shih et al. 2008), indicated by high levels of IL-5 and IL-13 in the microenvironment of the lamina propria (LP) T cells from UC patients (Fuss et al. 2004; Heller et al. 2005; Trinchieri 1994). Furthermore, mounting evidence implicates Th17 cells and their contribution to IL-17 levels to drive IBD pathogenesis (Shih et al. 2008; McGovern and Powrie 2007; Neurath 2014). Several genome-wide association studies (GWAS) have pinpointed genes involved in Th17 differentiation and expansion, including *IL-23R*, *IL-12B*, *JAK2*, *STAT3*, *CCR6*, and *TNFSF15*, as CD susceptibility loci with some overlap in UC (Franke et al. 2010; McGovern et al. 2010). These GWAS findings are corroborated by elevated frequencies of Th1, Th17, and generally RORγt$^+$ T cells in LP of CD and UC patients (Dambacher et al. 2009; Rovedatti et al. 2009; Sugihara et al. 2010).

Dysbiotic skewing of the microbial community toward elevated levels of pathobionts has been shown to translate into unrestrained T cell responses, finally resulting in intestinal inflammation (Elinav et al. 2011; Kamada et al. 2013). Several studies have confirmed that abnormal composition and activity of the intestinal microbiota drive IBD (Blander et al. 2012; Zechner 2017; Roy et al. 2017). However, how "dysbiosis" precisely drives effector Th cell differentiation in IBD patients remains an open question.

Memory and effector T cells function as permanent retainers of antigen-specific immune responses and support pro-inflammatory responses persistently. Their counterparts, regulatory T cells (Treg), can adjust and reduce these inflammatory reactions in the course of acute and chronic diseases. To balance the functions of effector T cells, Tregs are equipped to ameliorate intense immune reactions, globally, but also specifically within the T cell compartment (Sakaguchi et al. 2008). Importantly, peripherally induced Tregs (pTreg) are considered to harbor specificity toward harmless food-borne or microbiota-derived antigens, thereby limiting the development of pathogenic immune responses toward environmental antigens (Bach 2003; Vignali et al. 2008).

9.3 Characteristics and Contribution of Regulatory T Cells to Intestinal Immune Homeostasis

9.3.1 The Lineage Specification of Tregs

Tregs are a subset of $CD4^+$ T cells having fundamental functions not only to maintain immune homeostasis and peripheral tolerance but also in the prevention of overwhelming immune responses against invading pathogens (Smigiel et al. 2014). The hallmark transcription factor of the Treg lineage and their functional commitment is Forkhead box protein 3 (Foxp3) (Fontenot et al. 2003; Hori et al. 2003). Foxp3 is also required to maintain suppressive capacity of Tregs by maintaining the defined Treg-specific gene expression signature (Gavin et al. 2007; Hill et al. 2007). Work from others and us has demonstrated that epigenetic mechanisms play a key role in the stabilization of Foxp3 expression within the Treg lineage (Huehn et al. 2009; Ohkura et al. 2012). Particularly, the selective demethylation of the CpG-rich conserved noncoding sequence 2 (CNS-2) in the *Foxp3* locus, also known as Treg-specific demethylated region (TSDR), contributes to stable establishment of the Treg lineage identity (Floess et al. 2007; Kim and Leonard 2007; Polansky et al. 2008; Toker et al. 2013; Zheng et al. 2010).

9.3.2 Suppressive Functions of Tregs

Extensive studies have uncovered numerous suppressive mechanisms utilized by Tregs to balance immune responses. In general, Tregs exert suppression by secretion of inhibitory cytokines including IL-10, TGFβ, and IL-35. By that, Tregs shape the microenvironment of the respective niche (Banchereau et al. 2012; Shevach 2009). Moreover, their high CD25 expression enables Tregs to consume local IL-2 and therefore starve dividing effector T cells by depleting the IL-2 they need to survive and proliferate (Pandiyan et al. 2007). Tregs also confer immune suppression by releasing adenosine nucleosides via the ecto-5′-nucleotidase enzymes CD39 and CD73, which suppress effector T cell function through activation of the adenosine receptor 2A (Dwyer et al. 2007; Smyth et al. 2013). Additionally, Tregs can directly interact with effector T cells and/or APCs via molecules expressed on their surface such as CTLA-4, GITR, Nrp1, programmed death-1 (PD-1), lymphocyte activation gene 3 (LAG-3), fibrinogen-like protein-2 (FGL-2), and TNFR superfamily member 4 (OX40) (Park et al. 2015; Shevach 2009). Besides, Tregs were also shown to suppress effector T cell function directly by transferring the potent inhibitory second messenger cyclic AMP (cAMP) into effector T cells through membrane gap junctions (Bopp et al. 2007; Cao et al. 2007). How and to which extent microbiota and their constituents precisely shape the immune modulatory functional capacity of Tregs require additional studies.

In the past few years, mounting experimental evidence has suggested that Tregs are highly responsive to their local environment and distinct suppressor mechanisms prominently feature in particular tissue- and inflammation-dependent settings (Delacher et al. 2017; Josefowicz et al. 2012a). Upon instruction by the tissue environment, Tregs can induce expression of tissue-specific transcription factors whose cooperation with Foxp3 results in distinct tissue-specific Treg transcriptional and methylation signatures and functions and then supports Treg subset homeostasis in the respective tissues (Cretney et al. 2013; Delacher et al. 2017; Huehn and Beyer 2015).

9.3.3 Microbiota-Mediated Treg Accumulation in the Intestine

Abundant evidence shows that the intestinal microbiota affects the number and function of Tregs, but only few microbiota-derived constituents promoting Treg accumulation in the intestinal tissue have been identified. Predominant species in the microbiota that contribute to Treg accumulation belong to the class *Clostridia*. Already a mixture of 17 strains of human-derived *Clostridia* preferentially elevate the accumulation of *Clostridia* antigen-specific RORγt$^+$Helios$^-$ Tregs (see below) and can also facilitate the expression of IL-10, CTLA-4, and ICOS by colonic Tregs (Atarashi et al. 2011, 2013). The intestinal tissue hosts a dynamic community of microorganisms, which can influence the suppression apparatus and the differentiation of pTregs and thereby shape intestinal immune homeostasis (see below).

9.4 Gut-Draining Lymph Nodes as Initiation Points of Intestinal T Cell Responses

To prime effective adaptive immune responses, antigen-specific T cells need to meet their respective cognate antigen, a process that is preferentially taking place in the lymph nodes (LNs). The LNs are placed at strategic points throughout the body and drain a particular stretch of tissue, with its own peculiar environmental factors including tissue-specific microbiota. One of the key features of gut-draining LNs is its high Treg-inducing property. Gut-draining LNs include mesenteric LNs (mLNs), draining the small and large intestine, and the liver-draining celiac LN (cLN). Both mLN and cLN represent sites of superior pTreg generation (Cording et al. 2014; Feuerer et al. 2010; Siewert et al. 2007; Sun et al. 2007). In contrast, skin-draining LNs and also mandibular and lung-draining LN show substantially lower Treg induction capacities (own observations).

Importantly, the generation of functional intestinal pTregs as well as effector T cell responses is a successive stepwise process and involves the interconnection between the LN and the intestinal tissue (Pabst and Bernhardt 2013). Initially, CCR7-dependent migration of DCs from the LP into the mLN is required (Worbs et al. 2006). After migrating from the intestinal mucosa into the mLN, DCs promote pTreg generation in a TGFβ-, retinoic acid (RA)-, indoleamine 2,3-dioxygenase (IDO)-, and thymic stromal lymphopoietin (TSLP)-dependent manner (Fig. 9.2) (Coombes et al. 2007; Matteoli et al. 2010; Spadoni et al. 2012; Sun et al. 2007; Worbs et al. 2006). In particular CD103$^+$ DCs promote pTreg induction under steady-state conditions, whereas lymph-borne CD103$^-$ DCs promote the differentiation of effector T cells (Cerovic et al. 2013). CD103$^+$ DCs also promote the expression of gut-homing chemokine receptor CCR9 and integrin $\alpha_4\beta_7$ upon activation of naïve CD4$^+$ T cells (Fig. 9.2) (Agace 2006; Siewert et al. 2007). The establishment of effective pathogen-directed immune responses and mucosal tolerance requires subsequent homing of primed T cells from the mLN to the intestinal LP. The local expansion of pTregs under steady-state conditions depends on CX3CR1$^+$ gut-resident macrophages (Hadis et al. 2011; Zigmond et al. 2012). Hence, the initial differentiation of pTregs instigated in mLN is mitigated into a persistent tissue-specific state of tolerance (Pabst and Bernhardt 2013).

Under homeostatic, noninfectious conditions, the microenvironment of the intestinal tract favors

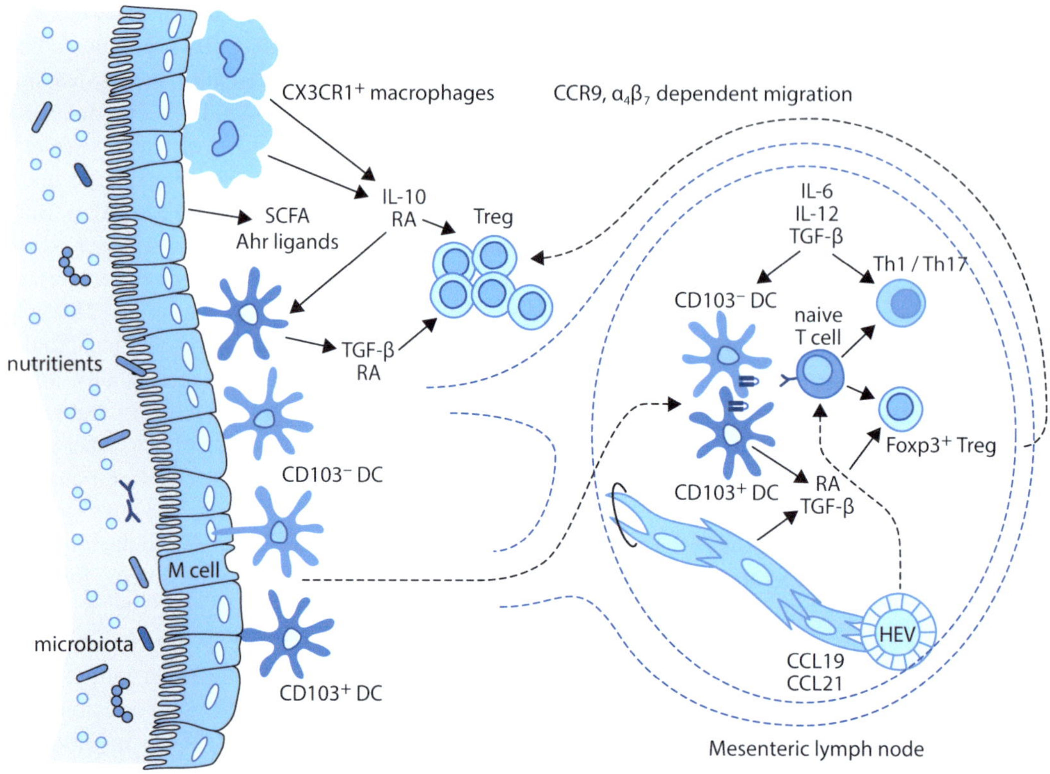

Fig. 9.2 T cell differentiation and expansion is a multistep process influenced by the local environment in a microbiota-dependent manner. T cell differentiation requires the encounter of naïve T cells with professional APCs presenting their cognate antigen. Based on the functional priming and ontogeny of the APC, being either preferentially tolerogenic CD103$^+$ DCs or non-tolerogenic CD103$^-$ DCs, pTreg or effector Th1/Th17 differentiation is instigated, respectively. Primary differentiation is taking place in the context of LN stromal cells that contribute to the tolerogenic environment of the mLN by contributing retinoic acid and TGFβ. Regardless of the type of T cell differentiation, the microenvironment of the mLN promotes the induction of gut-homing markers $\alpha_4\beta_7$ and CCR9 that allow for efficient T cell migration to intestinal tissues. Here, APCs continue to present the respective antigen and allow for the expansion and final differentiation of pTregs and T effector cells. The gut-tropic finalization of differentiation is strongly influenced by the microenvironment, including IL-10 provided by CX3CR1$^+$ macrophages and microbiota-derived SCFAs and Ahr ligands. *APC* antigen-presenting cell, *DC* dendritic cell, *LN* lymph node, *pTreg* peripherally induced Treg, *mLN* mesenteric LN, *SCFA* short-chain fatty acids, *HEV* high endothelial venule

pTreg generation, thereby contributing to the maintenance of intestinal homeostasis and establishment of oral tolerance against food-borne and microbiota-derived antigens. Under inflammatory conditions, however, APCs alter their phenotype to promote effector T cell responses. An interlocking system of unique APCs, immunomodulatory features of mLN stromal cells, and soluble immunoregulatory factors achieves the balance between tolerogenic and effector T cell responses within the high bacterial content environment of the intestine.

9.4.1 Antigen-Presenting Cells in T Cell Differentiation

Antigen presentation by DCs and macrophages is key for mounting immune responses. The effective presentation of antigens is required in both the draining LN and the intestinal tissues to prime and expand the pool of antigen-specific T cells. DCs are particularly required for efficient pTreg induction, as depletion of CD11c$^+$ DCs leads to decreased pTreg generation, Foxp3 expression, and Treg homeostasis (Cording et al. 2014;

Darrasse-Jeze et al. 2009). Importantly, the composition of costimulatory molecules on DCs modulates the Treg induction rate (Scott et al. 2011). For example, strong costimulation via CD28 inhibits pTreg generation (Benson et al. 2007), whereas the co-inhibitory molecule CTLA-4 has a positive effect on pTreg induction (Zheng et al. 2006). Importantly, costimulatory molecules exert their functional properties dependent on the stage of T cell differentiation, exemplified by the observation that CD30 and OX40 are not required for priming T cells in the mLN but for expansion within the LP (Nawaf et al. 2017). Furthermore, constitutive signaling via the CD27–CD70 axis attenuates induction of IL-17 and CCR6, ameliorates inflammation, and limits expansion of T cells in the intestinal LP (Coquet et al. 2013; Laouar et al. 2005). Remarkably, $CD70^{high}CD11c^{low}$ cells in the LP are triggered to produce IL-6, IL-23p19, and TGFβ by microbiota-derived ATP (Atarashi et al. 2011). These complementary observations underline the balance between microbiota-derived cues and interactions between T cells and APCs. Hence, fine-tuning of T cell responses in the intestinal tract requires modulatory interactions with different APCs, set within tissues and LNs.

Within the intestinal immune system, DCs and macrophages comprise the majority of APCs capable of priming or expanding T cells responses. Both DCs and macrophages coexpress CD11c and MHCII (Cerovic et al. 2014). To distinguish DCs from macrophages, the surface protein integrin α_E (CD103) is utilized (Merad et al. 2013), whereas high expression of CX3CR1 is typical for $F4/80^+CD64^+$ intestinal macrophages (Gross et al. 2015; Schulz et al. 2009). Based on the expression of CD103 and CD11b, intestinal DCs can be subdivided into four subsets. Within the LP and gut-draining LNs, the majority of DCs is $CD103^+$ and predominantly coexpresses CD11b, whereas only a small but functionally different $CD103^+CD11b^-CD8\alpha^+$ population is maintained (Cerovic et al. 2014; Fujimoto et al. 2011).

Generally, the maintenance of $CD103^+$ DCs with their unique tolerogenic and noninflammatory phenotype is supported by local factors. These include bile and dietary retinoids in the intestinal lumen but also vasoactive intestinal peptide (VIP), mucus glycoprotein, and prostaglandin (PG)E2 (Cerovic et al. 2014; Shan et al. 2013). Among these factors, RA is produced during vitamin A metabolism by the retinal dehydrogenase enzymes *Aldh1a2* and *Aldh1a3*, expressed at high levels in $CD103^+$ DCs (Coombes et al. 2007; Kang et al. 2007; Raverdeau and Mills 2014). RA can reduce cytokine production by effector T cells, finally resulting in elevated pTreg generation (Hill et al. 2008). Furthermore, RA can directly balance differentiation between Tregs and Th17 cells, by potentiating TGF-β-induced pTreg generation while antagonizing IL-6-driven Th17 differentiation (Mucida et al. 2007). In contrast to $CD103^+$ DCs, $CD103^-$ DC subsets have been shown to be mainly involved in the priming of Th1/Th17 cells (Cerovic et al. 2013). The non-tolerogenic potential of $CD103^-$ DCs is underlined by their ability to secrete IL-23, which promotes differentiation of Th17 cells, but also IL-22 production by ILCs (Guo et al. 2014; Kinnebrew et al. 2012). Furthermore, IL-22 production by different Th and ILC subsets contributes to contain microbiota by enforcing the epithelial cell barrier and elevating the production of mucus and antimicrobial peptides (Parks et al. 2015).

Generally, gut-resident macrophages are not able to instigate priming of naïve T cells as stand-alone. Particularly, $CD11b^+F4/80^+CD11c^-CX3CR1^+$ macrophages of the small intestinal LP promote the expansion of LN-primed Tregs by secreting high levels of IL-10 (Fig. 9.2) (Hadis et al. 2011; Murai et al. 2009). Nonetheless, depletion of $CD64^+$ macrophages, predominantly expressing CX3CR1, results in the abrogation of Th17 cell differentiation in the context of mice colonized with segmented filamentous bacteria (SFB) (Panea et al. 2015), indicating that intestinal macrophages are required to expand both pTreg and effector T cell responses. Under inflammatory conditions, $Ly6C^+$ monocytes rapidly accumulate in the intestinal LP (Bain and Mowat 2014). The incoming monocytes do not fully mature, remain highly responsive to TLR stimulation, and attain a pro-inflammatory cytokine profile including TNFα, IL-6, IL-12, and IL-23 secretion (Weber et al. 2011; Zigmond et al. 2012).

Thus, although intestinal macrophages have a diminished pro-inflammatory response capacity, they modulate T cell differentiation subsequent to priming in the gut-draining LNs appropriately to the current state of the intestinal environment.

9.4.2 Lymph Node Stromal Cells Shape T Cell Immune Responses

The LN stromal cells provide the lattice for interaction between naïve and effector T cells and incoming DCs that present the cognate antigen. These mesenchymal cells are predominantly known for their infrastructural functions. Lymphatic endothelial cells (LECs) establish a network of vessels transporting lymph from the tissue to and around the LN. Blood endothelial cells (BECs) make up the high endothelial venules (HEVs) that intersperse the LN and serve as the major entry gate for circulating lymphocytes within. The core of the LN consists of follicular dendritic cells (FDCs) and fibroblastic stromal cells (FSCs), dominating in the B cell follicles and T cell zones, respectively (Bajenoff 2012; Fletcher et al. 2015). LN stromal cells of the mesenteries are subject to the multitude of microbiota-derived signals, starting early during development, and vastly differ in their functional properties between skin- and gut-draining sites. These tissue-specific features define foci of T cell differentiation (Cording et al. 2014; Malhotra et al. 2012).

9.4.2.1 FSCs and Their Modulatory Function on T Cell Responses

FSCs, the dominant LN stromal cells within the T cell zone, contribute a vast battery of immune modulatory functions, which range from shaping T cell expansion to modulating peripheral tolerance (Baptista et al. 2014; Fletcher et al. 2010; Khan et al. 2011; Lukacs-Kornek et al. 2011; Siegert et al. 2011). FSCs are not able to prime T cell responses by themselves but express MHCII at low levels at homeostatic conditions. Upon inflammation, FSCs upregulate MHCII expression in an IFNγ-dependent manner (Abe et al. 2014; Dubrot et al. 2014). Furthermore, FSCs (and also LECs) can acquire MHCII complex originating from migratory DCs in a cell contact-dependent manner (Dubrot et al. 2014). Thus, FSCs have the ability to express and present self-antigens, potentially filtered from the conduits or directly processed by DCs from the periphery, and interact with cognate T cells in order to modulate T cell responses. Importantly, FSCs sheath the tube-like conduit system that is interspersing the LN cortex (Gretz et al. 2000; Sixt et al. 2005). Thus, FSCs are in direct contact to lymph-borne particles like chemokines, peptides, cytokines, and small metabolites derived from the tissue and microbiota. The content of the conduit network is sampled by DCs adjoined to the FSCs and could serve as a LN internal system to modulate T cell differentiation in the context of lymph-borne antigens originating from the intestinal tissue independent of migratory DCs (Roozendaal et al. 2009).

9.4.2.2 Location and Microbiota Define Lymph Node Stromal Cells to Impinge on T Cell Differentiation

In contrast to DCs, which have a short life expectancy after arriving at the LN (Cerovic et al. 2013; Kamath et al. 2002), stromal cells have a low turnover under homeostatic conditions and can provide a constant microenvironmental framework within LNs (Hammerschmidt et al. 2008; Molenaar et al. 2009). Importantly, DCs can be influenced by the LN environment and even change tissue-derived priming (Dudda et al. 2005; Hammerschmidt et al. 2008).

To dissect the immunological impact of the tissue-derived hematopoietic and LN-local stromal cell compartment, LN transplantation experiments can be utilized (Wolvers et al. 1999). Surgical resection of endogenous intestinal or skin-draining LN allows for the engraftment of transplanted LNs (Ahrendt et al. 2008). During engraftment, the hematopoietic compartment is replaced by migratory cells from the draining tissue. LN-resident cells, predominantly stromal cells of donor origin, are retained (Hammerschmidt et al. 2008).

Remarkably, transplanted LN stromal cells maintain immunomodulatory properties from their original site, best exemplified by the upregulation of $\alpha_4\beta_7$ on adoptively transferred ovalbumin (OVA)-specific, TCR-transgenic (OTII) naïve T cells in mLN transplanted into the skin-draining popliteal fossa upon immunization with OVA (Molenaar et al. 2009). Thus, stromal cells contribute to modulate T cell differentiation in a location-specific manner and can stably retain these properties upon transplantation. These findings implicate LN stromal cells as stable immunomodulatory elements in the context of the drained tissue.

It is evident that each tissue demands and shapes its own adaptive immune response tailored to its environmental stimuli (Matzinger and Kamala 2011). Hence, LN stromal cells, as the permanent infrastructural component in LNs, are likely modulated by their respective environment and in return impinge on tissue-specific immune responses. For example, FSCs from mLN express high levels of *Aldh1a2* and *Aldh1a3* (Hammerschmidt et al. 2008; Malhotra et al. 2012; Molenaar et al. 2011) and together with migratory CD103$^+$ DCs contribute to the high RA levels found in mLN. Importantly, FSCs from mLN produce substantially less IL-6 as compared to pLN (Malhotra et al. 2012) and potentially limit Th17 differentiation and upregulation of CCR6 (Pezoldt and Huehn 2016). Hence, FSCs significantly contribute to create a specific environment within LNs and shape effector T cell differentiation as well as pTreg induction.

To dissect the impact of LN stromal cells on tolerogenic T cell responses, we performed LN transplantation experiments in both gut- and skin-draining sites to investigate the role of LN stromal cells outside their endogenous position. We focused on the modulation of de novo Treg induction to dissect the impact of LN stromal cells to shape tolerogenic T cell responses. Remarkably, transplantation and engraftment of cLN and mLN into the skin-draining non-tolerogenic popliteal fossa retained an environment favoring Treg induction from OVA-specific, TCR-transgenic (DO11.10) naïve T cells upon i.v. injection of OVA peptide (Cording et al. 2014). These data suggest that LNs stably retain their distinct Treg-inducing capacities, a feature maintained by the stromal cell compartment.

Especially, gut-draining LNs, including the cLN and mLN, are exposed to specific tolerogenic environmental and microbiota-derived factors. The liver-draining cLN is subject to high levels of vitamin A (Winau et al. 2008), whereas the mLN is exposed to the intestinal microbiota. For both cLNs from vitamin A-deficient and mLNs from germ-free (GF) mice transplanted to the popliteal fossa, a reduced Treg-inducing capacity was observed as compared to cLN from normally fed or mLN from specific pathogen-free (SPF) mice, respectively (Cording et al. 2014). Thus, the functional stabilization of the tolerogenic phenotypes of gut-draining LNs is dependent on respective tissue-specific environmental cues. These findings suggest that LN stromal cells stably integrate tissue-specific signals and allow them to permanently impinge on T cell differentiation (Cording et al. 2014). Irrespective of the importance of resident LN stromal cells, DCs are indispensable for pTreg generation and are known to abundantly interact with FSCs (Acton et al. 2014; Cording et al. 2014; Gerner et al. 2012; Radtke et al. 2015). One could conclude that FSCs tweak incoming DCs toward an imprinted homeostatic condition to influence subsequent T cell differentiation. As FSCs are fixed location-wise and can "memorize" functional properties in a tissue-specific manner, they could contribute to shape antigen-specific T cell responses under steady-state conditions, thereby promoting induction of tolerance toward microbiota and food-borne antigens.

9.5 Microbiota and Their Metabolites Shape T Cell Differentiation

LNs are subjected to their surrounding local environment including tissue-dependent composition of the microbiota. Over the length of the intestine, the concentration of dietary antigens and the composition and density of the microbiota change. Hence, the different conditions along the intestinal tract require regionalized immune responses

(Mowat and Agace 2014). The dominant factor underlying antigenic content, metabolites, and pro-inflammatory mediators is dictated by the microbiota composition, whose alterations are a decisive factor for the development of intestinal inflammatory disorders, autoimmune diseases, allergy, and cancer (Belkaid and Hand 2014; Kosiewicz et al. 2011; Round and Mazmanian 2009). The increasing prevalence of these diseases emphasizes the importance of microbiota in shaping lastingly different adaptive immune responses. The key feature of a successful symbiotic relationship between the host and the colonizing bacteria is the establishment of a permanent but adaptable balance between protective effector responses strengthening the intestinal barrier and tolerogenic responses that limit the transgression of effector responses.

9.5.1 Intestinal Microbiota-Derived Metabolites Shape Effector T Cell and pTreg Balance

Microbiota are required to establish both tolerogenic pTreg and protective T cell effector responses. The requirement of intestinal colonization for balanced T cell responses was first indicated by several studies dissecting T cell composition in GF mice. These mice harbor less Helios^{-} Tregs, indicative of a reduced proportion of pTregs, particularly in the colon (Atarashi et al. 2011). Additionally, IL-17 and IFNγ production is compromised in GF mice, whereas the proportion of Th2 cells is increased (Hall et al. 2008; Ivanov et al. 2009; Mazmanian et al. 2005). Upon colonization, the proportion of pTregs rises, nonpathogenic Th17/Th1 responses equilibrate, and the proportion of Th2 cells ameliorates (Atarashi et al. 2011; Hall et al. 2008; Ivanov et al. 2009; Mazmanian et al. 2005). Further studies have highlighted the importance of particular bacteria dominantly located in the large intestine to influence T cell responses. These include *Bacteroides fragilis* expressing polysaccharide A, which promotes the expansion of IL-10 producing CD4^{+} T cells by limiting Th17 differentiation (Mazmanian et al. 2008). Additionally, different *Clostridiales* strains (*Clostridium* cluster XIV) promote pTreg expansion in the colon, attributed to short-chain fatty acids (SCFAs), including acetate, propionate, and butyrate (Arpaia et al. 2013; Furusawa et al. 2013; Smith et al. 2013). Furthermore, high concentrations of intestinal microbiota producing SCFAs are reached in the cecum and colon, where fermenting bacteria of the *Bacteroidetes* phylum produce SCFA during the breakdown of dietary fibers (Fig. 9.2) (Furusawa et al. 2013; Kuhn and Stappenbeck 2013; Mowat and Agace 2014).

SCFAs can shape T cell differentiation at multiple levels, including modulation of intestinal DCs, regulation of Treg migration, and instigating epigenetic modifications of T cells. The SCFA butyrate binds to Gpr109a, expressed by several colonic immune cells. DCs of Gpr109a-deficient mice show elevated levels of IL-17 and IL-6, reduced IL-10 and Aldh1a1, and concomitantly reduced Foxp3^{+} Treg frequencies in the colonic LP (Singh et al. 2014). The SCFA propionate can directly modulate the homing properties of Tregs, by binding to Gpr43, and upregulate chemoattractant receptor Gpr15 (Schaper et al. 2014; Smith et al. 2013), which is required for migration of Tregs to the colonic LP (Kim et al. 2013b). Finally, several studies have implicated the role of SCFAs in stabilizing the Treg lineage (Kamada et al. 2013). Consequently, these SCFAs provide an environment to allow imprinting of T cell differentiation depending on the current immune response required. In the context of infections and inflammations, effector T cell responses are impinged, whereas under homeostatic conditions, pTregs are supported (Kim et al. 2013a). This is best exemplified by the observation that under steady-state conditions, permissive histone acetylation within CNS-1 and CNS-3 of the *Foxp3* locus is elevated under SCFA treatment or colonization with SCFA-favoring microbiota, promoting increased Foxp3 expression, whereas other Th cell lineage specification factors like RORγt, T-bet, and GATA3 are not affected (Arpaia et al. 2013; Furusawa et al. 2013).

Although several microbiota-derived constituents have been shown to modulate T cell differentiation, stabilization, and expansion, there are no concise studies that distinguish the function of microbiota-derived metabolites with regard to the priming of the immune response in the gut-draining LNs and expansion of pre-differentiated T cells within the LP. A defined understanding of these interlinked processes and how they are influenced by microbiota and their constituents is required to precisely shape intestinal immune responses.

9.6 Treg Subsets Control Defined Effector T Cell Responses

9.6.1 The Myriads of Treg Subsets in the Intestinal Environment

9.6.1.1 Contribution of tTregs and pTregs in Gut

While Tregs are mainly generated within the thymus, the so-called tTregs, peripherally induced Foxp3^{+} pTregs complement the TCR repertoire to confer tolerance toward foreign antigens. Shortage of either tTregs or pTregs has been demonstrated to be detrimental to the host (Huehn et al. 2009).

The abundance of microbiota and food antigens in the intestine favors pTregs generation and maintenance. This is indicated by the low frequency of Helios and/or Neuropilin-1 (Nrp1) expressing Tregs in the large intestine. Thus, a major fraction of the Treg pool within the colonic tissue shows features of pTregs. However, tTregs seem to contribute to colonic Treg composition as the TCR repertoire is highly similar between thymus and colon-derived Tregs (Cebula et al. 2013), indicating that tTregs are a substantial source of intestinal Tregs. Neither Helios nor Nrp1 are reliable markers to distinguish tTregs from pTregs, particularly under inflammatory conditions (Gottschalk et al. 2012). Regardless, under homeostatic conditions, the level of complexity of microbial colonization defines the proportion of pTregs within the colonic tissue, where more complex microbial colonization resulted in higher frequencies of Helios^{-} Tregs, ergo pTregs (Yang et al. 2016). The peculiar requisites for pTreg generation are embedded within the locus of the *Foxp3* gene. The conserved noncoding sequence 1 (CNS-1) requires TGFβ/SMAD signaling and is thus crucial for pTreg generation but dispensable for tTreg development (Tone et al. 2008). BAC-transgenic ΔCNS-1 reporter mice allow to distinguish tTregs and pTregs based on CNS-1 requirement (Zhang et al. 2017). Own unpublished observations show that CNS-1-dependent Tregs are enriched in the colon, indicating that pTregs are the major Treg population in the intestine. If further considering that CNS-1-deficient mice specifically develop autoimmune diseases in the intestine (Josefowicz et al. 2012b) and that pTregs are required to prevent colitis (Haribhai et al. 2011), pTregs are highly relevant in maintaining the immunological balance toward foreign antigens within the intestine.

9.6.1.2 Functionality of Tregs Along the Intestine

GF mice and mice under antigen-free diets show strongly reduced Treg frequencies as compared to SPF-housed mice (Kim et al. 2016). These observations suggest that in addition to microbiota-derived antigens, food-derived antigens are another important trigger for establishing and maintaining Tregs in the intestine. Along the intestine, the composition and ratio of Tregs to effector T cells is increasing from the small intestine to the colon. This higher proportion of Tregs is due to elevated TGFβ and RA levels, which promote the generation of pTregs (Lathrop et al. 2011). Importantly, also, Tregs upregulate $\alpha_4\beta_7$ integrin, CCR9, and GPR15 when encountering high levels of RA and microbial metabolites, which allow for efficient homing to the intestine (Coombes et al. 2007; Siewert et al. 2007; Smith et al. 2013; Sun et al. 2007). Remarkably, tTregs migrate to the intestine early during ontogeny and are maintained in a niche, independent of IL-2 and MHCII antigen presentation (Korn et al. 2014; Torow et al. 2015), indicating that tTregs purposefully contribute to the suppressive microenvironment independent of foreign antigen recognition. The pTreg proportion, as indicated by lack of Helios expression, is affected by different environmental factors. While the availability of food-derived antigen defines pTreg content in the small intestine, microbial colonization dictates pTreg levels in the colon (Kim et al. 2016). Importantly, Helios$^+$ tTregs are maintained independent of antigenic diet content or colonization circumstance (Kim et al. 2016; Sefik et al. 2015). Thus, Treg distribution is altered along the intestine and depends on the environment provided by the host.

9.6.2 Treg Subsets Within the Intestinal Environment

9.6.2.1 Influence of Microbiota on Intestinal Treg Subsets

Beyond Treg segregation based on their origin as pTregs and tTregs, the expression of Th lineage specification factors in addition to Foxp3 is utilized to distinguish Treg subsets. Historically, studies have relied on key transcription factors identified in effector T cell populations. Among the intestinal Treg population, two transcription factors, namely, GATA3 and RORγt, are used to distinguish in total three dominant subpopulations. While intestinal Helios$^+$ Tregs also express GATA3, particularly, intestinal microbiota drive the expression of RORγt (Ohnmacht et al. 2015; Sefik et al. 2015; Yang et al. 2016).

GATA3$^+$ Tregs comprise up to 20% of the Treg pool within intestinal tissues, and the coexpression of Helios is indicative of their thymic origin (Wohlfert et al. 2011). Importantly, GATA3 expression by Tregs relies on IL-33 receptor signaling, and intestinal epithelial cell produced IL-33 (Schiering et al. 2014). Furthermore, GATA3$^+$ Tregs are present in GF mice and thus likely are independent of microbiota-derived cues (Sefik et al. 2015). The proportion of RORγt$^-$Helios$^-$Tregs, likely a pTreg-derived subpopulation, decreases from the small to large intestine, indicating that food-derived antigenic material might be required to maintain this Treg subset (Kim et al. 2016).

The dominant fraction of Tregs, with 35% in the small intestine and 65% in the colon, is RORγt$^+$Helios$^-$Tregs (Fig. 9.3A) (Ohnmacht et al. 2015; Sefik et al. 2015; Yang et al. 2016). Albeit expressing the hallmark Th17 lineage specification factor RORγt, these Tregs barely secrete IL-17 but on the contrary produce high levels of IL-10 (Ohnmacht et al. 2015). Interestingly, this Treg population is absent in GF or antibiotic-treated mice (Ohnmacht et al. 2015; Sefik et al. 2015). Thus, the extent of RORγt$^+$ Tregs in the intestinal tissues is highly dependent on the colonizing bacteria, emphasizing that RORγt$^+$ Helios$^-$Tregs are indeed microbiota-dependent. Strikingly, colonization with SFB, a bona fide RORγt$^+$ Th17-inducing bacterial strain in the intestine, is not the dominant stimulus for the generation and/or maintenance of RORγt$^+$Helios$^-$Tregs. Among others, *Clostridium ramosum* also promotes the proportion of RORγt$^+$ Tregs (Sefik et al. 2015). Importantly, RORγt$^+$ Tregs represent a stable population due to the consistently demethylated TSDR and their capacity to suppress intestinal inflammations and pathogenic Th2 responses (Ohnmacht et al. 2015; Yang et al. 2016). Furthermore, lack of RORγt$^+$ Tregs in the intestine correlates with exacerbated colitis induced by Th1-/Th17-mediated responses and is implicated in different autoimmune diseases and oncogenesis (Chellappa et al. 2016; Kluger et al. 2014, 2016; Tartar et al. 2010; Yang et al. 2016). Despite this accumulating evidence for a unique role of RORγt$^+$ Tregs in intestinal immune responses, their precise origin remains elusive.

9.6.2.2 Origin and Migration of RORγt$^+$ Tregs

Several key features required for the development of functional RORγt$^+$ Tregs have been dissected. Firstly, the canonical Th17-polarizing IL-6-/IL-23 signaling pathway is essential for inducing RORγt$^+$ Treg in vivo (Ohnmacht et al. 2015; Sefik et al. 2015). Induction of RORγt$^+$ Tregs is further supported by intestinal CX3CR1$^+$ APCs that integrate microbiota-derived stimuli in a myeloid differentiation primary response 88 (MyD88)-independent manner, while CD103$^+$ DCs can suppress induction via the CD40-CD40L pathway (Barthels et al. 2017; Solomon and Hsieh 2016).

Currently, RORγt$^+$ Tregs are viewed as being solely pTreg-derived, underlined by the finding that adoptively transferred clonal naïve CD4$^+$ T cells

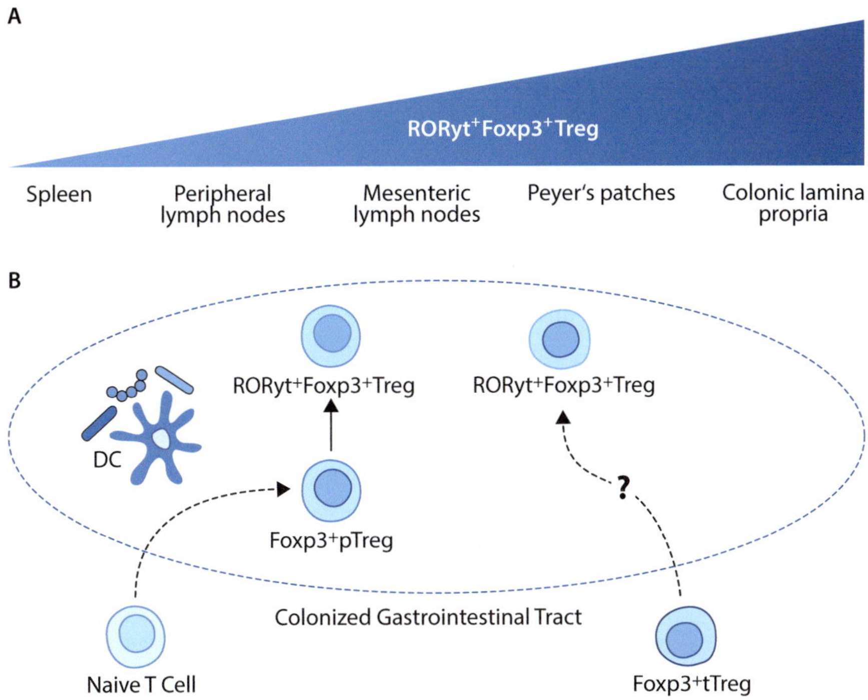

Fig. 9.3 RORγt⁺ Treg distribution and development. (A) The proportion of RORγt⁺ Tregs is influenced by the "proximity" to the intestine. The highest proportion is observed in colonic lamina propria and lowest in spleen. (B) Naïve CD4⁺ T cells differentiate into Foxp3⁺ pTregs and subsequently upregulate RORγt, a process that is promoted by microbiota-dependent cues and relies on DCs in the gastrointestinal tract. Whether Foxp3⁺ tTregs also contribute to the population of RORγt⁺ Tregs in the gastrointestinal tract is currently unclear

preferentially differentiate into Foxp3⁺RORγt⁺ cells (Solomon and Hsieh 2016). Nonetheless, considering the complexity of the intestinal ecosystem and the flexibility of tTregs to respond via upregulation of GATA3 to the proximal small intestinal environment, a functional involvement of RORγt⁺ tTregs is likely. The contradictory functionalities of RORγt⁺ Tregs range from fully suppressive in colon to partly pro-inflammatory in crescentic glomerulonephritis and pancreatic cancer (Chellappa et al. 2016; Kluger et al. 2014, 2016; Ohnmacht et al. 2015; Sefik et al. 2015; Tartar et al. 2010). Thus, RORγt⁺ Tregs might consist of varying subpopulations originating from different Treg progenitors (Fig. 9.3B). This notion is supported by the identification of two distinct RORγt⁺ Foxp3⁺ populations in humans (Halim et al. 2017). Own preliminary observations have revealed that tTregs are fully capable of upregulating RORγt both in vivo and in vitro. To derive a better perception of the modulatory composition of Treg functionality along the intestine, a more detailed understanding of Treg heterogeneity and the tTreg contribution to the pool of intestinal Tregs is required.

9.7 Outlook

Although the barrier functions of the intestine are set up to minimize the direct contact between bacteria and components of the immune system, a vast variety of T cell responses is initiated. Due to the necessary permeability of the system, essential for the absorption of nutrients, the immune system has to maintain a balance between effector T cell and tolerogenic Treg responses.

As the intestine hosts the largest compartment of T cells throughout the body, this clearly implicates that this energy-wise high burden for the host is required to keep the bacterial luminal content at bay without initiating pathogenic inflammations. The divergent variety of the T cell composition along the intestinal tract is an additional clear indicator that the intestinal immune system is adapted to the diverse environmental settings. Failure to provide an environment of intestinal balance results in localized IBDs affecting different parts of the intestine as observed for UC and CD with differently skewed T cell responses.

Thus, a better understanding of T cell differentiation and subsequent functionality along the complete stretch of intestinal segments is necessary to precisely modulate adaptive immune responses in the face of pathogenic inflammations driven by intestinal microbiota.

▶ Controversy

Tregs are essential to maintain intestinal tolerance to self-, food-, and microbiota-derived antigens. Aside from Foxp3-dependent epigenetic characteristics allowing for Treg lineage commitment and maintenance of suppressive functions, the intestinal microenvironment plays a central role for accumulating and differentiating Tregs. It is well known that a substantial proportion of Tregs in the intestine is derived via priming by CD103^{+} DC in mLN and subsequent migration to the intestinal tissues (Hadis et al. 2011). Whether the remaining colonic RORγt^{+} Tregs are also firstly educated in the mLNs or locally primed in the colonic LP is currently unknown. Several groups have studied a unique subpopulation of Tregs that coexpresses RORγt together with Foxp3 in the intestine of colonized adult mice (Ohnmacht et al. 2015; Sefik et al. 2015; Yang et al. 2016). The majority of studies on the origin of RORγt^{+} Tregs delineate them as being of pTreg origin, based on the requirement of microbiota for their generation and the lack of tTreg markers such as Helios. Importantly, two studies have assessed the TCR repertoire of colonic Tregs. While Lathrop et al. could show that the TCR repertoire of colonic Tregs vastly differs from tTregs (Lathrop et al. 2011), Cebula and colleagues observed, using TCR repertoire profiling on the single-cell level, that tTreg dominantly mediates tolerance to microbiota-produced antigens (Cebula et al. 2013). Despite the clear impact of microbiota on the proportion of RORγt^{+} Tregs in the intestine, it is currently unclear whether tTregs can acquire RORγt expression. Furthermore, Foxp3^{+}RORγt^{+} T cells have been shown to be both beneficial and detrimental to the host in a variety of different infection, inflammation, and infection models.

The delineation of the origin of Tregs in intestinal tissues and their contribution to peripheral tolerance are essential to enforce the establishment of homeostasis in IBDs and develop effective vaccination strategies.

History

From a historical perspective, the involvement of T cells in the context of intestinal immune regulation is novel. Only the advent of studying inflammatory bowel diseases (IBD), including Crohn's disease (CD) and ulcerative colitis (UC), directed the study of mucosal pathogenic alterations toward infection-independent disorders (Crohn et al. 1932). Although both UC and CD were idiopathic inflammations of the intestine lacking a direct bridge to a pathogen, numerous infection associations were established, ranging from *Pseudomonas aeruginosa* to *Chlamydia* and even different viruses (Mestecky et al. 2015). The first studies to link IBDs with microbiota were performed in the 1960s and 1970s using animal models inoculating rabbits with *E. coli*, a member of the intestinal microbiota together with Freud's adjuvants, establishing IBD-like syndromes (Halpern et al. 1967; Mee et al. 1979). Halpern and colleagues could also show that pre-immunization with *E. coli* prevented the development of IBDs (Halpern et al. 1967). From today's perspective, these findings point toward mucosal tolerance, a hitherto unresolved question. In 1978, the utilization of dinitrochlorobenzene for the induction of IBD and the inflammation ameliorating effect of the antibiotic metronidazole provided first clear evidence that microbiota contribute to the development of IBDs (Onderdonk et al. 1978). The concept that cytotoxic cells were in part responsible for IBD pathologies (Perlmann and Broberger 1963; Shorter et al. 1970) led to the first studies that T cells were dominant drivers of IBDs and that "suppressor" cell activity was reduced in IBDs (Elson et al. 1981; Hodgson et al. 1978). Importantly, effector T cells were identified to be key drivers of IBDs in the context of an adoptive transfer model of lymphopenic hosts (Powrie et al. 1993). The adoptive transfer of T cells with regulatory potential, originally identified as $CD4^{+}CD45RB^{low}$, and their protective role in different IBD models (Powrie et al. 1993, 1994) set the stage to study the cell population now known as regulatory T cells (Tregs).

Highlights

- Gut-draining lymph node stromal cells contribute to shape T cell priming.
- Microbial colonization is required to stabilize tolerogenic properties within gut-draining lymph node stromal cells.
- Microbial metabolites shape T cell expansion and differentiation in intestinal tissues.
- Intestinal homeostasis requires both pTregs and tTregs.

References

Abbas, A. K., Murphy, K. M., & Sher, A. (1996). Functional diversity of helper T lymphocytes. *Nature, 383*, 787–793.

Abe, J., Shichino, S., Ueha, S., Hashimoto, S., Tomura, M., Inagaki, Y., Stein, J. V., & Matsushima, K. (2014). Lymph node stromal cells negatively regulate antigen-specific $CD4^{+}$ T cell responses. *Journal of Immunology, 193*, 1636–1644.

Acton, S. E., Farrugia, A. J., Astarita, J. L., Mourao-Sa, D., Jenkins, R. P., Nye, E., Hooper, S., van Blijswijk, J., Rogers, N. C., Snelgrove, K. J., et al. (2014). Dendritic cells control fibroblastic reticular network tension and lymph node expansion. *Nature, 514*, 498–502.

Agace, W. W. (2006). Tissue-tropic effector T cells: Generation and targeting opportunities. *Nature Reviews Immunology, 6*, 682–692.

Ahrendt, M., Hammerschmidt, S. I., Pabst, O., Pabst, R., & Bode, U. (2008). Stromal cells confer lymph node-specific properties by shaping a unique microenvironment influencing local immune responses. *Journal of Immunology, 181*, 1898–1907.

Allen, J. E., & Maizels, R. M. (2011). Diversity and dialogue in immunity to helminths. *Nature Reviews Immunology, 11*, 375–388.

Amsen, D., Spilianakis, C. G., & Flavell, R. A. (2009). How are T(H)1 and T(H)2 effector cells made? *Current Opinion in Immunology, 21*, 153–160.

Arpaia, N., Campbell, C., Fan, X., Dikiy, S., van der Veeken, J., deRoos, P., Liu, H., Cross, J. R., Pfeffer, K., Coffer, P. J., & Rudensky, A. Y. (2013). Metabolites produced by commensal bacteria promote peripheral regulatory T-cell generation. *Nature, 504*, 451–455.

Atarashi, K., Tanoue, T., Shima, T., Imaoka, A., Kuwahara, T., Momose, Y., Cheng, G., Yamasaki, S., Saito, T., Ohba, Y., et al. (2011). Induction of colonic regulatory T cells by indigenous Clostridium species. *Science, 331*, 337–341.

Atarashi, K., Tanoue, T., Oshima, K., Suda, W., Nagano, Y., Nishikawa, H., Fukuda, S., Saito, T., Narushima, S., Hase, K., et al. (2013). Treg induction by a rationally selected mixture of Clostridia strains from the human microbiota. *Nature, 500*, 232–236.

Bach, J. F. (2003). Regulatory T cells under scrutiny. *Nature Reviews Immunology, 3*, 189–198.

Bain, C. C., & Mowat, A. M. (2014). Macrophages in intestinal homeostasis and inflammation. *Immunological Reviews, 260*, 102–117.

Bajenoff, M. (2012). Stromal cells control soluble material and cellular transport in lymph nodes. *Frontiers in Immunology, 3*, 304.

Banchereau, J., Pascual, V., & O'Garra, A. (2012). From IL-2 to IL-37: The expanding spectrum of anti-inflammatory cytokines. *Nature Immunology, 13*, 925–931.

Baptista, A. P., Roozendaal, R., Reijmers, R. M., Koning, J. J., Unger, W. W., Greuter, M., Keuning, E. D., Molenaar, R., Goverse, G., Sneeboer, M. M., et al. (2014). Lymph node stromal cells constrain immunity via MHC class II self-antigen presentation. *eLife, 3*, e04433.

Barthels, C., Ogrinc, A., Steyer, V., Meier, S., Simon, F., Wimmer, M., Blutke, A., Straub, T., Zimber-Strobl, U., Lutgens, E., et al. (2017). CD40-signalling abrogates induction of $ROR\gamma t^{+}$ Treg cells by intestinal $CD103^{+}$ DCs and causes fatal colitis. *Nature Communications, 8*, 14715.

Belkaid, Y., & Hand, T. W. (2014). Role of the microbiota in immunity and inflammation. *Cell, 157*, 121–141.

Benson, M. J., Pino-Lagos, K., Rosemblatt, M., & Noelle, R. J. (2007). All-trans retinoic acid mediates enhanced T reg cell growth, differentiation, and gut homing in the face of high levels of co-stimulation. *The Journal of Experimental Medicine, 204*, 1765–1774.

Blander, J. M., Torchinsky, M. B., & Campisi, L. (2012). Revisiting the old link between infection and autoimmune disease with commensals and T helper 17 cells. *Immunologic Research, 54*, 50–68.

Bopp, T., Becker, C., Klein, M., Klein-Hessling, S., Palmetshofer, A., Serfling, E., Heib, V., Becker, M., Kubach, J., Schmitt, S., et al. (2007). Cyclic adenosine monophosphate is a key component of regulatory T cell-mediated suppression. *The Journal of Experimental Medicine, 204*, 1303–1310.

Breese, E., Braegger, C. P., Corrigan, C. J., Walker-Smith, J. A., & Macdonald, T. T. (1993). Interleukin-2- and interferon-γ-secreting T cells in normal and diseased human intestinal mucosa. *Immunology, 78*, 5.

Cao, X., Cai, S. F., Fehniger, T. A., Song, J., Collins, L. I., Piwnica-Worms, D. R., & Ley, T. J. (2007). Granzyme B and perforin are important for regulatory T cell-mediated suppression of tumor clearance. *Immunity, 27*, 635–646.

Cebula, A., Seweryn, M., Rempala, G. A., Pabla, S. S., McIndoe, R. A., Denning, T. L., Bry, L., Kraj, P., Kisielow, P., & Ignatowicz, L. (2013). Thymus-derived regulatory T cells contribute to tolerance to commensal microbiota. *Nature, 497*, 258–262.

Cerovic, V., Houston, S. A., Scott, C. L., Aumeunier, A., Yrlid, U., Mowat, A. M., & Milling, S. W. (2013). Intestinal $CD103^-$ dendritic cells migrate in lymph and prime effector T cells. *Mucosal Immunology, 6*, 104–113.

Cerovic, V., Bain, C. C., Mowat, A. M., & Milling, S. W. (2014). Intestinal macrophages and dendritic cells: What's the difference? *Trends in Immunology, 35*, 270–277.

Chellappa, S., Hugenschmidt, H., Hagness, M., Line, P. D., Labori, K. J., Wiedswang, G., Tasken, K., & Aandahl, E. M. (2016). Regulatory T cells that co-express RORγt and FOXP3 are pro-inflammatory and immunosuppressive and expand in human pancreatic cancer. *Oncoimmunology, 5*, e1102828.

Ciofani, M., & Zuniga-Pflucker, J. C. (2010). Determining gd versus ab T cell development. *Nature Reviews Immunology, 10*, 657–663.

Coombes, J. L., Siddiqui, K. R., Arancibia-Carcamo, C. V., Hall, J., Sun, C. M., Belkaid, Y., & Powrie, F. (2007). A functionally specialized population of mucosal $CD103^+$ DCs induces $Foxp3^+$ regulatory T cells via a TGFb and retinoic acid-dependent mechanism. *The Journal of Experimental Medicine, 204*, 1757–1764.

Coquet, J. M., Middendorp, S., van der Horst, G., Kind, J., Veraar, E. A., Xiao, Y., Jacobs, H., & Borst, J. (2013). The CD27 and CD70 costimulatory pathway inhibits effector function of T helper 17 cells and attenuates associated autoimmunity. *Immunity, 38*, 53–65.

Cording, S., Wahl, B., Kulkarni, D., Chopra, H., Pezoldt, J., Buettner, M., Dummer, A., Hadis, U., Heimesaat, M., Bereswill, S., et al. (2014). The intestinal microenvironment imprints stromal cells to promote efficient Treg induction in gut-draining lymph nodes. *Mucosal Immunology, 7*, 359–368.

Cretney, E., Kallies, A., & Nutt, S. L. (2013). Differentiation and function of $Foxp3^+$ effector regulatory T cells. *Trends in Immunology, 34*, 74–80.

Crohn, B. B., Ginzburg, L., & Oppenheimer, G. D. (1932). Regional ileitis: A pathologic and clinical entity. *Journal of the American Medical Association, 99*, 1323–1329.

Cui, G., Zhang, Y., Gong, Z., Zhang, J. Z., & Zang, Y. Q. (2009). Induction of $CD4^+CD25^+Foxp3^+$ regulatory T cell response by glatiramer acetate in type 1 diabetes. *Cell Research, 19*, 574–583.

Dambacher, J., Beigel, F., Zitzmann, K., De Toni, E. N., Goke, B., Diepolder, H. M., Auernhammer, C. J., & Brand, S. (2009). The role of the novel Th17 cytokine IL-26 in intestinal inflammation. *Gut, 58*, 1207–1217.

Darrasse-Jeze, G., Deroubaix, S., Mouquet, H., Victora, G. D., Eisenreich, T., Yao, K. H., Masilamani, R. F., Dustin, M. L., Rudensky, A., Liu, K., & Nussenzweig, M. C. (2009). Feedback control of regulatory T cell homeostasis by dendritic cells in vivo. *The Journal of Experimental Medicine, 206*, 1853–1862.

Delacher, M., Imbusch, C. D., Weichenhan, D., Breiling, A., Hotz-Wagenblatt, A., Trager, U., Hofer, A. C., Kagebein, D., Wang, Q., Frauhammer, F., et al. (2017). Genome-wide DNA-methylation landscape defines specialization of regulatory T cells in tissues. *Nature Immunology, 18*(10), 1160–1172.

Dubrot, J., Duraes, F. V., Potin, L., Capotosti, F., Brighouse, D., Suter, T., LeibundGut-Landmann, S., Garbi, N., Reith, W., Swartz, M. A., & Hugues, S. (2014). Lymph node stromal cells acquire peptide-MHCII complexes from dendritic cells and induce antigen-specific $CD4^+$ T cell tolerance. *The Journal of Experimental Medicine, 211*, 1153–1166.

Dudda, J. C., Lembo, A., Bachtanian, E., Huehn, J., Siewert, C., Hamann, A., Kremmer, E., Forster, R., & Martin, S. F. (2005). Dendritic cells govern induction and reprogramming of polarized tissue-selective homing receptor patterns of T cells: Important roles for soluble factors and tissue microenvironments. *European Journal of Immunology, 35*, 1056–1065.

Dwyer, K. M., Deaglio, S., Gao, W., Friedman, D., Strom, T. B., & Robson, S. C. (2007). CD39 and control of cellular immune responses. *Purinergic Signalling, 3*, 171–180.

Ebbo, M., Crinier, A., Vely, F., & Vivier, E. (2017). Innate lymphoid cells: Major players in inflammatory diseases. *Nature Reviews Immunology, 17*(11), 665–678.

Elinav, E., Strowig, T., Kau, A. L., Henao-Mejia, J., Thaiss, C. A., Booth, C. J., Peaper, D. R., Bertin, J., Eisenbarth, S. C., Gordon, J. I., & Flavell, R. A. (2011). NLRP6 inflammasome regulates colonic microbial ecology and risk for colitis. *Cell, 145*, 745–757.

Elson, C. O., Graeff, A. S., James, S. P., & Strober, W. (1981). Covert suppressor T cells in Crohn's disease. *Gastroenterology, 80*, 1513–1521.

Eyerich, S., Eyerich, K., Pennino, D., Carbone, T., Nasorri, F., Pallotta, S., Cianfarani, F., Odorisio, T., Traidl-Hoffmann, C., Behrendt, H., et al. (2009). Th22 cells represent a distinct human T cell subset involved in epidermal immunity and remodeling. *The Journal of Clinical Investigation, 119*, 3573–3585.

Feuerer, M., Hill, J. A., Kretschmer, K., von Boehmer, H., Mathis, D., & Benoist, C. (2010). Genomic definition of multiple ex vivo regulatory T cell subphenotypes. *Proceedings of the National Academy of Sciences of the United States of America, 107*, 5919–5924.

Fletcher, A. L., Lukacs-Kornek, V., Reynoso, E. D., Pinner, S. E., Bellemare-Pelletier, A., Curry, M. S., Collier, A. R., Boyd, R. L., & Turley, S. J. (2010). Lymph node fibroblastic reticular cells directly present peripheral tissue antigen under steady-state and inflammatory conditions. *The Journal of Experimental Medicine, 207*, 689–697.

Fletcher, A. L., Acton, S. E., & Knoblich, K. (2015). Lymph node fibroblastic reticular cells in health and disease. *Nature Reviews Immunology, 15*, 350–361.

Floess, S., Freyer, J., Siewert, C., Baron, U., Olek, S., Polansky, J., Schlawe, K., Chang, H. D., Bopp, T., Schmitt, E., et al. (2007). Epigenetic control of the foxp3 locus in regulatory T cells. *PLoS Biology, 5*, e38.

Fontenot, J. D., Gavin, M. A., & Rudensky, A. Y. (2003). Foxp3 programs the development and function of CD4 $^{+}$CD25^{+} regulatory T cells. *Nature Immunology, 4*, 330–336.

Franke, A., McGovern, D. P., Barrett, J. C., Wang, K., Radford-Smith, G. L., Ahmad, T., Lees, C. W., Balschun, T., Lee, J., Roberts, R., et al. (2010). Genome-wide meta-analysis increases to 71 the number of confirmed Crohn's disease susceptibility loci. *Nature Genetics, 42*, 1118–1125.

Fujimoto, K., Karuppuchamy, T., Takemura, N., Shimohigoshi, M., Machida, T., Haseda, Y., Aoshi, T., Ishii, K. J., Akira, S., & Uematsu, S. (2011). A new subset of CD103^{+}CD8a^{+} dendritic cells in the small intestine expresses TLR3, TLR7, and TLR9 and induces Th1 response and CTL activity. *Journal of Immunology, 186*, 6287–6295.

Furusawa, Y., Obata, Y., Fukuda, S., Endo, T. A., Nakato, G., Takahashi, D., Nakanishi, Y., Uetake, C., Kato, K., Kato, T., et al. (2013). Commensal microbe-derived butyrate induces the differentiation of colonic regulatory T cells. *Nature, 504*, 446–450.

Fuss, I. J., Heller, F., Boirivant, M., Leon, F., Yoshida, M., Fichtner-Feigl, S., Yang, Z., Exley, M., Kitani, A., Blumberg, R. S., et al. (2004). Nonclassical CD1d-restricted NK T cells that produce IL-13 characterize an atypical Th2 response in ulcerative colitis. *The Journal of Clinical Investigation, 113*, 1490–1497.

Gavin, M. A., Rasmussen, J. P., Fontenot, J. D., Vasta, V., Manganiello, V. C., Beavo, J. A., & Rudensky, A. Y. (2007). Foxp3-dependent programme of regulatory T-cell differentiation. *Nature, 445*, 771–775.

Gerlach, K., Hwang, Y., Nikolaev, A., Atreya, R., Dornhoff, H., Steiner, S., Lehr, H. A., Wirtz, S., Vieth, M., Waisman, A., et al. (2014). TH9 cells that express the transcription factor PU.1 drive T cell-mediated colitis via IL-9 receptor signaling in intestinal epithelial cells. *Nature Immunology, 15*, 676–686.

Gerner, M. Y., Kastenmuller, W., Ifrim, I., Kabat, J., & Germain, R. N. (2012). Histo-cytometry: A method for highly multiplex quantitative tissue imaging analysis applied to dendritic cell subset microanatomy in lymph nodes. *Immunity, 37*, 364–376.

Gottschalk, R. A., Corse, E., & Allison, J. P. (2012). Expression of Helios in peripherally induced Foxp3^{+} regulatory T cells. *Journal of Immunology, 188*, 976–980.

Gretz, J. E., Norbury, C. C., Anderson, A. O., Proudfoot, A. E., & Shaw, S. (2000). Lymph-borne chemokines and other low molecular weight molecules reach high endothelial venules via specialized conduits while a functional barrier limits access to the lymphocyte microenvironments in lymph node cortex. *The Journal of Experimental Medicine, 192*, 1425–1440.

Gross, M., Salame, T. M., & Jung, S. (2015). Guardians of the Gut – murine intestinal macrophages and dendritic cells. *Frontiers in Immunology, 6*, 254.

Guo, X., Qiu, J., Tu, T., Yang, X., Deng, L., Anders, R. A., Zhou, L., & Fu, Y. X. (2014). Induction of innate lymphoid cell-derived interleukin-22 by the transcription factor STAT3 mediates protection against intestinal infection. *Immunity, 40*, 25–39.

Hadis, U., Wahl, B., Schulz, O., Hardtke-Wolenski, M., Schippers, A., Wagner, N., Muller, W., Sparwasser, T., Forster, R., & Pabst, O. (2011). Intestinal tolerance requires gut homing and expansion of FoxP3^{+} regulatory T cells in the lamina propria. *Immunity, 34*, 237–246.

Halim, L., Romano, M., McGregor, R., Correa, I., Pavlidis, P., Grageda, N., Hoong, S. J., Yuksel, M., Jassem, W., Hannen, R. F., et al. (2017). An Atlas of human regulatory T helper-like cells reveals features of Th2-like Tregs that support a tumorigenic environment. *Cell Reports, 20*, 757–770.

Hall, J. A., Bouladoux, N., Sun, C. M., Wohlfert, E. A., Blank, R. B., Zhu, Q., Grigg, M. E., Berzofsky, J. A., & Belkaid, Y. (2008). Commensal DNA limits regulatory T cell conversion and is a natural adjuvant of intestinal immune responses. *Immunity, 29*, 637–649.

Halpern, B., Zweibaum, A., Oriol Palau, R., & Monrard, J. C. (1967). Experimental immune ulcerative colitis. In: Miescher, P., & Grabar, P. (Eds.), *International symposium* (pp. 161–178). Basel.

Hammerschmidt, S. I., Ahrendt, M., Bode, U., Wahl, B., Kremmer, E., Forster, R., & Pabst, O. (2008). Stromal mesenteric lymph node cells are essential for the generation of gut-homing T cells in vivo. *The Journal of Experimental Medicine, 205*, 2483–2490.

Happel, K. I., Dubin, P. J., Zheng, M., Ghilardi, N., Lockhart, C., Quinton, L. J., Odden, A. R., Shellito, J. E., Bagby, G. J., Nelson, S., & Kolls, J. K. (2005). Divergent roles of IL-23 and IL-12 in host defense against Klebsiella pneumoniae. *The Journal of Experimental Medicine, 202*, 761–769.

Haribhai, D., Williams, J. B., Jia, S., Nickerson, D., Schmitt, E. G., Edwards, B., Ziegelbauer, J., Yassai, M., Li, S. H., Relland, L. M., et al. (2011). A requisite role for induced regulatory T cells in tolerance based on expanding antigen receptor diversity. *Immunity, 35*, 109–122.

He, Z., Ma, J., Wang, R., Zhang, J., Huang, Z., Wang, F., Sen, S., Rothenberg, E. V., & Sun, Z. (2017). A two-amino-acid substitution in the transcription factor RORγt disrupts its function in TH17 differentiation but not in thymocyte development. *Nature Immunology, 18*(10), 1128–1113.

Heller, F., Florian, P., Bojarski, C., et al. (2005). Interleukin-13 is the key effector Th2 cytokine in ulcerative colitis that affects epithelial tight junctions, apoptosis, and cell restitution. *Gastroenterology, 129*, 15.

Hill, J. A., Feuerer, M., Tash, K., Haxhinasto, S., Perez, J., Melamed, R., Mathis, D., & Benoist, C. (2007). Foxp3 transcription-factor-dependent and -independent regulation of the regulatory T cell transcriptional signature. *Immunity, 27*, 786–800.

Hill, J. A., Hall, J. A., Sun, C. M., Cai, Q., Ghyselinck, N., Chambon, P., Belkaid, Y., Mathis, D., & Benoist, C. (2008). Retinoic acid enhances Foxp3 induction indirectly by relieving inhibition from $CD4^{+}CD44^{hi}$ Cells. *Immunity, 29*, 758–770.

Hodgson, H. J., Wands, J. R., & Isselbacher, K. J. (1978). Decreased suppressor cell activity in inflammatory bowel disease. *Clinical and Experimental Immunology, 32*, 451–458.

Hogquist, K. A., Baldwin, T. A., & Jameson, S. C. (2005). Central tolerance: Learning self-control in the thymus. *Nature Reviews Immunology, 5*, 772–782.

Hori, S., Nomura, T., & Sakaguchi, S. (2003). Control of regulatory T cell development by the transcription factor Foxp3. *Science, 299*, 1057–1061.

Huehn, J., & Beyer, M. (2015). Epigenetic and transcriptional control of Foxp3 regulatory T cells. *Seminars in Immunology, 27*, 10–18.

Huehn, J., Polansky, J. K., & Hamann, A. (2009). Epigenetic control of FOXP3 expression: The key to a stable regulatory T-cell lineage? *Nature Reviews Immunology, 9*, 83–89.

Infante-Duarte, C., Horton, H. F., Byrne, M. C., & Kamradt, T. (2000). Microbial lipopeptides induce the production of IL-17 in Th cells. *The Journal of Immunology, 165*, 6107–6115.

Ivanov, I. I., Atarashi, K., Manel, N., Brodie, E. L., Shima, T., Karaoz, U., Wei, D., Goldfarb, K. C., Santee, C. A., Lynch, S. V., et al. (2009). Induction of intestinal Th17 cells by segmented filamentous bacteria. *Cell, 139*, 485–498.

Josefowicz, S. Z., Lu, L. F., & Rudensky, A. Y. (2012a). Regulatory T cells: Mechanisms of differentiation and function. *Annual Review of Immunology, 30*, 531–564.

Josefowicz, S. Z., Niec, R. E., Kim, H. Y., Treuting, P., Chinen, T., Zheng, Y., Umetsu, D. T., & Rudensky, A. Y. (2012b). Extrathymically generated regulatory T cells control mucosal TH2 inflammation. *Nature, 482*, 395–399.

Kamada, N., Seo, S. U., Chen, G. Y., & Nunez, G. (2013). Role of the gut microbiota in immunity and inflammatory disease. *Nature Reviews Immunology, 13*, 321–335.

Kamath, A. T., Henri, S., Battye, F., Tough, D. F., & Shortman, K. (2002). Developmental kinetics and lifespan of dendritic cells in mouse lymphoid organs. *Blood, 100*, 1734–1741.

Kang, S. G., Lim, H. W., Andrisani, O. M., Broxmeyer, H. E., & Kim, C. H. (2007). Vitamin A metabolites induce gut-homing $FoxP3^{+}$ regulatory T cells. *Journal of Immunology, 179*, 3724–3733.

Kaplan, M. H. (2013). Th9 cells: Differentiation and disease. *Immunological Reviews, 252*, 104–115.

Kaplan, M. H., Hufford, M. M., & Olson, M. R. (2015). The development and in vivo function of T helper 9 cells. *Nature Reviews Immunology, 15*, 295–307.

Khan, O., Headley, M., Gerard, A., Wei, W., Liu, L., & Krummel, M. F. (2011). Regulation of T cell priming by lymphoid stroma. *PLoS One, 6*, e26138.

Kim, H. P., & Leonard, W. J. (2007). CREB/ATF-dependent T cell receptor-induced FoxP3 gene expression: A role for DNA methylation. *The Journal of Experimental Medicine, 204*, 1543–1551.

Kim, M. H., Kang, S. G., Park, J. H., Yanagisawa, M., & Kim, C. H. (2013a). Short-chain fatty acids activate GPR41 and GPR43 on intestinal epithelial cells to promote inflammatory responses in mice. *Gastroenterology, 145*(396–406), e391–e310.

Kim, S. V., Xiang, W. V., Kwak, C., Yang, Y., Lin, X. W., Ota, M., Sarpel, U., Rifkin, D. B., Xu, R., & Littman, D. R. (2013b). GPR15-mediated homing controls immune homeostasis in the large intestine mucosa. *Science, 340*(6139), 1456–1459.

Kim, K. S., Hong, S. W., Han, D., Yi, J., Jung, J., Yang, B. G., Lee, J. Y., Lee, M., & Surh, C. D. (2016). Dietary antigens limit mucosal immunity by inducing regulatory T cells in the small intestine. *Science, 351* (6275), 858–863.

Kinnebrew, M. A., Buffie, C. G., Diehl, G. E., Zenewicz, L. A., Leiner, I., Hohl, T. M., Flavell, R. A., Littman, D. R., & Pamer, E. G. (2012). Interleukin 23 production by intestinal $CD103^{+}CD11b^{+}$ dendritic cells in response to bacterial flagellin enhances mucosal innate immune defense. *Immunity, 36*, 276–287.

Klein, L., Hinterberger, M., Wirnsberger, G., & Kyewski, B. (2009). Antigen presentation in the thymus for positive selection and central tolerance induction. *Nature Reviews Immunology, 9*, 833–844.

Kluger, M. A., Luig, M., Wegscheid, C., Goerke, B., Paust, H. J., Brix, S. R., Yan, I., Mittrucker, H. W., Hagl, B., Renner, E. D., et al. (2014). Stat3 programs Th17-specific regulatory T cells to control GN. *Journal of the American Society of Nephrology, 25*, 1291–1302.

Kluger, M. A., Meyer, M. C., Nosko, A., Goerke, B., Luig, M., Wegscheid, C., Tiegs, G., Stahl, R. A., Panzer, U.,

& Steinmetz, O. M. (2016). RORγt^{+}Foxp3^{+} cells are an independent bifunctional regulatory T cell lineage and mediate crescentic GN. *Journal of the American Society of Nephrology, 27*, 454–465.

Korn, L. L., Hubbeling, H. G., Porrett, P. M., Yang, Q., Barnett, L. G., & Laufer, T. M. (2014). Regulatory T cells occupy an isolated niche in the intestine that is antigen independent. *Cell Reports, 9*, 1567–1573.

Kosiewicz, M. M., Zirnheld, A. L., & Alard, P. (2011). Gut microbiota, immunity, and disease: A complex relationship. *Frontiers in Microbiology, 2*, 180.

Kuhn, K. A., & Stappenbeck, T. S. (2013). Peripheral education of the immune system by the colonic microbiota. *Seminars in Immunology, 25*, 364–369.

Laouar, A., Haridas, V., Vargas, D., Zhinan, X., Chaplin, D., van Lier, R. A., & Manjunath, N. (2005). CD70^{+} antigen-presenting cells control the proliferation and differentiation of T cells in the intestinal mucosa. *Nature Immunology, 6*, 698–706.

Lathrop, S. K., Bloom, S. M., Rao, S. M., Nutsch, K., Lio, C. W., Santacruz, N., Peterson, D. A., Stappenbeck, T. S., & Hsieh, C. S. (2011). Peripheral education of the immune system by colonic commensal microbiota. *Nature, 478*, 250–254.

LeibundGut-Landmann, S., Gross, O., Robinson, M. J., Osorio, F., Slack, E. C., Tsoni, S. V., Schweighoffer, E., Tybulewicz, V., Brown, G. D., Ruland, J., & Reis e Sousa, C. (2007). Syk- and CARD9-dependent coupling of innate immunity to the induction of T helper cells that produce interleukin 17. *Nature Immunology, 8*, 630–638.

Leung, J. M., Davenport, M., Wolff, M. J., Wiens, K. E., Abidi, W. M., Poles, M. A., Cho, I., Ullman, T., Mayer, L., & Loke, P. (2014). IL-22-producing CD4^{+} cells are depleted in actively inflamed colitis tissue. *Mucosal Immunology, 7*, 124–133.

Littman, D. R., & Rudensky, A. Y. (2010). Th17 and regulatory T cells in mediating and restraining inflammation. *Cell, 140*, 845–858.

Lukacs-Kornek, V., Malhotra, D., Fletcher, A. L., Acton, S. E., Elpek, K. G., Tayalia, P., Collier, A. R., & Turley, S. J. (2011). Regulated release of nitric oxide by nonhematopoietic stroma controls expansion of the activated T cell pool in lymph nodes. *Nature Immunology, 12*, 1096–1104.

Macatonia, S. E., Hosken, N. A., Litton, M., Vieira, P., Hsieh, C. S., Culpepper, J. A., Wysocka, M., Trinchieri, G., Murphy, K. M., & O'Garra, A. (1995). Dendritic cells produce IL-12 and direct the development of Th1 cells from naive CD4^{+} T cells. *Journal of Immunology, 154*, 5071–5079.

Malhotra, D., Fletcher, A. L., Astarita, J., Lukacs-Kornek, V., Tayalia, P., Gonzalez, S. F., Elpek, K. G., Chang, S. K., Knoblich, K., Hemler, M. E., et al. (2012). Transcriptional profiling of stroma from inflamed and resting lymph nodes defines immunological hallmarks. *Nature Immunology, 13*, 499–510.

Matteoli, G., Mazzini, E., Iliev, I. D., Mileti, E., Fallarino, F., Puccetti, P., Chieppa, M., & Rescigno, M. (2010). Gut CD103^{+} dendritic cells express indoleamine 2,3-dioxygenase which influences T regulatory/T effector cell balance and oral tolerance induction. *Gut, 59*, 595–604.

Matzinger, P., & Kamala, T. (2011). Tissue-based class control: The other side of tolerance. *Nature Reviews Immunology, 11*, 221–230.

Mazmanian, S. K., Liu, C. H., Tzianabos, A. O., & Kasper, D. L. (2005). An immunomodulatory molecule of symbiotic bacteria directs maturation of the host immune system. *Cell, 122*, 107–118.

Mazmanian, S. K., Round, J. L., & Kasper, D. L. (2008). A microbial symbiosis factor prevents intestinal inflammatory disease. *Nature, 453*, 620–625.

McGovern, D., & Powrie, F. (2007). The IL23 axis plays a key role in the pathogenesis of IBD. *Gut, 56*, 1333–1336.

McGovern, D. P., Gardet, A., Torkvist, L., Goyette, P., Essers, J., Taylor, K. D., Neale, B. M., Ong, R. T., Lagace, C., Li, C., et al. (2010). Genome-wide association identifies multiple ulcerative colitis susceptibility loci. *Nature Genetics, 42*, 332–337.

Medzhitov, R. (2007). Recognition of microorganisms and activation of the immune response. *Nature, 449*, 819–826.

Mee, A. S., McLaughlin, J. E., Hodgson, H. J. F., & Jewell, D. P. (1979). Chronic immune colitis in rabbits. *Gut Microbes, 20*(1), 1–5.

Merad, M., Sathe, P., Helft, J., Miller, J., & Mortha, A. (2013). The dendritic cell lineage: Ontogeny and function of dendritic cells and their subsets in the steady state and the inflamed setting. *Annual Review of Immunology, 31*, 563–604.

Mestecky, J., McGhee, J. R., Bienenstock, J., Lamm, M. E., Strober, W., Cebra, J. J., Mayer, L., Pearay, L. O., & Russel, M. W. (2015). Historical aspects of mucosal immunology, vol 4.

Molenaar, R., Greuter, M., van der Marel, A. P., Roozendaal, R., Martin, S. F., Edele, F., Huehn, J., Forster, R., O'Toole, T., Jansen, W., et al. (2009). Lymph node stromal cells support dendritic cell-induced gut-homing of T cells. *Journal of Immunology, 183*, 6395–6402.

Molenaar, R., Knippenberg, M., Goverse, G., Olivier, B. J., de Vos, A. F., O'Toole, T., & Mebius, R. E. (2011). Expression of retinaldehyde dehydrogenase enzymes in mucosal dendritic cells and gut-draining lymph node stromal cells is controlled by dietary vitamin A. *Journal of Immunology, 186*, 1934–1942.

Mosmann, T. R., & Coffman, R. L. (1989). TH1 and TH2 cells: Different patterns of lymphokine secretion lead to different functional properties. *Annual Review of Immunology, 7*, 145–173.

Mowat, A. M., & Agace, W. W. (2014). Regional specialization within the intestinal immune system. *Nature Reviews Immunology, 14*, 667–685.

Mucida, D., Park, Y., Kim, G., Turovskaya, O., Scott, I., Kronenberg, M., & Cheroutre, H. (2007). Reciprocal

TH17 and regulatory T cell differentiation mediated by retinoic acid. *Science, 317*, 256–260.

Murai, M., Turovskaya, O., Kim, G., Madan, R., Karp, C. L., Cheroutre, H., & Kronenberg, M. (2009). Interleukin 10 acts on regulatory T cells to maintain expression of the transcription factor Foxp3 and suppressive function in mice with colitis. *Nature Immunology, 10*, 1178–1184.

Nawaf, M. G., Ulvmar, M. H., Withers, D. R., McConnell, F. M., Gaspal, F. M., Webb, G. J., Jones, N. D., Yagita, H., Allison, J. P., & Lane, P. J. L. (2017). Concurrent OX40 and CD30 ligand blockade abrogates the CD4-driven autoimmunity associated with CTLA4 and PD1 blockade while preserving excellent anti-CD8 tumor immunity. *Journal of Immunology, 199*, 974–981.

Neurath, M. F. (2014). Cytokines in inflammatory bowel disease. *Nature Reviews Immunology, 14*, 329–342.

Ohkura, N., Hamaguchi, M., Morikawa, H., Sugimura, K., Tanaka, A., Ito, Y., Osaki, M., Tanaka, Y., Yamashita, R., Nakano, N., et al. (2012). T cell receptor stimulation-induced epigenetic changes and Foxp3 expression are independent and complementary events required for Treg cell development. *Immunity, 37*, 785–799.

Ohnmacht, C., Park, J. H., Cording, S., Wing, J. B., Atarashi, K., Obata, Y., Gaboriau-Routhiau, V., Marques, R., Dulauroy, S., Fedoseeva, M., et al. (2015). Mucosal immunology. The microbiota regulates type 2 immunity through RORγt$^+$ T cells. *Science, 349*, 989–993.

Onderdonk, A. B., Hermos, J. A., Dzink, J. L., & Bartlett, J. G. (1978). Protective effect of metronidazole in experimental ulcerative colitis. *Gastroenterology, 74*, 521–526.

Pabst, O., & Bernhardt, G. (2013). On the road to tolerance-generation and migration of gut regulatory T cells. *European Journal of Immunology, 43*, 1422–1425.

Pandiyan, P., Zheng, L., Ishihara, S., Reed, J., & Lenardo, M. J. (2007). CD4$^+$CD25$^+$Foxp3$^+$ regulatory T cells induce cytokine deprivation-mediated apoptosis of effector CD4$^+$ T cells. *Nature Immunology, 8*, 1353–1362.

Panea, C., Farkas, A. M., Goto, Y., Abdollahi-Roodsaz, S., Lee, C., Koscso, B., Gowda, K., Hohl, T. M., Bogunovic, M., & Ivanov, I. I. (2015). Intestinal monocyte-derived macrophages control commensal-specific Th17 responses. *Cell Reports, 12*, 1314–1324.

Park, H. J., Park, J. S., Jeong, Y. H., Son, J., Ban, Y. H., Lee, B. H., Chen, L., Chang, J., Chung, D. H., Choi, I., & Ha, S. J. (2015). PD-1 upregulated on regulatory T cells during chronic virus infection enhances the suppression of CD8$^+$ T cell immune response via the interaction with PD-L1 expressed on CD8$^+$ T cells. *Journal of Immunology, 194*, 5801–5811.

Parks, O. B., Pociask, D. A., Hodzic, Z., Kolls, J. K., & Good, M. (2015). Interleukin-22 signaling in the regulation of intestinal health and disease. *Frontiers in Cell and Development Biology, 3*, 85.

Paul, W. E., & Zhu, J. (2010). How are T(H)2-type immune responses initiated and amplified? *Nature Reviews Immunology, 10*, 225–235.

Perlmann, P., & Broberger, O. (1963). In vitro studies of ulcerative colitis. II. Cytotoxic action of white blood cells from patients on human fetal colon cells. *The Journal of Experimental Medicine, 117*, 717–733.

Pezoldt, J., & Huehn, J. (2016). Tissue-specific induction of CCR6 and Nrp1 during early CD4$^+$ T cell differentiation. *European Journal of Microbiology & Immunology, 6*, 219–226.

Polansky, J. K., Kretschmer, K., Freyer, J., Floess, S., Garbe, A., Baron, U., Olek, S., Hamann, A., von Boehmer, H., & Huehn, J. (2008). DNA methylation controls Foxp3 gene expression. *European Journal of Immunology, 38*, 1654–1663.

Powrie, F., Leach, M. W., Mauze, S., Caddle, L. B., & Coffman, R. L. (1993). Phenotypically distinct subsets of CD4$^+$ T cells induce or protect from chronic intestinal inflammation in C. B-17 scid mice. *International Immunology, 5*, 1461–1471.

Powrie, F., Correa Oliveira, R., Mauze, S., & Coffman, R. L. (1994). Regulatory interactions between CD45RB^{high} and CD45RB^{low} CD4$^+$ T cells are important for the balance between protective and pathogenic cell-mediated immunity. *The Journal of Experimental Medicine, 179*, 589–600.

Radtke, A. J., Kastenmuller, W., Espinosa, D. A., Gerner, M. Y., Tse, S. W., Sinnis, P., Germain, R. N., Zavala, F. P., & Cockburn, I. A. (2015). Lymph-node resident CD8a$^+$ dendritic cells capture antigens from migratory malaria sporozoites and induce CD8$^+$ T cell responses. *PLoS Pathogens, 11*, e1004637.

Raverdeau, M., & Mills, K. H. (2014). Modulation of T cell and innate immune responses by retinoic acid. *Journal of Immunology, 192*, 2953–2958.

Roozendaal, R., Mempel, T. R., Pitcher, L. A., Gonzalez, S. F., Verschoor, A., Mebius, R. E., von Andrian, U. H., & Carroll, M. C. (2009). Conduits mediate transport of low-molecular-weight antigen to lymph node follicles. *Immunity, 30*, 264–276.

Rothenberg, E. V., Moore, J. E., & Yui, M. A. (2008). Launching the T-cell-lineage developmental programme. *Nature Reviews Immunology, 8*, 9–21.

Round, J. L., & Mazmanian, S. K. (2009). The gut microbiota shapes intestinal immune responses during health and disease. *Nature Reviews Immunology, 9*, 313–323.

Rovedatti, L., Kudo, T., Biancheri, P., Sarra, M., Knowles, C. H., Rampton, D. S., Corazza, G. R., Monteleone, G., Di Sabatino, A., & Macdonald, T. T. (2009). Differential regulation of interleukin 17 and interferon-γ production in inflammatory bowel disease. *Gut, 58*, 1629–1636.

Roy, U., Galvez, E. J. C., Iljazovic, A., Lesker, T. R., Blazejewski, A. J., Pils, M. C., Heise, U., Huber, S., Flavell, R. A., & Strowig, T. (2017). Distinct microbial communities trigger colitis development upon intestinal barrier damage via innate or adaptive immune cells. *Cell Reports, 21*, 994–1008.

Sakaguchi, S., Yamaguchi, T., Nomura, T., & Ono, M. (2008). Regulatory T cells and immune tolerance. *Cell, 133*, 775–787.

Schaper, K., Kietzmann, M., & Baumer, W. (2014). Sphingosine-1-phosphate differently regulates the cytokine production of IL-12, IL-23 and IL-27 in activated murine bone marrow derived dendritic cells. *Molecular Immunology, 59*, 10–18.

Schiering, C., Krausgruber, T., Chomka, A., Frohlich, A., Adelmann, K., Wohlfert, E. A., Pott, J., Griseri, T., Bollrath, J., Hegazy, A. N., et al. (2014). The alarmin IL-33 promotes regulatory T-cell function in the intestine. *Nature, 513*, 564–568.

Schulz, O., Jaensson, E., Persson, E. K., Liu, X., Worbs, T., Agace, W. W., & Pabst, O. (2009). Intestinal $CD103^+$, but not $CX3CR1^+$, antigen sampling cells migrate in lymph and serve classical dendritic cell functions. *The Journal of Experimental Medicine, 206*, 3101–3114.

Scott, C. L., Aumeunier, A. M., & Mowat, A. M. (2011). Intestinal $CD103^+$ dendritic cells: Master regulators of tolerance? *Trends in Immunology, 32*, 412–419.

Sefik, E., Geva-Zatorsky, N., Oh, S., Konnikova, L., Zemmour, D., McGuire, A. M., Burzyn, D., Ortiz-Lopez, A., Lobera, M., Yang, J., et al. (2015). Mucosal immunology. Individual intestinal symbionts induce a distinct population of $ROR\gamma^+$ regulatory T cells. *Science, 349*, 993–997.

Shan, M., Gentile, M., Yeiser, J. R., Walland, A. C., Bornstein, V. U., Chen, K., He, B., Cassis, L., Bigas, A., Cols, M., et al. (2013). Mucus enhances gut homeostasis and oral tolerance by delivering immunoregulatory signals. *Science, 342*, 447–453.

Shevach, E. M. (2009). Mechanisms of $foxp3^+$ T regulatory cell-mediated suppression. *Immunity, 30*, 636–645.

Shih, D. Q., Targan, S. R., & McGovern, D. (2008). Recent advances in IBD pathogenesis: Genetics and immunobiology. *Current Gastroenterology Reports, 10*, 8.

Shorter, R. G., Huizenga, K. A., ReMine, S. G., & Spencer, R. J. (1970). Effects of preliminary incubation of lymphocytes with serum on their cytotoxicity for colonic epithelial cells. *Gastroenterology, 58*, 843–850.

Siegert, S., Huang, H. Y., Yang, C. Y., Scarpellino, L., Carrie, L., Essex, S., Nelson, P. J., Heikenwalder, M., Acha-Orbea, H., Buckley, C. D., et al. (2011). Fibroblastic reticular cells from lymph nodes attenuate T cell expansion by producing nitric oxide. *PLoS One, 6*, e27618.

Siewert, C., Menning, A., Dudda, J., Siegmund, K., Lauer, U., Floess, S., Campbell, D. J., Hamann, A., & Huehn, J. (2007). Induction of organ-selective $CD4^+$ regulatory T cell homing. *European Journal of Immunology, 37*, 978–989.

Singh, N., Gurav, A., Sivaprakasam, S., Brady, E., Padia, R., Shi, H., Thangaraju, M., Prasad, P. D., Manicassamy, S., Munn, D. H., et al. (2014). Activation of Gpr109a, receptor for niacin and the commensal metabolite butyrate, suppresses colonic inflammation and carcinogenesis. *Immunity, 40*, 128–139.

Sixt, M., Kanazawa, N., Selg, M., Samson, T., Roos, G., Reinhardt, D. P., Pabst, R., Lutz, M. B., & Sorokin, L. (2005). The conduit system transports soluble antigens from the afferent lymph to resident dendritic cells in the T cell area of the lymph node. *Immunity, 22*, 19–29.

Smigiel, K. S., Srivastava, S., Stolley, J. M., & Campbell, D. J. (2014). Regulatory T-cell homeostasis: Steady-state maintenance and modulation during inflammation. *Immunological Reviews, 259*, 40–59.

Smith, P. M., Howitt, M. R., Panikov, N., Michaud, M., Gallini, C. A., Bohlooly, Y. M., Glickman, J. N., & Garrett, W. S. (2013). The microbial metabolites, short-chain fatty acids, regulate colonic Treg cell homeostasis. *Science, 341*, 569–573.

Smyth, L. A., Ratnasothy, K., Tsang, J. Y., Boardman, D., Warley, A., Lechler, R., & Lombardi, G. (2013). CD73 expression on extracellular vesicles derived from $CD4^+$ $CD25^+Foxp3^+$ T cells contributes to their regulatory function. *European Journal of Immunology, 43*, 2430–2440.

Solomon, B. D., & Hsieh, C. S. (2016). Antigen-specific development of mucosal $Foxp3^+ROR\gamma t^+$ T cells from regulatory T cell precursors. *Journal of Immunology, 197*, 3512–3519.

Spadoni, I., Iliev, I. D., Rossi, G., & Rescigno, M. (2012). Dendritic cells produce TSLP that limits the differentiation of Th17 cells, fosters Treg development, and protects against colitis. *Mucosal Immunology, 5*, 184–193.

Sternberg, E. M. (2006). Neural regulation of innate immunity: A coordinated nonspecific host response to pathogens. *Nature Reviews Immunology, 6*, 318–328.

Sugihara, T., Kobori, A., Imaeda, H., Tsujikawa, T., Amagase, K., Takeuchi, K., Fujiyama, Y., & Andoh, A. (2010). The increased mucosal mRNA expressions of complement C3 and interleukin-17 in inflammatory bowel disease. *Clinical and Experimental Immunology, 160*, 386–393.

Sun, C. M., Hall, J. A., Blank, R. B., Bouladoux, N., Oukka, M., Mora, J. R., & Belkaid, Y. (2007). Small intestine lamina propria dendritic cells promote de novo generation of Foxp3 Treg cells via retinoic acid. *The Journal of Experimental Medicine, 204*, 1775–1785.

Szabo, S. J., Kim, S. T., Costa, G. L., Zhang, X., Fathman, C. G., & Glimcher, L. H. (2000). A novel transcription factor, T-bet, directs Th1 lineage commitment. *Cell, 100*, 655–669.

Tartar, D. M., VanMorlan, A. M., Wan, X., Guloglu, F. B., Jain, R., Haymaker, C. L., Ellis, J. S., Hoeman, C. M., Cascio, J. A., Dhakal, M., et al. (2010). $FoxP3^+ROR\gamma t^+$ T helper intermediates display suppressive function against autoimmune diabetes. *Journal of Immunology, 184*, 3377–3385.

Toker, A., Engelbert, D., Garg, G., Polansky, J. K., Floess, S., Miyao, T., Baron, U., Duber, S., Geffers, R., Giehr, P., et al. (2013). Active demethylation of the Foxp3 locus leads to the generation of stable regulatory T cells within the thymus. *Journal of Immunology, 190*, 3180–3188.

Tone, Y., Furuuchi, K., Kojima, Y., Tykocinski, M. L., Greene, M. I., & Tone, M. (2008). Smad3 and NFAT cooperate to induce Foxp3 expression through its enhancer. *Nature Immunology, 9*, 194–202.

Torow, N., Yu, K., Hassani, K., Freitag, J., Schulz, O., Basic, M., Brennecke, A., Sparwasser, T., Wagner, N., Bleich, A., et al. (2015). Active suppression of intestinal $CD4^{+}TCRab^{+}$ T-lymphocyte maturation during the postnatal period. *Nature Communications, 6*, 7725.

Trifari, S., Kaplan, C. D., Tran, E. H., Crellin, N. K., & Spits, H. (2009). Identification of a human helper T cell population that has abundant production of interleukin 22 and is distinct from T(H)-17, T(H)1 and T(H) 2 cells. *Nature Immunology, 10*, 864–871.

Trinchieri, G. (1994). Interleukin-12: A cytokine produced by antigen-presenting cells with immunoregulatory functions in the generation of T-helper cells type 1 and cytotoxic lymphocytes. *Blood, 84*, 20.

Vantourout, P., & Hayday, A. (2013). Six-of-the-best: Unique contributions of gdT cells to immunology. *Nature Reviews Immunology, 13*, 88–100.

Vignali, D. A., Collison, L. W., & Workman, C. J. (2008). How regulatory T cells work. *Nature Reviews Immunology, 8*, 523–532.

Weber, B., Saurer, L., Schenk, M., Dickgreber, N., & Mueller, C. (2011). CX3CR1 defines functionally distinct intestinal mononuclear phagocyte subsets which maintain their respective functions during homeostatic and inflammatory conditions. *European Journal of Immunology, 41*, 773–779.

Weinreich, M. A., & Hogquist, K. A. (2008). Thymic emigration: When and how T cells leave home. *Journal of Immunology, 181*, 2265–2270.

Winau, F., Quack, C., Darmoise, A., & Kaufmann, S. H. (2008). Starring stellate cells in liver immunology. *Current Opinion in Immunology, 20*, 68–74.

Wohlfert, E. A., Grainger, J. R., Bouladoux, N., Konkel, J. E., Oldenhove, G., Ribeiro, C. H., Hall, J. A., Yagi, R., Naik, S., Bhairavabhotla, R., et al. (2011). GATA3 controls $Foxp3^{+}$ regulatory T cell fate during inflammation in mice. *The Journal of Clinical Investigation, 121*, 4503–4515.

Wolvers, D. A., Coenen-de Roo, C. J., Mebius, R. E., van der Cammen, M. J., Tirion, F., Miltenburg, A. M., & Kraal, G. (1999). Intranasally induced immunological tolerance is determined by characteristics of the draining lymph nodes: Studies with OVA and human cartilage gp-39. *Journal of Immunology, 162*, 1994–1998.

Worbs, T., Bode, U., Yan, S., Hoffmann, M. W., Hintzen, G., Bernhardt, G., Forster, R., & Pabst, O. (2006). Oral tolerance originates in the intestinal immune system and relies on antigen carriage by dendritic cells. *The Journal of Experimental Medicine, 203*, 519–527.

Yang, B. H., Hagemann, S., Mamareli, P., Lauer, U., Hoffmann, U., Beckstette, M., Fohse, L., Prinz, I., Pezoldt, J., Suerbaum, S., et al. (2016). $Foxp3^{+}$ T cells expressing RORγt represent a stable regulatory T-cell effector lineage with enhanced suppressive capacity during intestinal inflammation. *Mucosal Immunology, 9*, 444–457.

Ye, P., Rodriguez, F. H., Kanaly, S., Stocking, K. L., Schurr, J., Schwarzenberger, P., Oliver, P., Huang, W., Zhang, P., Zhang, J., et al. (2001). Requirement of interleukin 17 receptor signaling for lung CXC chemokine and granulocyte colony-stimulating factor expression, neutrophil recruitment, and host defense. *The Journal of Experimental Medicine, 194*, 519–527.

Zechner, E. L. (2017). Inflammatory disease caused by intestinal pathobionts. *Current Opinion in Microbiology, 35*, 64–69.

Zhang, Z., Zhang, W., Guo, J., Gu, Q., Zhu, X., & Zhou, X. (2017). Activation and functional specialization of regulatory T cells lead to the generation of Foxp3 instability. *Journal of Immunology, 198*, 2612–2625.

Zheng, W., & Flavell, R. A. (1997). The transcription factor GATA-3 is necessary and sufficient for Th2 cytokine gene expression in CD4 T cells. *Cell, 89*, 587–596.

Zheng, S. G., Wang, J. H., Stohl, W., Kim, K. S., Gray, J. D., & Horwitz, D. A. (2006). TGF-beta requires CTLA-4 early after T cell activation to induce FoxP3 and generate adaptive $CD4^{+}CD25^{+}$ regulatory cells. *Journal of Immunology, 176*, 3321–3329.

Zheng, Y., Josefowicz, S., Chaudhry, A., Peng, X. P., Forbush, K., & Rudensky, A. Y. (2010). Role of conserved non-coding DNA elements in the Foxp3 gene in regulatory T-cell fate. *Nature, 463*, 808–812.

Zigmond, E., Varol, C., Farache, J., Elmaliah, E., Satpathy, A. T., Friedlander, G., Mack, M., Shpigel, N., Boneca, I. G., Murphy, K. M., et al. (2012). $Ly6C^{hi}$ monocytes in the inflamed colon give rise to proinflammatory effector cells and migratory antigen-presenting cells. *Immunity, 37*, 1076–1090.

Microbiome and Gut Immunity: B Cells 10

Oliver Pabst

Abstract

The intestinal mucosa is protected by secretory immunoglobulins (SIgs). SIgs are produced by the combined function of plasma cells and stroma cells, primarily the gut epithelial cells, and constitute the main type of antibody produced in mice and humans. After their release into the gut lumen, SIgs regulate the composition of the microbiota, neutralize toxins, and prevent infections. In this chapter, the organization of the intestinal B cell compartment, pathways of SIg generation, and the function of SIg in regulating the intestinal microbiota will be discussed.

10.1 Organization of the Intestinal B Cell Compartment

The intestinal immune system can be subdivided into inductive and effector compartments. Inductive compartments comprise the gut-associated lymphoid tissue (GALT) and the gut-draining mesenteric lymph nodes (mLN), whereas the gut lamina propria forms the major effector site (Fig. 10.1). With respect to the B cell compartment, naïve B cells and activated B cells undergoing somatic hypermutation and class switch recombination in germinal centers are characteristic of the inductive sites. In contrast, fully mature plasma cells producing antibodies are the major B cell population in the lamina propria (Tomasi et al. 1965).

GALT shows important differences compared to peripheral lymph nodes or mLN. Whereas lymph nodes are encapsulated structures and receive antigen via afferent lymphatics and blood, GALT structures lack a capsule, lack afferent lymphatics, and are separated from the intestinal lumen by a single epithelial layer, the follicle-associated epithelium (FAE). Within the FAE a unique type of epithelial cells, the M cells, together with myeloid cells (Lelouard et al. 2012) embedded in the FAE, confers the constitutive uptake of luminal antigen into the underlying lymphoid tissue.

The archetypical representatives of GALT are Peyer's patches (PP), secondary lymphoid organs of the small intestine named after the German anatomist Johann Conrad Peyer. PP form during gestation and consist of a number of aggregated lymphoid follicles, each covered by FAE and characterized by the constitutive presence of large germinal centers. In mice PP are macroscopically visible and distributed along the length of the small intestine. A single PP-like structure, the cecal patch, is found in the murine cecum, and 3–5 large lymphoid follicles are present in the colon (also referred to as colonic patches (Bunker

O. Pabst (✉)
Institute of Molecular Medicine, RWTH Aachen University, Aachen, Germany
e-mail: opabst@ukaachen.de

D. Haller (ed.), *The Gut Microbiome in Health and Disease*,
https://doi.org/10.1007/978-3-319-90545-7_10

et al. 2017). Besides PP, cecal and colonic patches, small-sized lymphoid clusters named isolated lymphoid follicles (ILF) are dispersed throughout the murine small intestine. ILF share key features of PP, including the presence of FAE and a characteristic B cell follicle, but in contrast to PP are much smaller and form only after birth (Mao et al. 2017).

Similar to the situation in mice, in humans PP form during gestation, but compared to mice show a more accentuated aggregation in the ileum. The ontogeny of human ILF has not yet been fully established, but similar to the situation in mice, small-sized lymphoid clusters have also been described in the human intestine.

GALT does not have a distinct border and diffuses gradually into the surrounding lamina propria. In consequence, cells isolated from lamina propria may easily be contaminated by cells isolated from adjacent GALT, in particular from ILF. Therefore, some controversy exists concerning the presence of naïve B cells and other non-plasma B cell populations in the lamina propria. Irrespective of this subtlety, the main population of B cells found in the lamina propria are fully matured antibody-producing plasma cells. Intestinal plasma cells are estimated to make up to 80% of all plasma cells and to produce several grams of antibody per day (Tomasi et al. 1965). This makes the intestinal mucosa the most active site of antibody production in the whole body.

10.2 Origin of Antibodies in the Gut Lumen

With respect to structure and generation, intestinal antibodies show important differences between mice and humans. In mice, by far the dominating isotype in the intestinal lamina propria is polymeric immunoglobulin A (pIgA). In contrast, humans have two IgA isotypes, IgA1 and IgA2, encoded by distinct constant regions as well as IgM-expressing plasma cells, which are more numerous compared to mice (Magri et al. 2017). This picture is characteristic of the healthy gut mucosa but dramatically changes during inflammation when in particular IgG-expressing plasma cells accumulate besides IgA- and IgM-secreting cells in the inflamed gut tissue.

IgA1 and IgA2 show unique biological properties. IgA1 possesses a hinge region that can be cleaved by bacterial proteases, whereas IgA2 lacks this region and is more resistant to proteolytic cleavage (Woof and Russell 2011). Consistently, the IgA2 isotype dominates in distal gut segments that are densely colonized by gut microbiota, whereas the proximal small intestine harbors about equal numbers of IgA1- and IgA2-producing plasma cells.

Polymeric Ig (pIg) complexes are tied together by the "joining" chain (J chain), a 15 kDa polypeptide not to be confused with the J gene segment encoded in the Ig loci. The J chain facilitates binding to the polymeric Ig receptor (pIgR) and enables the active transepithelial transport of pIg, thereby promoting active secretion into the gut lumen. In the gut lamina propria, J chain containing pIg binds to the pIgR present on the basolateral side of gut epithelial cells. Subsequently the complex of pIgR and pIg is internalized by receptor-mediated endocytosis and transported to the apical side of the epithelial cells. At the apical side, proteolytic cleavage of pIgR generates the secretory component, a fragment of the pIgR, and the complex of secretory component and pIg is released into the gut lumen (Fig. 10.2; Tomasi et al. 1965). Thus, secretory antibodies are effectively produced by the combined function of plasma cells and pIgR-expressing cells.

Besides the active pIgR-mediated transport by gut epithelial cells, additional mechanisms can facilitate the presence of antibodies in the gut lumen. Since SIg is also present in particular in proximal gut segments, bile-derived Ig contributes to the overall luminal Ig. Additionally, SIg is a crucial component of breast milk; hence, breastfeeding introduces a highly concentrated cocktail of antibodies into the neonate intestine. Finally, under some circumstances, antibodies can passively enter the gut lumen. While the healthy gut mucosa does not permit passive diffusion of antibodies, the situation is fundamentally different during inflammation and impaired barrier integrity.

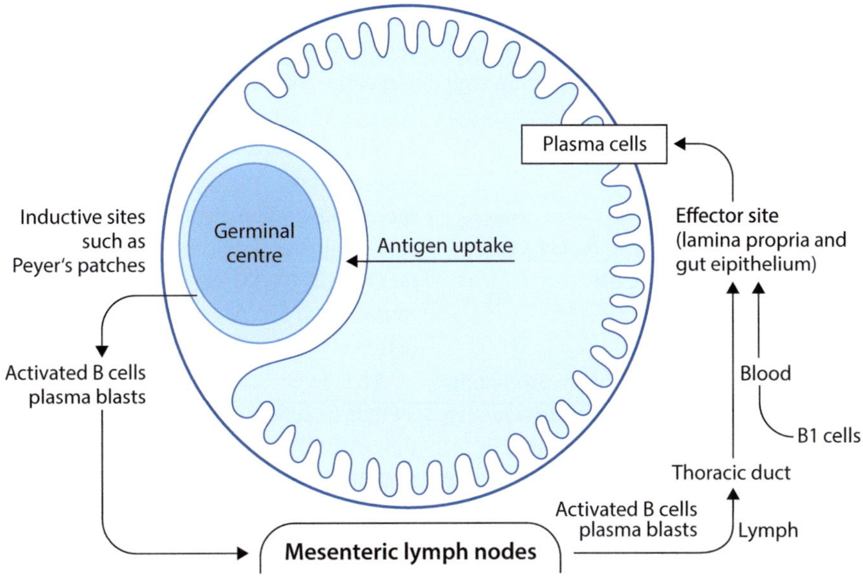

Fig. 10.1 Organization of the intestinal B cell immune system. Peyer's patches constitute a major inductive compartment of the gut immune system. PP constantly receive antigen from the gut lumen. Antigen presentation, lymphocyte activation, and cognate interaction between B and T cells generate activated B cells and plasma blasts. These activated cells leave the PP, enter mesenteric lymph nodes and via the thoracic duct they enter the central circulation. From blood plasma blasts enter the intestinal lamina propria. Alternative pathways including homing of B1 cells from circulation directly into the lamina propria have been suggested

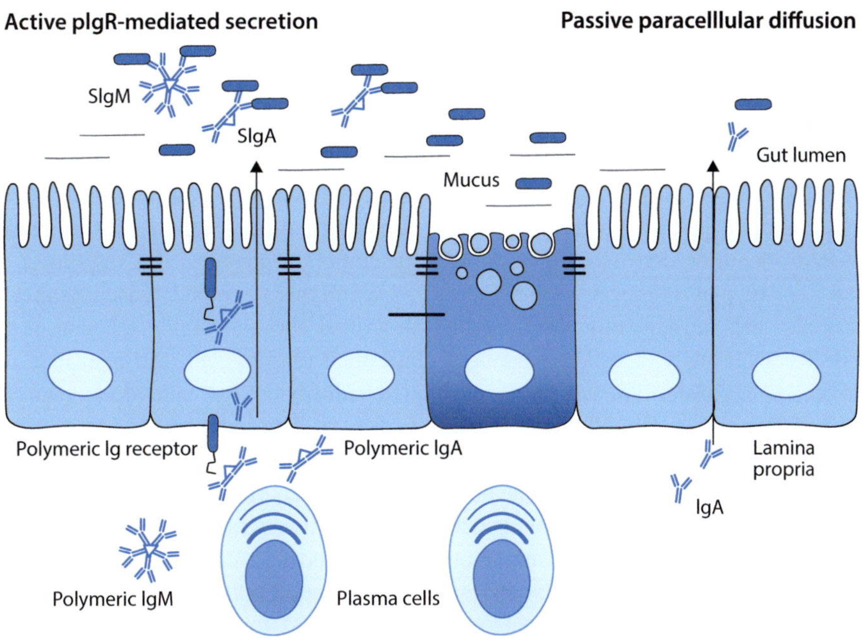

Fig. 10.2 Sources of intraluminal antibodies. Plasma cells in the lamina propria produce polymeric immunoglobulin, mostly dimeric IgA and pentameric IgM, both of which comprise a short polypeptide called the J chain. J chain-assisted binding of polymeric Ig to the polymeric Ig receptor enables transepithelial transport of the complex. At the apical side of epithelial cells, the complex of polymeric Ig and a fragment of the pIgR are released as secretory Ig (SIgA and SIgM). IgG present in the blood or produced by IgG-secreting plasma cells in the inflamed mucosa enters the gut lumen by paracellular diffusion

During inflammation, relevant numbers of IgG-expressing plasma cells accumulate in gut lamina propria and IgG produced locally as well as derived from circulation can enter the gut lumen.

10.3 Pathways of B Cell Activation and IgA Plasma Cell Differentiation

Pathways of how intestinal IgA-producing plasma cell are generated are controversially debated (Deng et al. 2017; Huang et al. 2017). Class switch to IgA can occur at different anatomical sites: within GALT germinal centers, inside GALT but independently of germinal centers, and in situ in the intestinal lamina propria. Moreover, T cell-dependent and T cell-independent pathways of IgA induction as well as an alternative origin of IgA-secreting plasma cells from B1 cells and B2 cells have been reported. I will firstly describe the archetypical pathway of T cell-dependent B cell activation in PP germinal centers before discussing alternative concepts of B cell activation.

There is a broad consensus that GALT serves the generation of IgA responses directed against the microbiota and pathogens. As described above, PP constitutively harbor germinal centers. These germinal centers form early after birth when luminal antigen is entering the "nascent" PP via the FAE, dendritic cells (DC) underlying the FAE, take up and present processed antigen to activate T cells that subsequently differentiate into effector T cells. Finally, T cell-dependent B cell activation and expansion initiates the formation of a germinal center. In adult PP the germinal center dominates the overall compartment and comprises newly activated B cells and memory B cell populations of different clonal origins that in parallel undergo selection and differentiation and class switch recombination. As such PP are considered prime compartments supporting the generation of IgA-secreting plasma cells in a T cell-dependent setting.

Germinal centers provide an environment supporting class switch recombination and somatic hypermutation. Class switch recombination enables a change of Ig isotype, which in PP typically, but not exclusively, entails a switch from IgM to IgA. During somatic hypermutation, B cells undergo successive rounds of mutations, consequently yielding selected B cells with increased antigen affinity. Both of these processes rely on the activity of activation-induced cytidine deaminase (AID), an enzyme that catalyzes the mutation of DNA and is transiently expressed in activated B cells.

The local environment in PP favors class switch to IgA and expression of the gut-homing factors CCR9 and $\alpha 4\beta 7$-integrin. IgA class switch is supported by the cytokine transforming growth factor β (TGFβ). Deficiency of the TGFβ receptor in mice results in almost complete lack of IgA, suggesting a critical role of this cytokine. However, the cellular sources of active TGFβ in PP still need to be defined. Moreover, DC can produce nitric oxide, the cytokine BAFF (B cell-activating factor, a member of the tumor necrosis factor family), and APRIL (a proliferation-inducing ligand) that all support IgA class switch recombination. Expression of gut-homing molecules can be induced by retinoic acid, and indeed DC in PP express high levels of retinoic acid-producing enzymes (Huang et al. 2017). However, in vivo the situation appears more complex, and various cell types including DC, stroma cells, and potentially epithelial cells cooperatively shape an environment in PP that facilitates the generation of intestinal IgA responses.

While there is broad consensus that the mode described above is highly efficient at generating intestinal plasma cells, the relevance of alternative pathways of IgA induction remain more controversial. The number of IgA-secreting plasma cells is unchanged in mice lacking germinal centers (Bemark et al. 2012), and also humans with genetically impaired ability to raise germinal centers have IgA. Indeed, AID expression has been reported outside of germinal centers in PP of mice and humans. Moreover, IgA-secreting plasma cells are strongly reduced but still detectable in T cell-deficient mice. Such T cell-independent class switch has been linked to the activity of BAFF and APRIL and might induce

AID-mediated class switch recombination inside the gut lamina propria. Moreover, besides B2 cells, the "conventional" B cell population populating lymphoid organs, B1 cells have been proposed to generate IgA-secreting plasma cells. The relevance of these pathways has not yet been resolved and may differ between species, with different developmental age and the particular type of antigen considered (Deng et al. 2017; Macpherson and McCoy 2015).

10.4 Plasma Blast Dissemination from GALT and Migration into the Lamina Propria

The germinal center response generates plasma cell precursors, also named plasma blasts. Plasma blasts exit the germinal center via efferent lymph or blood, disseminate to local effector sites, and finally differentiate into fully mature plasma cells. Migration from the circulation into the gut lamina propria is guided by gut-homing factors and their ligands present on intestinal venules. Accordingly, mice with defective β7-integrin, β7-integrin ligand, mucosal addressin cell adhesion molecule-1 (MadCAM-1), chemokine receptor CCR9, or its ligand CCL25 have impaired intestinal plasma cell populations. In the lamina propria, plasma cells can be maintained for prolonged time to secrete Ig.

Plasma blasts are thought to have similar homing properties with respect to different segments of the small but not the large intestine. Sequencing of the Ig-encoding gene loci identified clonally related plasma cells in proximal and distal small intestine. In contrast, fewer clonally related sequences were observed between plasma cells of the small and large intestine (Lindner et al. 2012). Yet, migration of GALT-derived plasma blasts is not limited to the gut, and clonally related populations of plasma cells have also been observed in gut and mammary glands (Lindner et al. 2015). Indirect evidence suggests that such notion may extent to the liver (Moro-Sibilot et al. 2016).

Occurrence of different plasma cell clones in small and large intestine indicates that distinct homing cues may allow the differential recruitment and/or retention of plasma cells in both compartments. One factor potentially involved in this is chemokine receptor CCR10. Plasma blasts activated in the cecal patch show higher expression of CCR10 compared to plasma blasts present in small intestinal PP. Additionally, B cells present in the cecal patch are enriched in colon compared to the small intestine (Masahata et al. 2014).

10.5 The Role of SIg in Maintaining Host-Microbiota Interactions

Various mechanisms of SIg-mediated mucosal protection have been described. For instance, SIg neutralize toxins in the gut lumen and limit infection with rotavirus or *Salmonella*. Yet, the precise functions of SIg in host-microbiota interaction are still emerging. Microbiota comprise a collection of eubacteria, archaea, viruses, fungi, and eukaryotes. After a highly dynamic early life phase, in adults composition of the microbiota stabilizes and shows a characteristic species composition that enables gut homeostasis and is essential for our well-being. A major deviation from this beneficial state of the microbiota is characteristic of various disease states and referred to as dysbiosis.

Global consequences of impaired SIg production have been studied in pIgR-deficient mice that fail to transport polymeric IgA into the gut lumen. pIgR-deficient mice show no overt immunopathology but instead significant differences in the composition of their microbiota compared to littermate controls. Similarly, B cell-deficient mice show reduced diversity of their gut microbiota compared to wild-type controls (Kawamoto et al. 2014). Accordingly, AID deficiency, marked by a defective class switch recombination and somatic hypermutation, results in intestinal dysbiosis, in particular in the overgrowth of *Firmicutes*. Interestingly, a related observation has been made in mice expressing a hypomorphic AID variant that preserves class switching but does not enable hypermutation (Wei et al. 2011). This indicates that not only the total amount but also the quality of IgA is relevant to sustain unperturbed host-microbiota interaction. In line with this concept, mice lacking PD-1 showed impaired effector T cell populations in PP that translated into a SIg-mediated intestinal dysbiosis (Kawamoto et al. 2012). Altogether, these observations support a general function of SIg in maintaining host-microbiota interactions in the intestine. Curiously, consequences of IgA deficiency on composition and function of the microbiota have not yet been systematically analyzed in humans.

Exciting new insights into the interplay between SIgA and the microbiota came from recent studies that investigated functional differences between bacteria targeted or bacteria non-targeted by IgA (Bunker et al. 2015; Kau et al. 2015; Palm et al. 2014). These studies relied on the purification of IgA-coated and non-coated bacteria from the microbiota and their subsequent identification by 16S rRNA gene sequencing. In common to all these studies, differential IgA coating was observed for distinct bacterial taxa (reviewed in Macpherson et al. 2015). SIg coating of gut bacteria was higher in proximal small intestine compared to colon. Moreover, bacteria that showed SIgA coating in the colon were typically also observed in small intestine whereas typical members of the colonic microbiota showed no detectable IgA coat (Bunker et al. 2015). This may indicate that small intestinal GALT plays a dominant role in generating microbiota-directed antibodies.

Notably, SIgA coating was preserved for some taxa in T cell-deficient mice suggesting that T cell-independent Ig responses can generate antibodies binding to gut bacteria. Accordingly, antibodies cloned from single murine plasma cells bound microbiota independent of somatic mutations (Bunker et al. 2017), a finding that contrasts with our own observations: We found that monoclonal antibodies derived from single human plasma cells typically lose their microbiota binding when these antibodies are expressed as germ-line variants (unpublished observations, J. Kabbert, O. Pabst). This in turn indicates that T cell-dependent somatic mutations might contribute to the generation of antibodies with efficient microbiota reactivity in humans. The situation might be even more complex in vivo, and the endogenous SIgA coat of single bacteria might comprise several antibodies of different specificities and affinities. Moreover, the secretory component in itself, as well as the glycosylated Fc part of IgA, can confer reactivity to microbiota (Mathias and Corthesy 2011). Therefore, the nature of the SIgA coat has to be explored in more detail as different pathways of IgA induction may predominate depending of the bacterial taxa looked at, the genetic layout of the host, and age.

10.6 Molecular Mechanism of SIg-Mediated Control of the Microbiota

How does SIgA coating of intestinal bacteria affect their function? Peterson and colleagues have studied this question in mice carrying single monoclonal antibodies directed against *Bacteroides thetaiotaomicron* (Peterson et al. 2007, 2015). One such antibody reduced proinflammatory responses and colonization levels of bacteria recognized by the respective antibody (Peterson et al. 2007). However, a second antibody did not show such an effect, and no reduction of bacteria fitness was observed (Peterson et al. 2015). Thus the precise mode of action, of how an antibody modulates the microbiota, might depend on the particular antibody. In fact, several mechanisms of SIg-mediated modulation of the microbiota have been proposed. In the following paragraph, I will discuss potential mechanisms of SIgA-mediated modulation of the microbiota. I will refer to these mechanisms as "immune exclusion," "enchainment," "fitness," "anchoring," and "sampling" (Fig. 10.3).

"Immune Exclusion" Antibody-mediated cross-linking of bacteria leads to agglutination and clump formation. Clumps of bacteria are thought to be excreted more efficiently compared to single bacteria by intestinal peristalsis in a process called immune exclusion (Fig. 10.3).

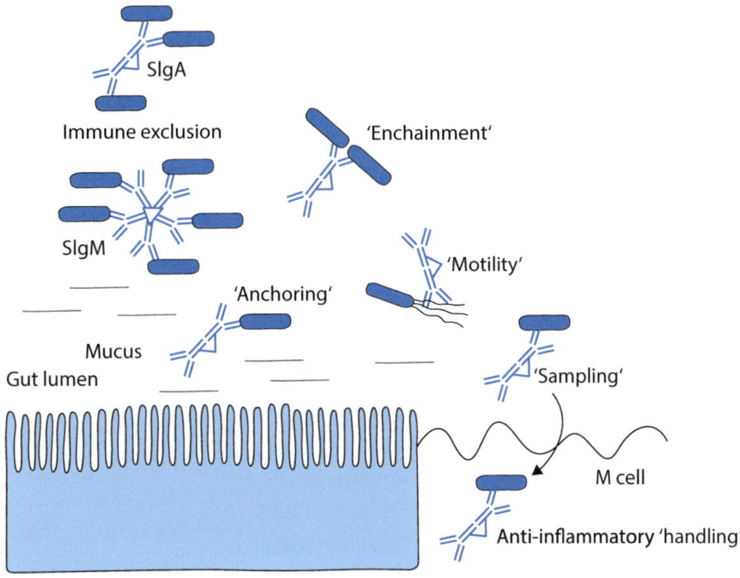

Fig. 10.3 Mechanisms of SIg-mediated modulation of the microbiota. SIgA and SIgM can agglutinate bacteria to form clumps that become secreted by gut peristalsis in a process called immune exclusion. "Enchainment" prevents separation of dividing bacteria and thereby leads to growth-dependent formation of bacterial clumps. Binding of bacteria by antibodies may affect their behavior directly such as antibodies can reduce bacterial fitness by binding to flagella. Antibody-coated bacteria may be anchored to the mucus and become more efficiently sampled by M cell compared to non-coated bacteria

Agglutination and immune exclusion is particularly efficient for multivalent antibodies such as SIgA and SIgM and is considered a major mechanism of SIg-mediated control of the microbiota.

"Enchainment" A variation of "immune exclusion" has recently been suggested to target growing bacteria (Moor et al. 2017). In a situation of low bacteria density, e.g., during the early phase of an infection, agglutination may be very inefficient. Nevertheless, SIgA has the ability to enchain dividing bacteria, thereby preventing their separation after division, resulting in growth-dependent clump formation.

"Fitness" Binding of antibodies to bacteria can directly modify their function. For example, flagella-binding antibodies have been shown to impair motility and potentially the virulence of *Salmonella*. Similarly, antibody binding of bacteria has been suggested to impact membrane integrity. Thus, antibodies may directly impact on bacterial fitness.

"Anchoring" SIgA loosely binds to mucus (Phalipon et al. 2002). Thus, SIgA coating of microbiota may anchor the bacteria within the mucus layer and thereby help to generate a niche that increases the competitive fitness of the bound bacteria, for instance in case of mucus degrading bacteria.

"Sampling" SIgA binding to luminal bacteria enhances their M cell-dependent sampling into PP (Boullier et al. 2009; Kadaoui and Corthesy 2007). In fact, coating of *Salmonella* with IgA facilitated its uptake and results in stronger adaptive immune responses (Fransen et al. 2015). Thus, coating with SIgA may drive a positive feedback loop that sustains antibody responses to certain bacterial species.

Cooperatively the abovementioned SIgA-mediated mechanisms might create niches for colonization and foster diversity of the microbiota. Agglutination might be particularly efficient for highly abundant bacteria, favor colonization by minorities and enhance microbial diversity. "Anchoring" in the mucus might retain bacteria within certain "habitats," and differential concentrations of SIgA along the crypt-villus axis may affect the competitive fitness of bacteria. Consequently, SIgA plays an important role in regulating beneficial and stable host-microbe interaction in the intestine.

Controversy

What determines whether or not bacteria become SIgA-coated? IgA coating has been suggested to target "inflammatory commensals that preferentially drive intestinal disease" (Palm et al. 2014). Additionally, the anatomical location (of the respective microbes but also the site of IgA induction) has been suggested to contribute. I speculate that the propensity of bacteria to acquire an IgA coat might primarily reflect their anatomical localization. Bacteria that occupy niches close to the gut epithelium are more likely to be sampled by the immune system and to induce antibody responses. At the same time, proximity might favor detrimental effects, such as the induction of colitogenic immune responses, but also beneficial effects (such as the stimulation of lymphoid organogenesis).

What is the contribution of alternative pathways of SIgA induction? Alternative pathways of SIg induction have been proposed but we do not know how these pathways cooperate. This question relates to the role of T cell-dependent versus T cell-independent pathways of IgA induction, the site of class switch recombination (e.g., in GALT or in situ in the lamina propria), the B cell populations involved (contribution of B1 and B2 cells), and the role of germinal center-dependent versus germinal center-independent processes. I propose that microbiota-targeted therapies may benefit from in depth understanding of these mechanisms.

What is the specificity of SIg? Different members of the microbiota show a differential propensity to be coated with IgA, but the molecular structures recognized by SIg are largely unknown. I speculate that SIg might be targeted against molecular epitopes spread across phylogenetically unrelated bacteria.

History

In 1965 "an immunological system which is characteristic of certain external secretions" was described, and it was suggested that this system might "play a significant role in the body's defense mechanisms against allergens and microorganisms" (Tomasi et al. 1965). Subsequently, intensive research established the pathways that enable transepithelial transport and secretion of polymeric immunoglobulin, polymeric IgA and IgM, into body fluids (Brandtzaeg and Prydz 1984), and the concept of compartmentalized mucosal immunity that was thought to structurally and functionally differ from systemic immunity was developed (a concept which based on current knowledge seems to be less strict as compared to its initial formulation). During these years research on secretory immunoglobulin dominated the field of mucosal immunology. However, eventually other aspects of mucosal immunity such as immune regulation, dendritic cells in mucosal tissues, and most recently innate lymphocyte functions moved into the lime light, and the IgA system received less attention. Now this situation seems to change again and with the growing interest in microbiota, research on mucosal B cells and their function is on the rise again. It will be exciting to follow the field that with the availability of new technologies and the integration of research on microbiota and mucosal immunity might rapidly move forward.

Highlights

- Immunoglobulin A is the main antibody isotype produced in humans and mice.
- IgA induction comprises T cell-dependent and T cell-independent pathways as well as different anatomical sites.
- Secretory antibodies protect mucosal surfaces from pathogen invasion and toxins.
- Secretory IgA coats distinct members of the microbiota and regulates their interaction with the host.

References

Bemark, M., Boysen, P., & Lycke, N. Y. (2012). Induction of gut IgA production through T cell-dependent and T cell-independent pathways. *Annals of the New York Academy of Sciences, 1247*, 97–116.

Boullier, S., Tanguy, M., Kadaoui, K. A., Caubet, C., Sansonetti, P., Corthesy, B., & Phalipon, A. (2009). Secretory IgA-mediated neutralization of Shigella flexneri prevents intestinal tissue destruction by down-regulating inflammatory circuits. *Journal of Immunology, 183*, 5879–5885.

Brandtzaeg, P., & Prydz, H. (1984). Direct evidence for an integrated function of J chain and secretory component in epithelial transport of immunoglobulins. *Nature, 311*, 71–73.

Bunker, J. J., Erickson, S. A., Flynn, T. M., Henry, C., Koval, J. C., Meisel, M., Jabri, B., Antonopoulos, D. A., Wilson, P. C., & Bendelac, A. (2017). Natural polyreactive IgA antibodies coat the intestinal microbiota. *Science, 358*. https://doi.org/10.1126/science.aan6619.

Bunker, J. J., Flynn, T. M., Koval, J. C., Shaw, D. G., Meisel, M., McDonald, B. D., Ishizuka, I. E., Dent, A. L., Wilson, P. C., Jabri, B., et al. (2015). Innate and adaptive humoral responses coat distinct commensal bacteria with immunoglobulin A. *Immunity, 43*, 541–553.

Deng, S., Kain, L., Pereira, C. S., Mata, S., Macedo, M. F., Bendelac, A., Teyton, L., & Savage, P. B. (2017). Psychosine variants as antigens for natural killer T cells. *Chemical Science, 8*, 2204–2208.

Fransen, F., Zagato, E., Mazzini, E., Fosso, B., Manzari, C., El Aidy, S., Chiavelli, A., D'Erchia, A. M., Sethi, M. K., Pabst, O., et al. (2015). BALB/c and C57BL/6 mice differ in polyreactive IgA abundance, which impacts the generation of antigen-specific IgA and microbiota diversity. *Immunity, 43*, 527–540.

Huang, Y., Dalal, S., Antonopoulos, D., Hubert, N., Raffals, L. H., Dolan, K., Weber, C., Messer, J. S., Jabri, B., Bendelac, A., et al. (2017). Early transcriptomic changes in the ileal pouch provide insight into the molecular pathogenesis of pouchitis and ulcerative colitis. *Inflammatory Bowel Diseases, 23*, 366–378.

Kadaoui, K. A., & Corthesy, B. (2007). Secretory IgA mediates bacterial translocation to dendritic cells in mouse Peyer's patches with restriction to mucosal compartment. *Journal of Immunology, 179*, 7751–7757.

Kau, A. L., Planer, J. D., Liu, J., Rao, S., Yatsunenko, T., Trehan, I., Manary, M. J., Liu, T. C., Stappenbeck, T. S., Maleta, K. M., et al. (2015). Functional characterization of IgA-targeted bacterial taxa from undernourished Malawian children that produce diet-dependent enteropathy. *Science Translational Medicine, 7*, 276ra224.

Kawamoto, S., Maruya, M., Kato, L. M., Suda, W., Atarashi, K., Doi, Y., Tsutsui, Y., Qin, H., Honda, K., Okada, T., et al. (2014). Foxp3(+) T cells regulate immunoglobulin a selection and facilitate diversification of bacterial species responsible for immune homeostasis. *Immunity, 41*, 152–165.

Kawamoto, S., Tran, T. H., Maruya, M., Suzuki, K., Doi, Y., Tsutsui, Y., Kato, L. M., & Fagarasan, S. (2012). The inhibitory receptor PD-1 regulates IgA selection and bacterial composition in the gut. *Science, 336*, 485–489.

Lelouard, H., Fallet, M., de Bovis, B., Meresse, S., & Gorvel, J. P. (2012). Peyer's patch dendritic cells sample antigens by extending dendrites through M cell-specific transcellular pores. *Gastroenterology, 142*, 592–601 e593.

Lindner, C., Thomsen, I., Wahl, B., Ugur, M., Sethi, M. K., Friedrichsen, M., Smoczek, A., Ott, S., Baumann, U., Suerbaum, S., et al. (2015). Diversification of memory B cells drives the continuous adaptation of secretory antibodies to gut microbiota. *Nature Immunology, 16*, 880–888.

Lindner, C., Wahl, B., Fohse, L., Suerbaum, S., Macpherson, A. J., Prinz, I., & Pabst, O. (2012). Age, microbiota, and T cells shape diverse individual IgA repertoires in the intestine. *The Journal of Experimental Medicine, 209*, 365–377.

Macpherson, A. J., Koller, Y., & McCoy, K. D. (2015). The bilateral responsiveness between intestinal microbes and IgA. *Trends in Immunology, 36*, 460–470.

Macpherson, A. J., & McCoy, K. D. (2015). Independence day for IgA. *Immunity, 43*, 416–418.

Magri, G., Comerma, L., Pybus, M., Sintes, J., Llige, D., Segura-Garzon, D., Bascones, S., Yeste, A., Grasset, E. K., Gutzeit, C., et al. (2017). Human secretory IgM emerges from plasma cells clonally related to gut memory B cells and targets highly diverse commensals. *Immunity, 47*, 118–134 e118.

Mao, A. P., Ishizuka, I. E., Kasal, D. N., Mandal, M., & Bendelac, A. (2017). A shared Runx1-bound Zbtb16 enhancer directs innate and innate-like lymphoid lineage development. *Nature Communications, 8*, 863.

Masahata, K., Umemoto, E., Kayama, H., Kotani, M., Nakamura, S., Kurakawa, T., Kikuta, J., Gotoh, K., Motooka, D., Sato, S., et al. (2014). Generation of colonic IgA-secreting cells in the caecal patch. *Nature Communications, 5*, 3704.

Mathias, A., & Corthesy, B. (2011). Recognition of gram-positive intestinal bacteria by hybridoma- and colostrum-derived secretory immunoglobulin A is mediated by carbohydrates. *The Journal of Biological Chemistry, 286*, 17239–17247.

Moor, K., Diard, M., Sellin, M. E., Felmy, B., Wotzka, S. Y., Toska, A., Bakkeren, E., Arnoldini, M., Bansept, F., Co, A. D., et al. (2017). High-avidity IgA protects the intestine by enchaining growing bacteria. *Nature, 544*, 498–502.

Moro-Sibilot, L., Blanc, P., Taillardet, M., Bardel, E., Couillault, C., Boschetti, G., Traverse-Glehen, A., Defrance, T., Kaiserlian, D., & Dubois, B. (2016).

Mouse and human liver contain immunoglobulin A-secreting cells originating from Peyer's patches and directed against intestinal antigens. *Gastroenterology, 151*, 311–323.

Palm, N. W., de Zoete, M. R., Cullen, T. W., Barry, N. A., Stefanowski, J., Hao, L., Degnan, P. H., Hu, J., Peter, I., Zhang, W., et al. (2014). Immunoglobulin A coating identifies colitogenic bacteria in inflammatory bowel disease. *Cell, 158*, 1000–1010.

Peterson, D. A., McNulty, N. P., Guruge, J. L., & Gordon, J. I. (2007). IgA response to symbiotic bacteria as a mediator of gut homeostasis. *Cell Host & Microbe, 2*, 328–339.

Peterson, D. A., Planer, J. D., Guruge, J. L., Xue, L., Downey-Virgin, W., Goodman, A. L., Seedorf, H., & Gordon, J. I. (2015). Characterizing the interactions between a naturally primed immunoglobulin A and its conserved Bacteroides thetaiotaomicron species-specific epitope in gnotobiotic mice. *The Journal of Biological Chemistry, 290*, 12630–12649.

Phalipon, A., Cardona, A., Kraehenbuhl, J. P., Edelman, L., Sansonetti, P. J., & Corthesy, B. (2002). Secretory component: A new role in secretory IgA-mediated immune exclusion in vivo. *Immunity, 17*, 107–115.

Tomasi, T. B., Jr., Tan, E. M., Solomon, A., & Prendergast, R. A. (1965). Characteristics of an immune system common to certain external secretions. *The Journal of Experimental Medicine, 121*, 101–124.

Wei, M., Shinkura, R., Doi, Y., Maruya, M., Fagarasan, S., & Honjo, T. (2011). Mice carrying a knock-in mutation of Aicda resulting in a defect in somatic hypermutation have impaired gut homeostasis and compromised mucosal defense. *Nature Immunology, 12*, 264–270.

Woof, J. M., & Russell, M. W. (2011). Structure and function relationships in IgA. *Mucosal Immunology, 4*, 590–597.

11 Microbiome and Diseases: Inflammatory Bowel Diseases

Kai Hildner, Nadine Waldschmitt, and Dirk Haller

Abstract

Inflammatory bowel diseases, such as Crohn's disease and ulcerative colitis, are chronically relapsing, immune-mediated disorders of the gastrointestinal tract that have been steadily increasing over the past decades. The hallmark of IBD is an uncontrolled manifestation of intestinal and extraintestinal inflammation within genetically susceptible individuals. Herein, compelling research on host genetics pave the way to a better understanding of disease pathogenesis. Over 200 genetic risk factors have been identified showing a disturbed cross talk of the immune epithelial cell microbiota axis. Additionally, epidemiologic studies pointed toward Western lifestyle and habits as part of central environmental factors contributing to both development and maintenance of intestinal inflammation. In this regard, the gut microbiota is thought to play a decisive role in disease progression. The intestinal microbiota was unequivocally shown to be indispensable in orchestrating the development and functionality of the immune system further having a critical impact on both intestinal homeostasis and inflammation in preclinical models. Even though profound changes in the composition of the intestinal microbiota have been frequently observed in human IBD, unraveling cause and consequences of intestinal dysbiosis need further understanding of the interaction between host genetics, microbial ecosystems, and environmental triggers. In this chapter we will discuss different aspects of the etiology of intestinal inflammation and particularly address the role of host-microbe interaction in disease development, progression, and intervention.

K. Hildner
Department of Medicine 1, University Hospital Erlangen, University of Erlangen-Nuremberg, Erlangen, Germany
e-mail: Kai.Hildner@uk-erlangen.de

N. Waldschmitt
Chair of Nutrition and Immunology, Technical University of Munich, Munich, Germany

D. Haller (✉)
Chair of Nutrition and Immunology, Technical University of Munich, Munich, Germany

ZIEL – Institute for Food & Health, Technical University of Munich, Munich, Germany
e-mail: dirk.haller@tum.de

11.1 Introduction

Inflammatory bowel diseases (IBD) represent an umbrella term for a number of conditions characterized by chronically active and/or relapsing inflammatory episodes and largely consist of the two main subtypes Crohn's disease (CD) and ulcerative colitis (UC). Gastrointestinal tract (GIT) manifestations in IBD are typically restricted to selected anatomic regions: In UC, the intestinal manifestation is mainly restricted to the colon,

D. Haller (ed.), *The Gut Microbiome in Health and Disease*,
https://doi.org/10.1007/978-3-319-90545-7_11

except for 10–20% of the patients with additionally ileal manifestations (so-called backwash ileitis). In CD, however, intestinal inflammation can occur throughout the entire GIT from the mouth to the anus in a discontinuous, i.e., segmental, manner, while the terminal ileum and the colon are most frequently affected. Disease initiation, progression, and persistence are assumed to be triggered by environmental factors in genetically susceptible individuals (Podolsky 2002). Both IBD entities are often complicated by extraintestinal manifestations most frequently involving the skin, eyes, and joints. Genome-wide association studies (GWAS) and animal data provide evidence for an interrelated role of the intestinal microbiota, epithelial interface, and immune system driving the pathogenesis of IBD (Hormannsperger et al. 2015; Khor et al. 2011). This chapter provides an overview of the current state of research on the role of the intestinal microbiota in CD and UC. In addition, approaches to the treatment of IBD by the modulation of the intestinal microbiota are discussed.

11.2 Epidemiology of IBD

IBD frequently occurs in populations of the northern hemisphere (Podolsky 2002). Here, highest IBD rates are found in Northern and Western Europe, respectively, and North America (Molodecky et al. 2012). Published data on IBD incidences range from 5 to 20 IBD cases within 100,000 individuals (Kaplan 2015). Among those, CD cases have continued to rise over the past decades due to so far poorly understood reasons, while the prevalence for UC seems to have reached a plateau or has even decreased. However, estimations of an overall prevalence are methodically compromised due to the challenge of a nonhomogenous regional distribution of IBD, including countries with enhanced IBD rates (Molodecky et al. 2012).

The increased risk of IBD in populations with a Western lifestyle nurtured the hypothesis that common environmental signals, such as dietary habits (e.g., high-fat, high-salt, and high-sugar contents) and/or consumption of industrialized food (e.g., highly processed food products), confer a higher susceptibility to develop IBD. In this regard, the enhanced susceptibility could be driven by direct or indirect effects, such as the modulation of immune cell functionality and changes of the intestinal microbiota compositions and/or functionality, respectively, thereby promoting an overall pro-inflammatory micromilieu. The idea that environmental triggers could abet IBD manifestation gained further attention by the observation that individuals moving from low- to high-incidence countries and vice versa, especially before or latest during adolescence, adapt to the incidence rates of their new environment. Consistent with these findings, economically developing countries (e.g., China, India) that progressively change to Western lifestyle habits are noticing a steady increase in IBD cases, thus providing further evidence for an impact of environmental signals in disease progression (Kaplan 2015).

The onset of clinical manifestations of IBD is usually peaking during young adulthood, i.e., between the second and fourth decade of life, while 10–15% of the patients represent children (Podolsky 2002; Kelsen and Baldassano 2008). In UC, male individuals were described to numerically prevail over female. In contrast, female patients preferentially suffer from CD especially in countries with a high socioeconomic standard, while males appear to outnumber female patients in low-incidence countries. Conversely to adults, pediatric CD patients are predominately boys, while the majority of pediatric UC patients are females (Kelsen and Baldassano 2008).

Heritability of IBD has been suspected since and first described at the beginning of the twentieth century. However, details on the complex genetic traits associated with IBD could not be identified and confirmed until modern times with the invention and broad application of GWAS and *next-generation sequencing* (NGS) technologies. Here, studies analyzing familial occurrence of IBD clearly indicated that familial cases of IBD confer the highest risk for developing IBD (Orholm et al. 1991). While there is a pronounced interfamilial heterogeneity of the clinical presentation of IBD cases, intrafamilial IBD manifestations are usually surprisingly homogenous, further supporting the interpretation that genetic cues underlie the manifestation of specific familial IBD phenotypes (Satsangi et al. 1996; Colombel et al. 1996). The importance of genetics gained further attention in the light of study cohorts that compared both IBD incidences and phenotypic manifestations among mono- and dizygotic twins. Here, numerous studies with the first published in the late 1980s showed a higher concordance rate in monozygotic compared to dizygotic twins especially in CD (Tysk et al. 1988). One conclusion of these studies is that genetic traits are more affecting CD than UC pathology. Nevertheless, drawing conclusion on twin studies are difficult to interpret since results are confined by the fact that besides the common genetic pool twins usually share similar environmental signals acquired both, during pregnancy and/or during infancy, which might critically impact IBD susceptibility. Regardless, however, population-centered genetic IBD research over the last five decades built the basis for more in-depth genetics-based approaches that have been performed over the past 10–20 years.

11.3 Genetic Susceptibility in IBD

Genetic linkage studies generally aim at identifying abnormal chromosomal regions that are enriched in populations suffering from complex diseases such as IBD when compared to matched healthy individuals. Hence, this approach represents a largely hypothesis-free, however, valuable tool to decipher the genetic basis of complex, non-Mendelian diseases. The identification of disease-associated chromosomal regions usually reflects the first step that is followed by a more detailed characterization of the precise nature of the mutation within the allelic variant. Finally, the description of the affected gene or genetic region should be complemented by attempts to experimentally prove that the variant functionally contributes to an increased risk for the disease of choice. Historically, HLA class II genes were among the first described genetic alterations that had been associated with an increased risk for IBD and other immune mediated disorders (Stokkers et al. 1999). However, first specific associations between distinct chromosomal loci and IBD development resulted from first-generation genetic linkage studies in the mid- to late 1990s which identified a total of nine IBD (IBD1–9) loci (Ahmad et al. 2001).

One of these genes, located within the IBD1 locus, initially found by GWAS is the protein *nucleotide-binding oligomerization domain-containing protein 2* (NOD2), an intracellular sensor for microbial-derived components (Hugot et al. 2001; Ogura et al. 2001). NOD2 is specifically linked to CD but also to acute graft versus host disease in patients undergoing allogeneic stem cell transplantation, whereas associations to UC were not found. Consecutive studies helped to molecularly define a number of distinct single nucleotide polymorphisms (SNPs) that were shown to increase disease susceptibility, including allele variants of the *NOD2* gene or within the *NOD2* locus, respectively (Hugot et al. 2001; Ogura et al. 2001). Notably, carriage of one risk-conferring allele leads to a two- to threefold increased risk of CD, while homozygosity as well as the heterozygous presence of two risk alleles particularly enhances the risk up to 20- to 40-fold (Lesage et al. 2002). Studies on NOD2 biology in the context of IBD revealed important mechanisms that putatively contribute to IBD pathology and paved the way to better understand gene-environment interactions linking microbial triggers to the immune system (Fig. 11.1a). NOD2 is a pattern recognition receptor that is broadly expressed within both immune and epithelial cells. Three of the most abundant SNPs were shown to result in functionally hampered NOD2 activity subsequently leading to diminished nuclear factor κB (NF-κB) activation upon the presence of the NOD2 ligand *muramyl dipeptide* (MDP) that displays a cell wall component of most bacteria. In mice, however, *Nod2* deficiency leads to changes of the microbial composition in the intestine under physiological conditions, while the bacterial clearance was compromised in the case of an intestinal infection (Petnicki-Ocwieja et al. 2009). Hence, one hypothesis deduced from these preclinical data is that impaired microbial-host interactions in patients

Figure 1A

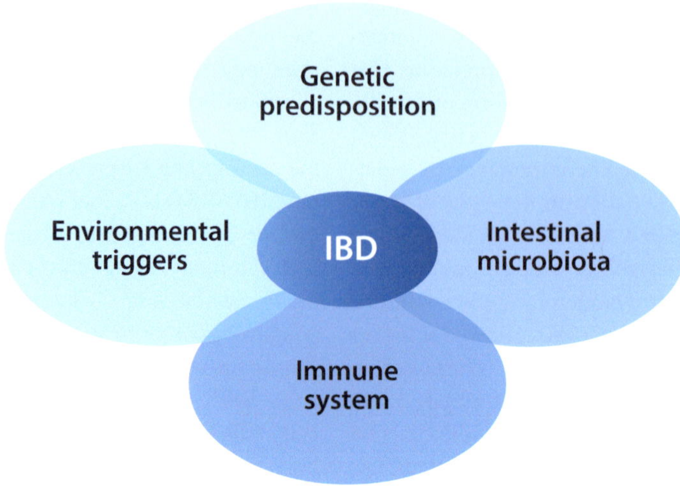

Figure 1B

Inflammatory bowel diseases (IBD)	Ulcerative Colitis (UC)	Crohn's Disease (CD)
Epidemiology	***Diseases of the Northern hemisphere*** - highest annual indicences (about 10-15 / 100.000) in North America & Europe - increasing incidences in selected Asian countries are paralleling western life-style adaptations Peak incidence between ***second and fourth decade of life***	
	m ≥ f	f ≥ m
Clinical presentation & anatomic distribution	Chronic relapsing ***large bowel inflammation*** Small bowel rarely affected (backwash-ileitis) ***Obligatory*** involving the ***procto-rectal region with ascending dissemination***	***Chronic inflammation*** putatively of the ***entire GIT*** ***Small bowel (terminal ileum) > 80%*** affected About 30% of patients have perianal disease while in 20% of patients the disease is limited to the colon
Pathology - macroscopic	Inflammatory affections limited to the ***mucosa*** Usually ***circumferrential & continuous lesions***	***Transmural*** inflammation with ***fistula*** formation Usually ***discontinuous ("skip") lesion***
- microscopic	***Crypt abscessess*** (UC >> CD)	***Granulomas*** (CD >> UC)
Clinical manifestations - intestinal	***<u>Diarrhoe</u>***: frequently bloody, hemorrhagy	***<u>Diarrhoe</u>***: more commonly non-bloody, watery
	<u>Abdominal pain</u>: less typical rarely fistulas	***<u>Abdominal pain</u>***: typically observed frequently fistulas -> abscess
	<u>Upper GIT lesions</u>: absent	***<u>Upper GIT lesions</u>***: e.g. oral aphtous ulcers in 5-15%
- extraintestinal	*Joint and connective tissue affections* (e.g. ankylosing spondylitis) *Eye involvement* (e.g. uveitis) *Skin lesions* (e.g. erythema nodosum or pyoderma gangrenosum) *Cholangiopathies* (e.g. primary sclerosing cholangitis [PSC]) *Hypercoagulability* (e.g. venous thrombembolism) *Kidney lesions* (e.g. renal stones) *Pulmonary manifestations* (e.g. Bronchiolitis obliterans with organizing pneumonia, BOOP)	
Complications	***Intestinal Hemorrhagy*** ***Intestinal fibrosis, strictures, short bowel syndrom*** ***Colorectal cancer*** (CU > CD), ***cholangiocarcinoma***	***Phlegmona or abscesses*** with sepsis -> loss of enteral functionality, malabsorption (esp. if PSC+); ***small bowel cancer*** (CD)

Fig. 11.1 Etiologic, epidemiologic, and clinical features of IBD. (**a**) IBD describe a number of disease states that manifest in genetically susceptible individuals. IBD are characterized by recurrent episodes of intestinal inflammation resulting from an uncontrolled immune response that is considered to be initiated and/or promoted by both

carrying NOD2-associated risk factors for CD could result in a harmful modulation of the composition and/or location of the intestinal microbiota. In the occasional case of a disruption of the intestinal epithelial barrier (e.g., in the course of an intestinal infection), increased bacterial translocation and/or hampered elimination of those bacteria might result in an over-shooting, inadequate immune stimulation that is frequently observed in IBD patients. However, *NOD2* alleles that confer an increased risk of CD are common variants found in >5% of the population (Hugot et al. 2001). Additionally, *NOD2* mutations are predominately found in CD sub-cohorts displaying an ileal manifestation and are almost exclusively detected in Caucasian but not in Asian populations. Conclusively, accumulating knowledge on NOD2 genetics and biology are valuable in the understanding of host-microbe interactions, but, however, can be only applied to subsets of CD patients, vividly illustrating the shortcoming of genetics-based approaches to decipher complex disease mechanisms in the pathogenesis of the vast majority of adult-onset IBD forms.

Up to now, more than 200 genetic loci associated with an increased risk for IBD have been identified by GWAS considering cohorts both from European and non-European populations (Jostins et al. 2012; Liu et al. 2015). Within these risk loci, candidate genes coding for proteins were identified that crucially impact the innate (CARD9, IL18RAP, NOD2, etc.) and adaptive immune response (IL-23R, Stat3, Jak2, CCR6, RORC, IRF4, NFkB1, IL-21, PRDM1, Bach2, etc.), the epithelial barrier function, integrity, and repair (NKX2-3, Stat3, NOD2, Rel, MUC19, etc.) as well as the autophagy response (ATG16L1, etc.). Strikingly, these results particularly illustrate to what extent a largely hypothesis-free approach can unravel numerous disease-associated pathways, thus providing enormous insight into the diversity of affected cellular compartments and important signaling mechanisms in IBD (Liu and Stappenbeck 2016). Nevertheless, these studies further showed that the majority of the identified loci are not specifically associated with CD and/or UC, respectively, but is further linked to the occurrence of other immune-mediated disorders, such as psoriasis and arthritis.

In summary, genetic approaches, such as GWAS, exponentially enlarged the knowledge on the mechanisms and pathways underlying IBD. However, in line with a non-Mendelian inheritance at least in adult-onset IBD, an allelic variant alone, i.e., a single IBD susceptibility locus, in average merely affects the individual probability to develop IBD but of note can specifically entail a high risk with regard to a homozygous carriage of one allelic variant, such as the NOD2 risk allele (Ogura et al. 2001). In addition, the currently known susceptibility loci for IBD are predicted to only account for ~13% of the CD and ~8% of the UC cases, again underscoring the multifactorial pathogenesis and especially the impact of environmental aspects for IBD manifestation (Jostins et al. 2012). Finally, as a matter of fact, the causal genomic regions or even the involved genes have remained unknown for the majority of the identified genomic risk loci. Hence, future studies should take advantage of novel technologies, such as *whole genome* and/or *whole exome sequencing* (WES), respectively, to cover uncommon alleles and molecularly pinpoint genetic alteration within single IBD patients beyond the level of resolution that is reached by GWAS-based analyses. However, as *a conditio sine qua non*, functional studies need to flank genetics-based approaches including NGS-derived data sets to evaluate the pathogenetic relevance of mutations in IBD.

Fig. 11.1 (continued) environmental and microbial signals. (**b**) The two major subtypes of IBD, ulcerative colitis (UC) and Crohn's disease (CD), are characterized by chronic or frequently relapsing episodes of intestinal inflammation. Many disease-related features are shared by both forms of IBD, while only very few epidemiological, pathological, or clinical aspects are indicative for the presence of a certain IBD subset

In contrast to the non-Mendelian inheritance of adult-onset IBD, very early- and early-onset forms of IBD that is observed in pediatric patients often represent monogenic Mendelian disorders. They are frequently associated with a positive family history for IBD-like disorders (Kelsen and Baldassano 2008). Rare, monogenic diseases are usually missed by classical GWAS. However, as NGS technologies (e.g., WES) emerged, a series of single mutants could be identified and functionally validated to be causatively involved in these early manifesting pediatric IBD forms. Mutations identified by this approach include genes that usually account for the integrity of the epithelial barrier function (dysregulated upon deletion of, e.g., *ADAM17*) (Blaydon et al. 2011) and prevent hyper-inflammatory states due to the proper regulation of apoptosis (dysregulated upon deletion of, e.g., *XIAP*) (Worthey et al. 2011) or due to the uncompromised functionality of regulatory T cells (e.g., dysregulated upon mutations of *FOXP3*) (Okou et al. 2014).

In summary, genetic findings, i.e., allelic variants which are detectable within adult-onset IBD patient cohorts, have strongly informed and inspired the field of IBD research since these studies confirmed previous findings but further uncovered so far less recognized pathways (e.g., autophagy, cell stress) to be relevant in IBD pathogenesis. However, the overall power of individual chromosomal aberrations is increasingly seen to be limited and superseded by the impact of environmental signals. In contrast, however, genetics-based approaches are a preserve to particularly evaluate rare forms, but not exclusively in pediatric IBD patients with early-onset forms. Here, a monogenic etiology often results in the manifestation of immune-mediated disorders including an IBD-like disease. Hence, NGS-based approaches of individual IBD cases will be the method of choice within the near future, hereby simultaneously integrating both host-derived genetic and environmental (e.g., from intestinal microbiota) triggers and behavioral pattern (e.g., diet).

11.4 Pathology and Clinical Manifestation

The clinical presentation of IBD patients depends on the obligatory presence of inflammatory manifestations within the gastrointestinal tract, while extraintestinal manifestations are frequently but variably observed (Podolsky 2002). Intestinal manifestations of UC are characterized by chronically persisting and/or relapsing ulcers within the large bowel that are morphologically superficial, i.e., limited to the intestinal mucosa. In 10–20% of the cases, however, inflammatory lesions can also be found in the terminal ileum, which is described by the so-called backwash ileitis. In general, the rectum is obligatorily afflicted in UC. From there, the colitis manifestation continuously ascends in a retrograde circumferential and in terms of longitudinal extent variable fashion.

In contrast to UC, inflammation in CD occurs within the entire gastrointestinal tract, i.e., from the oral cavity to the anus. Here, more than 80% of CD patients show a terminal ileitis. CD lesions are characterized by a segmental, discontinuous distribution of typically aphthous or confluent deep linear ulcers. These lesions are also termed "skip lesions" due to the patchy, adjacent coexistence of healthy and highly inflamed areas within the same anatomic location giving the macroscopic impression of a cobblestone pattern. Another critical feature of CD-associated lesions is the transmural extent of the inflammatory affection of the bowel wall. Transmural inflammation is assumed to represent the preexisting condition that facilitates and precedes the development of fistulas, which is especially found in the small bowel and the perianal region. Here, transmural inflammatory lesions rather randomly inflict adjacent tissues and organs resulting in the formation of penetrating fistulas. Both interenteric (i.e., between two intestinal loops), entero-genital, entero-vesical, and entero-cutaneous fistulas are frequently observed and often accompanied by the formation of abscesses and septic complications (Gupta et al. 2010). Another complication associated with chronic intestinal inflammation represents the development of intestinal fibrosis and—if fibrosis is restricted to a distinct anatomic location—of strictures overall hampering the functionality of the gut, which further increases the risk for additional complications as, e.g., ileus formation (Rieder and Fiocchi 2008). Altogether, macroscopic pathological manifestations often guide the way to diagnose IBD and to identify the subtype on the basis of the endoscopic and imaging-based (e.g., ultrasound or MRI of the small intestine) distribution pattern of intestinal inflammation. However, none of the visible signs of intestinal inflammation is sufficient to unequivocally diagnose IBD and/or discriminate between IBD subtypes (Fig. 11.1b).

In this regard, histopathological assessments of samples that are obtained by endoscopy-guided biopsies or surgical resection are considered to be potentially helpful in characterizing intestinal tissue manifestations and subtyping IBD (Magro et al. 2013), but are not sufficient to unequivocally diagnose IBD. Histopathological correlates typical for UC are represented by widespread crypt architectural distortion and depletion of mucin as well as by a diffuse transmucosal inflammatory infiltrate with basal plasmacytosis that putatively triggers cryptitis and the formation of crypt abscesses (Magro et al. 2013). In delineation from UC, focal chronic inflammation together with crypt irregularities and granulomas that are not associated with crypt injuries is assumed to reflect CD-associated colonic and less accurately ileal manifestation. Overall, however, discrimination of CD and UC entities that are based on detailed histopathological criteria has not been reliably established and reproducibly achieved, even among experts in the field of gastrointestinal pathology.

While the onset of IBD is assumed to represent the starting point of a lifelong and per se incurable disease state, type, characteristics, and severity of clinical presentations among IBD, patients are highly variable. Moreover, even within the same IBD patient, the distribution pattern and the longitudinal dynamics of the clinical manifestation and symptomatology are highly volatile, thereby indicating a high degree of intraindividual phenotypical adaptations over time, which is presumably due to so far poorly understood changes in immunological processes underlying IBD. Despite these limitations, all IBD patients usually show at least one of the two hallmark symptoms reflecting intestinal manifestations: (a) irregularities/changes of the stool habits (i.e., mostly diarrhea but also obstipation, increase of stool frequency, bloody feces, hemorrhagy) and/or (b) abdominal pain sensations. Any of these symptoms justify the initiation of further diagnostic testing (e.g., microbiological stool analyses, endoscopy of the upper and lower gastrointestinal tract with histopathologic evaluation of biopsies, ultrasound of the small and large bowel) that aims to confirm or dismiss the diagnosis of IBD. From a clinical point of view, UC patients typically develop more frequently bloody diarrhea, while abdominal pain episodes are less prevalent in UC (Podolsky 2002). However, the most abundant symptom in CD represents abdominal pain episodes, which are putatively due to and/or indicative for the affection of the small bowel. The increased burden of abdominal pain among CD patients is also partly dependent on the appearance of penetrating fistulas and abscesses. Furthermore, CD patients who are describing either upper abdominal and chest pain, respectively, or seem to have difficulties to chew or swallow might be screened for the presence of upper gastrointestinal tract lesions, especially in the oral cavity, the esophagus, stomach, or small intestine.

In the majority of IBD patients, IBD are associated with the occurrence of extraintestinal disease manifestations over time (Levine and Burakoff 2011). Here, similar to the intestinal inflammation, the underlying pathomechanisms are largely unknown. However, it is assumed that extraintestinal manifestations are directly or indirectly affected by the pro-inflammatory micromilieu. Although extraintestinal manifestations can be virtually found in all organ systems, the predominately affected sites represent the musculoskeletal (e.g., arthritis), the dermal (e.g., erythema nodosum), the ocular (e.g., uveitis), the hepatopancreatobiliary (e.g., primary sclerosing cholangitis, PSC), the pulmonary (e.g., obstructive lung disease) and the renal systems (e.g., calcium oxalate kidney stones), or the vasculature (e.g., venous thromboembolism) (Levine and Burakoff 2011). In contrast to the clinical typology of intestinal manifestations, there is a less stringent assignment of extraintestinal disease manifestations to either IBD subtypes, although in clinical practice and according to epidemiologic studies, certain disease states appear to be more frequently associated with CD (e.g., erythema nodosum) and UC (e.g., pyoderma gangrenosum), respectively, as the underlying IBD condition.

Finally, both intestinal and extraintestinal disease manifestations promote the development of long-term complications, such as intestinal dysfunction due to fibrosis, strictures, or short bowel syndrome in resected patients (Rieder and Fiocchi 2008), cancer development within the large (UC and CD patients with colon involvement) and less frequently small bowel (in CD) (Grivennikov et al. 2010), occurrence of carcinomas arising from fistulas, and cholangiocarcinoma (Gulamhusein et al. 2016). Together with the multifaceted spectrum of clinical symptoms directly elicited by chronic intestinal and extraintestinal inflammatory processes, long-term complications contribute to and add another level of morbidity to the preexisting disease burden prevalent in IBD patients.

11.5 Microbiome Changes in Human IBD

The human gastrointestinal (GI) tract represents the most densely colonized organ of the body, with the highest microbial load of 10^{11} bacteria/mL content in the colon (Sartor 2008). Bacteria dominate the microbial ecosystem in the GI tract, with more than 90% belonging to the phyla *Bacteroidetes* and the *Firmicutes*, but archaea, viruses, yeasts, and protozoa contribute to the diversity of this microbial ecosystem (Eckburg et al. 2005; Human Microbiome Project 2012; Lozupone et al. 2012). At the species level, high-throughput 16S rRNA gene sequencing and metagenomic analysis allow the detection of approximately 100–300 operational taxonomic units (OTU) per individual and overall, more than 1500 different species of bacteria were detected in human fecal samples and biopsies (Li et al. 2014). Nevertheless, interindividual variation of microbiota composition and spatial differences that occur along the proximal to the distal part of the intestine need to be considered in the evaluation of microbiome signatures relevant for the pathogenesis of IBD (Knights et al. 2011). Thus, a profound functional understanding of the microbiome is crucial in order to explain the complex intestinal ecosystem and its interaction with the host. The evidence that many chronic disorders, such as allergies and autoimmune and metabolic diseases, are associated with compositional shifts of the gut microbiota in clinical and/or animal studies underlies the therapeutic importance of this field of research.

The identification of genetic risk factors in IBD paved the way for a better mechanistic understanding of the pathogenesis (Jostins et al. 2012; Liu et al. 2015). Nevertheless, the presence of environmental triggers is thought to play the dominant role in the etiology of IBD (Kaplan 2015; Renz et al. 2011). Fecal stream diversion in subsets of patients with active Crohn's disease provided first clinical evidence for a central role of bacteria in the pathogenesis of this chronic inflammatory disorder (Rutgeerts et al. 1991). In addition, GWAS identified a variety of target genes that point toward a disruption of microbe-host interactions, including genetic loci associated with microbial sensing and clearance as well as resilience mechanisms to cope with accumulating cell stress. Paneth cells that affect epithelial stem cell homeostasis and antimicrobial defense in the small intestinal crypt region and genetic variations in *NOD2*, *ATG16L1*, and *XBP1* support the hypothesis that disturbed Paneth cell activity contributes to the pathogenesis at least in a subset of patients with ileal phenotype (Fritz et al. 2011) (Fig. 11.2). High-throughput sequencing analysis of microbial communities was used to identify risk patterns associated with distinct IBD phenotypes (Frank et al. 2007), first-degree relatives (Joossens et al. 2011), and geographic distribution of disease manifestation (Rehman et al. 2016). One representative study

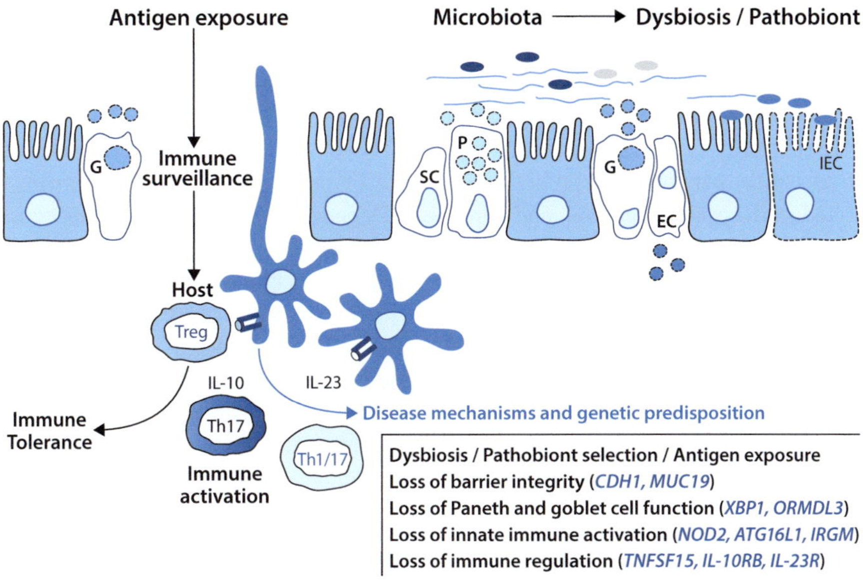

Fig. 11.2 Barrier function and immune surveillance at the epithelial interface—linking genetic susceptibility to IBD pathogenesis. The single cell layer (10 μm) of intestinal epithelial cells (IECs) is dynamically renewed by stem cells (SC) at the crypt basis giving rise to absorptive enterocytes (IEC), Paneth cells (P), goblet cells (G), and enteroendocrine cells (EC). Collectively, the intestinal interface maintains barrier function and orchestrates immune surveillance to luminal antigens leading to immune tolerance under normal conditions. In IBD, barrier integrity of the epithelium is disrupted at all levels, e.g., the secretion of antimicrobial peptides by Paneth cells and mucus production by goblet cells, leading to chronic inflammation. Microbiota changes including dysbiosis or pathobiont selection contribute to IBD pathogenesis. Genetic susceptibility (examples of risk genes are indicated in blue) shifts intestinal tissue and immune responses toward chronically relapsing, inflammatory processes

demonstrated that bacterial diversity and composition of mucosa-associated and fecal samples substantially differed between Crohn's disease phenotypes, and the most pronounced effects were shown for patients with ileal compared to colonic disease involvement (Willing et al. 2010). A recent analysis of a large cohort (RISK) including 447 treatment-naïve new-onset pediatric CD and 221 non-IBD control samples confirmed the association between disease severity and low species richness (alpha diversity) (Gevers et al. 2014), and microbiome signatures might even predict clinical responses to biology therapy. Interestingly, overall community structures only differed between patients and controls when correlated with ileal gene expression, suggesting that individual patterns of microbiota composition or function are linked to host responses including disease phenotype, activity, and location. Despite the availability of data from these larger cohorts, consensus about specific disease-relevant taxa in IBD is still hampered. Meta-analyses of combined 16S sequence data sets from all cross-sectional studies might help to increase sample size; however, knowledge extraction from this approach is heavily confounded by technical differences in sample collection, storage, and extraction as well as age, geographic location, medication, and disease phenotypes/activity of the various study individuals (Lozupone et al. 2013). Due to this noise in the data sets, it seems unlikely that in the absence of mechanistic understanding, the sole description of microbial communities, including their gene repertoires will identify IBD-relevant phylotypes or disease-conditioning bacterial networks across a broader range of patients. Host genomic loci might have an impact on the composition and functionality of the bacterial community, thus making twin cohort studies an excellent tool to limit interindividual differences and to control the environment. In patients with ileal Crohn's disease, Willing and co-workers showed significantly lower levels of *Faecalibacterium* and *Roseburia* and higher levels of *Enterobacteriaceae* and *Ruminococcus* compared to their healthy twins (Willing et al. 2010). In a cohort of ulcerative colitis patients, discordant twins showed a reduced species richness and alpha diversity. At the compositional level, *Actinobacteria* (mostly *Rhodococcus* genus) and *Proteobacteria* (mainly *Enterobacteriaceae*, i.e., *Shigella/Escherichia*) were enriched in diseased twins compared to their healthy siblings (Lepage et al. 2011). This evidence is of particular importance because it indicates that the gut microbiota composition is affected much more by the disease state rather than the genetic component in the context of ileitis and colitis, emphasizing that familial aggregation, and thus genetic background and environment, may promote low microbial diversity. How these factors favor the development of IBD in some individuals and not in their siblings is still a forum of speculations. In conclusion, some bacterial species seem to be consistently associated with protective effects, such as *Faecalibacterium prausnitzii* (Sokol et al. 2008), while others are linked to adverse clinical situations, including *Pasteurellacaea* and *Enterobacteriaceae*, *Escherichia coli, and* specifically adherent-invasive *E. coli* (*AIEC*) (Frank et al. 2007; Rehman et al. 2010). AIEC and *Mycobacterium avium subsp. paratuberculosis* are frequently associated with the pathogenesis of inflammation in Crohn's disease patients leading to the speculation that transient infection or specific pathobionts rather than complex microbial signatures are causally linked to IBD pathogenesis.

Dysbiosis is considered as an alteration in microbiota community structure and/or function, capable of causing/driving a detrimental distortion of microbe-host homeostasis that specifically initiates or propagates disease. Loss of species richness correlates with disease activity in various studies and subsets of IBD patients; however, the "egg or hen" question related to the cause or consequence in the context of inflammation-driven changes in the microbiota remains unanswered. In addition and similar to IBD, other immune-mediated pathologies such as type 1 diabetes (T1D) also show low species richness at disease-onset questioning the specificity of this readout in IBD (Kostic et al. 2015). We previously showed in a clinical intervention trial using oral or intravenous iron replacement therapy in IBD patients that treatment-induced changes in the community structure and metabolic landscape of the microbiota are not necessarily associated with alterations in disease activity (Lee et al. 2017). Thus, the need to define the functional relevance and specificity of single bacteria (pathobiont) or bacterial networks (dysbiosis) is most important for the understanding of microbe-host interactions in the pathogenesis of IBD (Fig. 11.2).

11.6 Pathogenesis of Microbe-Host Interactions

Despite considerable progress the functional complexity of the microbiome is still unresolved, and to date, mechanisms of microbe-host interactions involve a pleiotropic network of immune, metabolic, and trophic functions. Studies in germ-free animals recognized the essential role played by the intestinal microbiome in the development and regulation of the mucosal immune system during early life (Macpherson and Harris 2004). While many organisms have been shown to fulfill protective functions in the GI tract and are critical for host physiology, complex shifts in the community structure and abundance of certain microbes have been associated with the onset of inflammatory diseases.

11.6.1 IBD Mouse Models

A substantial number of different mouse models develop IBD-like pathologies (Fig. 11.3). Although none of them completely replicate the human disease phenotype, at least 74 genetically engineered mice spontaneously develop intestinal inflammation, and 20 of these experimental models carry IBD-relevant susceptibility genes (Mizoguchi et al. 2016). Various broad-spectrum

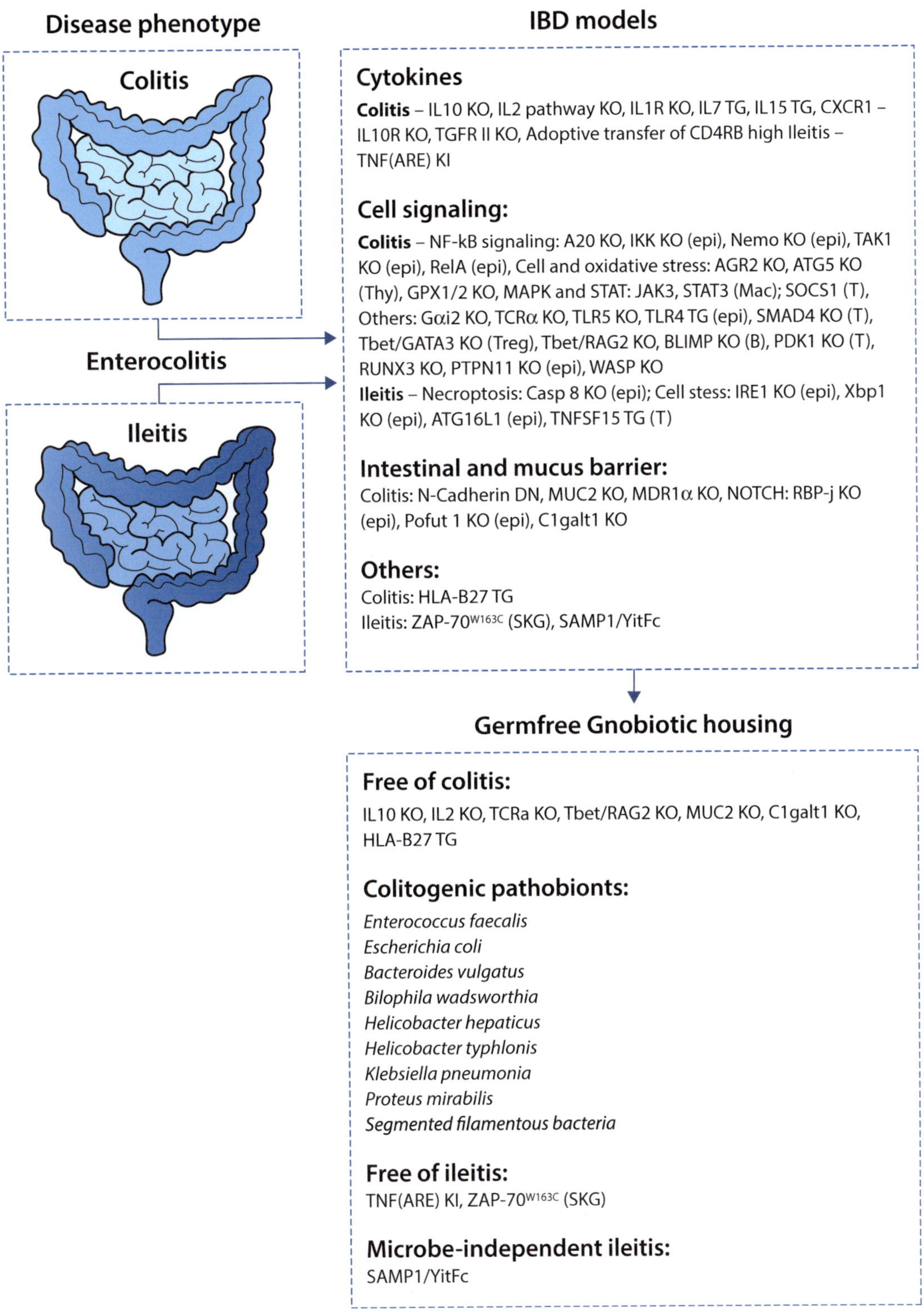

Fig. 11.3 Overview about IBD-relevant animal models. Complete- and tissue-specific genetic manipulation (KO, knockout; KI, knock-in; TG, transgenic) mouse and rat (e.g., HLA-B27 TG) models are summarized according to

antibiotics have a profound impact on the incidence and severity of disease in an array of immune-related experimental disease models, suggesting that the intestinal microbiota is an important disease modulator, for example, in experimental IBD, T1D, and experimental autoimmune encephalopathy (EAE). However, one important way to investigate the role of the (intestinal) microbiota in the context of specific host functions or complex diseases is the use of germ-free experimental disease models. Mono-association studies or studies using selected consortia of microorganisms allow important mechanistic insights into specific microbe-microbe and microbe-host interactions under highly controlled and standardized conditions regarding genotype, diet, and microbial colonization (Hormannsperger et al. 2015).

As introduced above, *NOD2* is an important susceptibility gene for Crohn's disease associated with a loss of function mutation (Hugot et al. 2001; Ogura et al. 2001); however, *Nod2*$^{-/-}$ mice fail to spontaneously develop inflammation but show increased susceptibility to *Listeria monocytogenes* infection (Kobayashi et al. 2005). Most IBD animal models show reduced or lack of disease under germ-free conditions, indicating that the intestinal microbiota is involved in the development of most complex diseases. At present, the majority of IBD-related animal models are suitable to study chronic inflammation in the colon (Magro et al. 2013), but only a few models develop Crohn's disease-like pathology, including transmural inflammation in the ileum, cobblestone lesions in the mucosa as well as extraintestinal disease manifestations such as arthritis (Mizoguchi and Mizoguchi 2010). SAMP1/YitFc and *Tnf* $^{\Delta ARE/+}$ mice spontaneously develop chronic inflammation in the ileum and combine most of the Crohn's disease-like pathological manifestations (Rodriguez-Palacios et al. 2015). In genetically susceptible *Tnf* $^{\Delta ARE/+}$ mice, inflammation-associated changes in the microbiota caused transmissible disease in the genetically susceptible host (TNF$^{\Delta ARE}$ mice) but not in wild-type (WT) recipients, while germ-free mice remain disease-free, demonstrating that microbial communities, although developed under the same experimental conditions, may undergo progressive changes in their microbial environment, leading to a more aggressive phenotype over time (Schaubeck et al. 2016). Interestingly, ZAP-70 ^{W163C}–mutant (SKG) mice which develop spondyloarthritis and ileitis upon intraperitoneal injection of β-1,3-glucan (curdlan) remain completely disease-free under germ-free conditions (Rehaume et al. 2014), suggesting a role for the microbiota also in joint inflammation. In addition, targeted gene deletion and inducible Crohn's disease models such as *Caspase8*$^{\Delta IEC}$ (Gunther et al. 2011) and *Xbp1*$^{\Delta IEC}$ (X *Atg16l1* $^{\Delta IEC}$ or *Atg7*$^{\Delta IEC}$) (Adolph et al. 2013) have been recently developed. However, mechanisms of microbe-host interactions in these models are not well established due to the lack of systematic studies with germ-free mice.

11.6.2 Bacterial Sensing at the Epithelial Interface

Loss of epithelial barrier function and innate immunity are fundamental to the pathogenesis of inflammatory and infectious diseases (Fig. 11.2). The intestinal immune system has the challenge of responding to pathogens while remaining tolerant to food antigens and the commensal microbiota. The intestinal epithelium executes a compartmentalization between the lumen and the host, simultaneously acting as a selectively permeable first line of defense to fulfill its function of absorption while maintaining an

Fig. 11.3 (continued) the anatomical site of inflammation (ileitis, colitis) and the cellular/molecular targets of genetic manipulation (cytokine expression, signaling molecules, barrier integrity). Most of the IBD models are disease-free under germ-free housing conditions. Gnotobiotic mice colonized with selective pathobionts reestablish disease

effective barrier against the intestinal microbiota, antigens, and toxins. Intestinal epithelial cells (IECs) express pro-inflammatory cytokines in response to infectious invasive bacteria but largely ignore nonpathogenic commensals. Certain intestinal pathogens and opportunistic commensals, however, can evade this first line of defense and enter IECs, suggesting that the existence of IEC-intrinsic immune mechanisms for bacterial detection and limitation is essential. One key cell-autonomous mechanism of antibacterial defense is IEC autophagy, shown to be activated following bacterial invasion through adaptor protein myeloid differentiation primary response gene 88 (MyD88) cell-intrinsic signaling, with autophagy deficiency in mice causing increased dissemination of invasive bacteria (Benjamin et al. 2013), indicating that autophagy could have a broader role in inflammatory diseases. IECs and innate immune cells of the lamina propria are able to differentiate self from nonself through a selective spatial and cellular expression of pattern recognition receptors (PRRs). Classically the detection of pathogen-associated molecular patterns (PAMPs) allows the intestinal epithelium to activate signaling pathways that induce the early host response to infection. The role of microbe-associated molecular patterns (MAMPs) in mediating innate recognition of the commensal "noninfectious" microbiota remains controversial. Paradoxically, recent progress in understanding the IBD pathogenesis suggests that a defective innate immune system predisposes the host toward chronic inflammation (Liu et al. 2015; Khor et al. 2011; Jostins et al. 2012), supporting a protective role of PRR signaling in maintaining intestinal tissue homeostasis. Early work related to the activation of inflammation-related transcription factors, such as NF-κB, suggested a hormetic adaptation of the epithelium in response to commensal bacteria (Haller et al. 2002, 2003), with elegant studies related to IEC-specific inhibition of NF-κB activation validating the importance of this signaling pathway in maintaining tissue homeostasis (Nenci et al. 2007). This paradigm shift was supported by Medzhitov and colleagues, demonstrating that microbiota-derived signals via the toll-like receptor (TLR)-related adaptor protein MyD88 protect mice from the development of colitis (Rakoff-Nahoum et al. 2004) and intestinal tumor formation (Rakoff-Nahoum and Medzhitov 2007). Thus, bacteria (dead or alive) and their metabolites form key mediators for the cross talk between IECs and other mucosal cell types, through the interaction with host PRRs (Table 11.1).

In spite of the symbiotic nature of the microbe-host relationship, the close proximity of bacteria to intestinal tissue poses a considerable health challenge. An effective and dynamic intestinal epithelial barrier is therefore crucial to conserve a compartmentalized microbe-host interaction and tissue homeostasis. In the healthy organ, the epithelium maintains a distinct anatomical barrier relevant for a constant state of homeostasis while being exposed to a myriad of environmental stimuli that include, but are not limited to, microbes, dietary products, and inorganic materials. A single cell layer of IECs forms a continuous physical barrier, with tight junctions connecting adjacent IECs and associating with cytoplasmic actin and myosin networks that regulate intestinal permeability. Long-lived pluripotent stem cells located at the base of intestinal crypts continuously produce tissue-specific precursor cells that transit through a differentiation pathway that gives rise to absorptive lineage cells (enterocyte/colonocyte) or secretory lineage cells (goblet, Paneth, enteroendocrine, and tuft cells) (Sood et al. 2009). IECs represent not merely a physical barrier but contribute to intestinal health through the production of mucus (goblet cells) and the secretion of antimicrobial peptides (AMPs) (Paneth cells) (Fig. 11.2). Goblet cells secrete mucin glycoproteins, of which MUC2 is the main constituent of the approximately 150 μm thick (in the mouse) colonic mucus layer. In the colon, two structurally distinct mucus layers are formed: an inner mucus layer that is devoid of bacteria and an outer mucus layer that forms a habitat for a large number of bacteria (Johansson et al. 2008). Further, MUC2 secretion in small intestinal tissue has been shown to impact antigen sampling by dendritic cells, thereby enhancing gut homeostasis (Shan et al. 2013). Intestinal

Table 11.1 Mammalian TLRs with their associated ligands and adaptors

TLR	Ligand(s)	Adaptor(s)
TLR1	Lipopeptides, modulin	MyD88/MAL
TLR2	Lipoteichoic acid, lipoproteins, peptidoglycan, yeast mannans, lipoarabinomannan	MyD88/MAL
TLR3	dsRNA	TRIF
TLR4	Lipopolysaccharide, envelope proteins, glycosyl-phosphatidylinositols	MyD88/MAL/TRIF/TRAM
TLR5	Flagellin	MyD88
TLR6	Modulin	MyD88/MAL
TLR7	ssViral RNA	MyD88
TLR8	ssViral RNA	MyD88
TLR9	CpG-containing DNA	MyD88
TLR10	Undefined	Unknown
TLR11	*Toxoplasma gondii* profilin-like protein	MyD88
TLR12	*Toxoplasma gondii* profilin-like protein	MyD88
TLR13	Bacterial ribosomal RNA sequence "CGGAAAGACC"	MyD88/TAK-1

dsRNA double-stranded RNA, *ssRNA* single-stranded RNA, *MyD88* myeloid differentiation primary response gene 88, *MAL* MyD88-adapter-like, *TRIF* TIR-domain-containing adapter-inducing interferon-β, *TRAM* TRIF-related adaptor molecule, *TAK-1* transforming growth factor-β activated kinase 1

Paneth cells are the main source of AMPs that function in host defense and in establishing and maintaining the intestinal microbiota (Vangay et al. 2015). Secretory immunoglobulin A (sIgA) directed against intestinal bacteria and produced by lamina propria plasma cells binds the polymeric immunoglobulin receptor (pIgR) and transcytoses across the epithelium to prevent microbial translocation across the epithelial barrier (Macpherson and Uhr 2004). This concerted interplay between plasma cells and IECs provides an adaptive immune element to the intestinal epithelial barrier. Also found scattered throughout the lamina propria are T cells, stromal cells, and antigen-presenting cells such as dendritic cells (DCs) and macrophages. Specialized IECs, called microfold (M) cells, and goblet cells facilitate the transport of luminal antigens and bacteria across the intestinal epithelial barrier to DCs, with macrophages sampling through transepithelial dendrites (Rescigno et al. 2001). Under steady-state conditions, the intestinal immune system detects commensal bacteria and provides basal signals without full activation of immune responses.

The intestinal microbiota forms part of the intestinal barrier by limiting bacterial colonization and stimulating epithelial turnover. For example, *Bifidobacteria* species produce high concentrations of the short-chain fatty acid (SCFA) acetate and can thereby prevent enteropathogenic *Escherichia coli* (EHEC) infection and its Shiga toxin release (Fukuda et al. 2011). Similarly, butyrate-producing *Faecalibacterium prausnitzii*, *Eubacterium rectale*, and *Roseburia* species directly target virulence gene expression to prevent bacterial infection (Bibiloni et al. 2005). Studies have demonstrated that bacteria-dependent signals regulate the intestinal epithelial barrier and contribute to its effective functioning. Experiments in germ-free mice have shown that mucus layer thickness is reduced compared to conventionally housed mice and that stimulation with lipopolysaccharide (LPS) and peptidoglycan (PGN) can reverse this to specific pathogen-free (SPF)-like levels of mucus thickness. Similarly, AMP production depends on and is enhanced by the presence of intestinal microbial signals. TLR, NOD-like receptor (NLR), RIG-like receptor (RLR), and C-type lectin receptor (CLR) family members provide distinct microbial signaling pathways in the intestinal epithelium (Abreu et al. 2005; Tursi et al. 2010). Despite evidence from mouse models deficient in PRRs and signaling adaptors (Nenci et al. 2007; Zocco et al. 2006), there is further need for epithelial-specific PRR knockout mice to fully comprehend the role of bacteria-derived signals in intestinal epithelial homeostasis and repair.

The mono-association of germ-free mice with the prominent gut commensal *Bacteroides fragilis* revealed that this bacterium specifically signals through TLR2 on regulatory T cells via its polysaccharide A (Samouilidou et al. 2012) symbiosis factor, to enable its niche-specific mucosal colonization (Furrie et al. 2005). Similarly, the colonization of mice with *B. fragilis* protects against experimental colitis in a TLR2-dependent manner. Mono-colonization in germ-free rats with the commensal *Bifidobacterium lactis* was shown to cause TLR2-mediated MAPK and NF-κB pathway activation in IECs (Ruiz et al. 2005). Furthermore, the colonization of germ-free rodents with *Enterococcus faecalis* or *Bacteroides vulgatus* activates NF-κB signaling and induces chemokine expression in colonic IECs through TLR2 and TLR4 signaling, respectively (Ruiz et al. 2006; Haller et al. 2003). A study in TLR5-deficient mice showed that the cecal microbiota differed from WT littermates in more than 100 bacterial phylotypes (Vijay-Kumar et al. 2010), indicating that TLR signaling has implications in the regulation of the intestinal microbiota. This was also shown in MyD88-deficient mice that demonstrated higher-level cecal *Rikenellaceae* and *Porphyromonadaceae* families (Wen et al. 2008). In the healthy state, mice deficient in TLR signaling (MyD88 deficient, TLR4 deficient, MyD88/TRIF knockouts) do not show any differences in proliferation and IEC barrier function compared to WT mice (Slack et al. 2009; Oliva et al. 2012). Under conditions of injury, however, MyD88-, TLR2-, and TLR4-deficient mice show increased susceptibility to dextran sodium sulfate (DSS)-induced colitis. Despite the importance of PRRs in the regulation of the intestinal microbiota, studies in PRR-deficient mice have shown that only those deficient in TLR5, NLRP6, or RIG-I develop spontaneous intestinal inflammation (Vijay-Kumar et al. 2007; Whelan and Quigley 2013; Steed et al. 2010). This may suggest a major role of compensatory mechanisms, where PAMPs are recognized by multiple host PRRs.

Abnormal PRR signaling is implicated in the development of chronic intestinal inflammation. The cytosolic NLR NOD2 (also known as CARD15) recognizes bacterial peptidoglycan-derived muramyl peptide (MDP) to elicit NF-κB-mediated pro-inflammatory responses and AMP synthesis (Kobayashi et al. 2005). *Nod2*-deficient mice harbor an elevated load of commensal resident bacteria, display dysbiosis, and show a reduced ability to prevent intestinal pathogen colonization (Rehman et al. 2011; Petnicki-Ocwieja et al. 2009). In turn, NOD2 expression depends on the intestinal microbiota, suggesting a feedback mechanism in the maintenance of intestinal homeostasis. In line with the above findings, *NOD2* gene mutations were identified in patients with CD (Hugot et al. 2001; Ogura et al. 2001), suggesting an association with changes in the commensal microbiota that may facilitate disease progression.

The genetically engineered interleukin-10-deficient mouse (IL-$10^{-/-}$) provides a model of spontaneous intestinal inflammation (Kuhn et al. 1993) and has been extensively used to dissect IBD etiology. Evidence for the requirement of resident enteric bacteria for the development of colitis in *IL10*$^{-/-}$ mice stemmed from studies in germ-free animals, where colitis development was not observed (Sellon et al. 1998). It has been shown that the Gram-positive intestinal bacterium *E. faecalis* drives distal colonic inflammation in IL-$10^{-/-}$ mice following mono-association (Balish and Warner 2002). Findings from our own group identified that the virulence factor gelatinase E (GelE) partially impairs intestinal epithelial barrier integrity in IL-$10^{-/-}$ mice (Steck et al. 2011) and that the colitogenic activity of *E. faecalis* was partially and almost completely abrogated when deficient for the enterococcal polysaccharide antigen (Δ*epaB*) and lipoproteins (Δ*lgt*) envelope structures, respectively (Ocvirk et al. 2015). Mono-association of IL-$10^{-/-}$ mice with the commensal bacteria *E. faecalis*, *E. coli*, or *Pseudomonas fluorescens* demonstrated that different commensal species selectively initiate distinct immune-mediated intestinal inflammation in the same host (Kim et al. 2005). Such results invite the hypothesis that particular microbial effectors, or a combination of effectors from different bacteria, are required to elicit pathogenesis or maintain the necessary barrier function for intestinal homeostasis. Additionally, not only the specific bacterium but the susceptibility of the host plays a major role in disease progression, as shown by the induction of colitis by *Bacteroides vulgatus* in HLA/B27-β2m transgenic rats, but not in IL-$2^{-/-}$ mice (Kim et al. 2005; Waidmann et al. 2003).

Identifying bacterial gene products that drive protective rather than pathogenic inflammation in the intestine is crucial to rebalance homeostasis in inflammatory diseases and malignancies. *Lactobacillus* species, such as *L. acidophilus*, are normal inhabitants of the intestinal microbiota and have received considerable attention as beneficial ecosystem members. *L. acidophilus* stimulates DCs through TLR2 via lipoteichoic acid (LTA) to trigger the production of inflammatory and regulatory cytokines. Deletion of the phosphoglycerol transferase gene (LBA0447) that synthesizes LTA generated an *L. acidophilus* derivative (NCK2025) that diminishes colitis when administered orally in a murine colitis model (Mohamadzadeh et al. 2011), confirming the role of LTA in inducing inflammation. In another example, *L. paracasei*, a single strain derived from the VSL#3 bacterial mixture (Sood et al. 2009), was found to secrete the prtP-encoded protease lactocepin with anti-inflammatory effects via the degradation of pro-inflammatory chemokines (von Schillde et al. 2012).

Collectively, the above findings support the notion that the colitogenic activity can be assigned to specific bacterial structures and that such characterizations are indispensable in understanding host-microbe interactions relevant for the development of intestinal inflammation.

11.7 Therapeutic Intervention

Despite the strong correlative evidence of changes in the gut microbial ecosystem and disease activity, functional prove for the causative nature of microbe-host interactions especially the role of complex changes in community structure (dysbiosis) and related clinical adaptation is still lacking in IBD. At present, microbial therapies are intensively studied with the aim to influence the disease activity by modulating the intestinal milieu. Although a variety of therapeutic intention strategies have been applied to modulate the intestinal microbiota including the administration of antibiotics, prebiotics (i.e., dietary components that promote the growth and metabolic activity of beneficial bacteria), probiotics (i.e., beneficial bacteria), or fecal microbiota transplantation (FMT), the clinical relevance in IBD is still not conclusive. Recently, two initial double-blind randomized control trials showed opposite efficacy of FMT in UC patients. While the study from Moayyedi et al. reported secondary efficacy end points of FMT in active UC (Moayyedi et al. 2015), the trial carried out by the Ponsioen group showed no benefits for IBD patients by FMT (Rossen et al. 2015). The contrasting results may be explained by the different approach adopted by the two research teams, including mode of administration and the composition and the dose of the donor microbiota used. However, a recently published controlled trial efficiently used a multi-donor and multi-dosing FMT strategy to induce steroid-free clinical and endoscopic remission in active ulcerative colitis (Paramsothy et al. 2017), suggesting dose-response mechanisms in successfully treat chronic intestinal inflammation.

11.8 Concluding Remarks

Compelling evidence from human and mouse studies support the view that the microbiome plays an important role in the pathogenesis of IBD. However, the multifactorial nature of gene-environment interactions and the existence of a variety of confounding factors in human studies (e.g., drugs, disease location, and behavior) hamper a clear conclusion on how microbiome risk profiles should look like in order to be clinically implemented for prognostic and therapeutic purposes. It is still unclear to what extent disease risk is predicably based on characteristic changes in bacterial community structure and/or function and whether microbiome changes have a causal role in disease onset. The development of standards is essential to ensure reproducibility and comparability between human studies. The scientific field needs to validate the specificity and selectivity of microbiome signatures considering different mechanisms in the onset and progression of Crohn's disease and ulcerative colitis. To reach this goal, research strategies need to go beyond metagenomics sequencing, aiming at the identification of functional mediators and most importantly, addressing mechanistic links between microbiota-derived signals and host sensors and finally, leading to evidence-based intervention strategies selectively targeted to specific human diseases.

Controversy

Deciphering the pathogenesis of inflammatory bowel diseases is particularly challenging with regard to the multifactorial interaction between host genetics, intestinal microbiota, and environmental triggers. Here, a first milestone can be claimed on the association of *NOD2* mutations with the susceptibility to Crohn's disease, thereby identifying host-microbe interactions to be compromised in CD patients (Hugot et al. 2001; Ogura et al. 2001). Following genome-wide association studies on CD and UC patients highly improved our knowledge in IBD pathogenesis, consequently promoting distinct pathway analysis research, such as Paneth cell signaling, autophagy, and the Th17/Th23 pathway (Jostins et al. 2012). However, based on the investigations of host genetics, it became increasingly evident that environmental factors must play a dominant role in triggering disease. In consideration of the inherited risk to develop IBD, genetic factors contribute to approximately 30% prevalence of intestinal inflammation reflecting the need of environmental triggers to cause disease development (Bennett et al. 1991). In this context, the microbial ecosystem is well known to be crucially involved in gut health and disease, while a precise evaluation of the consequences of dysbiosis in human IBD remains difficult. Even though the abundance of several taxa, including adherent-invasive *Escherichia coli* and *Mycobacterium avium subsp. Paratuberculosis*, was highly associated with IBD phenotypes, dysbiosis without known pathobionts was further observed in intestinal inflammation (Frank et al. 2007; Rehman et al. 2010). In this regard, microbial changes that are characterized by a reduced anti-inflammatory capacity (e.g., by low abundant *Faecalibacterium prausnitzii*) are suggested to negatively impact host tolerance toward the intestinal microbiota, thereby promoting host-derived inflammatory responses (Sokol et al. 2008). Contrarily, levels of *Faecalibacterium prausnitzii* have been observed rather increased in a pediatric cohort study, thus complicating the interpretation of the role of individual bacterial taxa in an overall altered microbial ecosystem (Hansen et al. 2012). However, it is not clear whether any changes of the gut microbiota precedes IBD, since a genetically driven influx of immune cells in the intestinal mucosa could rather induce alterations of the microbial composition in the gut, which in turn result in an inflamed gut. Interindividual differences and alterations in the microbiome during the disease progression hamper a clear separation of microbiome signatures. Conclusively, it still remains elusive how an inherited risk affects the impact of environmental triggers and *vice versa* in IBD.

History

Due to the incomplete understanding of the underlying mechanisms inducing, promoting, resolving, and/or failing to resolve intestinal inflammation, until now a precise definition of IBD subtypes, accurate diagnostic tools of discrimination, and curative approaches and therapeutics of either IBD manifestation are lacking. Although disease states associated with diarrhea as the most important symptom of IBD were mentioned in historic documents dating back to Greek age (Baillie 1793), it took until 1859 that the physician Sir Samuel Wilks coined the term "ulcerative colitis" (UC). In 1909 a larger cohort of UC patients was broadly discussed in the scientific community (Wilks 1859) presumably for the first time. CD was first described as a distinct, separate IBD entity ("regional ileitis") by Burril Bernhard Crohn (Crohn et al. 1932) although several previous case reports may have unintentionally described clinical cases of CD falsely as UC cases given the presence of "regional ileitis" (Crohn et al. 1932).

Ever since, microbial cues ("germs") have been implicated to be a putatively critical component contributing to the IBD pathogenesis. This hypothesis was presumably further promoted at the end of the nineteenth and beginning of the twentieth century by the perception that infections frequently represent the underlying trigger of diarrhea (Paulson 1928). Consecutively, upon invention of antibiotics in the late 1930s, IBD patients were broadly treated with antibiotics. However, these treatment regimens yielded expectedly variable results and failed to induce stable clinical long-term remission.

This phase was followed by the steroid era in the 1950s yielding for the first time dramatically positive therapeutic effects that established a sustainable success of steroids in IBD treatment (Truelove and Witts 1955). Consecutively, the scientific focus steadily shifted to the immune system as a worthwhile module to be targeted in IBD patients. Hence the era of anti-inflammatory (e.g., sulfasalazine) (Svartz 1948) and immunosuppressive (e.g., 6-mercaptopurine) (Bean 1962) drugs started its successful march through the clinic and represents still a mainstay in IBD treatment. However, the dissection of the underlying immunologic mechanisms with modern technologies resulted in the identification of cytokines as putative therapeutic targets. Initially developed with the intention to control sepsis, TNF-alpha antibodies showed marked efficacy to restrain non-septic tissue inflammation as initially observed in murine arthritis models and also rheumatoid arthritis patients (Knight et al. 1993; Elliott et al. 1994; Williams et al. 1992). Consecutively, TNF-alpha antibodies proved to be safe and effective in CD (van Dullemen et al. 1995) and consecutively also in UC patients (Rutgeerts et al. 2005). Until now, TNF-alpha targeting represents one of the therapeutic mainstays to induce and maintain remission in IBD patients.

The genomic era generated awareness for the heritability of IBD, while especially genome-wide association studies managed to identify a plethora of disease risk susceptibility loci. These loci contain genes like *NOD2* that is assumed to provide a critical link between bacterial, immunological, and epithelial abnormalities observed in IBD (Hugot et al. 2001). However, novel effective therapeutic inventions derived from these data sets are still awaited.

Highlights

- An increased risk of IBD is observed in populations with a Western lifestyle.
- Association of NOD2 mutations with CD gave first evidence of disease-relevant implications of the intestinal microbiota.
- GWAS on IBD patient cohorts identified over 200 genetic factors showing a genetically driven disruption of distinct host-microbe interactions.
- Fecal stream diversion in subsets of IBD patients provided first clinical evidence for a central role of bacteria in the pathogenesis of Crohn's disease.
- Germ-free animal models are mostly disease-free indicating a fundamental relevance for microbial triggers in IBD pathogenesis.

References

Abreu, M. T., Fukata, M., & Arditi, M. (2005). TLR signaling in the gut in health and disease. *Journal of Immunology, 174*, 4453–4460.

Adolph, T. E., Tomczak, M. F., Niederreiter, L., Ko, H. J., Bock, J., Martinez-Naves, E., Glickman, J. N., Tschurtschenthaler, M., Hartwig, J., Hosomi, S., Flak, M. B., Cusick, J. L., Kohno, K., Iwawaki, T., Billmann-Born, S., Raine, T., Bharti, R., Lucius, R., Kweon, M. N., Marciniak, S. J., Choi, A., Hagen, S. J., Schreiber, S., Rosenstiel, P., Kaser, A., & Blumberg, R. S. (2013). Paneth cells as a site of origin for intestinal inflammation. *Nature, 503*, 272–276.

Ahmad, T., Satsangi, J., Mcgovern, D., Bunce, M., & Jewell, D. P. (2001). Review article: The genetics of inflammatory bowel disease. *Alimentary Pharmacology & Therapeutics, 15*, 731–748.

Baillie, M. (1793). *The morbid anatomy of some of the most important parts of the human body*. London: J. Johnson and G. Nicol.

Balish, E., & Warner, T. (2002). Enterococcus faecalis induces inflammatory bowel disease in interleukin-10 knockout mice. *The American Journal of Pathology, 160*, 2253–2257.

Bean, R. H. (1962). The treatment of chronic ulcerative colitis with 6-mercaptopurine. *The Medical Journal of Australia, 49*(2), 592–593.

Benjamin, J. L., Sumpter, R., Jr., Levine, B., & Hooper, L. V. (2013). Intestinal epithelial autophagy is essential for host defense against invasive bacteria. *Cell Host & Microbe, 13*, 723–734.

Bennett, R. A., Rubin, P. H., & Present, D. H. (1991). Frequency of inflammatory bowel disease in offspring of couples both presenting with inflammatory bowel disease. *Gastroenterology, 100*, 1638–1643.

Bibiloni, R., Fedorak, R. N., Tannock, G. W., Madsen, K. L., Gionchetti, P., Campieri, M., de Simone, C., & Sartor, R. B. (2005). VSL#3 probiotic-mixture induces remission in patients with active ulcerative colitis. *The American Journal of Gastroenterology, 100*, 1539–1546.

Blaydon, D. C., Biancheri, P., Di, W. L., Plagnol, V., Cabral, R. M., Brooke, M. A., van Heel, D. A., Ruschendorf, F., Toynbee, M., Walne, A., O'toole, E. A., Martin, J. E., Lindley, K., Vulliamy, T., Abrams, D. J., Macdonald, T. T., Harper, J. I., & Kelsell, D. P. (2011). Inflammatory skin and bowel disease linked to ADAM17 deletion. *The New England Journal of Medicine, 365*, 1502–1508.

Colombel, J. F., Grandbastien, B., Gower-Rousseau, C., Plegat, S., Evrard, J. P., Dupas, J. L., Gendre, J. P., Modigliani, R., Belaiche, J., Hostein, J., Hugot, J. P., van Kruiningen, H., & Cortot, A. (1996). Clinical characteristics of Crohn's disease in 72 families. *Gastroenterology, 111*, 604–607.

Crohn, B. B., Ginzburg, L., & Oppenheimer, G. D. (1932). Regional ileitis: A pathologic and clinical entity. *JAMA, 99*(16), 1323–1329.

Eckburg, P. B., Bik, E. M., Bernstein, C. N., Purdom, E., Dethlefsen, L., Sargent, M., Gill, S. R., Nelson, K. E., & Relman, D. A. (2005). Diversity of the human intestinal microbial flora. *Science, 308*, 1635–1638.

Elliott, M. J., Maini, R. N., Feldmann, M., Long-Fox, A., Charles, P., Bijl, H., & Woody, J. N. (1994). Repeated therapy with monoclonal antibody to tumour necrosis factor alpha (cA2) in patients with rheumatoid arthritis. *Lancet, 344*, 1125–1127.

Frank, D. N., St Amand, A. L., Feldman, R. A., Boedeker, E. C., Harpaz, N., & Pace, N. R. (2007). Molecular-phylogenetic characterization of microbial community imbalances in human inflammatory bowel diseases. *Proceedings of the National Academy of Sciences of the United States of America, 104*, 13780–13785.

Fritz, T., Niederreiter, L., Adolph, T., Blumberg, R. S., & Kaser, A. (2011). Crohn's disease: NOD2, autophagy and ER stress converge. *Gut, 60*, 1580–1588.

Fukuda, S., Toh, H., Hase, K., Oshima, K., Nakanishi, Y., Yoshimura, K., Tobe, T., Clarke, J. M., Topping, D. L., Suzuki, T., Taylor, T. D., Itoh, K., Kikuchi, J., Morita, H., Hattori, M., & Ohno, H. (2011). Bifidobacteria can protect from enteropathogenic infection through production of acetate. *Nature, 469*, 543–547.

Furrie, E., Macfarlane, S., Kennedy, A., Cummings, J. H., Walsh, S. V., O'neil, D. A., & Macfarlane, G. T. (2005). Synbiotic therapy (Bifidobacterium longum/ Synergy 1) initiates resolution of inflammation in patients with active ulcerative colitis: A randomised controlled pilot trial. *Gut, 54*, 242–249.

Gevers, D., Kugathasan, S., Denson, L. A., Vazquez-Baeza, Y., van Treuren, W., Ren, B., Schwager, E., Knights, D., Song, S. J., Yassour, M., Morgan, X. C., Kostic, A. D., Luo, C., Gonzalez, A., Mcdonald, D., Haberman, Y., Walters, T., Baker, S., Rosh, J., Stephens, M., Heyman, M., Markowitz, J., Baldassano, R., Griffiths, A., Sylvester, F., Mack, D., Kim, S., Crandall, W., Hyams, J., Huttenhower, C., Knight, R., & Xavier, R. J. (2014). The treatment-naive microbiome in new-onset Crohn's disease. *Cell Host & Microbe, 15*, 382–392.

Grivennikov, S. I., Greten, F. R., & Karin, M. (2010). Immunity, inflammation, and cancer. *Cell, 140*, 883–899.

Gulamhusein, A. F., Eaton, J. E., Tabibian, J. H., Atkinson, E. J., Juran, B. D., & Lazaridis, K. N. (2016). Duration of inflammatory bowel disease is associated with increased risk of cholangiocarcinoma in patients with primary sclerosing cholangitis and IBD. *The American Journal of Gastroenterology, 111*, 705–711.

Gunther, C., Martini, E., Wittkopf, N., Amann, K., Weigmann, B., Neumann, H., Waldner, M. J., Hedrick, S. M., Tenzer, S., Neurath, M. F., & Becker, C. (2011). Caspase-8 regulates TNF-alpha-induced epithelial necroptosis and terminal ileitis. *Nature, 477*, 335–339.

Gupta, N., Bostrom, A. G., Kirschner, B. S., Ferry, G. D., Gold, B. D., Cohen, S. A., Winter, H. S., Baldassano, R. N., Abramson, O., Smith, T., & Heyman, M. B. (2010). Incidence of stricturing and penetrating complications of Crohn's disease diagnosed in pediatric patients. *Inflammatory Bowel Diseases, 16*, 638–644.

Haller, D., Holt, L., Kim, S. C., Schwabe, R. F., Sartor, R. B., & Jobin, C. (2003). Transforming growth factor-beta 1 inhibits non-pathogenic Gram negative bacteria-induced NF-kappa B recruitment to the interleukin-6 gene promoter in intestinal epithelial cells through modulation of histone acetylation. *The Journal of Biological Chemistry, 278*, 23851–23860.

Haller, D., Russo, M. P., Sartor, R. B., & Jobin, C. (2002). IKK beta and phosphatidylinositol 3-kinase/Akt participate in non-pathogenic Gram-negative enteric bacteria-induced RelA phosphorylation and NF-kappa B activation in both primary and intestinal epithelial cell lines. *The Journal of Biological Chemistry, 277*, 38168–38178.

Hansen, R., Russell, R. K., Reiff, C., Louis, P., Mcintosh, F., Berry, S. H., Mukhopadhya, I., Bisset, W. M., Barclay, A. R., Bishop, J., Flynn, D. M., Mcgrogan, P., Loganathan, S., Mahdi, G., Flint, H. J., El-Omar, E. M., & Hold, G. L. (2012). Microbiota of de-novo pediatric IBD: Increased Faecalibacterium prausnitzii and reduced bacterial diversity in Crohn's but not in ulcerative colitis. *The American Journal of Gastroenterology, 107*, 1913–1922.

Hormannsperger, G., Schaubeck, M., & Haller, D. (2015). Intestinal microbiota in animal models of inflammatory diseases. *ILAR Journal, 56*, 179–191.

Hugot, J. P., Chamaillard, M., Zouali, H., Lesage, S., Cezard, J. P., Belaiche, J., Almer, S., Tysk, C., O'Morain, C. A., Gassull, M., Binder, V., Finkel, Y., Cortot, A., Modigliani, R., Laurent-Puig, P., Gower-Rousseau, C., Macry, J., Colombel, J. F., Sahbatou, M., & Thomas, G. (2001). Association of NOD2 leucine-rich repeat variants with susceptibility to Crohn's disease. *Nature, 411*, 599–603.

Human Microbiome Project, C. (2012). Structure, function and diversity of the healthy human microbiome. *Nature, 486*, 207–214.

Johansson, M. E., Phillipson, M., Petersson, J., Velcich, A., Holm, L., & Hansson, G. C. (2008). The inner of the two Muc2 mucin-dependent mucus layers in colon is devoid of bacteria. *Proceedings of the National Academy of Sciences of the United States of America, 105*, 15064–15069.

Joossens, M., Huys, G., Cnockaert, M., de Preter, V., Verbeke, K., Rutgeerts, P., Vandamme, P., & Vermeire, S. (2011). Dysbiosis of the faecal microbiota in patients with Crohn's disease and their unaffected relatives. *Gut, 60*, 631–637.

Jostins, L., Ripke, S., Weersma, R. K., Duerr, R. H., Mcgovern, D. P., Hui, K. Y., Lee, J. C., Schumm, L. P., Sharma, Y., Anderson, C. A., Essers, J., Mitrovic, M., Ning, K., Cleynen, I., Theatre, E., Spain, S. L., Raychaudhuri, S., Goyette, P., Wei, Z., Abraham, C., Achkar, J. P., Ahmad, T., Amininejad, L., Ananthakrishnan, A. N., Andersen, V., Andrews, J. M., Baidoo, L., Balschun, T., Bampton, P. A., Bitton, A., Boucher, G., Brand, S., Buning, C., Cohain, A., Cichon, S., D'Amato, M., de Jong, D., Devaney, K. L., Dubinsky, M., Edwards, C., Ellinghaus, D., Ferguson, L. R., Franchimont, D., Fransen, K., Gearry, R., Georges, M., Gieger, C., Glas, J., Haritunians, T., Hart, A., Hawkey, C., Hedl, M., Hu, X., Karlsen, T. H., Kupcinskas, L., Kugathasan, S., Latiano, A., Laukens, D., Lawrance, I. C., Lees, C. W., Louis, E., Mahy, G., Mansfield, J., Morgan, A. R., Mowat, C., Newman, W., Palmieri, O., Ponsioen, C. Y., Potocnik, U., Prescott, N. J., Regueiro, M., Rotter, J. I., Russell, R. K., Sanderson, J. D., Sans, M., Satsangi, J., Schreiber, S., Simms, L. A., Sventoraityte, J., Targan, S. R., Taylor, K. D., Tremelling, M., Verspaget, H. W., de Vos, M., Wijmenga, C., Wilson, D. C., Winkelmann, J., Xavier, R. J., Zeissig, S., Zhang, B., Zhang, C. K., Zhao, H., International, IBDGC, Silverberg, M. S., Annese, V., Hakonarson, H., Brant, S. R., Radford-Smith, G., Mathew, C. G., Rioux, J. D., et al. (2012). Host-microbe interactions have shaped the genetic architecture of inflammatory bowel disease. *Nature, 491*, 119–124.

Kaplan, G. G. (2015). The global burden of IBD: From 2015 to 2025. *Nature Reviews Gastroenterology & Hepatology, 12*, 720–727.

Kelsen, J., & Baldassano, R. N. (2008). Inflammatory bowel disease: The difference between children and adults. *Inflammatory Bowel Diseases, 14*(Suppl 2), S9–S11.

Khor, B., Gardet, A., & Xavier, R. J. (2011). Genetics and pathogenesis of inflammatory bowel disease. *Nature, 474*, 307–317.

Kim, S. C., Tonkonogy, S. L., Albright, C. A., Tsang, J., Balish, E. J., Braun, J., Huycke, M. M., & Sartor, R. B. (2005). Variable phenotypes of enterocolitis in interleukin 10-deficient mice monoassociated with two different commensal bacteria. *Gastroenterology, 128*, 891–906.

Knight, D. M., Trinh, H., Le, J., Siegel, S., Shealy, D., Mcdonough, M., Scallon, B., Moore, M. A., Vilcek, J., Daddona, P., et al. (1993). Construction and initial characterization of a mouse-human chimeric anti-TNF antibody. *Molecular Immunology, 30*, 1443–1453.

Knights, D., Parfrey, L. W., Zaneveld, J., Lozupone, C., & Knight, R. (2011). Human-associated microbial signatures: Examining their predictive value. *Cell Host & Microbe, 10*, 292–296.

Kobayashi, K. S., Chamaillard, M., Ogura, Y., Henegariu, O., Inohara, N., Nunez, G., & Flavell, R. A. (2005). Nod2-dependent regulation of innate and adaptive immunity in the intestinal tract. *Science, 307*, 731–734.

Kostic, A. D., Gevers, D., Siljander, H., Vatanen, T., Hyotylainen, T., Hamalainen, A. M., Peet, A., Tillmann, V., Poho, P., Mattila, I., Lahdesmaki, H., Franzosa, E. A., Vaarala, O., de Goffau, M., Harmsen, H., Ilonen, J., Virtanen, S. M., Clish, C. B., Oresic, M., Huttenhower, C., Knip, M., Group, DS, & Xavier, R. J. (2015). The dynamics of the human infant gut microbiome in development and in progression toward type 1 diabetes. *Cell Host & Microbe, 17*, 260–273.

Kuhn, R., Lohler, J., Rennick, D., Rajewsky, K., & Muller, W. (1993). Interleukin-10-deficient mice develop chronic enterocolitis. *Cell, 75*, 263–274.

Lee, T., Clavel, T., Smirnov, K., Schmidt, A., Lagkouvardos, I., Walker, A., Lucio, M., Michalke, B., Schmitt-Kopplin, P., Fedorak, R., & Haller,

D. (2017). Oral versus intravenous iron replacement therapy distinctly alters the gut microbiota and metabolome in patients with IBD. *Gut, 66*, 863–871.

Lepage, P., Hasler, R., Spehlmann, M. E., Rehman, A., Zvirbliene, A., Begun, A., Ott, S., Kupcinskas, L., Dore, J., Raedler, A., & Schreiber, S. (2011). Twin study indicates loss of interaction between microbiota and mucosa of patients with ulcerative colitis. *Gastroenterology, 141*, 227–236.

Lesage, S., Zouali, H., Cezard, J. P., Colombel, J. F., Belaiche, J., Almer, S., Tysk, C., O'Morain, C., Gassull, M., Binder, V., Finkel, Y., Modigliani, R., Gower-Rousseau, C., Macry, J., Merlin, F., Chamaillard, M., Jannot, A. S., Thomas, G., Hugot, J. P., & Group, E-I, Group, E & Group, G. (2002). CARD15/NOD2 mutational analysis and genotype-phenotype correlation in 612 patients with inflammatory bowel disease. *American Journal of Human Genetics, 70*, 845–857.

Levine, J. S., & Burakoff, R. (2011). Extraintestinal manifestations of inflammatory bowel disease. *Gastroenterology and Hepatology (NY), 7*, 235–241.

Li, J., Jia, H., Cai, X., Zhong, H., Feng, Q., Sunagawa, S., Arumugam, M., Kultima, J. R., Prifti, E., Nielsen, T., Juncker, A. S., Manichanh, C., Chen, B., Zhang, W., Levenez, F., Wang, J., Xu, X., Xiao, L., Liang, S., Zhang, D., Zhang, Z., Chen, W., Zhao, H., Al-Aama, J. Y., Edris, S., Yang, H., Wang, J., Hansen, T., Nielsen, H. B., Brunak, S., Kristiansen, K., Guarner, F., Pedersen, O., Dore, J., Ehrlich, S. D., Bork, P., Wang, J., & Meta, HITC. (2014). An integrated catalog of reference genes in the human gut microbiome. *Nature Biotechnology, 32*, 834–841.

Liu, J. Z., van Sommeren, S., Huang, H., Ng, S. C., Alberts, R., Takahashi, A., Ripke, S., Lee, J. C., Jostins, L., Shah, T., Abedian, S., Cheon, J. H., Cho, J., Dayani, N. E., Franke, L., Fuyuno, Y., Hart, A., Juyal, R. C., Juyal, G., Kim, W. H., Morris, A. P., Poustchi, H., Newman, W. G., Midha, V., Orchard, T. R., Vahedi, H., Sood, A., Sung, J. Y., Malekzadeh, R., Westra, H. J., Yamazaki, K., Yang, S. K., International Multiple Sclerosis Genetics, C, International, IBDGC, Barrett, J. C., Alizadeh, B. Z., Parkes, M., Bk, T., Daly, M. J., Kubo, M., Anderson, C. A., & Weersma, R. K. (2015). Association analyses identify 38 susceptibility loci for inflammatory bowel disease and highlight shared genetic risk across populations. *Nature Genetics, 47*, 979–986.

Liu, T. C., & Stappenbeck, T. S. (2016). Genetics and pathogenesis of inflammatory bowel disease. *Annual Review of Pathology, 11*, 127–148.

Lozupone, C. A., Stombaugh, J., Gonzalez, A., Ackermann, G., Wendel, D., Vazquez-Baeza, Y., Jansson, J. K., Gordon, J. I., & Knight, R. (2013). Meta-analyses of studies of the human microbiota. *Genetical Research, 23*, 1704–1714.

Lozupone, C. A., Stombaugh, J. I., Gordon, J. I., Jansson, J. K., & Knight, R. (2012). Diversity, stability and resilience of the human gut microbiota. *Nature, 489*, 220–230.

Macpherson, A. J., & Harris, N. L. (2004). Interactions between commensal intestinal bacteria and the immune system. *Nature Reviews Immunology, 4*, 478–485.

Macpherson, A. J., & Uhr, T. (2004). Induction of protective IgA by intestinal dendritic cells carrying commensal bacteria. *Science, 303*, 1662–1665.

Magro, F., Langner, C., Driessen, A., Ensari, A., Geboes, K., Mantzaris, G. J., Villanacci, V., Becheanu, G., Borralho Nunes, P., Cathomas, G., Fries, W., Jouret-Mourin, A., Mescoli, C., De Petris, G., Rubio, C. A., Shepherd, N. A., Vieth, M., Eliakim, R., European Society of, P, European, CS, & Colitis, O. (2013). European consensus on the histopathology of inflammatory bowel disease. *Journal of Crohn's & Colitis, 7*, 827–851.

Mizoguchi, A., & Mizoguchi, E. (2010). Animal models of IBD: Linkage to human disease. *Current Opinion in Pharmacology, 10*, 578–587.

Mizoguchi, A., Takeuchi, T., Himuro, H., Okada, T., & Mizoguchi, E. (2016). Genetically engineered mouse models for studying inflammatory bowel disease. *The Journal of Pathology, 238*, 205–219.

Moayyedi, P., Surette, M. G., Kim, P. T., Libertucci, J., Wolfe, M., Onischi, C., Armstrong, D., Marshall, J. K., Kassam, Z., Reinisch, W., & Lee, C. H. (2015). Fecal microbiota transplantation induces remission in patients with active ulcerative colitis in a randomized controlled trial. *Gastroenterology, 149*, 102–109 e6.

Mohamadzadeh, M., Pfeiler, E. A., Brown, J. B., Zadeh, M., Gramarossa, M., Managlia, E., Bere, P., Sarraj, B., Khan, M. W., Pakanati, K. C., Ansari, M. J., O'Flaherty, S., Barrett, T., & Klaenhammer, T. R. (2011). Regulation of induced colonic inflammation by Lactobacillus acidophilus deficient in lipoteichoic acid. *Proceedings of the National Academy of Sciences of the United States of America, 108*(Suppl 1), 4623–4630.

Molodecky, N. A., Soon, I. S., Rabi, D. M., Ghali, W. A., Ferris, M., Chernoff, G., Benchimol, E. I., Panaccione, R., Ghosh, S., Barkema, H. W., & Kaplan, G. G. (2012). Increasing incidence and prevalence of the inflammatory bowel diseases with time, based on systematic review. *Gastroenterology, 142*, 46–54 e42 quiz e30.

Nenci, A., Becker, C., Wullaert, A., Gareus, R., van Loo, G., Danese, S., Huth, M., Nikolaev, A., Neufert, C., Madison, B., Gumucio, D., Neurath, M. F., & Pasparakis, M. (2007). Epithelial NEMO links innate immunity to chronic intestinal inflammation. *Nature, 446*, 557–561.

Ocvirk, S., Sava, I. G., Lengfelder, I., Lagkouvardos, I., Steck, N., Roh, J. H., Tchaptchet, S., Bao, Y., Hansen, J. J., Huebner, J., Carroll, I. M., Murray, B. E., Sartor, R. B., & Haller, D. (2015). Surface-associated lipoproteins link enterococcus faecalis virulence to colitogenic activity in IL-10-deficient mice independent of their expression levels. *PLoS Pathogens, 11*, e1004911.

Ogura, Y., Bonen, D. K., Inohara, N., Nicolae, D. L., Chen, F. F., Ramos, R., Britton, H., Moran, T., Karaliuskas, R., Duerr, R. H., Achkar, J. P., Brant, S. R., Bayless, T. M., Kirschner, B. S., Hanauer, S. B., Nunez, G., & Cho, J. H. (2001). A frameshift mutation in NOD2 associated with susceptibility to Crohn's disease. *Nature, 411*, 603–606.

Okou, D. T., Mondal, K., Faubion, W. A., Kobrynski, L. J., Denson, L. A., Mulle, J. G., Ramachandran, D., Xiong, Y., Svingen, P., Patel, V., Bose, P., Waters, J. P., Prahalad, S., Cutler, D. J., Zwick, M. E., & Kugathasan, S. (2014). Exome sequencing identifies a novel FOXP3 mutation in a 2-generation family with inflammatory bowel disease. *Journal of Pediatric Gastroenterology and Nutrition, 58*, 561–568.

Oliva, S., di Nardo, G., Ferrari, F., Mallardo, S., Rossi, P., Patrizi, G., Cucchiara, S., & Stronati, L. (2012). Randomised clinical trial: The effectiveness of Lactobacillus reuteri ATCC 55730 rectal enema in children with active distal ulcerative colitis. *Alimentary Pharmacology & Therapeutics, 35*, 327–334.

Orholm, M., Munkholm, P., Langholz, E., Nielsen, O. H., Sorensen, T. I., & Binder, V. (1991). Familial occurrence of inflammatory bowel disease. *The New England Journal of Medicine, 324*, 84–88.

Paramsothy, S., Kamm, M. A., Kaakoush, N. O., Walsh, A. J., van den Bogaerde, J., Samuel, D., Leong, R. W. L., Connor, S., Ng, W., Paramsothy, R., Xuan, W., Lin, E., Mitchell, H. M., & Borody, T. J. (2017). Multidonor intensive faecal microbiota transplantation for active ulcerative colitis: A randomised placebo-controlled trial. *Lancet, 389*, 1218–1228.

Paulson, M. (1928). Chronic ulcerative colitis with reference to a bacterial etiology. Experimental studies. *Archives of Internal Medicine, 41*(1), 75–96.

Petnicki-Ocwieja, T., Hrncir, T., Liu, Y. J., Biswas, A., Hudcovic, T., Tlaskalova-Hogenova, H., & Kobayashi, K. S. (2009). Nod2 is required for the regulation of commensal microbiota in the intestine. *Proceedings of the National Academy of Sciences of the United States of America, 106*, 15813–15818.

Podolsky, D. K. (2002). Inflammatory bowel disease. *The New England Journal of Medicine, 347*, 417–429.

Rakoff-Nahoum, S., & Medzhitov, R. (2007). Regulation of spontaneous intestinal tumorigenesis through the adaptor protein MyD88. *Science, 317*, 124–127.

Rakoff-Nahoum, S., Paglino, J., Eslami-Varzaneh, F., Edberg, S., & Medzhitov, R. (2004). Recognition of commensal microflora by toll-like receptors is required for intestinal homeostasis. *Cell, 118*, 229–241.

Rehaume, L. M., Mondot, S., Aguirre De Carcer, D., Velasco, J., Benham, H., Hasnain, S. Z., Bowman, J., Ruutu, M., Hansbro, P. M., Mcguckin, M. A., Morrison, M., & Thomas, R. (2014). ZAP-70 genotype disrupts the relationship between microbiota and host, leading to spondyloarthritis and ileitis in SKG mice. *Arthritis and Rheumatism, 66*, 2780–2792.

Rehman, A., Lepage, P., Nolte, A., Hellmig, S., Schreiber, S., & Ott, S. J. (2010). Transcriptional activity of the dominant gut mucosal microbiota in chronic inflammatory bowel disease patients. *Journal of Medical Microbiology, 59*, 1114–1122.

Rehman, A., Rausch, P., Wang, J., Skieceviciene, J., Kiudelis, G., Bhagalia, K., Amarapurkar, D., Kupcinskas, L., Schreiber, S., Rosenstiel, P., Baines, J. F., & Ott, S. (2016). Geographical patterns of the standing and active human gut microbiome in health and IBD. *Gut, 65*, 238–248.

Rehman, A., Sina, C., Gavrilova, O., Hasler, R., Ott, S., Baines, J. F., Schreiber, S., & Rosenstiel, P. (2011). Nod2 is essential for temporal development of intestinal microbial communities. *Gut, 60*, 1354–1362.

Renz, H., von Mutius, E., Brandtzaeg, P., Cookson, W. O., Autenrieth, I. B., & Haller, D. (2011). Gene-environment interactions in chronic inflammatory disease. *Natural Immunity, 12*, 273–277.

Rescigno, M., Urbano, M., Valzasina, B., Francolini, M., Rotta, G., Bonasio, R., Granucci, F., Kraehenbuhl, J. P., & Ricciardi-Castagnoli, P. (2001). Dendritic cells express tight junction proteins and penetrate gut epithelial monolayers to sample bacteria. *Natural Immunity, 2*, 361–367.

Rieder, F., & Fiocchi, C. (2008). Intestinal fibrosis in inflammatory bowel disease—Current knowledge and future perspectives. *Journal of Crohn's & Colitis, 2*, 279–290.

Rodriguez-Palacios, A., Kodani, T., Kaydo, L., Pietropaoli, D., Corridoni, D., Howell, S., Katz, J., Xin, W., Pizarro, T. T., & Cominelli, F. (2015). Stereomicroscopic 3D-pattern profiling of murine and human intestinal inflammation reveals unique structural phenotypes. *Nature Communications, 6*, 7577.

Rossen, N. G., Fuentes, S., van der Spek, M. J., Tijssen, J. G., Hartman, J. H., Duflou, A., Lowenberg, M., van den Brink, G. R., Mathus-Vliegen, E. M., de Vos, W. M., Zoetendal, E. G., D'Haens, G. R., & Ponsioen, C. Y. (2015). Findings from a randomized controlled trial of fecal transplantation for patients with ulcerative colitis. *Gastroenterology, 149*, 110–118 e4.

Ruiz, P. A., Hoffmann, M., Szcesny, S., Blaut, M., & Haller, D. (2005). Innate mechanisms for Bifidobacterium lactis to activate transient pro-inflammatory host responses in intestinal epithelial cells after the colonization of germ-free rats. *Immunology, 115*, 441–450.

Ruiz, P. A., Shkoda, A., Kim, S. C., Sartor, R. B., & Haller, D. (2006). IL-10 gene-deficient mice lack TGF-beta/Smad-mediated TLR2 degradation and fail to inhibit proinflammatory gene expression in intestinal epithelial cells under conditions of chronic inflammation. *Annals of the New York Academy of Sciences, 1072*, 389–394.

Rutgeerts, P., Goboes, K., Peeters, M., Hiele, M., Penninckx, F., Aerts, R., Kerremans, R., & Vantrappen, G. (1991). Effect of faecal stream diversion on recurrence of Crohn's disease in the neoterminal ileum. *Lancet, 338*, 771–774.

Rutgeerts, P., Sandborn, W. J., Feagan, B. G., Reinisch, W., Olson, A., Johanns, J., Travers, S., Rachmilewitz, D., Hanauer, S. B., Lichtenstein, G. R., de Villiers, W. J., Present, D., Sands, B. E., & Colombel, J. F. (2005). Infliximab for induction and maintenance therapy for ulcerative colitis. *The New England Journal of Medicine, 353*, 2462–2476.

Samouilidou, E. C., Karpouza, A. P., Kostopoulos, V., Bakirtzi, T., Pantelias, K., Petras, D., Tzanatou-Exarchou, H., & grapsa, E. J. (2012). Lipid abnormalities and oxidized LDL in chronic kidney disease patients on hemodialysis and peritoneal dialysis. *Renal Failure, 34*, 160–164.

Sartor, R. B. (2008). Microbial influences in inflammatory bowel diseases. *Gastroenterology, 134*, 577–594.

Satsangi, J., Parkes, M., Louis, E., Hashimoto, L., Kato, N., Welsh, K., Terwilliger, J. D., Lathrop, G. M., Bell, J. I., & Jewell, D. P. (1996). Two stage genome-wide search in inflammatory bowel disease provides evidence for susceptibility loci on chromosomes 3, 7 and 12. *Nature Genetics, 14*, 199–202.

Schaubeck, M., Clavel, T., Calasan, J., Lagkouvardos, I., Haange, S. B., Jehmlich, N., Basic, M., Dupont, A., Hornef, M., von Bergen, M., Bleich, A., & Haller, D. (2016). Dysbiotic gut microbiota causes transmissible Crohn's disease-like ileitis independent of failure in antimicrobial defence. *Gut, 65*, 225–237.

Sellon, R. K., Tonkonogy, S., Schultz, M., Dieleman, L. A., Grenther, W., Balish, E., Rennick, D. M., & Sartor, R. B. (1998). Resident enteric bacteria are necessary for development of spontaneous colitis and immune system activation in interleukin-10-deficient mice. *Infection and Immunity, 66*, 5224–5231.

Shan, M., Gentile, M., Yeiser, J. R., Walland, A. C., Bornstein, V. U., Chen, K., He, B., Cassis, L., Bigas, A., Cols, M., Comerma, L., Huang, B., Blander, J. M., Xiong, H., Mayer, L., Berin, C., Augenlicht, L. H., Velcich, A., & Cerutti, A. (2013). Mucus enhances gut homeostasis and oral tolerance by delivering immunoregulatory signals. *Science, 342*, 447–453.

Slack, E., Hapfelmeier, S., Stecher, B., Velykoredko, Y., Stoel, M., Lawson, M. A., Geuking, M. B., Beutler, B., Tedder, T. F., Hardt, W. D., Bercik, P., Verdu, E. F., Mccoy, K. D., & Macpherson, A. J. (2009). Innate and adaptive immunity cooperate flexibly to maintain host-microbiota mutualism. *Science, 325*, 617–620.

Sokol, H., Pigneur, B., Watterlot, L., Lakhdari, O., Bermudez-Humaran, L. G., Gratadoux, J. J., Blugeon, S., Bridonneau, C., Furet, J. P., Corthier, G., Grangette, C., Vasquez, N., Pochart, P., Trugnan, G., Thomas, G., Blottiere, H. M., Dore, J., Marteau, P., Seksik, P., & Langella, P. (2008). Faecalibacterium prausnitzii is an anti-inflammatory commensal bacterium identified by gut microbiota analysis of Crohn disease patients. *Proceedings of the National Academy of Sciences of the United States of America, 105*, 16731–16736.

Sood, A., Midha, V., Makharia, G. K., Ahuja, V., Singal, D., Goswami, P., & Tandon, R. K. (2009). The probiotic preparation, VSL#3 induces remission in patients with mild-to-moderately active ulcerative colitis. *Clinical Gastroenterology and Hepatology, 7*, 1202–1209 e1.

Steck, N., Hoffmann, M., Sava, I. G., Kim, S. C., Hahne, H., Tonkonogy, S. L., Mair, K., Krueger, D., Pruteanu, M., Shanahan, F., Vogelmann, R., Schemann, M., Kuster, B., Sartor, R. B., & Haller, D. (2011). Enterococcus faecalis metalloprotease compromises epithelial barrier and contributes to intestinal inflammation. *Gastroenterology, 141*, 959–971.

Steed, H., Macfarlane, G. T., Blackett, K. L., Bahrami, B., Reynolds, N., Walsh, S. V., Cummings, J. H., & Macfarlane, S. (2010). Clinical trial: The microbiological and immunological effects of synbiotic consumption—a randomized double-blind placebo-controlled study in active Crohn's disease. *Alimentary Pharmacology & Therapeutics, 32*, 872–883.

Stokkers, P. C., Reitsma, P. H., Tytgat, G. N., & van Deventer, S. J. (1999). HLA-DR and -DQ phenotypes in inflammatory bowel disease: A meta-analysis. *Gut, 45*, 395–401.

Svartz, N. (1948). The treatment of rheumatic polyarthritis with acid azo compounds. *Rheumatism, 4*, 180–185.

Truelove, S. C., & Witts, L. J. (1955). Cortisone in ulcerative colitis; Final report on a therapeutic trial. *British Medical Journal, 2*, 1041–1048.

Tursi, A., Brandimarte, G., Papa, A., Giglio, A., Elisei, W., Giorgetti, G. M., Forti, G., Morini, S., Hassan, C., Pistoia, M. A., Modeo, M. E., Rodino, S., D'Amico, T., Sebkova, L., Sacca, N., di Giulio, E., Luzza, F., Imeneo, M., Larussa, T., di Rosa, S., Annese, V., Danese, S., & Gasbarrini, A. (2010). Treatment of relapsing mild-to-moderate ulcerative colitis with the probiotic VSL#3 as adjunctive to a standard pharmaceutical treatment: A double-blind, randomized, placebo-controlled study. *The American Journal of Gastroenterology, 105*, 2218–2227.

Tysk, C., Lindberg, E., Jarnerot, G., & Floderus-Myrhed, B. (1988). Ulcerative colitis and Crohn's disease in an unselected population of monozygotic and dizygotic twins. A study of heritability and the influence of smoking. *Gut, 29*, 990–996.

van Dullemen, H. M., van Deventer, S. J., Hommes, D. W., Bijl, H. A., Jansen, J., Tytgat, G. N., & Woody, J. (1995). Treatment of Crohn's disease with anti-tumor necrosis factor chimeric monoclonal antibody (cA2). *Gastroenterology, 109*, 129–135.

Vangay, P., Ward, T., Gerber, J. S., & Knights, D. (2015). Antibiotics, pediatric dysbiosis, and disease. *Cell Host & Microbe, 17*, 553–564.

Vijay-Kumar, M., Aitken, J. D., Carvalho, F. A., Cullender, T. C., Mwangi, S., Srinivasan, S., Sitaraman, S. V., Knight, R., Ley, R. E., & Gewirtz, A. T. (2010). Metabolic syndrome and altered gut microbiota in mice lacking Toll-like receptor 5. *Science, 328*, 228–231.

Vijay-Kumar, M., Sanders, C. J., Taylor, R. T., Kumar, A., Aitken, J. D., Sitaraman, S. V., Neish, A. S., Uematsu, S., Akira, S., Williams, I. R., & Gewirtz, A. T. (2007).

Deletion of TLR5 results in spontaneous colitis in mice. *The Journal of Clinical Investigation, 117*, 3909–3921.

von Schillde, M. A., Hormannsperger, G., Weiher, M., Alpert, C. A., Hahne, H., Bauerl, C., van Huynegem, K., Steidler, L., Hrncir, T., Perez-Martinez, G., Kuster, B., & Haller, D. (2012). Lactocepin secreted by Lactobacillus exerts anti-inflammatory effects by selectively degrading proinflammatory chemokines. *Cell Host & Microbe, 11*, 387–396.

Waidmann, M., Bechtold, O., Frick, J. S., Lehr, H. A., Schubert, S., Dobrindt, U., Loeffler, J., Bohn, E., & Autenrieth, I. B. (2003). Bacteroides vulgatus protects against *Escherichia coli*-induced colitis in gnotobiotic interleukin-2-deficient mice. *Gastroenterology, 125*, 162–177.

Wen, L., Ley, R. E., Volchkov, P. Y., Stranges, P. B., Avanesyan, L., Stonebraker, A. C., Hu, C., Wong, F. S., Szot, G. L., Bluestone, J. A., Gordon, J. I., & Chervonsky, A. V. (2008). Innate immunity and intestinal microbiota in the development of type 1 diabetes. *Nature, 455*, 1109–1113.

Whelan, K., & Quigley, E. M. (2013). Probiotics in the management of irritable bowel syndrome and inflammatory bowel disease. *Current Opinion in Gastroenterology, 29*, 184–189.

Wilks, S. (1859). Morbid appearances in the intestines of Miss Bankes. *London Medical Gazette, 2*, 264–265.

Williams, R. O., Feldmann, M., & Maini, R. N. (1992). Anti-tumor necrosis factor ameliorates joint disease in murine collagen-induced arthritis. *Proceedings of the National Academy of Sciences of the United States of America, 89*, 9784–9788.

Willing, B. P., Dicksved, J., Halfvarson, J., Andersson, A. F., Lucio, M., Zheng, Z., Jarnerot, G., Tysk, C., Jansson, J. K., & Engstrand, L. (2010). A pyrosequencing study in twins shows that gastrointestinal microbial profiles vary with inflammatory bowel disease phenotypes. *Gastroenterology, 139*, 1844–1854 e1.

Worthey, E. A., Mayer, A. N., Syverson, G. D., Helbling, D., Bonacci, B. B., Decker, B., Serpe, J. M., Dasu, T., Tschannen, M. R., Veith, R. L., Basehore, M. J., Broeckel, U., Tomita-Mitchell, A., Arca, M. J., Casper, J. T., Margolis, D. A., Bick, D. P., Hessner, M. J., Routes, J. M., Verbsky, J. W., Jacob, H. J., & Dimmock, D. P. (2011). Making a definitive diagnosis: Successful clinical application of whole exome sequencing in a child with intractable inflammatory bowel disease. *Genetics in Medicine, 13*, 255–262.

Zocco, M. A., Dal Verme, L. Z., Cremonini, F., Piscaglia, A. C., Nista, E. C., Candelli, M., Novi, M., Rigante, D., Cazzato, I. A., Ojetti, V., Armuzzi, A., Gasbarrini, G., & Gasbarrini, A. (2006). Efficacy of Lactobacillus GG in maintaining remission of ulcerative colitis. *Alimentary Pharmacology & Therapeutics, 23*, 1567–1574.

Microbiome and Diseases: Allergy

12

Florian Wölbing, Martin Köberle, Thomas Volz, and Tilo Biedermann

Abstract

The increasing prevalence of allergies and atopic diseases is contrasted by a yet unmet goal of prevention and treatment of allergic diseases. Allergies and atopic diseases develop and manifest at surface organs, which are inhabited by a specific composition of a microbiome. Consequently, the interdependence of microbial colonization of surface organs and the development of allergic/atopic diseases orchestrating either immune tolerance or immune hyperreactivity are a focus of research. This chapter will focus on what is known about the gut microbiota regulating atopic/allergic diseases and shortly allude to a possible role of microbiomes beyond the gut in regulating these diseases. Three prototypic diseases will be covered: food allergy, atopic dermatitis, and allergic asthma. As many functional analyses were performed in the murine system, these will be discussed and complemented with studies in humans, in some cases with appropriate interventional trials.

F. Wölbing · M. Köberle · T. Volz · T. Biedermann (✉)
Klinik und Poliklinik für Dermatologie und Allergologie am Biederstein, Technische Universität München, München, Germany
e-mail: florian.woelbing@tum.de; tilo.biedermann@tum.de

An ongoing increase of the prevalence of allergies and atopic diseases is well documented, and therefore these diseases are often referred to as the epidemic of developed countries (Platts-Mills 2015). They are associated with significant loss of the quality of life. Permanent medication and frequent sick leaves burden public health budgets. Preventing onset or progression of disease is the unmet goal for the treatment of allergic diseases. However, in-depth understanding of underlying mechanisms is the basis for the development of novel strategies for prevention and treatment. Of note, allergies and atopic diseases are illnesses that develop and manifest at surface organs. Consequently, among the new frontiers explored to better understand and intervene with allergies and atopic diseases, the role and influence of the composition of the microbiome at different surfaces are increasingly studied (Belkaid and Harrison 2017).

Allergic reactions can be subgrouped into different variants based on the immune cells and the reaction pattern involved (Coombs and Gells 1963). Environmental allergens elicit type I allergic reactions, and those will be the focus of this chapter. The majority of the other variants develop in response to drugs, topical medication, and stimuli ranging from cosmetics to occupational substances and are less likely to be fundamentally influenced by microbiomes. Type I allergic reactions develop on the basis of allergen-specific Th2 cells that orchestrate the switch of B cells to produce allergen-specific

D. Haller (ed.), *The Gut Microbiome in Health and Disease*,
https://doi.org/10.1007/978-3-319-90545-7_12

IgE antibodies (Kay 2001). Allergen-specific antibodies bind to the high affinity receptor FcεRI on mast cells and basophils. Upon subsequent contact to the allergen, cross-linking of FcεRI-bound IgE by these allergens leads to the so-called immediate or type I immune reaction that is the degranulation of mast cells or basophils (Fig. 12.1). This degranulation sets free various mediators, among them histamine, inducing symptoms within minutes. Typical elicitors of type I allergic reactions are proteins derived from pollen, from house dust mites or from animal fur (Traidl-Hoffmann et al. 2009). Importantly, type I allergies can develop in patients with and without atopy. Atopy describes the predisposition to develop the atopic diseases, which are atopic dermatitis, allergic asthma, allergic rhinoconjunctivitis as well as food allergy. Prototypic type I allergies that can be found in patients independent of being atopic are type I allergy to bee or wasp venom. The most severe type I allergic reaction is anaphylaxis (Wölbing and Biedermann 2013). It is classified into four levels of severity of symptoms that include urticaria of the skin, asthmatic reactions, diarrhea, and the drop of blood pressure that can be followed by unconsciousness and even cardiac arrest (Ring et al. 2004). Importantly, many of the presentations of the atopic diseases cannot be solely explained by IgE-mediated degranulation of mast cells and basophils but rather involve the cellular compartment of the adaptive immune system. Especially Th2 cells are the basis of cutaneous inflammation in atopic dermatitis and are dominant contributors to inflammation in allergic asthma (Biedermann et al. 2004; Cohn et al. 2004).

It is evident that allergic/atopic inflammatory diseases develop predominantly on surface organs of the body, such as the gut, the skin, and the respiratory system. As these surfaces are inhabited by a specifically composed microbiota, the interdependence of microbial colonization of surface organs and the development of allergic/atopic diseases have gained tremendous interest (Belkaid and Harrison 2017; Biedermann et al. 2015). Especially, the immune system of the gut is acknowledged to be orchestrating tolerance to, e.g., food antigens. Thus, disturbance of the interface between the microbes, allergens such as food antigens, and the intestinal barrier and its immune system is a focus of research to better understand also systemic immune responses. Today we understand that a similar concept is also functional at the respiratory surfaces and the skin, orchestrating immune tolerance as well as immune responses to antigenic constituents they are exposed to (Fig. 12.2).

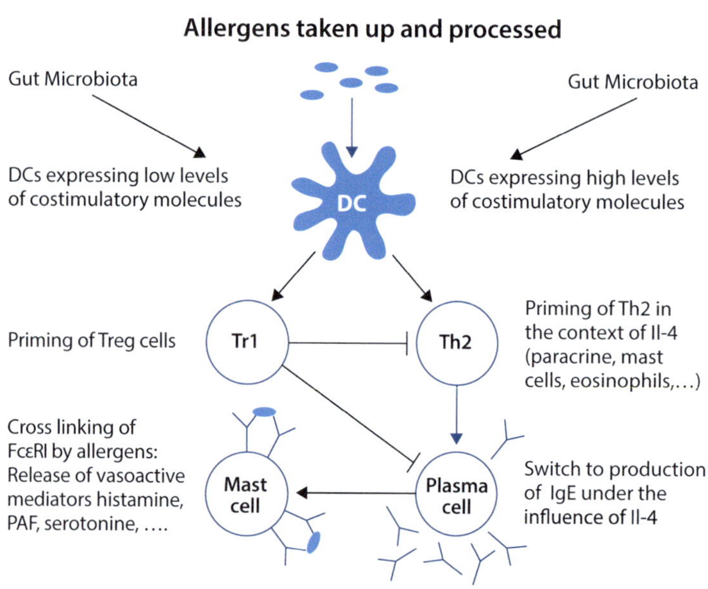

Fig. 12.1 Gut microbiota shapes the context of allergen presentation. Gut antigens are presented to T cells by dendritic cells. The microbiota regulates the behavior and immune profile such as the expression of co-stimulatory molecules of dendritic cells and thus contributes to skewing the differentiation of T cells primed to gut antigens. The result may be a pro-allergic Th2 or a tolerogenic Treg response with local and possibly systemic immune consequences

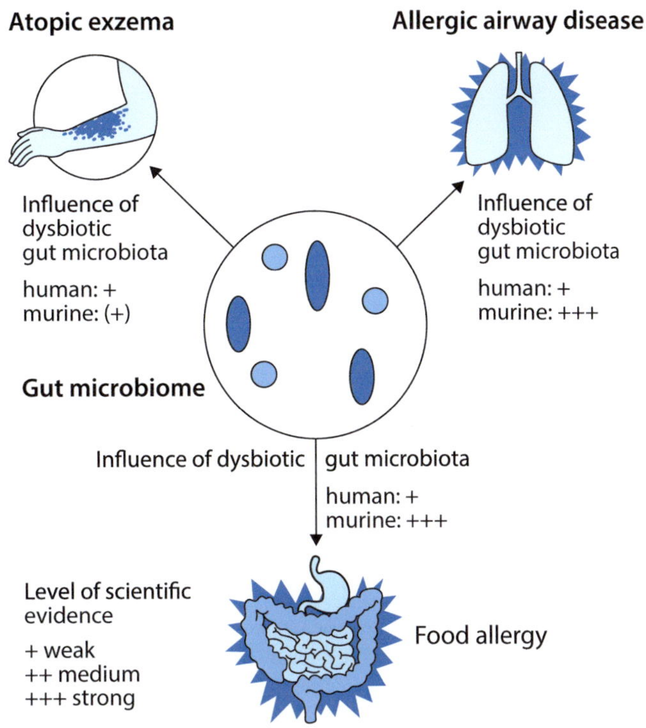

Fig. 12.2 Gut microbiota influences sensitization and allergic diseases at interface organs. The gut microbiota is believed to shape immune responses at interface organs. There is accumulating evidence that dysbiotic gut microbiota can promote sensitization to environmental antigens and development of allergic/atopic diseases at the skin, the airways, and the gut. The level of scientific evidence of these processes varies between the different interface organs and the human and murine immune system and is therefore indicated (+ and +++)

This chapter will focus on what is known about the gut microbiota regulating atopic/allergic diseases and shortly allude to a possible role of microbiomes beyond the gut in regulating allergic/atopic diseases in effector organs (Fig. 12.3). The most obvious interdependence between the intestinal microbiome and allergies may be anticipated for food allergy (Plunkett and Nagler 2017).

12.1 Gut Microbiome and Food Allergy

12.1.1 Allergies to Food and Tolerance Induction in the Intestine

Food allergy has substantially increased in the last two decades with up to 11% of children, and 5% of adults reported to suffer from adverse reactions to food (Peters et al. 2017; Rinaldi et al. 2012; Sicherer and Sampson 2014). The highest prevalence was found in infants at the age of 1, and a decrease over time until the age of 4 has been observed (Peters et al. 2017). Nevertheless, emergency department visits caused by food allergy-related anaphylaxis are reported to be approximately 100,000/year in the United States demonstrating the socioeconomic and personal burden of this disease (Clark et al. 2011).

Under physiological circumstances, ingested antigens are tolerated by the intestinal immune system for which the term "oral tolerance" has been coined (Weiner et al. 2011). In recent years it could be demonstrated that food antigens are not neglected by the immune system but rather induce tolerogenic immune responses (Weiner et al. 2011). Underlying mechanisms have been subsequently elucidated, and a multistep model for oral tolerance has been proposed (Pabst and Mowat 2012). In brief, $CD103^+$ DC of the lamina propria take up antigens from the intestine and migrate to the mesenteric lymph nodes (mLN) (Schulz et al. 2009). In mLN these DC induce differentiation of peripheral $FoxP3^+$ regulatory T cells (iTregs) under the influence of retinoic acid and TGF-β. In a third step, iTregs home back to

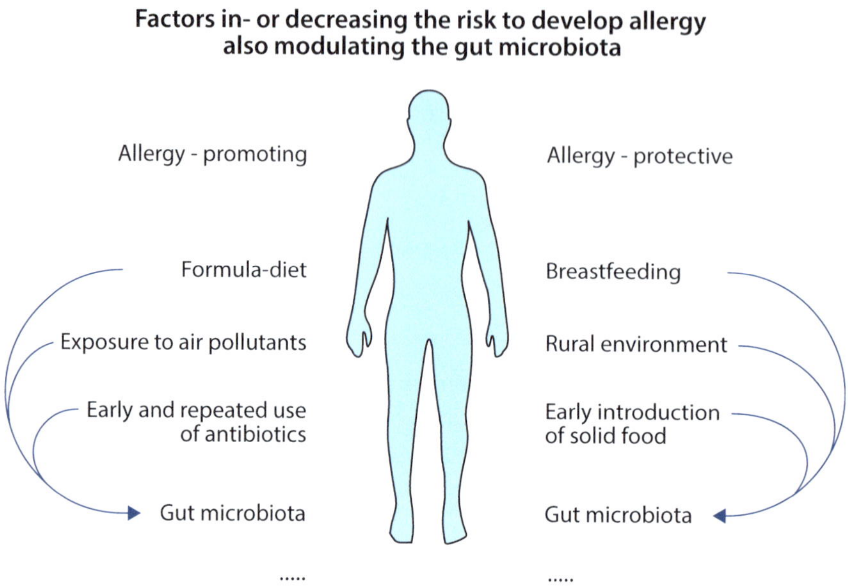

Fig. 12.3 Environmental conditions modulating the risk to develop allergic diseases and their interplay. There is a high evidence level that certain environmental conditions can either increase or decrease the risk to develop allergic diseases. All these factors can also modulate the gut microbiota, which, dependent on its composition, maybe crucially involved in regulating tolerance development. However, general mechanisms still need to be identified

the lamina propria and are expanded depending on IL-10 producing CX3CR1^{+} macrophages (Hadis et al. 2011). In addition other pathways for induction of tolerance in the gut exist, e.g., by induction of regulatory Tr1 cells that may team up with FoxP3^{+} iTregs (Weiner et al. 2011). The breakdown of oral tolerance has been suggested to account for the development of sensitization to food resulting in food allergy, although the detailed mechanisms are unknown yet. A crucial role for FoxP3^{+} Tregs in establishing and maintaining oral tolerance was demonstrated in murine models either by depletion of Tregs or by transfer of CD4^{+} CD25^{+} Tregs to sensitized hosts (Curotto de Lafaille et al. 2008; Yamashita et al. 2012).

12.1.2 Role of Intestinal Microbiota Orchestrating Food Allergy or Tolerance

Functional analyses from murine studies demonstrated that intestinal microbiota induces a subset of regulatory T cells expressing the transcription factor RORγt controlling Th2 responses (Ohnmacht et al. 2015; Sefik et al. 2015). As the development of the intestinal immune system in early life is paralleled by microbial colonization of the gut, perturbations of the gut microbiota may directly influence shaping of tolerogenic immune responses, hampering induction of oral tolerance (Maynard et al. 2012; Plunkett and Nagler 2017). The majority of the investigations regarding the influence of the intestinal microbiome and its relationship to food allergy have been undertaken in the murine system, but there is also increasing evidence from human studies.

An early report showed that oral administration of the antibiotic kanamycin, targeting mainly Gram-negative bacteria, to 3-week-old mice, resulted in increased IgE and IgG1 serum levels as well as predominant production of the Th2 hallmark cytokine IL-4 while suppressing the Th1 cytokine IFN-γ in splenic lymphocytes (Oyama et al. 2001). In line with this finding, mice devoid of TLR4 were shown to be prone evolving predominant Th2 responses and enhanced IgE levels after intragastric sensitization to peanut extract (Bashir et al. 2004). Depletion of the intestinal microbiota by antibiotics in wild-type animals mimicked the results obtained with TLR4-deficient mice demonstrating a pivotal role for the gut microbiota in maintaining oral tolerance and inhibiting food allergy (Bashir et al. 2004). Interestingly the susceptibility to develop food allergy in this model was restricted to specific genetic backgrounds such as BALB/c mice, whereas C3H or C57Bl/6 mice did not develop anaphylaxis (Berin et al. 2006). Analysis of germ-free mice revealed significantly elevated IgE levels in the absence of commensal microbiota compared to conventionally reared animals (Cahenzli et al. 2013; Hill et al. 2012; Rodriguez et al. 2011). Moreover, after sensitization, germ-free mice also showed more severe anaphylaxis compared to control animals with normal microbiome (Cahenzli et al. 2013; Rodriguez et al. 2011; Stefka et al. 2014). Strikingly, administration of a low-complexity microbiota harboring only 40 phylotypes completely protected germ-free mice from aberrant IgE production (Cahenzli et al. 2013). These observations clearly highlight the important role of the commensal intestinal microbiota to maintain oral tolerance and to prevent a detrimental type 2 immune response to orally ingested antigens.

Food allergy-prone mice with a gain-of-function mutation in the IL-4 receptor α-chain (*Il4raF709*) displayed a severely altered microbiome compared to their wild-type littermates (Noval Rivas et al. 2013). A significant decrease in the relative abundance of the bacterial family *Erysipelotrichaceae*, which belongs to the *Firmicutes* phylum as well as the *Enterobacteriaceae* family (*Proteobacteria* phylum), was observed in sensitized mice (Noval Rivas et al. 2013). By transfer of the microbiota of *IL4raF709* mice to germ-free mice, food-induced anaphylaxis after oral sensitization could be elicited demonstrating a crucial role for a dysbiotic microbiota in triggering food allergy.

Detailed analysis of the intestinal microbiota in humans has made tremendous advances due to state-of-the-art techniques such as bacterial 16S rRNA gene phylotyping and metagenomics. However, few studies have analyzed the microbiome by means of 16S rRNA gene sequencing in humans in regard to food allergy. Comparing the intestinal microbiome of food allergic and non-allergic children, a small but pioneering study found increased levels of the genera *Clostridium sensu stricto* spp. and *Anaerobacter* spp. and decreased levels of *Bacteroides* and *Clostridium* cluster XVIII in infants with IgE-mediated food allergy (Ling et al. 2014). These findings are of utmost importance as it could be shown that certain *Clostridium* species devoid of toxin expression, including *Clostridium* cluster XVIII species, induce FoxP3^{+} regulatory T cells, thereby dampening IgE production and severity in an ovalbumin-induced allergic diarrhea model (Atarashi et al. 2011, 2013). Moreover, colonization of germ-free mice with *Clostridia* attenuated anaphylactic reactions in orally sensitized mice highlighting a specific role for *Clostridia* in establishing oral tolerance and preventing food allergy (Stefka et al. 2014).

A recently published study in infants with food sensitization in early life demonstrated an altered fecal microbiota with lower microbiota diversity compared to healthy controls (Chen et al. 2016). Children with food sensitization had significantly increased numbers of *Sphingomonas* spp., *Sutterella* spp., *Bifidobacterium* spp., *Collinsella* spp., *Clostridium sensu stricto* spp., *Clostridium* IV spp., *Enterococcus* spp., *Lactobacillus* spp., *Roseburia* spp., *Faecalibacterium* spp., *Ruminococcus* spp., *Subdoligranulum* spp., and *Akkermansia* spp.; *Bacteroides* spp., *Parabacteroides* spp., *Prevotella* spp., *Alistipes* spp., *Streptococcus* spp., and *Veillonella* spp. were decreased in children with sensitization to food (Chen et al. 2016). Comparability of the results reported by Ling et al. and Chen et al. is limited as the latter reported on a food-sensitized cohort of children on the basis of specific IgE levels. In contrast, Ling et al. investigated infants with challenge-proven food allergy. A specific signature in the microbiome of children with cow's milk allergy during infancy predictive for resolution at the age of 8 years was recently reported (Bunyavanich et al. 2016). Resolution of cow's milk allergy until the age of 8 years was associated with an enrichment of *Clostridia* and other *Firmicutes* in the microbiome at the age of 3–6 months (Bunyavanich et al. 2016). Taken together, results from murine and human investigations show that development of food allergy is associated with a dysbiotic microbiome. It still needs to be studied in more detail how diets based on food avoidance interfere with these results. In addition, further research is needed to characterize more precisely the impact of specific bacterial species or strains on the development or resolution of food allergy.

12.1.3 Prevention and Therapy of Food Allergy by Modulating the Microbiota

Given the association of a dysbiotic gut microbiome with the development of food allergy, probiotic bacteria have been evaluated for therapy or prevention of food allergy (Aitoro et al. 2017; du Toit et al. 2016). In a recently published meta-analysis on beneficial effects of probiotics to prevent food allergy, these were only observed when given pre- and postnatally to mother and child. Intervention limited either to prenatal administration to the pregnant mother or to postnatal application to the child displayed no significant effects (Zhang et al. 2016).

Studies investigating therapeutic efficacy of probiotics in resolution of food allergy have obtained conflicting results. Administration of *Lactobacillus casei* CRL 431 and *Bifidobacterium lactis* Bb12 to children with cow's milk allergy failed to induce significantly enhanced allergy resolution and acquisition of oral tolerance to milk (Hol et al. 2008). In contrast supplementation of a highly hydrolyzed casein formula diet with the popular and well-studied probiotic *Lactobacillus rhamnosus* GG revealed significantly augmented acquisition of tolerance to cow's milk protein after a 12-month period (Berni Canani et al. 2012). More recently it has been shown that administration of probiotic *Lactobacillus rhamnosus* CGMCC 1.3724 together with peanut in a placebo-controlled randomized clinical trial induced tolerance in >80% of peanut allergic children (Tang et al. 2015).

12.2 Microbiomes Beyond the Gut

12.2.1 The Vaginal Microbiome

In contrast to other human microbiome communities, which typically display a high degree of diversity under healthy steady state conditions, the vaginal microbiome is typically dominated by one of the *Lactobacillus* species *L. iners, L. crispatus, L. gasseri,* or *L. jensenii*, but more diverse communities with higher numbers of anaerobic bacteria also can be found. It seems to be stable during reproductive age but is strongly influenced by human behavior (e.g., hygiene, contraception) and ethnic group membership (Ravel et al. 2011). The production of lactic acid by vaginal bacteria causes the low vaginal pH. In recent years, attention to the vaginal microbiome was boosted by studies suggesting it a prominent role in shaping the microbiomes of the prior sterile newborn, but improved evidence suggests that differences in the microbiomes correlating with the mode of delivery may be rather caused by the medical conditions leading to the caesarean section (Chu et al. 2017).

12.2.2 Respiratory Microbiome

Nasal, lung, and sometimes oropharyngeal microbial communities are summarized as the respiratory microbiome. Traditionally, the healthy lung has been regarded as a sterile compartment. But recent NGS-based approaches identified bacteria of the *Proteobacteria*, *Firmicutes*, and *Bacteroidetes* phyla, with the predominant genera being *Pseudomonas*, *Streptococcus*, *Prevotella*, *Fusobacteria, and Veillonella*, although some caveats are issued concerning possible contaminations by nasal or faucal microbes (Beck et al. 2012).

The nasal microbiome was found to be dominated by *Actinobacteria*, *Firmicutes*, and *Proteobacteria*, with *Corynebacteriaceae* and *Propionibacteriaceae* being the most important *Actinobacteria* families, while *Staphylococcus epidermidis* or *S. aureus* are the most frequent *Firmicutes* (Bassis et al. 2014; Liu et al. 2015).

The respiratory microbiome receives increasing attention in regard to its possible role in the development of atopic asthma, but more research is required for definite answers (Holt 2015).

12.2.3 Skin Microbiome

Compared to other body sites harboring microbiomes, the skin is rather dry and cool. However, important factors like pH, moisture, and sebum quantities vary a lot between different skin locations and also in terms of age and sex. *Propionibacterium* spp. (and *Malassezia* spp.) are the dominant colonizers of the sebaceous skin (e.g., on the forehead or back) and moist areas (mostly the body's vaults and creases) and attract *Staphylococcus* spp. and *Corynebacterium* species, while dry locations show the diverse colonization with bacteria belonging to the *Actinobacteria*, *Proteobacteria*, *Firmicutes*, and *Bacteroidetes* phyla (Grice and Segre 2011).

Despite the skin is readily accessible for microbial exchange with the environment, the skin microbiome seems to be remarkably stable over time and predominantly shaped by host factors and not so much by environmental microbes. While it is very evident that noninfectious diseases can drastically reshape the skin microbiome, leading to the dominance of pathologic species, it is less evident how specific skin commensals can contribute to a healthy steady state, in part because of overlapping metabolic capabilities of different species (Oh et al. 2016).

12.3 Microbiome Shaping of Atopic Dermatitis

12.3.1 Introduction into Atopic Dermatitis

The other large surface organ affected by an atopy-related disease is the skin. Atopic dermatitis (AD) is a chronic recurrent skin disease. Primarily affected sites are the flexures of the limbs, face, and neck. AD often affects children (up to 18% prevalence) and can persist for life. Dominant symptoms are highly irritable, itching skin with erythemato-squamous plaques, papulovesicles, and lichenification especially in chronic lesions (Werfel 2015).

The pathogenesis of AD is the subject of intense research and assumed to be multifactorial (Eyerich et al. 2015). These factors include a compromised skin barrier and microbial dysbiosis within the lesions of AD.

Skin barrier defects in AD can be based on mutations of central proteins of the cutaneous barrier in affected patients. Alternatively, immune responses dominated by so-called type 2 cytokines such as interleukin (IL-)4 and IL-13 or microbial dysbiosis can secondarily reduce cutaneous barrier function (Agrawal and Woodfolk 2014; Biedermann 2006). Findings include reduced levels of tight junction protein claudin 1 (Tokumasu et al. 2016), enhanced levels of kallikrein protease, or changes in the composition of lipids (Egawa and Kabashima 2016). The most prominent genetic predisposition is the loss of function mutations of the structural protein filaggrin, determining the quality of the most outer layer of the skin, the stratum corneum, representing a key protein of the skin barrier (Irvine et al. 2011; Palmer et al. 2006). The aforementioned barrier defects result in increased epidermal water loss and decreased physical protection against microbes (Agrawal and Woodfolk 2014). They also reduce protection against sensitization to allergens, leading to a correlation of AD with food allergies and atopic asthma (atopic march) (Spergel 2010).

AD is characterized by the dominance of type 2 immune responses associated with elevated secretion of the Th2 cytokines IL-4 and IL-13 and elevated frequencies of Th2 cells (Eyerich et al. 2015).

Several polymorphisms in pathogen recognition receptors, most notable TLR2, may compromise microbial sensing in AD (Kuo et al. 2013). Skin migration or function of cells of the innate immune system, especially neutrophils, is often impaired (De Benedetto et al. 2009), and type 2 cytokines suppress the IL-23-IL-17 axis by type 2 immune responses and/or microbial dysbiosis (Guenova et al. 2015).

On the interface between skin barrier and skin immunity, antimicrobial peptides (AMP) play an important role in the early defense of microbial pathogens and the orchestration of the microbiome composition. The lesional skin of AD patients has been shown to express significantly lower amounts of antimicrobial peptides like HBD-2 and LL-37 than the skin from psoriatic lesions (Ong et al. 2002), possibly due to inhibition by Th2 cytokines. But many antimicrobial peptides, including HBD-2, are upregulated in both the atopic and psoriatic skin when compared to the healthy skin (Schröder 2011). Taken together, these alterations associated with AD not only boost allergic inflammation and facilitate sensitization to allergens but also hamper proper antimicrobial immune responses and facilitate microbial dysbiosis.

12.3.2 The Intestinal Microbiome in Atopic Dermatitis

In some older studies, atopic infants were found to be colonized less with *Bacteroides* and *Bifidobacterium* species as well as having a lower ratio of bifidobacteria to *Clostridia* when developing AD (Kalliomäki et al. 2001; Watanabe et al. 2003). However, many of the subsequent studies examining larger collectives have not been able to fully reproduce these findings (Abrahamsson et al. 2012) or to show alterations of the intestinal microbiome in AD at all (Adlerberth et al. 2007). Notably, patient collectives differed between these studies, especially with regard to patients who also suffered from food allergy. Maybe a critical weakness of many hitherto performed studies is their limited phylogenetic resolution. Typically, the genus to species level is resolved, but a recent study suggests that a subspecies of *Faecalibacterium prausnitzii* may correlate with AD (Song et al. 2016).

Basically the same holds true regarding the question, whether probiotics are suitable for the prevention or therapy of AD. Some studies found that the application of probiotics is effective. *Bifidobacteria* spp. or *Lactobacilli* were found to ameliorate AD when given after weaning (Isolauri et al. 2000; Penders et al. 2013). Others found no effects (Brouwer et al. 2006) or found probiotics only effective in atopic children with food allergy (Sistek et al. 2006). A recent meta-analysis found probiotic treatment effective only when administered both pre- and postnatally (Panduru et al. 2015), as has been suggested for food allergy (Zhang et al. 2016).

Because of the heterogeneous results of human epidemiological studies and the lack of evidence created by mouse model research, it is hard to decide whether risk factors for atopic dermatitis may act via the gut microbiome (Figs. 12.2 and 12.3).

12.3.3 The Skin Microbiome in Atopic Dermatitis

AD lesions display a drastically changed skin microbiome, with a strong expansion of *Staphylococcus aureus*, which is responsible both for the loss of microbial diversity and the exacerbation of the disease by intensifying inflammation. Other staphylococcal populations may be expanded as well, especially *S. epidermidis*. Notably *S. aureus* is also expanded in the nares. After successful treatment, microbial diversity rises again with increased proportions of *Streptococcus*, *Propionibacterium*, and *Corynebacterium* species. Microbial diversity can be effectively reinforced by anti-inflammatory treatment but also by treatment with emollients alone (Kong et al. 2012; Seite et al. 2014). Interestingly, most recent data indicate that the composition of epidermal lipids determines the bacterial composition, in particular *Propionibacterium* and *Corynebacterium* species (Baurecht et al. 2018). Key mechanisms by which staphylococci increase and prolong AD inflammation are thought to be staphylococcal superantigens, host TLR2 and NOD/CARD pathogen sensing, IL-36 driven T-cell activation, and the appearance of IL-17 and associated transcription signatures (Biedermann 2006; Volz et al. 2010; Liu et al. 2017; Eyerich et al. 2009). We were able to show that the exposure to TLR2 ligands such as from *Staphylococcus aureus* from the inflamed skin due to Th2 cell activation leads to a significant inhibition of the IL-10 response and subsequently to exaggerated and tremendously prolonged cutaneous inflammation (Kaesler et al. 2014; Volz et al. 2014). The fungal opportunistically pathogenic *Malassezia* species can also contribute to AD inflammation, both by stimulating IgE production and Th2 cells and by delivering ligands (Baker 2006; Glatz et al. 2015). Similar to the intestinal microbiome, grafting beneficial bacteria to the skin could ameliorate AD symptoms. In a recent study, autologously transferred commensal staphylococci decreased *S. aureus* burden by secretion of antimicrobial lantibiotics (Nakatsuji et al. 2017).

Interestingly, application of lysates of the nonpathogenic bacterium *Vitreoscilla filiformis* in a cream significantly ameliorated mild atopic dermatitis compared to placebo (Gueniche et al. 2008). Subsequent mechanistic analyses displayed that TLR2 ligands within these lysates upregulate IL-10 production in innate immune cells such as dendritic cells and subsequently in adaptive immune cells (Tr1 cells) (Volz et al. 2014). Thus, the loss of IL-10 through the concomitant signaling through the IL-4 receptor and TLR2 can be reconstituted by appropriate signals from beneficial bacteria. Importantly, the latest treatments targeting type 2 immunity by blocking the IL-4 receptor α-chain (dupilumab) demonstrated efficacy in moderate to severe atopic dermatitis, reducing the inflammatory signatures in the skin, and ongoing research investigates the consequences on microbial dysbiosis and the role for cutaneous microbes in this type of targeted treatment (Beck et al. 2014; Hamilton et al. 2014).

12.4 Asthma and Gut Microbiota

12.4.1 Introduction into Asthma

Asthma is a chronic inflammatory obstructive disease of the airways. It belongs to the most relevant chronic diseases worldwide, affecting about 300 million people with economic costs exceeding those of infectious diseases such as tuberculosis or HIV (Accordini et al. 2013). Yet, the incidence is continuously increasing since decades suggesting an environmental etiology (Masoli et al. 2004). It can be classified as allergic, non-allergic, or a combination of both as well as by severity. The most common form of asthma is at least initially caused by type I allergies to aeroallergens. In type I allergy, the Th2 cytokines IL-4, IL-5, and IL-13 trigger a class switch in plasma cells from production of allergen-specific IgG to IgE which binds to FcεRI on effector cells like mast cells and basophils and mediates their activation in case of allergen contact (Fig. 12.1). Recurrent inflammation results in airway remodeling and allergen-independent bronchial hyperreactivity. Other cells involved in the pathogenesis of asthma are Treg cells. Hartl et al. found a reduction in Treg cells in adult asthma patients (Hartl et al. 2007). Exacerbation of allergic airway disease in mice depleted of Treg cells suggests that Treg cells might crucially regulate the severity of the disease (Lewkowich et al. 2005). Obviously, the initiation of allergic asthma is the result of a complex interplay of pro-inflammatory, especially Th2-prone conditions and tolerance-promoting factors as well as local conditions like, e.g., the epithelial barrier and airway remodeling.

12.4.2 Intestinal Microbiota and the Development of Allergic Sensitization and Asthma

There is increasing evidence that the composition of the gut microbiota very early in life helps to finally shape the development of the immune system which might also determine the pattern of immune responses later in life. Strikingly, virtually all conditions already known to reduce the risk to develop type I allergy and asthma have in common to increase the exposure to microbes and thereby most likely to have an impact on the diversity and the composition of the gut microbiota in infancy. Well known such conditions even mentioned in guidelines are vaginal birth (Kolokotroni et al. 2012), breast-feeding (Oddy 2009), close contact to dogs or certain farm animals, having multiple older siblings (Ball et al. 2000), raw milk consumption (Waser et al. 2007), early day-care attendance (Ball et al. 2000), or even in a broader sense, growing up in more rural and less industrialized countries (ISAAC 1998) (Fig. 12.3). Certain foods, especially those more frequently consumed in Western countries, were also shown to be associated with an increased risk to develop asthma which most likely depends on the effects of food ingredients on the gut microbiota. Based on these observations and the knowledge about the role of the gut microbiota for shaping patterns of immune responses, the current model of asthma pathogenesis includes the concept of a cross talk between mucosal immune compartments, in recent publications often called "gut-lung axis" (Fig. 12.2).

The initial microbial colonization of the gut directly after birth by the so-called pioneer microbiome has lifelong implications. Next to its impact on the developing immune system, it also determines the conditions for subsequent colonization by other microbes. It is therefore shaping the developing gut microbiome that becomes relatively stable in adulthood. The route of delivery at birth, contact to the mother's skin and to food, provides the first contact of the gut to microbes. Children born by caesarean section were found to have a reduced "bacterial richness and diversity" (Azad et al. 2013). Although evidence is in parts conflicting, summarized published data show that birth by caesarean section seems to be associated with an increased risk to develop "atopic sensitization" and asthma in childhood (Renz-Polster et al. 2005). A strain enriched in the human vaginal tract immediately before birth is *Lactobacillus johnsonii* (Aagaard et al. 2012). Fujimura et al. reported that mice with an altered gut microbiome enriched in *Lactobacillus johnsonii* were protected from ovalbumin-induced allergic airway disease (AAD), which is the commonly used model for asthma in mice. They further demonstrated that supplementation of mice with this strain was sufficient to protect them from AAD (Fujimura et al. 2014). This suggests that early colonization with *Lactobacillus johnsonii* might crucially modulate the microbial colonization pattern of the gut. This is in line with the effects observed by breast-feeding since breast milk has a unique microbiota composition including staphylococci, streptococci, lactobacilli and bifidobacteria (Gomez-Gallego et al. 2016). A recent meta-analysis again confirmed that breast-feeding has a protective role with regard to asthma development (Lodge et al. 2015). Interestingly, even among breast-fed children, differences in the gut microbiota composition can be found. The feces of those later on developing allergies contain other bifidobacteria species than the feces of non-allergic children (Ouwehand et al. 2001).

Another possibility to explain the association between the risk to develop asthma and the route of delivery at birth is the routine antibiotic treatment given before caesarean section. Keski-Nisula et al. could show that intrapartum administration of antibiotics was associated with a reduced transmission of the vaginal *Lactobacillus* spp. dominated microbiota to the neonates (Keski-Nisula et al. 2013). Furthermore, a number of studies in different countries consistently prove that even short-term antibiotic treatment within the first year of life can result in a long-term shift of the gut microbiome composition. For example, treatment with amoxicillin was shown to especially deplete *Lactobacillus* spp., *Bifidobacterium* spp., and *Acidophilus* spp. A study from Finland shows that the use of macrolides results in a long-lasting dysbiosis with predominant depletion of *Actinobacteria* and a significantly increased risk to develop asthma in the age of 2–7 (Korpela et al. 2016). However, there are also large observational studies and sibling analyses reporting no relationship between asthma and treatment with antibiotics. Most likely the observed discrepancies depend on the group of antibiotics used for treatment. This interpretation is supported by studies performed in mice. Treatment of neonatal mice with vancomycin, but not streptomycin, resulted in a reduced microbial diversity and alteration of the gut microbiota composition as well as increased serum- and surface-bound IgE, enhanced susceptibility to AAD, and decreased Treg cell numbers in the colon. Most interestingly, these effects could not be observed in similarly treated adult mice (Russell et al. 2012). This proves the existence and relevance of a gut-lung axis. Interestingly, albeit weak, there is evidence that vice versa exposure of the lung to LPS within 24 hours has an influence on the gut microbiota resulting in a significantly increased bacterial count in the cecum (Sze et al. 2014). This implies that the gut-lung axis can act in both directions.

12.4.3 Concepts and Interventional Studies on Intestinal Microbiota and the Development of Allergic Sensitization and Asthma

However, although the abovementioned observations provide strong evidence that conditions interfering with the microbial colonization of the gut especially early in life can modulate the risk to develop asthma, it remained unclear whether it is the gut microbiota composition itself that modulates the risk to develop asthma or AAD in mice. To address this question, the Canadian Healthy Infant Longitudinal Development (CHILD) study analyzed and compared the gut microbiome in 319 children between 3 and 12 months of age (Arrieta et al. 2015). Healthy controls were compared to children either suffering from atopy, wheezing, or atopy plus wheezing; the latter were expected to have the highest risk to develop asthma later in life. In this group, the genera *Faecalibacterium*, *Lachnospira*, *Rothia*, and *Veillonella* displayed lower abundance exclusively at 3 months of age. To figure out if this causes a higher risk to develop asthma, germ-free mice were either reconstituted with feces of a 3-month-old patient from this group or with the same feces but additionally supplemented with the mentioned genera. Intriguingly, the comparison of both groups showed relatively reduced ovalbumin-induced AAD in the latter group, proving a direct impact of not only composition but rather abundance of certain gut microbiota on the development of AAD. In accordance with an earlier study published by Atarashi et al. who were able to show that "oral inoculation of a cocktail of *Clostridium* species exclusively during the early life of conventionally reared mice" results in significantly reduced ovalbumin-specific IgE following sensitization. This suggests that colonization of the gut with certain microbes very early in life can reduce type I allergy and related AAD risk (Atarashi et al. 2011).

12.4.4 Short-Chain Fatty Acids, Microbiota, and the Development of Allergic Sensitization and Asthma

Another important finding of the CHILD study was decreased levels of acetate in the 3-month feces samples of those children more prone to asthma. Acetate like butyrate, propionate, and formate is a metabolic product of fermentation of nondigestible carbohydrates, so-called short-chain fatty acid (SCFA). The level of SCFA depends on the interplay between diet and gut microbiota composition. SCFA production, especially of acetate, is widely distributed among bacteria. Thorburn et al. have shown that "feeding mice a high-fibre diet results in shaping a distinctive gut microbiota composition" accompanied by increased acetate levels. Both the diet and direct feeding of acetate led to "marked suppression of house dust mite extract induced AAD" (Thorburn et al. 2015). Strikingly, a study by De Filippo et al. comparing cohorts of 1 to 6-year-old children from Italy and Burkina Faso, expected to represent low- or high-fiber nutrition, respectively, disclosed that the gut microbiota associated with a high-fiber diet included higher numbers of *Bacteroidetes* but a reduction in *Firmicutes* (De Filippo et al. 2010). Concordantly, in mice a high-fiber diet also changes the ratio of *Firmicutes* to *Bacteroidetes* resulting in increased circulating SCFA concentrations. Mice fed a high-fiber diet were protected against house dust mite extract-induced allergic inflammation of the lung. In contrast, mice fed a low-fiber diet had increased IL-4, IL-5, and IL-13 cytokine levels in the lung tissue, increased airway mucus production, and increased serum IgE (Trompette et al. 2014). In conclusion, a certain gut microbiota composition might be able to reduce the risk to develop asthma by altering the metabolism of nutrients. This might explain why certain foods more prevalent in industrialized countries have been associated with an increased risk to develop asthma. For example, pasteurized milk contains significantly less SCFA than unpasteurized milk (Velez et al. 2010). Further evidence supporting this hypothesis comes from a study by Berthon et al. showing that "subjects with severe persistent asthma consumed less fibre (32 g/day ± 11 g/day) as compared with healthy controls (37 g/day ± 13 g/day)" (Berthon et al. 2013).

For the SCFA butyrate, acetate, and propionate, it could be shown that their luminal concentration directly regulates the number of colonic Treg cells (Furusawa et al. 2013; Smith et al. 2013). This corresponds to reports of Atarashi et al. showing that colonization of germ-free mice with *Clostridium* species led to a strong upregulation of colonic Treg cells and of Geuking et al., who observed proliferation of Treg cells following colonization of germ-free mice with altered Schaedler flora, a defined mixture of eight species also including *Clostridium* species (Atarashi et al. 2013; Geuking et al. 2011). Vice versa, iTreg deficiency was shown to be associated with a dysbiotic gut microbiota (Josefowicz et al. 2012). A reduced number of Treg cells as well as accumulation of invariant natural killer T (iNKT) cells in the lung and colonic lamina propria soon after birth is characteristic for germ-free mice and at least one reason why they are more susceptible to AAD. However, according to Olszak et al. "colonization of neonatal germ-free mice with a conventional microbiota protected the animals from mucosal iNKT accumulation" and ovalbumin-driven allergic asthma. Of note, these effects were absent in similarly treated adult mice (Olszak et al. 2012). Taken together, this implies that the microbial colonization of the gut very early in life is crucial for induction of Treg cells, and this most likely also mediates reduced susceptibility to AAD.

In conclusion, the composition of the gut microbiota in early childhood, most likely within the first 3 months of life, may play an orchestrating role in the development of type I sensitization and allergic asthma. The composition of the gut microbiome in this period of life is largely dependent on the route of delivery at birth and the environmental microbiome. Its interplay with diet and inherited factors like the strength of the epithelial barrier via bacterial metabolites crucially regulates the balance between pro-inflammatory and tolerance-mediating immune cells, primarily in the gut. Observational studies supported by mouse model experiments showing that it also determines the susceptibility to asthma or allergic airway disease however, suggest a role far beyond the gut.

12.5 Conclusion

In recent years it has become evident that microbes have an enormous capacity to educate and fine-tune our immune system, and allergic diseases are no exception. Diversity of the human microbiomes has been recognized as an important factor associated with reduced manifestation of allergic diseases. This is most evident in food allergy. Conversely, dysbiotic microbial colonization boosts inflammation and thus symptom severity of allergic diseases. Further studies providing high-resolution microbiome data will lead to a better understanding of the contribution of specific microbes to incidence or severity of different allergic diseases (Fig. 12.2).

▶ **Controversy**

Atopic diseases and allergies develop at organ surfaces colonized by a multitude of microorganisms, the microbiome. Thus, it is obvious that "the microbiome" somehow interferes with inflammation in atopic diseases and allergic reactions.

However, controversies regarding specific interactions and consequences remain:

- How much can microbiomes orchestrate immune reactions and barrier function across surfaces? It is well accepted that the gut microbiome governs immune reactions functional within the gut and the skin microbiome conditions cutaneous immunity toward effective and specific reactions of immune modulation and tolerance. The constantly increasing understanding of tissue-specific immune memory (T_{RM} cells) further strengthens the belief that compartmentalized immunity is a default pathway in immune orchestration and regulation. Consequently, local microbiomes will directly influence local immunity and may only under exceptional circumstances drive systemic immune profiles and reactions. Contrariwise, the gastrointestinal barrier function and the development of the intestinal immune system especially early in life impacts systemic immune cell development and reactivity arguing for a role of the gut microbiome also for immune reactions at other surfaces than the intestinal mucosa. Further research will dissect if this influence is restricted to certain windows of opportunity or whether there is a constant chance to shape immune reactions through the gut microbiome, possibly allowing therapeutic interventions.
- How can we distinguish between epiphenomena from causal findings regarding the microbiome and atopic diseases and allergies? As pointed out, atopic inflammation and allergies develop at surfaces, which are colonized by "the microbiome." We understand that atopic diseases and allergies develop based on a complex genetic trait, including genes responsible for barrier functions, innate/adaptive immune responses or nonimmune cells "inflammation." Microbial constituents can therefore function as triggers, but changes in microbial composition will also be a consequence of a dysfunctional barrier, innate or adaptive immune reactions or inflammatory pathways. The latter may result in microbial dysbiosis based on "surface instability" tempting us to call this an epiphenomenon of the susceptibility for atopic diseases or allergies. However, this dysbiosis may well act as switch to transform initial immune dysregulation into overt inflammation and disease. In addition, dysbiosis may result of subtle changes in surface response patterns, which then impact disease development directly, representing a primary cause rather than an epiphenomenon. As the latter also implies that intervention at an early stage of disease development/predisposition may be feasible, further research is highly desirable.

In addition, controversies remain regarding specific aspects of investigation:

- Human studies by nature deliver less causal insights compared to animal experiments and can remain speculative in regard to mechanistic interpretations. For example, there is controversial evidence regarding the role of caesarean sections for changes of the composition of the gut microbiota on the risk to develop allergy, atopic dermatitis, or asthma, which cannot be resolved with animal experiments.
- Loss of microbial diversity due to a westernized lifestyle with higher hygiene standards, less infectious diseases, smaller family sizes, and less contact to rural environments (biodiversity hypothesis) has well been shown to be associated with allergic sensitization and asthma, not so well with atopic dermatitis or food allergy. Compared to the rise of allergic asthma and atopic dermatitis, the increase of food allergy in the last two decades can only in part be explained by the biodiversity hypothesis as its rise occurred approximately >30 years later. Thus additional yet unknown susceptibility factors with direct effects on the induction of type 2 immunity in the gut or with indirect effects shaping the composition of the intestinal microbiota are likely to exist.

History

- Allergies and atopic diseases are not new to mankind and have been described since ancient times. But their etiology has been poorly understood before the concept of an immune system was introduced in the late nineteenth century. Therefore, earlier reports can hardly be related to modern disease classifications (Bergmann and Ring 2014).
- 1873: Charles Blackley found hay fever to be less frequent among farmers' children (no consideration of microbes at the time, though).
- Beginning of the twentieth century: Several physicians independently recognized that infectious diseases symptoms did not necessarily have to be caused by the pathogens. Instead, they were largely the results of the pro-inflammatory immune reactions (because symptoms and incubation times were similar with infections by various pathogens). Furthermore, impressive, sometimes fatal consequences were noted with patients repeatedly treaded with the recently developed antisera (Igea 2013).
- 1902: Portier and Richet coined the term "anaphylaxis."
- 1906: von Pirquet coined the term "Allergie."
- 1933: Wise and Sulzberger proposed the term "atopic dermatitis" for a recurrent eczema with family history. Scientific discussion about the influence of food allergy on AD begins.
- 1974: Leyden et al. found *Staphylococcus aureus* to be the dominant microbe on atopic lesions (Leyden et al. 1974).
- 1989: Strachan introduced the "hygiene hypothesis" in a study correlating asthma incidence with big, unhygienic households (Strachan 1989).
- 1997: Majamaa and Isolauri reported the first successful use of a probiotic (*Lactobacillus rhamnosus* GG) on children with cow's milk allergy and AD after lactobacilli were reported to strengthen the intestinal barrier (Majamaa and Isolauri 1997).
- 2000: Kalliomäki et al. demonstrated distinct microbiomes in healthy children and children with atopic dermatitis by non-culture methods. Bacterial culture was shown to be not sensitive enough, hinting on potential of NGS investigations (Kalliomäki et al. 2001).
- 2008: Chen and Blaser reported inverse correlation between *Helicobacter pylori* infection and asthma in children (Chen and Blaser 2008).
- 2012: Kong et al. provide high-resolution NGS investigation of microbiome shifts in AD (Kong et al. 2012).

Highlights

- Investigations in the murine and human system using 16S rRNA gene phylotyping and metagenomics have demonstrated an association of food allergy with an altered (dysbiotic) intestinal microbiota (Chen et al. 2016; Ling et al. 2014; Noval Rivas et al. 2013).
- Research in the last decade has elucidated cellular and molecular mechanisms of oral tolerance and the role of the intestinal microbiota in shaping tolerogenic immune responses in regard to orally ingested antigens (Cahenzli et al. 2013; Hadis et al. 2011; Pabst and Mowat 2012).
- Gut microbiota composition and levels of SCFA at age 3 months correlate with the risk to develop asthma later in life. Observations made in humans confirmed by animal experiments (Arrieta et al. 2015).
- High-fiber diet in animal experiments changes the ratio of *Firmicutes* to *Bacteroidetes* in the gut resulting in increased circulating SCFA concentrations associated with protection against house dust mite extract-

(continued)

induced allergic lung inflammation (Trompette et al. 2014).
- The controversy on a role of the gut microbiome an atopic dermatitis and consecutively on probiotics as possible intervention strategy could be attributed to the window of opportunity: A recent meta-analysis found probiotic treatment effective only when administered both pre- and postnatally (Panduru et al. 2015).

References

Aagaard, K., Riehle, K., Ma, J., Segata, N., Mistretta, T. A., Coarfa, C., Raza, S., et al. (2012). A metagenomic approach to characterization of the vaginal microbiome signature in pregnancy. *PLoS One, 7*, e36466.

Abrahamsson, T. R., Jakobsson, H. E., Andersson, A. F., Bjorksten, B., Engstrand, L., & Jenmalm, M. C. (2012). Low diversity of the gut microbiota in infants with atopic eczema. *Journal of Allergy and Clinical Immunology, 129*, 434–440 440 e431–432.

Accordini, S., Corsico, A. G., Braggion, M., Gerbase, M. W., Gislason, D., Gulsvik, A., Heinrich, J., et al. (2013). The cost of persistent asthma in Europe: An international population-based study in adults. *International Archives of Allergy and Immunology, 160*, 93–101.

Adlerberth, I., Strachan, D. P., Matricardi, P. M., Ahrne, S., Orfei, L., Aberg, N., Perkin, M. R., et al. (2007). Gut microbiota and development of atopic eczema in 3 European birth cohorts. *The Journal of Allergy and Clinical Immunology, 120*, 343–350.

Agrawal, R., & Woodfolk, J. A. (2014). Skin barrier defects in atopic dermatitis. *Current Allergy and Asthma Reports, 14*, 433.

Aitoro, R., Paparo, L., Amoroso, A., Di Costanzo, M., Cosenza, L., Granata, V., Di Scala, C., et al. (2017). Gut microbiota as a target for preventive and therapeutic intervention against food allergy. *Nutrients, 9*, E672.

Arrieta, M. C., Stiemsma, L. T., Dimitriu, P. A., Thorson, L., Russell, S., Yurist-Doutsch, S., Kuzeljevic, B., et al. (2015). Early infancy microbial and metabolic alterations affect risk of childhood asthma. *Science Translational Medicine, 7*, 307ra152.

Atarashi, K., Tanoue, T., Oshima, K., Suda, W., Nagano, Y., Nishikawa, H., Fukuda, S., et al. (2013). Treg induction by a rationally selected mixture of Clostridia strains from the human microbiota. *Nature, 500*, 232–236.

Atarashi, K., Tanoue, T., Shima, T., Imaoka, A., Kuwahara, T., Momose, Y., Cheng, G., et al. (2011). Induction of colonic regulatory T cells by indigenous Clostridium species. *Science, 331*, 337–341.

Azad, M. B., Konya, T., Maughan, H., Guttman, D. S., Field, C. J., Sears, M. R., Becker, A. B., et al. (2013). Infant gut microbiota and the hygiene hypothesis of allergic disease: Impact of household pets and siblings on microbiota composition and diversity. *Allergy, Asthma and Clinical Immunology, 9*, 15.

Baker, B. S. (2006). The role of microorganisms in atopic dermatitis. *Clinical and Experimental Immunology, 144*, 1–9.

Ball, T. M., Castro-Rodriguez, J. A., Griffith, K. A., Holberg, C. J., Martinez, F. D., & Wright, A. L. (2000). Siblings, day-care attendance, and the risk of asthma and wheezing during childhood. *The New England Journal of Medicine, 343*, 538–543.

Bashir, M. E., Louie, S., Shi, H. N., & Nagler-Anderson, C. (2004). Toll-like receptor 4 signaling by intestinal microbes influences susceptibility to food allergy. *Journal of Immunology, 172*, 6978–6987.

Bassis, C. M., Tang, A. L., Young, V. B., & Pynnonen, M. A. (2014). The nasal cavity microbiota of healthy adults. *Microbiome, 2*, 27.

Baurecht, H., Rühlemann, M. C., Rodríguez, E., Thielking, F., Harder, I., Erkens, A. S., Stölzl, D., Ellinghaus, E., Hotze, M., Lieb, W., Wang, S., Heinsen, F. A., Franke, A., & Weidinger, S. (2018). Epidermal lipid composition, barrier integrity and eczematous inflammation are associated with skin microbiome configuration. *The Journal of Allergy and Clinical Immunology*. pii: S0091-6749(18) 30198-2. https://doi.org/10.1016/j.jaci.2018.01.019.

Beck, T. C., Gomes, A. C., Cyster, J. G., & Pereira, J. P. (2014). CXCR4 and a cell-extrinsic mechanism control immature B lymphocyte egress from bone marrow. *The Journal of Experimental Medicine, 211*, 2567–2581.

Beck, J. M., Young, V. B., & Huffnagle, G. B. (2012). The microbiome of the lung. *Translational Research, 160*, 258–266.

Belkaid, Y., & Harrison, O. J. (2017). Homeostatic immunity and the microbiota. *Immunity, 46*, 562–576.

Bergmann, K.-C., & Ring, J. (2014). History of allergy. *Chemical Immunology and Allergy, 100*, 54–61.

Berin, M. C., Zheng, Y., Domaradzki, M., Li, X. M., & Sampson, H. A. (2006). Role of TLR4 in allergic sensitization to food proteins in mice. *Allergy, 61*, 64–71.

Berni Canani, R., Nocerino, R., Terrin, G., Coruzzo, A., Cosenza, L., Leone, L., & Troncone, R. (2012). Effect of Lactobacillus GG on tolerance acquisition in infants with cow's milk allergy: A randomized trial. *Journal of Allergy and Clinical Immunology, 129*, 580–582.e585.

Berthon, B. S., Macdonald-Wicks, L. K., Gibson, P. G., & Wood, L. G. (2013). Investigation of the association between dietary intake, disease severity and airway inflammation in asthma. *Respirology, 18*, 447–454.

Biedermann, T. (2006). Dissecting the role of infections in atopic dermatitis. *Acta Dermato-Venereologica, 86*, 99–109.

Biedermann, T., Röcken, M., & Carballido, J. M. (2004). TH1 and TH2 lymphocyte development and regulation of TH cell-mediated immune responses of the skin. *The Journal of Investigative Dermatology. Symposium Proceedings, 9*, 5–14.

Biedermann, T., Skabytska, Y., Kaesler, S., & Volz, T. (2015). Regulation of T cell immunity in atopic dermatitis by microbes: The Yin and Yang of cutaneous inflammation. *Frontiers in Immunology, 6*, 353.

Brouwer, M. L., Wolt-Plompen, S. A., Dubois, A. E., van der Heide, S., Jansen, D. F., Hoijer, M. A., Kauffman, H. F., et al. (2006). No effects of probiotics on atopic dermatitis in infancy: A randomized placebo-controlled trial. *Clinical and Experimental Allergy, 36*, 899–906.

Bunyavanich, S., Shen, N., Grishin, A., Wood, R., Burks, W., Dawson, P., Jones, S. M., et al. (2016). Early-life gut microbiome composition and milk allergy resolution. *The Journal of Allergy and Clinical Immunology, 138*, 1122–1130.

Cahenzli, J., Koller, Y., Wyss, M., Geuking, M. B., & McCoy, K. D. (2013). Intestinal microbial diversity during early-life colonization shapes long-term IgE levels. *Cell Host & Microbe, 14*, 559–570.

Chen, Y., & Blaser, M. J. (2008). Helicobacter pylori colonization is inversely associated with childhood asthma. *The Journal of Infectious Diseases, 198*, 553–560.

Chen, C. C., Chen, K. J., Kong, M. S., Chang, H. J., & Huang, J. L. (2016). Alterations in the gut microbiotas of children with food sensitization in early life. *Pediatric Allergy and Immunology, 27*, 254–262.

Chu, D. M., Ma, J., Prince, A. L., Antony, K. M., Seferovic, M. D., & Aagaard, K. M. (2017). Maturation of the infant microbiome community structure and function across multiple body sites and in relation to mode of delivery. *Nature Medicine, 23*, 314–326.

Clark, S., Espinola, J., Rudders, S. A., Banerji, A., & Camargo, C. A., Jr. (2011). Frequency of US emergency department visits for food-related acute allergic reactions. *The Journal of Allergy and Clinical Immunology, 127*, 682–683.

Cohn, L., Elias, J. A., & Chupp, G. L. (2004). Asthma: Mechanisms of disease persistence and progression. *Annual Review of Immunology, 22*, 789–815.

Coombs, R. R. A., & Gells, P. G. H. (1963). The classification of allergic reactions underlying disease. In R. R. A. Coombs & P. G. H. Gells (Eds.), *Clinical aspects of immunology*. Oxford: Blackwell Science.

Curotto de Lafaille, M. A., Kutchukhidze, N., Shen, S., Ding, Y., Yee, H., & Lafaille, J. J. (2008). Adaptive Foxp3+ regulatory T cell-dependent and -independent control of allergic inflammation. *Immunity, 29*, 114–126.

De Benedetto, A., Agnihothri, R., McGirt, L. Y., Bankova, L. G., & Beck, L. A. (2009). Atopic dermatitis: A disease caused by innate immune defects? *The Journal of Investigative Dermatology, 129*, 14–30.

De Filippo, C., Cavalieri, D., Di Paola, M., Ramazzotti, M., Poullet, J. B., Massart, S., Collini, S., et al. (2010). Impact of diet in shaping gut microbiota revealed by a comparative study in children from Europe and rural Africa. *Proceedings of the National Academy of Sciences of the United States of America, 107*, 14691–14696.

du Toit, G., Tsakok, T., Lack, S., & Lack, G. (2016). Prevention of food allergy. *The Journal of Allergy and Clinical Immunology, 137*, 998–1010.

Egawa, G., & Kabashima, K. (2016). Multifactorial skin barrier deficiency and atopic dermatitis: Essential topics to prevent the atopic march. *The Journal of Allergy and Clinical Immunology, 138*, 350–358.e351.

Eyerich, K., Eyerich, S., & Biedermann, T. (2015). The multi-modal immune pathogenesis of atopic eczema. *Trends in Immunology, 36*, 788–801.

Eyerich, K., Pennino, D., Scarponi, C., Foerster, S., Nasorri, F., Behrendt, H., Ring, J., Traidl-Hoffmann, C., Albanesi, C., & Cavani, A. (2009). IL-17 in atopic eczema: Linking allergen-specific adaptive and microbial-triggered innate immune response. *The Journal of Allergy and Clinical Immunology, 123*, 59–66.

Fujimura, K. E., Demoor, T., Rauch, M., Faruqi, A. A., Jang, S., Johnson, C. C., Boushey, H. A., et al. (2014). House dust exposure mediates gut microbiome Lactobacillus enrichment and airway immune defense against allergens and virus infection. *Proceedings of the National Academy of Sciences of the United States of America, 111*, 805–810.

Furusawa, Y., Obata, Y., Fukuda, S., Endo, T. A., Nakato, G., Takahashi, D., Nakanishi, Y., et al. (2013). Commensal microbe-derived butyrate induces the differentiation of colonic regulatory T cells. *Nature, 504*, 446–450.

Geuking, M. B., Cahenzli, J., Lawson, M. A., Ng, D. C., Slack, E., Hapfelmeier, S., McCoy, K. D., et al. (2011). Intestinal bacterial colonization induces mutualistic regulatory T cell responses. *Immunity, 34*, 794–806.

Glatz, M., Bosshard, P. P., Hoetzenecker, W., & Schmid-Grendelmeier, P. (2015). The role of Malassezia spp. in atopic dermatitis. *Journal of Clinical Medicine, 4*, 1217–1228.

Gomez-Gallego, C., Garcia-Mantrana, I., Salminen, S., & Collado, M. C. (2016). The human milk microbiome and factors influencing its composition and activity. *Seminars in Fetal and Neonatal Medicine, 21*, 400–405.

Grice, E. A., & Segre, J. A. (2011). The skin microbiome. *Nature Reviews. Microbiology, 9*, 244–253.

Gueniche, A., Knaudt, B., Schuck, E., Volz, T., Bastien, P., Martin, R., Röcken, M., et al. (2008). Effects of nonpathogenic gram-negative bacterium Vitreoscilla filiformis lysate on atopic dermatitis: A prospective, randomized, double-blind, placebo-controlled clinical study. *The British Journal of Dermatology, 159*, 1357–1363.

Guenova, E., Skabytska, Y., Hoetzenecker, W., Weindl, G., Sauer, K., Tham, M., Kim, K. W., et al. (2015). IL-4 abrogates T(H)17 cell-mediated inflammation by selective silencing of IL-23 in antigen-presenting cells. *Proceedings of the National Academy of Sciences of the United States of America, 112*, 2163–2168.

Hadis, U., Wahl, B., Schulz, O., Hardtke-Wolenski, M., Schippers, A., Wagner, N., Muller, W., et al. (2011). Intestinal tolerance requires gut homing and expansion of FoxP3+ regulatory T cells in the lamina propria. *Immunity, 34*, 237–246.

Hamilton, J. D., Suarez-Farinas, M., Dhingra, N., Cardinale, I., Li, X., Kostic, A., Ming, J. E., et al. (2014). Dupilumab improves the molecular signature in skin of patients with moderate-to-severe atopic dermatitis. *The Journal of Allergy and Clinical Immunology, 134*, 1293–1300.

Hartl, D., Koller, B., Mehlhorn, A. T., Reinhardt, D., Nicolai, T., Schendel, D. J., Griese, M., et al. (2007). Quantitative and functional impairment of pulmonary CD4+CD25hi regulatory T cells in pediatric asthma. *The Journal of Allergy and Clinical Immunology, 119*, 1258–1266.

Hill, D. A., Siracusa, M. C., Abt, M. C., Kim, B. S., Kobuley, D., Kubo, M., Kambayashi, T., et al. (2012). Commensal bacteria-derived signals regulate basophil hematopoiesis and allergic inflammation. *Nature Medicine, 18*, 538–546.

Hol, J., van Leer, E. H., Elink Schuurman, B. E., de Ruiter, L. F., Samsom, J. N., Hop, W., Neijens, H. J., et al. (2008). The acquisition of tolerance toward cow's milk through probiotic supplementation: A randomized, controlled trial. *The Journal of Allergy and Clinical Immunology, 121*, 1448–1454.

Holt, P. G. (2015). The mechanism or mechanisms driving atopic asthma initiation: The infant respiratory microbiome moves to center stage. *The Journal of Allergy and Clinical Immunology, 136*, 15–22.

Igea, J. M. (2013). The history of the idea of allergy. *Allergy, 68*, 966–973.

Irvine, A. D., McLean, W. H., & Leung, D. Y. (2011). Filaggrin mutations associated with skin and allergic diseases. *The New England Journal of Medicine, 365*, 1315–1327.

ISAAC. (1998). Worldwide variation in prevalence of symptoms of asthma, allergic rhinoconjunctivitis, and atopic eczema: ISAAC. The International Study of Asthma and Allergies in Childhood (ISAAC) Steering Committee. *Lancet, 351*, 1225–1232.

Isolauri, E., Arvola, T., Sutas, Y., Moilanen, E., & Salminen, S. (2000). Probiotics in the management of atopic eczema. *Clinical and Experimental Allergy, 30*, 1604–1610.

Josefowicz, S. Z., Niec, R. E., Kim, H. Y., Treuting, P., Chinen, T., Zheng, Y., Umetsu, D. T., et al. (2012). Extrathymically generated regulatory T cells control mucosal TH2 inflammation. *Nature, 482*, 395–399.

Kaesler, S., Volz, T., Skabytska, Y., Köberle, M., Hein, U., Chen, K. M., Guenova, E., et al. (2014). Toll-like receptor 2 ligands promote chronic atopic dermatitis through IL-4-mediated suppression of IL-10. *The Journal of Allergy and Clinical Immunology, 134*, 92–99.

Kalliomäki, M., Kirjavainen, P., Eerola, E., Kero, P., Salminen, S., & Isolauri, E. (2001). Distinct patterns of neonatal gut microflora in infants in whom atopy was and was not developing. *The Journal of Allergy and Clinical Immunology, 107*, 129–134.

Kay, A. B. (2001). Allergy and allergic diseases. First of two parts. *The New England Journal of Medicine, 344*, 30–37.

Keski-Nisula, L., Kyynarainen, H. R., Karkkainen, U., Karhukorpi, J., Heinonen, S., & Pekkanen, J. (2013). Maternal intrapartum antibiotics and decreased vertical transmission of Lactobacillus to neonates during birth. *Acta Paediatrica, 102*, 480–485.

Kolokotroni, O., Middleton, N., Gavatha, M., Lamnisos, D., Priftis, K. N., & Yiallouros, P. K. (2012). Asthma and atopy in children born by caesarean section: Effect modification by family history of allergies – a population based cross-sectional study. *BMC Pediatrics, 12*, 179.

Kong, H. H., Oh, J., Deming, C., Conlan, S., Grice, E. A., Beatson, M. A., Nomicos, E., et al. (2012). Temporal shifts in the skin microbiome associated with disease flares and treatment in children with atopic dermatitis. *Genome Research, 22*, 850–859.

Korpela, K., Salonen, A., Virta, L. J., Kekkonen, R. A., Forslund, K., Bork, P., & de Vos, W. M. (2016). Intestinal microbiome is related to lifetime antibiotic use in Finnish pre-school children. *Nature Communications, 7*, 10410.

Kuo, I. H., Yoshida, T., De Benedetto, A., & Beck, L. A. (2013). The cutaneous innate immune response in patients with atopic dermatitis. *The Journal of Allergy and Clinical Immunology, 131*, 266–278.

Lewkowich, I. P., Herman, N. S., Schleifer, K. W., Dance, M. P., Chen, B. L., Dienger, K. M., Sproles, A. A., et al. (2005). CD4+CD25+ T cells protect against experimentally induced asthma and alter pulmonary dendritic cell phenotype and function. *The Journal of Experimental Medicine, 202*, 1549–1561.

Leyden, J. J., Marples, R. R., & Kligman, A. M. (1974). Staphylococcus aureus in the lesions of atopic dermatitis. *The British Journal of Dermatology, 90*, 525–530.

Ling, Z., Li, Z., Liu, X., Cheng, Y., Luo, Y., Tong, X., Yuan, L., et al. (2014). Altered fecal microbiota composition associated with food allergy in infants. *Applied and Environmental Microbiology, 80*, 2546–2554.

Liu, H., Archer, N. K., Dillen, C. A., Wang, Y., Ashbaugh, A. G., Ortines, R. V., Kao, T., Lee, S. K., Cai, S. S., Miller, R. J., Marchitto, M. C., Zhang, E., Riggins, D. P., Plaut, R. D., Stibitz, S., Geha, R. S., & Miller, L. S. (2017). Staphylococcus aureus epicutaneous exposure drives skin inflammation via IL-36-mediated T cell responses. *Cell Host & Microbe, 22*, 653–666.

Liu, C. M., Price, L. B., Hungate, B. A., Abraham, A. G., Larsen, L. A., Christensen, K., Stegger, M., et al. (2015). Staphylococcus aureus and the ecology of the nasal microbiome. *Science Advances, 1*, e1400216.

Lodge, C. J., Tan, D. J., Lau, M. X., Dai, X., Tham, R., Lowe, A. J., Bowatte, G., et al. (2015). Breastfeeding and asthma and allergies: A systematic review and meta-analysis. *Acta Paediatrica, 104*, 38–53.

Majamaa, H., & Isolauri, E. (1997). Probiotics: A novel approach in the management of food allergy. *The Journal of Allergy and Clinical Immunology, 99*, 179–185.

Masoli, M., Fabian, D., Holt, S., & Beasley, R. (2004). The global burden of asthma: Executive summary of the GINA Dissemination Committee report. *Allergy, 59*, 469–478.

Maynard, C. L., Elson, C. O., Hatton, R. D., & Weaver, C. T. (2012). Reciprocal interactions of the intestinal microbiota and immune system. *Nature, 489*, 231–241.

Nakatsuji, T., Chen, T. H., Narala, S., Chun, K. A., Two, A. M., Yun, T., Shafiq, F., et al. (2017). Antimicrobials from human skin commensal bacteria protect against Staphylococcus aureus and are deficient in atopic dermatitis. *Science Translational Medicine, 9*. https://doi.org/10.1126/scitranslmed.aah4680

Noval Rivas, M., Burton, O. T., Wise, P., Zhang, Y. Q., Hobson, S. A., Garcia Lloret, M., Chehoud, C., et al. (2013). A microbiota signature associated with experimental food allergy promotes allergic sensitization and anaphylaxis. *The Journal of Allergy and Clinical Immunology, 131*, 201–212.

Oddy, W. H. (2009). The long-term effects of breastfeeding on asthma and atopic disease. *Advances in Experimental Medicine and Biology, 639*, 237–251.

Oh, J., Byrd, A. L., Park, M., Program, N. C. S., Kong, H. H., & Segre, J. A. (2016). Temporal stability of the human skin microbiome. *Cell, 165*, 854–866.

Ohnmacht, C., Park, J. H., Cording, S., Wing, J. B., Atarashi, K., Obata, Y., Gaboriau-Routhiau, V., et al. (2015). Mucosal Immunology. The microbiota regulates type 2 immunity through RORgammat(+) T cells. *Science, 349*, 989–993.

Olszak, T., An, D., Zeissig, S., Vera, M. P., Richter, J., Franke, A., Glickman, J. N., et al. (2012). Microbial exposure during early life has persistent effects on natural killer T cell function. *Science, 336*, 489–493.

Ong, P. Y., Ohtake, T., Brandt, C., Strickland, I., Boguniewicz, M., Ganz, T., Gallo, R. L., et al. (2002). Endogenous antimicrobial peptides and skin infections in atopic dermatitis. *The New England Journal of Medicine, 347*, 1151–1160.

Ouwehand, A. C., Isolauri, E., He, F., Hashimoto, H., Benno, Y., & Salminen, S. (2001). Differences in Bifidobacterium flora composition in allergic and healthy infants. *The Journal of Allergy and Clinical Immunology, 108*, 144–145.

Oyama, N., Sudo, N., Sogawa, H., & Kubo, C. (2001). Antibiotic use during infancy promotes a shift in the T(H)1/T(H)2 balance toward T(H)2-dominant immunity in mice. *The Journal of Allergy and Clinical Immunology, 107*, 153–159.

Pabst, O., & Mowat, A. M. (2012). Oral tolerance to food protein. *Mucosal Immunology, 5*, 232–239.

Palmer, C. N., Irvine, A. D., Terron-Kwiatkowski, A., Zhao, Y., Liao, H., Lee, S. P., Goudie, D. R., et al. (2006). Common loss-of-function variants of the epidermal barrier protein filaggrin are a major predisposing factor for atopic dermatitis. *Nature Genetics, 38*, 441–446.

Panduru, M., Panduru, N. M., Salavastru, C. M., & Tiplica, G. S. (2015). Probiotics and primary prevention of atopic dermatitis: A meta-analysis of randomized controlled studies. *Journal of the European Academy of Dermatology and Venereology, 29*, 232–242.

Penders, J., Gerhold, K., Stobberingh, E. E., Thijs, C., Zimmermann, K., Lau, S., & Hamelmann, E. (2013). Establishment of the intestinal microbiota and its role for atopic dermatitis in early childhood. *The Journal of Allergy and Clinical Immunology, 132*, 601–607e608.

Peters, R. L., Koplin, J. J., Gurrin, L. C., Dharmage, S. C., Wake, M., Ponsonby, A. L., Tang, M. L. K., et al. (2017). The prevalence of food allergy and other allergic diseases in early childhood in a population-based study: HealthNuts age 4-year follow-up. *The Journal of Allergy and Clinical Immunology, 140*, 145–153e148.

Platts-Mills, T. A. (2015). The allergy epidemics: 1870-2010. *The Journal of Allergy and Clinical Immunology, 136*, 3–13.

Plunkett, C. H., & Nagler, C. R. (2017). The influence of the microbiome on allergic sensitization to food. *Journal of Immunology, 198*, 581–589.

Ravel, J., Gajer, P., Abdo, Z., Schneider, G. M., Koenig, S. S., McCulle, S. L., Karlebach, S., et al. (2011). Vaginal microbiome of reproductive-age women. *Proceedings of the National Academy of Sciences of the United States of America, 108*(Suppl 1), 4680–4687.

Renz-Polster, H., David, M. R., Buist, A. S., Vollmer, W. M., O'Connor, E. A., Frazier, E. A., & Wall, M. A. (2005). Caesarean section delivery and the risk of allergic disorders in childhood. *Clinical and Experimental Allergy, 35*, 1466–1472.

Rinaldi, M., Harnack, L., Oberg, C., Schreiner, P., St Sauver, J., & Travis, L. L. (2012). Peanut allergy diagnoses among children residing in Olmsted County, Minnesota. *The Journal of Allergy and Clinical Immunology, 130*, 945–950.

Ring, J., Brockow, K., & Behrendt, H. (2004). History and classification of anaphylaxis. *Novartis Foundation Symposium, 257*, 6–16 discussion 16–24, 45–50, 276–285.

Rodriguez, B., Prioult, G., Bibiloni, R., Nicolis, I., Mercenier, A., Butel, M. J., & Waligora-Dupriet, A. J. (2011). Germ-free status and altered caecal subdominant microbiota are associated with a high susceptibility to cow's milk allergy in mice. *FEMS Microbiology Ecology, 76*, 133–144.

Russell, S. L., Gold, M. J., Hartmann, M., Willing, B. P., Thorson, L., Wlodarska, M., Gill, N., et al. (2012). Early life antibiotic-driven changes in microbiota enhance susceptibility to allergic asthma. *EMBO Reports, 13*, 440–447.

Schröder, J. M. (2011). Antimicrobial peptides in healthy skin and atopic dermatitis. *Allergology International, 60*, 17–24.

Schulz, O., Jaensson, E., Persson, E. K., Liu, X., Worbs, T., Agace, W. W., & Pabst, O. (2009). Intestinal CD103+, but not CX3CR1+, antigen sampling cells migrate in lymph and serve classical dendritic cell functions. *The Journal of Experimental Medicine, 206*, 3101–3114.

Sefik, E., Geva-Zatorsky, N., Oh, S., Konnikova, L., Zemmour, D., McGuire, A. M., Burzyn, D., et al. (2015). Mucosal Immunology. Individual intestinal symbionts induce a distinct population of RORgamma(+) regulatory T cells. *Science, 349*, 993–997.

Seite, S., Flores, G. E., Henley, J. B., Martin, R., Zelenkova, H., Aguilar, L., & Fierer, N. (2014). Microbiome of affected and unaffected skin of patients with atopic dermatitis before and after emollient treatment. *Journal of Drugs in Dermatology, 13*, 1365–1372.

Sicherer, S. H., & Sampson, H. A. (2014). Food allergy: Epidemiology, pathogenesis, diagnosis, and treatment. *The Journal of Allergy and Clinical Immunology, 133*, 291–307 quiz 308.

Sistek, D., Kelly, R., Wickens, K., Stanley, T., Fitzharris, P., & Crane, J. (2006). Is the effect of probiotics on atopic dermatitis confined to food sensitized children? *Clinical and Experimental Allergy, 36*, 629–633.

Smith, P. M., Howitt, M. R., Panikov, N., Michaud, M., Gallini, C. A., Bohlooly, Y. M., Glickman, J. N., et al. (2013). The microbial metabolites, short-chain fatty acids, regulate colonic Treg cell homeostasis. *Science, 341*, 569–573.

Song, H., Yoo, Y., Hwang, J., Na, Y. C., & Kim, H. S. (2016). Faecalibacterium prausnitzii subspecies-level dysbiosis in the human gut microbiome underlying atopic dermatitis. *The Journal of Allergy and Clinical Immunology, 137*, 852–860.

Spergel, J. M. (2010). From atopic dermatitis to asthma: The atopic march. *Annals of Allergy, Asthma, & Immunology, 105*, 99–106 quiz 107–109, 117.

Stefka, A. T., Feehley, T., Tripathi, P., Qiu, J., McCoy, K., Mazmanian, S. K., Tjota, M. Y., et al. (2014). Commensal bacteria protect against food allergen sensitization. *Proceedings of the National Academy of Sciences of the United States of America, 111*, 13145–13150.

Strachan, D. P. (1989). Hay fever, hygiene, and household size. *BMJ, 299*, 1259–1260.

Sze, M. A., Tsuruta, M., Yang, S. W., Oh, Y., Man, S. F., Hogg, J. C., & Sin, D. D. (2014). Changes in the bacterial microbiota in gut, blood, and lungs following acute LPS instillation into mice lungs. *PLoS One, 9*, e111228.

Tang, M. L., Ponsonby, A. L., Orsini, F., Tey, D., Robinson, M., Su, E. L., Licciardi, P., et al. (2015). Administration of a probiotic with peanut oral immunotherapy: A randomized trial. *The Journal of Allergy and Clinical Immunology, 135*, 737–744e738.

Thorburn, A. N., McKenzie, C. I., Shen, S., Stanley, D., Macia, L., Mason, L. J., Roberts, L. K., et al. (2015). Evidence that asthma is a developmental origin disease influenced by maternal diet and bacterial metabolites. *Nature Communications, 6*, 7320.

Tokumasu, R., Yamaga, K., Yamazaki, Y., Murota, H., Suzuki, K., Tamura, A., Bando, K., et al. (2016). Dose-dependent role of claudin-1 in vivo in orchestrating features of atopic dermatitis. *Proceedings of the National Academy of Sciences of the United States of America, 113*, E4061–E4068.

Traidl-Hoffmann, C., Jakob, T., & Behrendt, H. (2009). Determinants of allergenicity. *The Journal of Allergy and Clinical Immunology, 123*, 558–566.

Trompette, A., Gollwitzer, E. S., Yadava, K., Sichelstiel, A. K., Sprenger, N., Ngom-Bru, C., Blanchard, C., et al. (2014). Gut microbiota metabolism of dietary fiber influences allergic airway disease and hematopoiesis. *Nature Medicine, 20*, 159–166.

Velez, M. A., Perotti, M. C., Wolf, I. V., Hynes, E. R., & Zalazar, C. A. (2010). Influence of milk pretreatment on production of free fatty acids and volatile compounds in hard cheeses: Heat treatment and mechanical agitation. *Journal of Dairy Science, 93*, 4545–4554.

Volz, T., Nega, M., Buschmann, J., Kaesler, S., Guenova, E., Peschel, A., Röcken, M., et al. (2010). Natural Staphylococcus aureus-derived peptidoglycan fragments activate NOD2 and act as potent costimulators of the innate immune system exclusively in the presence of TLR signals. *The FASEB Journal, 24*, 4089–4102.

Volz, T., Skabytska, Y., Guenova, E., Chen, K. M., Frick, J. S., Kirschning, C. J., Kaesler, S., et al. (2014). Nonpathogenic bacteria alleviating atopic dermatitis inflammation induce IL-10-producing dendritic cells and regulatory Tr1 cells. *The Journal of Investigative Dermatology, 134*, 96–104.

Waser, M., Michels, K. B., Bieli, C., Floistrup, H., Pershagen, G., von Mutius, E., Ege, M., et al. (2007). Inverse association of farm milk consumption with asthma and allergy in rural and suburban populations across Europe. *Clinical and Experimental Allergy, 37*, 661–670.

Watanabe, S., Narisawa, Y., Arase, S., Okamatsu, H., Ikenaga, T., Tajiri, Y., & Kumemura, M. (2003). Differences in fecal microflora between patients with atopic dermatitis and healthy control subjects. *The Journal of Allergy and Clinical Immunology, 111*, 587–591.

Weiner, H. L., da Cunha, A. P., Quintana, F., & Wu, H. (2011). Oral tolerance. *Immunological Reviews, 241*, 241–259.

Werfel, T. (2015). Atopische dermatitis. In T. Biedermann, W. Heppt, H. Renz, & M. Röcken (Eds.), *Allergologie* (pp. 249–259). Berlin: Springer.

Wölbing, F., & Biedermann, T. (2013). Anaphylaxis: Opportunities of stratified medicine for diagnosis and risk assessment. *Allergy, 68*, 1499–1508.

Yamashita, H., Takahashi, K., Tanaka, H., Nagai, H., & Inagaki, N. (2012). Overcoming food allergy through acquired tolerance conferred by transfer of Tregs in a murine model. *Allergy, 67*, 201–209.

Zhang, G. Q., Hu, H. J., Liu, C. Y., Zhang, Q., Shakya, S., & Li, Z. Y. (2016). Probiotics for prevention of atopy and food hypersensitivity in early childhood: A PRISMA-compliant systematic review and meta-analysis of randomized controlled trials. *Medicine (Baltimore), 95*, e2562.

13 Microbiome and Diseases: Graft-Versus-Host Disease

D. Weber and E. Holler

Abstract

Acute graft-versus-host disease (GvHD) of the gastrointestinal (GI) tract is still a major cause of severe morbidity and mortality following allogeneic stem cell transplantation (ASCT). The intestinal "microflora" has been in the focus of the pathophysiology of acute GI GvHD since many years. In 1974 van Bekkum and colleagues reported a possible role of microbiota as they observed that mice kept under germ-free conditions did not develop acute GI GvHD. Clinical studies showed a reduction of GvHD severity with total or selective gut decontamination and isolation which became standard for years. However, with deeper insights obtained by new molecular techniques, our understanding of the association between microbiota and GI GvHD has changed. During ASCT an early loss of intestinal microbiome diversity especially with regard to commensal *Clostridiales* and a shift toward an enteropathogenic flora was observed which associated with increased GvHD-related mortality. A major risk factor for the loss of commensal bacteria is the use of systemic broad-spectrum antibiotics for prophylaxis and therapy of neutropenic infections which occur frequently in ASCT patients. Both the kind of antibiotics and the timing of antibiotic treatment influence the outcome of patients after ASCT. The damage and destruction of Paneth cells by GvHD itself reduce antimicrobial peptides further aggravating intestinal dysbiosis during GvHD. These results indicate that microbiota manipulation could be a promising future approach not only for prophylaxis and treatment of acute GI GvHD but also for eradication of a colonization with multiple antibiotic-resistant pathogens to prevent infections or even to prevent relapse.

13.1 Introduction

Acute and chronic graft-versus-host disease and associated immunodeficiencies are still the major causes of transplant-related mortality (TRM) and long-term morbidity in patients receiving allogeneic stem cell transplantation (SCT) for hematological malignancies such as acute leukemias (Ferrara et al. 2009): Although improvements in supportive care and reduction of toxicity have overall improved the results of SCT, still 15–20% of patients experience 1-year TRM, and more than 50% depend on long-term immunosuppression with all negative side effects and long-term complications (Gooley et al. 2010). Activation of donor T lymphocytes co-transplanted with

D. Weber (✉) · E. Holler
Experimentelle allogene KMT – Innere Medizin III, Universitätsklinikum Regensburg, Regensburg, Germany
e-mail: daniela.weber@ukr.de; ernst.holler@ukr.de

D. Haller (ed.), *The Gut Microbiome in Health and Disease*,
https://doi.org/10.1007/978-3-319-90545-7_13

the stem cell graft by major and minor HLA differences is the crucial step in the pathophysiology of GvHD; however, this activation is strongly regulated by the tissue microenvironment. Main target organs of acute GvHD are the skin, gut, and liver, whereas chronic GvHD additionally involves eyes, mouth, lung, or urogenital manifestations. A common denominator of all these targets is the exposure and extensive interaction of these organs with the respective microbiota, already indicating a major role of microbiota in the pathophysiology of GvHD- and SCT-related complications.

13.2 The Early Times: van Bekkum and the Clinical Impact of Decontamination

The role of microbiota in intestinal GvHD was described in the early period of experimental SCT. Van Bekkum reported that mice receiving the bone marrow did not experience GvHD at all, if they were housed under germ-free conditions for at least 40 days after transplantation. Stepwise reduction of the germ-free period resulted in stepwise increase of GvHD-related mortality, and almost more than 90% of mice transplanted under conventional housing died from GvHD. This effect of the "intestinal flora" could also be observed when he transplanted gut from germ-free versus conventionally housed mice into recipients and then induced GvHD: Again, the histological GvHD score was highly increased in the gut of mice exposed to a conventional flora (van Bekkum et al. 1974; van Bekkum and Knaan 1977). These observations initiated experiments in dogs and monkeys confirming the role of maximal decontamination in prevention of GvHD, and early reports of patients with aplastic anemia and children transplanted under maximal decontamination in protective environments supported these conceptions (Vriesendorp et al. 1982; Storb et al. 1983; Schmeiser et al. 1985). Later on, protective environments and decontamination were combined with prophylactic application of antibiotics such as quinolones in order to reduce the incidence and mortality from neutropenic infections (Gafter-Gvili et al. 2005). The Essen group performed the only randomized study and compared prophylactic ciprofloxacin with ciprofloxacin and metronidazole, and the combined approach was associated with far less acute GvHD and improved outcome (Beelen et al. 1999). Up to now, centers have different standards of decontamination and prophylactic antibiotics ranging from maximal to almost no prophylaxis, as further studies were missing or nonconclusive (Whangbo et al. 2017).

13.3 Indirect Evidence for a More Complex Role of Microbiota

In the 1990s, experimental and clinical studies indicated that T cell activation by antigen-presenting cells required inflammatory signals to induce antigen expression, and proinflammatory cytokines like TNF (Piguet et al. 1987; Holler et al. 1990) or IL1 (Ferrara et al. 2009) were shown to be involved both in the initiation and effector phase of GvHD finally resulting in a cytokine storm. LPS translocated via the damaged intestinal mucosa was recognized as a major trigger of inflammation as confirmed by experiments using LPS-sensitive and LPS-resistant mice (Cooke et al. 2001, 2002). With the description of Toll-like and NOD-like receptors and their respective ligands, it became soon clear that multiple ligands derived either from pathogens like bacteria and viruses (so-called PAMPs) or endogenous ligands derived from damaged tissue, so-called DAMPs (Zeiser et al. 2011; Apostolova and Zeiser 2016), can activate and initiate this inflammation and subsequently GvHD. Parallels to the pathophysiology of inflammatory bowel diseases were described: *NOD2/CARD15* and *ATG16L1* gene polymorphisms, which have been described as risk factors of IBD, identified also patients at higher risk of GvHD, and a direct role of NOD2/CARD15 in GvHD has been proven in murine ko-models (Holler et al. 2006; Penack et al. 2009). As *NOD2/CARD15* is involved in the regulation of production of antibacterial peptides by Paneth cells (Bevins et al. 2009), these data indicated a strong interaction of GvHD-related inflammation and antibacterial defense in the target tissues of GvHD like the gastrointestinal tract and intestinal bacteria. Interestingly, the prognostic significance of *NOD2/CARD15* polymorphisms showed high variations between different transplant centers, and the type of gastrointestinal decontamination used in different centers seemed to explain at least partially this heterogeneity of results (Holler et al. 2006).

13.4 Microbiome Analysis Reveals Severe Dysbiosis Early After SCT

While early data on decontamination always depended on cultural analyses of intestinal and stool sample bacteria and culture-dependent analyses up to now show a benefit of decontamination (Vossen et al. 2014), these studies may have missed a large proportion of bacteria, especially commensal bacteria: The introduction of 16s rRNA sequencing within the Human Microbiome Project opened a new level of understanding and knowledge also in the setting of SCT. The Memorial Sloan Kettering group around Eric Pamer and Marcel van den Brink was the first (Jenq et al. 2012) to report microbiota sequencing both in murine models of GvHD and in patients and reported a strong association of dysbiosis and loss of diversity with GvHD. In the murine model, modulation of microbiota had a strong impact on survival, as elimination of *Lactobacillales* aggravated GvHD, while restoration of this genus improved outcome. Similar changes were observed in patients with GvHD. Our group applied 16s rRNA sequencing in serial samples in 32 patients, and in line with Jenq's observation, we saw a strong loss of diversity early after SCT with a strong abundance of enterococci, especially in patients developing acute gastrointestinal GvHD (Holler et al. 2014). Taur reported at the same time a direct association of diversity early after engraftment with mortality (Taur et al. 2014). Subsequently, we were able to confirm this association of early microbiota disruption and loss of diversity with increased TRM and GvHD by applying an indirect approach of urine analysis of indoxyl sulfate (IS) in a large series of patients: As shown in 2014 and 2015, urinary IS results from bacterial indoles produced by commensal GI bacteria and conjugated in the liver. Patients with low IS levels on the first 10 days after SCT had worse outcome, and low IS was an independent risk factor of NRM (Weber et al. 2015). The protective role of commensal anaerobic bacteria, especially *Blautia* spp., was also confirmed by R. Jenq (Jenq et al. 2015), as patients with abundance of *Blautia* at the time of engraftment had a low TRM of 5%, while patients with loss of *Blautia* had a TRM almost approaching 30%. In the meantime, several groups confirmed these associations of early dysbiosis with more GvHD and inferior outcome (Golob et al. 2017; Doki et al. 2017; Kaysen et al. 2017; Bilinski et al. 2016).

13.5 Causes of Early Dysbiosis

13.5.1 Antibiotic Prophylaxis and Treatment

In all experimental and clinical analyses reported so far, antibiotic prophylaxis as used for infection prevention and decontamination and broad-spectrum antibiotics given for neutropenic fever and infections in SCT patients were driving factors of early dysbiosis (Jenq et al. 2012; Holler et al. 2014; Weber et al. 2015). Shono and colleagues clearly showed that antibiotics strongly suppressing commensal bacteria like carbapenems and piperacillin/tazobactam induce more GvHD in experimental and in clinical SCT, whereas antibiotics with some commensal and anaerobic sparing effect like cefepime or aztreonam seem to be more protective (Shono et al. 2016). In an analysis of more than 600 patients transplanted at the MSKCC in New York and in Regensburg, we recently demonstrated that not only the application of these broad-spectrum antibiotics in general but also the timing is critical, as patients with start of these antibiotics prior to the day of transplantation had increased NRM as compared to patients with later initiation of antibiotic treatment. The best survival was observed in the unfortunately so far small subgroup of about 8% of patients who never required systemic antibiotic treatment beyond prophylaxis (Weber et al. 2017b). Early initiation of systemic antibiotics resulted in earlier loss of *Clostridia* and in deeper suppression of early IS levels again connecting early suppression of commensals with more GvHD. Surprisingly, the early use of broad-spectrum antibiotics was an independent risk factor of outcome, as even low-risk patients expected to have a superior outcome showed this strong negative effect (Fig. 13.1).

Randomized prospective studies on antibiotic prophylaxis are missing. Our group stopped the prophylactic use of ciprofloxacin and metronidazole as we observed the high abundance of enterococci and concomitantly also a high prevalence of vancomycin-resistant *Enterococcus* (VRE) in patients and switched to prophylaxis with low doses of the nonabsorbable antibiotic rifaximin, as it is widely used for treatment of travelers' diarrhea. In a retrospective analysis, patients on rifaximin prophylaxis had higher urinary IS levels and a decreased incidence of severe GvHD (Weber et al. 2016). In line with a negative impact of antibiotic prophylaxis is a recent observation from Canadian centers (Routy et al. 2016), as patients with antibiotic prophylaxis had an almost twofold increased incidence of severe GI GvHD (21% vs. 11%, *p* 0.001) and a reduced overall survival as compared to patients not receiving any prophylaxis.

13.5.2 Paneth Cell Damage in GvHD

In an attempt to define biomarkers of GvHD, a proteomic approach is identified among others, Reg3α, the human homologue of murine Reg3y, as a human biomarker of intestinal GvHD (Paczesny et al. 2009). The specificity was confirmed in a study performed with samples from Ann Arbor, Regensburg, and Kyushu (Ferrara et al. 2011). Interestingly, Reg3α turned out to be an antimicrobial peptide produced by Paneth cells, and in a subsequent study on biopsies from intestinal GvHD, we could demonstrate Paneth cell damage and destruction as a sensitive histological hallmark of human GvHD (Levine et al. 2013). Paneth cells are located in the crypts beneath epithelial stem cells which are sensitive targets of intestinal GvHD as well. This is in line with Paneth cell damage and subsequent dysbiosis in murine GvHD (Eriguchi et al. 2012, 2013, 2015). More recently, we reported loss of expression of all major defensins (HBD5,6 and Reg3alpha) in severe GvHD in ileal biopsies from patients with severe GvHD which paralleled increases in systemic Reg3α levels (Weber et al. 2017a). As Reg3α levels rise early in the course of GvHD frequently preceding clinical symptoms of GI GvHD, these data indicate that Paneth cell damage is an early event in GvHD and thus enhances dysbiosis during GvHD.

13.6 Potential Mechanisms Explaining a Pathophysiological Role of Dysbiosis in GvHD

So far, damaging mechanisms associated with dysbiosis in GvHD are not different from mechanisms reported for microbiota-dependent diseases in general. Therefore, we focus here on specific findings in the setting of SCT.

13.6.1 Microbial Metabolites and GvHD

Based on the observation that histone deacetylase inhibitors modulate the immunosuppressive enzyme indoleamine dioxygenase and subsequently GvHD (Reddy et al. 2008), the group of Pavan Reddy became interested in short-chain fatty acids, especially butyrate, as a natural HDAC inhibitor derived from bacterial metabolites. They recently showed reduction of butyrate levels in intestinal walls of mice suffering from GvHD; restoration of commensal *Clostridia* in a murine BMT model not only increased butyrate levels but also ameliorated

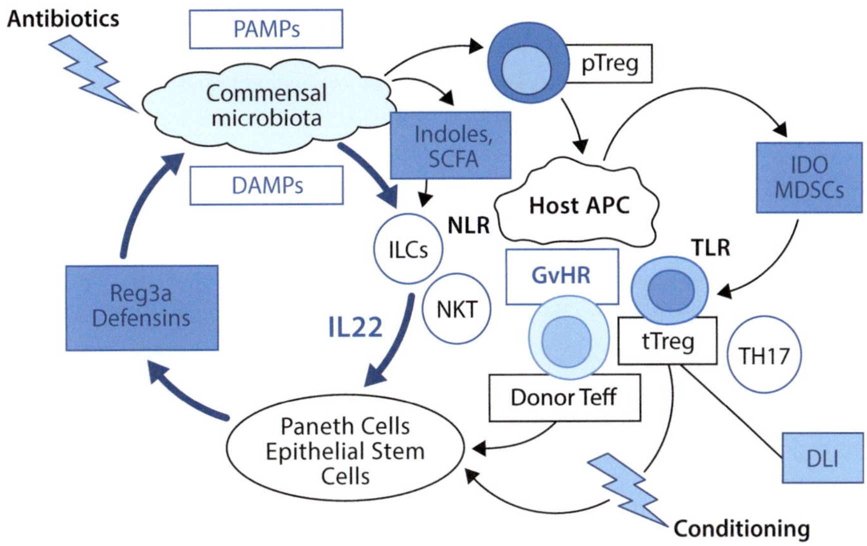

Fig. 13.1 Regulation of graft-versus-host reactions by microbiota

GvHD-related mortality, and similar effects could be observed by butyrate gavage (Mathewson et al. 2016). Reddy and colleagues recently summarized the potential impact of bacterial metabolites on GvHD (Riwes and Reddy 2017) and indicated that bacterial metabolites [short-chain fatty acids like butyrate, bile acids, polyamines, and indoles binding to aryl hydrocarbon receptor (AHR)] involved in epithelial and immunological regulation may be beneficial in the setting of GvHD. In line with this, we recently used the microbiome biomarker and metabolite indoxyl sulfate, derived from commensal bacteria and hydroxylated in the liver, in in vitro cultures of dendritic cells and mixed lymphocyte reactions. Indeed, IS induced in physiologically observed concentration is an anti-inflammatory cytokine profile in dendritic cells with low IL12 and IL6 and high IL10, which translated in diminished IFNy release in mixed lymphocyte reaction (Ghimire et al. 2018).

13.6.2 Mucus Layer and GvHD

Antibiotics have repeatedly been shown to worsen dysbiosis and subsequent GvHD. Shono et al. (2016) reported a higher incidence of GvHD and related mortality in patients and mice treated with imipenem as compared to fourth-generation cephalosporins. They could not detect differences in regulatory or effector T cells in imipenem-treated mice but observed increased expression of IL23 and neutrophil influx in the colon. Notably, imipenem treatment of mice with GVHD led to loss of the protective mucus lining of the colon and the compromising of intestinal barrier function as shown by increased FITC-dextran uptake. Sequencing of mouse stool specimen showed an increase in *Akkermansia muciniphila*, a gut bacterium with mucus-degrading capabilities, raising the possibility that mucus degradation may contribute to murine GVHD.

13.6.3 Microbiota-Dependent Effector Cells

Interleukin 22 produced by recipient innate lymphoid cells has been reported to prevent epithelial stem cell apoptosis in the GI tract and protect mice from lethal GvHD (Hanash et al. 2012). IL22 release by ILCs is under strong regulation of binding of metabolites (like indoles) from commensal bacteria to their AHR and thus may be a further mechanism contributing to epithelial instability in GvHD. Besides ILCs, neutrophils are increasingly recognized as amplifiers of GvHD-related inflammation. The group of R. Zeiser observed increased accumulation of neutrophils in the ileum of GvHD mice, which was not present in germ-free mice. Depletion of neutrophils by antibodies reduced GvHD severity and mortality indicating a microbiota-neutrophil recruitment pathway of damage in GvHD (Schwab et al. 2014), which can be explained by increased inflammation via reactive oxygen species (ROS). IL17 produced by certain T cell populations may exacerbate GvHD but also modulate inflammation dependent on the involved cellular subtypes. Recently, the group of Hill et al. reported protection by IL17, as mice unable to signal via the IL17 receptor developed hyperactive GvHD: Interestingly, the susceptibility to GvHD could be transferred by the microbiota, as wild-type mice cohoused with IL17 RA/RC mice also developed accelerated and more severe GvHD (Varelias et al. 2017). The authors speculate that host-derived IL17 contributes to increased antimicrobial peptide expression (e.g., Reg3g) in these mice.

13.6.4 Specific Immune Responses: T Cells and IgA, Donor Versus Recipient

Specific immune responses within the adaptive immune system are in the center of GvHD: As shown in a recent study in mice, antibiotic treatment has profound effects on almost all specific immune cell populations including regulatory T cells and B cells, which can be partially restored by experimental FMT. Functionally, these cells substantially alter their cytokine profile, especially also cytokines which are important for intestinal homeostasis like IL17 and IL22 (Ekmekciu et al. 2017). However, antigen specificity has not been addressed in this study. Recent data obtained in healthy individuals and in patients with IBD show a broad and polyclonal T cell response against commensals and therefore support this hypothesis. The immunological situation in HSCT is even more complex due to the exchange of recipient by donor immune cells: A yet unanswered question is whether the different T cell and B cell repertoire of the stem cell donor against microbiota contributes to the pathophysiology of GvHD. Given the important role of IgA in the control of commensals versus pathogens, a role of specific immunity is likely, but has not been addressed to date. So far only one experimental study analyzed the effects of microbiota from germ-free versus conventional versus SPF donor mice on T cell reactivity in a murine BMT model and reported no impact (Tawara et al. 2013). A very recent clinical study comparing recipient and donor microbiota diversity confirmed the increased mortality in patients with low diversity but also reported a lower incidence of GvHD in patients who received stem cells from donors with high diversity (Liu et al. 2017). Whether these reflect an optimized T cell repertoire against intestinal microbiota remains to be clarified in future prospective studies.

13.7 Microbiota Manipulation as a Prophylactic or Therapeutic Approach

Experimental studies which used either lactobacilli to modulate GvHD in mice (Gerbitz et al. 2004) or to modulate GvHD in mice pretreated with antibiotics (Jenq et al. 2012) or consortia of *Clostridia* to increase production of intestinal butyrate (Mathewson et al. 2016) clearly indicate that microbiota modulation has a potential impact on the outcome following SCT. The numerous retrospective analyses on antibiotic prophylaxis and treatment in patients further support this concept, but prospective studies are currently running, and the results are pending. The potential candidates (pre-, pro-, postbiotics) have been recently summarized (Peled et al. 2016). So far, only one prospective randomized probiotic trial using lactobacilli in patients has been published with almost no effect, but similar to a previously performed prophylaxis trial with *Lactobacillus rhamnosus* in our unit, probiotics were only applied starting from engraftment (Gorshein et al. 2017). As almost all analyses revealed dysbiosis occurring prior to engraftment as the most critical damage enhancing GvHD, these findings are not unexpected as probiotic manipulation may have started too late. There are concerns using probiotics during the neutropenic period as *Lactobacillus* spp. septicemias have been described in leukemia patients, but a recent safety trial using lactobacilli from day-8 pretransplant until recovery from neutropenia in children revealed no single episode of *Lactobacillus* bacteremia in 30 patients (Ladas et al. 2016). Similar concerns regarding infectious risks were raised with regard to performing fecal microbiota transplantation (FMT) in HSCT patients. FMT for *Clostridium difficile* however has been reported to be safe for immunosuppressed patients (Mandalai et al. 2016) and also in seven patients after HSCT (Webb et al. 2016) with five of them being on immunosuppression. In 2016, two groups made the next step and used FMT to treat steroid refractory GvHD. In Kakihana's study (2016), four patients were treated and three complete and one partial response was observed without major side effects, and a potential modulation of circulating regulatory T cells was observed. Similarly, Spindelboeck and colleagues treated three patients refractory GvHD with repeated FMT and observed comparable responses (2017). To prove these promising first reports in GvHD, a European multicenter phase II trial is currently prepared, and prophylactic pilot trials using either pretransplant autologous fecal microbiota or third-party donor FMT are also currently performed. Diet and nutrition is another important factor maintaining microbiota diversity. So far, only one randomized study compared the impact of enteric versus parenteral nutrition on GvHD and reported an almost 50% reduction of severe GvHD in the group receiving enteral nutrition (Gonzales et al. 2017). Several groups report a reduced incidence of GvHD in patients receiving outpatient or homecare transplantation. An interesting but yet not analyzed hypothesis for this observation could be both a better enteral nutrition and continuous exposure to their home microbiota, but of course, other factors such as a selection bias of patients with lower risk and less antibiotic history might explain this observation as well (Svahn et al. 2002).

13.8 Impact of Dysbiosis on Further Complications

13.8.1 Role in Infections

A major finding in allogeneic SCT recipients but also leukemia patients is the rapid replacement of a diverse microbiota by abundance of one or very few species representing extensive colonization. This abundance is a major risk factor of subsequent bloodstream infections and septicemia, as these bacteria are translocated through the damaged intestinal barrier: Enterococcal domination was shown to increase the risk of enterococcal bloodstream infection (BSI) ninefold, similarly *Proteobacteria* domination conferred a fivefold risk (Taur et al. 2012; Lee et al. 2017). A French group even developed a BSI-risk index based on pretreatment microbiota composition predicting infections with a 90% specificity (Montassier et al. 2016). The lung is another critical target organ with regard to infections following allogeneic SCT: So far, one small study reported an association of shifts in the oral microbiome with subsequent pulmonary infections (Ames et al. 2012), but it seems worth and needed to further analyze these interactions. As generally discussed it seems also possible that pulmonary infections are again triggered by intestinal microbiota; this is at least suggested by the recent study from Harris (Harris et al. 2016) where gamma-*Proteobacteria* domination of the intestinal microbiome predicted pulmonary complications and resulting mortality.

Colonization with multiple antibiotic-resistant pathogens is an increasing thread to hematological patients. The recent reports on the use of healthy donor FMT to eliminate these bacteria rely on the natural ability of commensals to restore colonization resistance and have been reported to have an unexpected high success rate close to 80% in leukemia and HSCT patients (Bilinski et al. 2017).

Given the lack of new antibiotics and the potential risk of further induction of resistance genes, FMT may become an attractive alternative to restore a diverse microbiome in these patients (Manges et al. 2016).

13.8.2 Relapse and Microbiota

The opposing and interacting principles of allogeneic SCT, GvHD, and graft-versus-leukemia (GvL) effects suggest that any principle interfering with GvHD may also affect GvL effects. So far, only one large analysis addressed this question in detail: Interestingly, Peled et al. were able to identify a single bacterial species, *Eubacterium limosum*, which conferred protection against relapse in a larger series of patients (2017). This observation suggests more specific mechanisms beyond less or more alloreaction and indicates that specific bacteria may promote antileukemic activity by yet unknown immunological mechanisms. However, these data are in line with the recent experimental and clinical observation that antitumoral and antileukemic activity of specific anticancer treatment strategies are under the control of microbiota (Pflug et al. 2016; Routy et al. 2017; Zitvogel et al 2017).

13.9 Future Aspects and Unsolved Issues

- As for the vast majority of dysbiosis-associated diseases, the exact impact of microbiota modulation on prevention and treatment of GvHD is currently rather speculative. As almost all major organs of GvHD both in acute and also in chronic GvHD are organs with epithelia exposed to their respective microbiota, we can expect a huge expansion on knowledge of interactions in these organs and potential therapeutic applications. Seeing the profound impact of commensal microbiota on function of both, innate and specific immune cells, it is likely that in almost all targets of GvHD epithelial tissue, recipient and donor immune cells and the respective microbiota have to be considered as an interacting "menage a trois."
- A still larger unknown world is the role of the virome and mycobiome in the setting of transplantation. While the mycobiome might be less heterogenous and future research might help to understand why certain patients are more susceptible to fungal infections (Shelburne et al. 2015), a recent longitudinal analysis of the virome in 44 HSCT patients revealed an exciting burst of vertebrate viral genomes in the period of T cell deficiency early after SCT: Especially, *Picobirnaviruses* correlated highly with more severe gastrointestinal GvHD (Legoff et al. 2017), and a comparable increase in knowledge can be expected regarding the role of the virome.
- Modern assessment of clinical interactions by "omic" approaches brings a new challenge to translational research as each new finding needs to be confirmed and validated in prospective studies. Large consortia dedicated to prospective clinical data and sample collection will be needed to provide adequate clinical instruments to meet these diagnostic and scientific challenges, as it was started within the biomarker "MAGIC" consortium for acute GvHD (MAGIC Manuscript, JCI Insight 2017).

Controversy

As in other disease models, the major unresolved issue is the exact pathophysiological contribution of dysbiosis and microbiota damage to GvHD: One can consider dysbiosis just as a factor changing the threshold for initiation of inflammation. However, it could well be that specific bacteria contribute specific pathophysiological mechanisms such as destruction of the mucus layer or epithelial destruction by strain-specific epitheliolysins. Here, clearly deeper analyses are needed. The same question holds true for the association of microbiota shifts with regard to relapse—are there specific mechanisms how certain bacteria stimulate antitumoral immune responses? Even one step more, it is up to now unclear whether specific immune responses on the T and B cell side (Th17, follicular T helper cells, IgA-producing plasma cells) are involved in the observed pathophysiological associations. The debate is ongoing and sometimes highly emotional as the observations question dogmas present in clinical practice for decades: Should this translate in altered clinical strategies—are we ready to change antibiotic prophylaxis and treatment or even to induce FMT in heavily immunosuppressed patients?

History

Allogeneic stem cell transplantation is a standard treatment in patients with high-risk leukemias but associates with the risk of immunological complications raising from activation of donor lymphocytes. Therefore, graft-versus-host disease is the major complication and cause of treatment-related mortality following allogeneic stem cell transplantation. Based on early studies showing a reduced incidence of GvHD in animals growing under germ-free conditions or receiving maximal isolation and decontamination, and supported by clinical trials on reduction of neutropenic infections by prophylactic antibiotics, the concept of maximal decontamination has become standard of care for a long time. Increasing knowledge about the interaction of innate and adaptive immunity and lessons learned by the application of modern molecular microbiota sequencing, however, lead to reevaluation of these strategies, as dysbiosis rather than decontamination is achieved in the majority of patients.

Highlights

Dysbiosis induced by prophylactic and therapeutic antibiotics is new and independent risk factor of acute GvHD:

- GvHD itself potentiates dysbiosis as Paneth cells are the most sensitive targets of intestinal GvHD resulting in diminished release of antimicrobial peptides.
- As a consequence, increased systemic levels of Reg3alpha, an antimicrobial Paneth cell peptide, are among the most sensitive and predictive biomarkers of intestinal GvHD.
- Maintaining and protecting or reinstitution of commensal bacteria seems to be a new option to modulate GvHD.

References

Ames, N. J., Sulima, P., Ngo, T., Barb, J., Munson, P. J., Paster, B. J., & Hart, T. C. (2012). A characterization of the oral microbiome in allogeneic stem cell transplant patients. *PLoS One, 7*(10), e47628. https://doi.org/10.1371/journal.pone.0047628. Epub 2012 Oct 29. PubMed PMID: 23144704; PubMed Central PMCID: PMC3483166.

Apostolova, P., & Zeiser, R. (2016). The role of purine metabolites as DAMPs in acute graft-versus-host disease. *Frontiers in Immunology, 7*, 439. eCollection 2016.

Beelen, D. W., Elmaagacli, A., Müller, K. D., Hirche, H., & Schaefer, U. W. (1999). Influence of intestinal bacterial decontamination using metronidazole and ciprofloxacin or ciprofloxacin alone on the development of acute graft-versus-host disease after marrow transplantation in patients with hematologic malignancies: Final results and long-term follow-up of an open-label prospective randomized trial. *Blood, 93*(10), 3267–3275.

Bevins, C. L., Stange, E. F., & Wehkamp, J. (2009). Decreased Paneth cell defensin expression in ileal Crohn's disease is independent of inflammation, but linked to the NOD2 1007fs genotype. *Gut, 58*(6), 882–883. discussion 883–4. PubMed PMID: 19433600.

Bilinski, J., Grzesiowski, P., Sorensen, N., Madry, K., Muszynski, J., Robak, K., Wroblewska, M., Dzieciatkowski, T., Dulny, G., Dwilewicz-Trojaczek, J., Wiktor-Jedrzejczak, W., & Basak, G. W. (2017). Fecal microbiota transplantation in patients with blood disorders inhibits gut colonization with antibiotic-resistant bacteria: Results of a prospective, single-center study. *Clinical Infectious Diseases, 65*(3), 364–370. https://doi.org/10.1093/cid/cix252. PubMed PMID: 28369341.

Bilinski, J., Robak, K., Peric, Z., Marchel, H., Karakulska-Prystupiuk, E., Halaburda, K., Rusicka, P., Swoboda-

Kopec, E., Wroblewska, M., Wiktor-Jedrzejczak, W., & Basak, G. W. (2016). Impact of gut colonization by antibiotic-resistant bacteria on the outcomes of allogeneic hematopoietic stem cell transplantation: A retrospective, single-center study. *Biology of Blood and Marrow Transplantation, 22*(6), 1087–1093. https://doi.org/10.1016/j.bbmt.2016.02.009. Epub 2016 Feb 18. PubMed PMID: 26900084.

Cooke, K. R., Gerbitz, A., Crawford, J. M., Teshima, T., Hill, G. R., Tesolin, A., Rossignol, D. P., & Ferrara, J. L. (2001). LPS antagonism reduces graft-versus-host disease and preserves graft-versus-leukemia activity after experimental bone marrow transplantation. *The Journal of Clinical Investigation, 107*(12), 1581–1589.

Cooke, K. R., Olkiewicz, K., Erickson, N., & Ferrara, J. L. (2002). The role of endotoxin and the innate immune response in the pathophysiology of acute graft versus host disease. *Journal of Endotoxin Research, 8*(6), 441–448.

Doki, N., Suyama, M., Sasajima, S., Ota, J., Igarashi, A., Mimura, I., Morita, H., Fujioka, Y., Sugiyama, D., Nishikawa, H., Shimazu, Y., Suda, W., Takeshita, K., Atarashi, K., Hattori, M., Sato, E., Watakabe-Inamoto, K., Yoshioka, K., Najima, Y., Kobayashi, T., Kakihana, K., Takahashi, N., Sakamaki, H., Honda, K., & Ohashi, K. (2017). Clinical impact of pre-transplant gut microbial diversity on outcomes of allogeneic hematopoietic stem cell transplantation. *Annals of Hematology, 96*(9), 1517–1523. https://doi.org/10.1007/s00277-017-3069-8. Epub 2017 Jul 21. PubMed PMID: 28733895.

Ekmekciu, I., von Klitzing, E., Fiebiger, U., Escher, U., Neumann, C., Bacher, P., Scheffold, A., Kühl, A. A., Bereswill, S., & Heimesaat, M. M. (2017). Immune responses to broad-spectrum antibiotic treatment and fecal microbiota transplantation in mice. *Frontiers in Immunology, 8*, 397. https://doi.org/10.3389/fimmu.2017.00397.

Eriguchi, Y., Nakamura, K., Hashimoto, D., Shimoda, S., Shimono, N., Akashi, K., Ayabe, T., & Teshima, T. (2015). Decreased secretion of Paneth cell α-defensins in graft-versus-host disease. *Transplant Infectious Disease, 17*(5), 702–706.

Eriguchi, Y., Takashima, S., Oka, H., Shimoji, S., Nakamura, K., Uryu, H., Shimoda, S., Iwasaki, H., Shimono, N., Ayabe, T., Akashi, K., & Teshima, T. (2012). Graft-versus-host disease disrupts intestinal microbial ecology by inhibiting Paneth cell production of α-defensins. *Blood, 120*(1), 223–231.

Eriguchi, Y., Uryu, H., Nakamura, K., Shimoji, S., Takashima, S., Iwasaki, H., Miyamoto, T., Shimono, N., Hashimoto, D., Akashi, K., Ayabe, T., & Teshima, T. (2013). Reciprocal expression of enteric antimicrobial proteins in intestinal graft-versus-host disease. *Biology of Blood and Marrow Transplantation, 19*(10), 1525–1529.

Ferrara, J. L., Harris, A. C., Greenson, J. K., Braun, T. M., Holler, E., Teshima, T., Levine, J. E., Choi, S. W., Huber, E., Landfried, K., Akashi, K., Vander Lugt, M., Reddy, P., Chin, A., Zhang, Q., Hanash, S., & Paczesny, S. (2011). Regenerating islet-derived 3-alpha is a biomarker of gastrointestinal graft-versus-host disease. *Blood, 118*(25), 6702–6708.

Ferrara, J. L., Levine, J. E., Reddy, P., & Holler, E. (2009). Graft-versus-host disease. *Lancet, 373*(9674), 1550–1561. https://doi.org/10.1016/S0140-6736(09)60237-3.

Gafter-Gvili, A., Fraser, A., Paul, M., van de Wetering, M., Kremer, L., & Leibovici, L. (2005). Antibiotic prophylaxis for bacterial infections in afebrile neutropenic patients following chemotherapy. *Cochrane Database Systematic Review*, (4), CD004386. Review. Update in: Cochrane Database Syst Rev. 2012;1:CD004386.

Gerbitz, A., Schultz, M., Wilke, A., Linde, H. J., Schölmerich, J., Andreesen, R., & Holler, E. (2004). Probiotic effects on experimental graft-versus-host disease: Let them eat yogurt. *Blood, 103*(11), 4365–4367.

Ghimire, S., Matos, C., Caioni, M., Weber, D., Peter, K., Holler, E., Kreutz, M., & Renner, K. (2018). Indoxyl 3-sulfate inhibits maturation and activation of human monocyte-derived dendritic cells. *Immunobiology, 223*(2), 239–245.

Golob, J. L., Pergam, S. A., Srinivasan, S., Fiedler, T. L., Liu, C., Garcia, K., Mielcarek, M., Ko, D., Aker, S., Marquis, S., Loeffelholz, T., Plantinga, A., Wu, M. C., Celustka, K., Morrison, A., Woodfield, M., & Fredricks, D. N. (2017). The stool microbiota at neutrophil recovery is predictive for severe acute graft versus host disease after hematopoietic cell transplantation. *Clinical Infectious Diseases*. https://doi.org/10.1093/cid/cix699. [Epub ahead of print] PubMed PMID: 29020185.

Gonzales, F., Bruno, B., Alarcón Fuentes, M., De Berranger, E., Guimber, D., Behal, H., Gandemer, V., Spiegel, A., Sirvent, A., Yakoub-Agha, I., Nelken, B., Duhamel, A., & Seguy, D. (2017). Better early outcome with enteral rather than parenteral nutrition in children undergoing MAC allo-SCT. *Clinical Nutrition*. https://doi.org/10.1016/j.clnu.2017.10.005. pii: S0261-5614(17)31365-1. [Epub ahead of print] PubMed PMID: 29097037.

Gooley, T. A., Chien, J. W., Pergam, S. A., Hingorani, S., Sorror, M. L., Boeckh, M., Martin, P. J., Sandmaier, B. M., Marr, K. A., Appelbaum, F. R., Storb, R., & McDonald, G. B. (2010). Reduced mortality after allogeneic hematopoietic-cell transplantation. *The New England Journal of Medicine, 363*(22), 2091–2101.

Gorshein, E., Wei, C., Ambrosy, S., Budney, S., Vivas, J., Shenkerman, A., Manago, J., McGrath, M. K., Tyno, A., Lin, Y., Patel, V., Gharibo, M., Schaar, D., Jenq, R. R., Khiabanian, H., & Strair, R. (2017). Lactobacillus rhamnosus GG probiotic enteric regimen does not appreciably alter the gut microbiome or provide protection against GVHD after allogeneic hematopoietic stem cell transplantation. *Clinical Transplantation, 31*(5). https://doi.org/10.1111/ctr.12947. Epub 2017 Mar 31.

Hanash, A. M., Dudakov, J. A., Hua, G., O'Connor, M. H., Young, L. F., Singer, N. V., West, M. L., Jenq, R. R., Holland, A. M., Kappel, L. W., Ghosh, A., Tsai, J. J., Rao, U. K., Yim, N. L., Smith, O. M., Velardi, E., Hawryluk, E. B., Murphy, G. F., Liu, C., Fouser, L. A., Kolesnick, R., Blazar, B. R., & van den Brink, M. R. (2012). Interleukin-22 protects intestinal stem cells from immune-mediated tissue damage and regulates sensitivity to graft versus host disease. *Immunity, 37*(2), 339–350.

Harris, B., Morjaria, S. M., Littmann, E. R., Geyer, A. I., Stover, D. E., Barker, J. N., Giralt, S. A., Taur, Y., & Pamer, E. G. (2016). Gut microbiota predict pulmonary infiltrates after allogeneic hematopoietic cell transplantation. *American Journal of Respiratory and Critical Care Medicine, 194*(4), 450–463.

Holler, E., Butzhammer, P., Schmid, K., Hundsrucker, C., Koestler, J., Peter, K., Zhu, W., Sporrer, D., Hehlgans, T., Kreutz, M., Holler, B., Wolff, D., Edinger, M., Andreesen, R., Levine, J. E., Ferrara, J. L., Gessner, A., Spang, R., & Oefner, P. J. (2014). Metagenomic analysis of the stool microbiome in patients receiving allogeneic stem cell transplantation: Loss of diversity is associated with use of systemic antibiotics and more pronounced in gastrointestinal graft-versus-host disease. *Biology of Blood and Marrow Transplantation, 20*(5), 640–645.

Holler, E., Kolb, H. J., Möller, A., Kempeni, J., Liesenfeld, S., Pechumer, H., Lehmacher, W., Ruckdeschel, G., Gleixner, B., Riedner, C., et al. (1990). Increased serum levels of tumor necrosis factor alpha precede major complications of bone marrow transplantation. *Blood, 75*(4), 1011–1016. PubMed PMID: 2405918.

Holler, E., Rogler, G., Brenmoehl, J., Hahn, J., Herfarth, H., Greinix, H., Dickinson, A. M., Socié, G., Wolff, D., Fischer, G., Jackson, G., Rocha, V., Steiner, B., Eissner, G., Marienhagen, J., Schoelmerich, J., & Andreesen, R. (2006). Prognostic significance of NOD2/CARD15 variants in HLA-identical sibling hematopoietic stem cell transplantation: Effect on long-term outcome is confirmed in 2 independent cohorts and may be modulated by the type of gastrointestinal decontamination. *Blood, 107*(10), 4189–4193.

Jenq, R. R., Taur, Y., Devlin, S. M., Ponce, D. M., Goldberg, J. D., Ahr, K. F., Littmann, E. R., Ling, L., Gobourne, A. C., Miller, L. C., Docampo, M. D., Peled, J. U., Arpaia, N., Cross, J. R., Peets, T. K., Lumish, M. A., Shono, Y., Dudakov, J. A., Poeck, H., Hanash, A. M., Barker, J. N., Perales, M. A., Giralt, S. A., Pamer, E. G., & van den Brink, M. R. (2015). Intestinal blautia is associated with reduced death from graft-versus-host disease. *Biology of Blood and Marrow Transplantation, 21*(8), 1373–1383.

Jenq, R. R., Ubeda, C., Taur, Y., Menezes, C. C., Khanin, R., Dudakov, J. A., Liu, C., West, M. L., Singer, N. V., Equinda, M. J., Gobourne, A., Lipuma, L., Young, L. F., Smith, O. M., Ghosh, A., Hanash, A. M., Goldberg, J. D., Aoyama, K., Blazar, B. R., Pamer, E. G., & van den Brink, M. R. (2012). Regulation of intestinal inflammation by microbiota following allogeneic bone marrow transplantation. *The Journal of Experimental Medicine, 209*(5), 903–911.

Kakihana, K., Fujioka, Y., Suda, W., Najima, Y., Kuwata, G., Sasajima, S., Mimura, I., Morita, H., Sugiyama, D., Nishikawa, H., Hattori, M., Hino, Y., Ikegawa, S., Yamamoto, K., Toya, T., Doki, N., Koizumi, K., Honda, K., & Ohashi, K. (2016). Fecal microbiota transplantation for patients with steroid-resistant acute graft-versus-host disease of the gut. *Blood, 128*(16), 2083–2088.

Kaysen, A., Heintz-Buschart, A., Muller, E. E. L., Narayanasamy, S., Wampach, L., Laczny, C. C., Graf, N., Simon, A., Franke, K., Bittenbring, J., Wilmes, P., & Schneider, J. G. (2017). Integrated meta-omic analyses of the gastrointestinal tract microbiome in patients undergoing allogeneic hematopoietic stem cell transplantation. *Translational Research, 186*, 79–94.e1. https://doi.org/10.1016/j.trsl.2017.06.008. Epub 2017 Jun 20. PubMed PMID: 28686852.

Ladas, E. J., Bhatia, M., Chen, L., Sandler, E., Petrovic, A., Berman, D. M., Hamblin, F., Gates, M., Hawks, R., Sung, L., & Nieder, M. (2016). The safety and feasibility of probiotics in children and adolescents undergoing hematopoietic cell transplantation. *Bone Marrow Transplantation, 51*(2), 262–266.

Lee, Y. J., Arguello, E. S., Jenq, R. R., Littmann, E., Kim, G. J., Miller, L. C., Ling, L., Figueroa, C., Robilotti, E., Perales, M. A., Barker, J. N., Giralt, S., van den Brink, M. R. M., Pamer, E. G., & Taur, Y. (2017). Protective factors in the intestinal microbiome against Clostridium difficile infection in recipients of allogeneic hematopoietic stem cell transplantation. *The Journal of Infectious Diseases, 215*(7), 1117–1123. https://doi.org/10.1093/infdis/jix011. PubMed PMID: 28498996; PubMed Central PMCID: PMC5426375.

Legoff, J., Resche-Rigon, M., Bouquet, J., Robin, M., Naccache, S. N., Mercier-Delarue, S., Federman, S., Samayoa, E., Rousseau, C., Piron, P., Kapel, N., Simon, F., Socié, G., & Chiu, C. Y. (2017). The eukaryotic gut virome in hematopoietic stem cell transplantation: New cluesin enteric graft-versus-host disease. *Nature Medicine, 23*(9), 1080–1085. https://doi.org/10.1038/nm.4380. Epub 2017 Jul 31. PubMed PMID: 28759053.

Levine, J. E., Huber, E., Hammer, S. T., Harris, A. C., Greenson, J. K., Braun, T. M., Ferrara, J. L., & Holler, E. (2013). Low Paneth cell numbers at onset of gastrointestinal graft-versus-host disease identify patients at high risk for nonrelapse mortality. *Blood, 122*(8), 1505–1509.

Liu, C., Frank, D. N., Horch, M., Chau, S., Ir, D., Horch, E. A., Tretina, K., van Besien, K., Lozupone, C. A., & Nguyen, V. H. (2017). Associations between acute gastrointestinal GvHD and the baseline gut microbiota of allogeneic hematopoietic stem cell transplant recipients and donors. *Bone Marrow Transplantation.*

https://doi.org/10.1038/bmt.2017.200. [Epub ahead of print].

Mandalia, A., Ward, A., Tauxe, W., Kraft, C. S., & Dhere, T. (2016). Fecal transplant is as effective and safe in immunocompromised as non-immunocompromised patients for Clostridium difficile. *International Journal of Colorectal Disease, 31*(5), 1059–1060.

Manges, A. R., Steiner, T. S., & Wright, A. J. (2016). Fecal microbiota transplantation for the intestinal decolonization of extensively antimicrobial-resistant opportunistic pathogens: A review. *Infectious Diseases (London), 48*(8), 587–592.

Mathewson, N. D., Jenq, R., Mathew, A. V., Koenigsknecht, M., Hanash, A., Toubai, T., Oravecz-Wilson, K., Wu, S. R., Sun, Y., Rossi, C., Fujiwara, H., Byun, J., Shono, Y., Lindemans, C., Calafiore, M., Schmidt, T. C., Honda, K., Young, V. B., Pennathur, S., van den Brink, M., & Reddy, P. (2016). Gut microbiome-derived metabolites modulate intestinal epithelial cell damage and mitigate graft-versus-host disease. *Nature Immunology, 17*(5), 505–513.

Montassier, E., Al-Ghalith, G. A., Ward, T., Corvec, S., Gastinne, T., Potel, G., Moreau, P., de la Cochetiere, M. F., Batard, E., & Knights, D. (2016). Pretreatment gut microbiome predicts chemotherapy-related bloodstream infection. *Genome Medicine, 8*(1), 49. https://doi.org/10.1186/s13073-016-0301-4. Erratum in: Genome Med. 2016;8(1):61. PubMed PMID: 27121964; PubMed Central PMCID: PMC4848771.

Paczesny, S., Krijanovski, O. I., Braun, T. M., Choi, S. W., Clouthier, S. G., Kuick, R., Misek, D. E., Cooke, K. R., Kitko, C. L., Weyand, A., Bickley, D., Jones, D., Whitfield, J., Reddy, P., Levine, J. E., Hanash, S. M., & Ferrara, J. L. (2009). A biomarker panel for acute graft-versus-host disease. *Blood, 113*(2), 273–278.

Peled, J. U., Devlin, S. M., Staffas, A., Lumish, M., Khanin, R., Littmann, E. R., Ling, L., Kosuri, S., Maloy, M., Slingerland, J. B., Ahr, K. F., Porosnicu Rodriguez, K. A., Shono, Y., Slingerland, A. E., Docampo, M. D., Sung, A. D., Weber, D., Alousi, A. M., Gyurkocza, B., Ponce, D. M., Barker, J. N., Perales, M. A., Giralt, S. A., Taur, Y., Pamer, E. G., Jenq, R. R., & van den Brink, M. R. M. (2017). Intestinal microbiota and relapse after hematopoietic-cell transplantation. *Journal of Clinical Oncology, 35*(15), 1650–1659.

Peled, J. U., Jenq, R. R., Holler, E., & van den Brink, M. R. (2016). Role of gut flora after bone marrow transplantation. *Nature Microbiology, 1*, 16036. https://doi.org/10.1038/nmicrobiol.2016.36.

Penack, O., Smith, O. M., Cunningham-Bussel, A., Liu, X., Rao, U., Yim, N., Na, I. K., Holland, A. M., Ghosh, A., Lu, S. X., Jenq, R. R., Liu, C., Murphy, G. F., Brandl, K., & van den Brink, M. R. (2009). NOD2 regulates hematopoietic cell function during graft-versus-host disease. *The Journal of Experimental Medicine, 206*(10), 2101–2110. https://doi.org/10.1084/jem.20090623. Epub 2009 Sep 8. PubMed PMID: 19737867; PubMed Central PMCID: PMC2757869.

Pflug, N., Kluth, S., Vehreschild, J. J., Bahlo, J., Tacke, D., Biehl, L., Eichhorst, B., Fischer, K., Cramer, P., Fink, A. M., von Bergwelt-Baildon, M., Stilgenbauer, S., Hallek, M., Cornely, O. A., & Vehreschild, M. J. (2016). Efficacy of antineoplastic treatment is associated with the use of antibiotics that modulate intestinal microbiota. *Oncoimmunology, 5*(6), e1150399. https://doi.org/10.1080/2162402X.2016.1150399. eCollection 2016 Jun. PubMed PMID: 27471619; PubMed Central PMCID: PMC4938364.

Piguet, P. F., Grau, G. E., Allet, B., & Vassalli, P. (1987). Tumor necrosis factor/cachectin is an effector of skin and gut lesions of the acute phase of graft-vs.-host disease. *The Journal of Experimental Medicine, 166*(5), 1280–1289.

Reddy, P., Sun, Y., Toubai, T., Duran-Struuck, R., Clouthier, S. G., Weisiger, E., Maeda, Y., Tawara, I., Krijanovski, O., Gatza, E., Liu, C., Malter, C., Mascagni, P., Dinarello, C. A., & Ferrara, J. L. (2008). Histone deacetylase inhibition modulates indoleamine 2,3-dioxygenase-dependent DC functions and regulates experimental graft-versus-host disease in mice. *The Journal of Clinical Investigation, 118*(7), 2562–2573. Review.

Riwes, M., & Reddy, P. (2017). Microbial metabolites and graft versus host disease. *American Journal of Transplantation*. https://doi.org/10.1111/ajt.14443. [Epub ahead of print] Review. PubMed PMID: 28742948.

Routy, B., Le Chatelier, E., Derosa, L., Duong, C. P. M., Alou, M. T., Daillère, R., Fluckiger, A., Messaoudene, M., Rauber, C., Roberti, M. P., Fidelle, M., Flament, C., Poirier-Colame, V., Opolon, P., Klein, C., Iribarren, K., Mondragón, L., Jacquelot, N., Qu, B., Ferrere, G., Clémenson, C., Mezquita, L., Masip, J. R., Naltet, C., Brosseau, S., Kaderbhai, C., Richard, C., Rizvi, H., Levenez, F., Galleron, N., Quinquis, B., Pons, N., Ryffel, B., Minard-Colin, V., Gonin, P., Soria, J. C., Deutsch, E., Loriot, Y., Ghiringhelli, F., Zalcman, G., Goldwasser, F., Escudier, B., Hellmann, M. D., Eggermont, A., Raoult, D., Albiges, L., Kroemer, G., & Zitvogel, L. (2017). Gut microbiome influences efficacy of PD-1-based immunotherapy against epithelial tumors. *Science*. https://doi.org/10.1126/science.aan3706. pii: eaan3706. [Epub ahead of print] PubMed PMID: 29097494.

Routy, B., Letendre, C., Enot, D., Chénard-Poirier, M., Mehraj, V., Séguin, N. C., Guenda, K., Gagnon, K., Woerther, P. L., Ghez, D., & Lachance, S. (2016). The influence of gut-decontamination prophylactic antibiotics on acute graft-versus-host disease and survival following allogeneic hematopoietic stem cell transplantation. *Oncoimmunology, 6*(1), e1258506. https://doi.org/10.1080/2162402X.2016.1258506. eCollection 2017.

Schmeiser, T., Kurrle, E., Arnold, R., Heit, W., Krieger, D., Kubanek, B., & Heimpel, H. (1985). The influence of the microbial flora on incidence of graft-versus-host disease in human allogeneic bone marrow transplantation. *Progress in Clinical and Biological Research, 181*, 433–435.

Schwab, L., Goroncy, L., Palaniyandi, S., Gautam, S., Triantafyllopoulou, A., Mocsai, A., Reichardt, W., Karlsson, F. J., Radhakrishnan, S. V., Hanke, K., Schmitt-Graeff, A., Freudenberg, M., von Loewenich, F. D., Wolf, P., Leonhardt, F., Baxan, N., Pfeifer, D., Schmah, O., Schönle, A., Martin, S. F., Mertelsmann, R., Duyster, J., Finke, J., Prinz, M., Henneke, P., Häcker, H., Hildebrandt, G. C., Häcker, G., & Zeiser, R. (2014). Neutrophil granulocytes recruited upon translocation of intestinal bacteria enhance graft-versus-host disease via tissue damage. *Nature Medicine, 20*(6), 648–654.

Shelburne, S. A., Ajami, N. J., Chibucos, M. C., Beird, H. C., Tarrand, J., Galloway-Peña, J., Albert, N., Chemaly, R. F., Ghantoji, S. S., Marsh, L., Pemmaraju, N., Andreeff, M., Shpall, E. J., Wargo, J. A., Rezvani, K., Alousi, A., Bruno, V. M., Futreal, P. A., Petrosino, J. F., & Kontoyiannis, D. P. (2015). Implementation of a pan-genomic approach to investigate holobiont-infecting microbe interaction: A case report of a leukemic patient with invasive mucormycosis. *PLoS One, 10*(11), e0139851.

Shono, Y., Docampo, M. D., Peled, J. U., Perobelli, S. M., Velardi, E., Tsai, J. J., Slingerland, A. E., Smith, O. M., Young, L. F., Gupta, J., Lieberman, S. R., Jay, H. V., Ahr, K. F., Porosnicu Rodriguez, K. A., Xu, K., Calarfiore, M., Poeck, H., Caballero, S., Devlin, S. M., Rapaport, F., Dudakov, J. A., Hanash, A. M., Gyurkocza, B., Murphy, G. F., Gomes, C., Liu, C., Moss, E. L., Falconer, S. B., Bhatt, A. S., Taur, Y., Pamer, E. G., van den Brink, M. R. M., & Jenq, R. R. (2016). Increased GVHD-related mortality with broad-spectrum antibiotic use after allogeneic hematopoietic stem cell transplantation in human patients and mice. *Science Translational Medicine, 8*(339), 339ra71.

Spindelboeck, W., Schulz, E., Uhl, B., Kashofer, K., Aigelsreiter, A., Zinke-Cerwenka, W., Mulabecirovic, A., Kump, P. K., Halwachs, B., Gorkiewicz, G., Sill, H., Greinix, H., Högenauer, C., & Neumeister, P. (2017). Repeated fecal microbiota transplantations attenuate diarrhea and lead to sustained changes in the fecal microbiota in acute, refractory gastrointestinal graft-versus-host-disease. *Haematologica, 102*(5), e210-e213.

Storb, R., Prentice, R. L., Buckner, C. D., Clift, R. A., Appelbaum, F., Deeg, J., Doney, K., Hansen, J. A., Mason, M., Sanders, J. E., Singer, J., Sullivan, K. M., Witherspoon, R. P., & Thomas, E. D. (1983). Graft-versus-host disease and survival in patients with aplastic anemia treated by marrow grafts from HLA-identical siblings. Beneficial effect of a protective environment. *The New England Journal of Medicine, 308*(6), 302–307.

Svahn, B. M., Remberger, M., Myrbäck, K. E., Holmberg, K., Eriksson, B., Hentschke, P., Aschan, J., Barkholt, L., & Ringdén, O. (2002). Home care during the pancytopenic phase after allogeneic hematopoietic stem cell transplantation is advantageous compared with hospital care. *Blood, 100*(13), 4317–4324.

Taur, Y., Jenq, R. R., Perales, M. A., Littmann, E. R., Morjaria, S., Ling, L., No, D., Gobourne, A., Viale, A., Dahi, P. B., Ponce, D. M., Barker, J. N., Giralt, S., van den Brink, M., & Pamer, E. G. (2014). The effects of intestinal tract bacterial diversity on mortality following allogeneic hematopoietic stem cell transplantation. *Blood, 124*(7), 1174–1182.

Taur, Y., Xavier, J. B., Lipuma, L., Ubeda, C., Goldberg, J., Gobourne, A., Lee, Y. J., Dubin, K. A., Socci, N. D., Viale, A., Perales, M. A., Jenq, R. R., van den Brink, M. R., & Pamer, E. G. (2012). Intestinal domination and the risk of bacteremia in patients undergoing allogeneic hematopoietic stem cell transplantation. *Clinical Infectious Diseases, 55*(7), 905–914.

Tawara, I., Liu, C., Tamaki, H., Toubai, T., Sun, Y., Evers, R., Nieves, E., Mathewson, N., Nunez, G., & Reddy, P. (2013). Influence of donor microbiota on the severity of experimental graft-versus-host-disease. *Biology of Blood and Marrow Transplantation, 19*(1), 164–168.

van Bekkum, D. W., & Knaan, S. (1977). Role of bacterial microflora in development of intestinal lesions from graft-versus-host reaction. *Journal of the National Cancer Institute, 58*(3), 787–790.

van Bekkum, D. W., Roodenburg, J., Heidt, P. J., & van der Waaij, D. (1974). Mitigation of secondary disease of allogeneic mouse radiation chimeras by modification of the intestinal microflora. *Journal of the National Cancer Institute, 52*(2), 401–404.

Varelias, A., Ormerod, K. L., Bunting, M. D., Koyama, M., Gartlan, K. H., Kuns, R. D., Lachner, N., Locke, K. R., Lim, C. Y., Henden, A. S., Zhang, P., Clouston, A. D., Hasnain, S. Z., McGuckin, M. A., Blazar, B. R., MacDonald, K. P., Hugenholtz, P., & Hill, G. R. (2017). Acute graft-versus-host disease is regulated by an IL-17-sensitive microbiome. *Blood, 129*(15), 2172–2185.

Vossen, J. M., Guiot, H. F., Lankester, A. C., Vossen, A. C., Bredius, R. G., Wolterbeek, R., Bakker, H. D., & Heidt, P. J. (2014). Complete suppression of the gut microbiome prevents acute graft-versus-host disease following allogeneic bone marrow transplantation. *PLoS One, 9*(9), e105706. https://doi.org/10.1371/journal.pone.0105706. eCollection 2014. PubMed PMID: 25180821; PubMed Central PMCID: PMC4152127.

Vriesendorp, H. M., Klapwijk, W. M., Visser, T. P., Zurcher, C., & van Bekkum, D. W. (1982). Alternatives to donor matching for control of graft-versus-host disease. *Immunogenetics, 15*(1), 79–94.

Webb, B. J., Brunner, A., Ford, C. D., Gazdik, M. A., Petersen, F. B., & Hoda, D. (2016). Fecal microbiota transplantation for recurrent Clostridium difficile infection in hematopoietic stem cell transplant recipients. *Transplant Infectious Disease, 18*(4), 628–633.

Weber, D., Frauenschläger, K., Ghimire, S., Peter, K., Panzer, I., Hiergeist, A., Weber, M., Kutny, D., Wolff, D., Grube, M., Huber, E., Oefner, P., Gessner, A., Hehlgans, T., Herr, W., & Holler, E. (2017a). The association between acute graft-versus-host disease and antimicrobial peptide expression in the gastrointestinal tract after allogeneic stem cell transplantation. *PLoS One, 12*(9), e0185265. https://doi.org/10.1371/journal.pone.0185265. eCollection 2017.

Weber, D., Jenq, R. R., Peled, J. U., Taur, Y., Hiergeist, A., Koestler, J., Dettmer, K., Weber, M., Wolff, D., Hahn, J., Pamer, E. G., Herr, W., Gessner, A., Oefner, P. J., van den Brink, M. R., & Holler, E. (2017b). Microbiota disruption induced by early use of broad-spectrum antibiotics is an independent risk factor of outcome after allogeneic stem cell transplantation. *Biology of Blood and Marrow Transplantation, 23*(5), 845–852.

Weber, D., Oefner, P. J., Dettmer, K., Hiergeist, A., Koestler, J., Gessner, A., Weber, M., Stämmler, F., Hahn, J., Wolff, D., Herr, W., & Holler, E. (2016). Rifaximin preserves intestinal microbiota balance in patients undergoing allogeneic stem cell transplantation. *Bone Marrow Transplantation, 51*(8), 1087–1092.

Weber, D., Oefner, P. J., Hiergeist, A., Koestler, J., Gessner, A., Weber, M., Hahn, J., Wolff, D., Stämmler, F., Spang, R., Herr, W., Dettmer, K., & Holler, E. (2015). Low urinary indoxyl sulfate levels early after transplantation reflect a disrupted microbiome and are associated with poor outcome. *Blood, 126*(14), 1723–1728.

Whangbo, J., Ritz, J., & Bhatt, A. (2017). Antibiotic-mediated modification of the intestinal microbiome in allogeneic hematopoietic stem cell transplantation. *Bone Marrow Transplantation, 52*(2), 183–190. https://doi.org/10.1038/bmt.2016.206. Epub 2016 Aug 15. Review. PubMed PMID: 27526283.

Zeiser, R., Penack, O., Holler, E., & Idzko, M. (2011). Danger signals activating innate immunity in graft-versus-host disease. *Journal of Molecular Medicine (Berlin), 89*(9), 833–845.

Zitvogel, L., Daillère, R., Roberti, M. P., Routy, B., & Kroemer, G. (2017). Anticancer effects of the microbiome and its products. *Nature Reviews. Microbiology, 15*(8), 465–478. https://doi.org/10.1038/nrmicro.2017.44. Epub 2017 May 22. Review. PubMed PMID: 28529325.

Microbiome and Diseases: Pathogen Infection

14

Christine Josenhans and Guntram A. Grassl

Abstract

The host and the intestinal microbiota contribute in manifold ways to an immune balance and defense against diseases and pathogens. Infectious diseases of the intestine are major diseases which severely threaten individual and global health. Enteropathogenic agents causing infectious diseases have evolved specific tactics how they interact with the host and with the microbiota simultaneously. This chapter characterizes the main players of the intestinal niche, which contribute to the well-being of the host and can ward off or limit pathogenic invaders. It also illustrates, using a few prominent and well-studied examples, how intestinal pathogenic agents can interact with both the host and the microbiota in order to promote their own expansion, to overcome the defenses by the resident microbiota and the mucosal barrier, and how they finally cause disease. This chapter also introduces novel ways how to treat intestinal infections by addressing the microbiota.

C. Josenhans (✉)
Institute for Medical Microbiology and Hospital Epidemiology, Hannover Medical School, Hannover, Germany

DZIF, Partner Site Hannover-Braunschweig, Braunschweig, Germany

DZIF, Partner Site Munich, Munich, Germany

Max von Pettenkofer Institute, LMU Munich, Munich, Germany
e-mail: Josenhans.christine@mh-hannover.de

G. A. Grassl
Institute for Medical Microbiology and Hospital Epidemiology, Hannover Medical School, Hannover, Germany

DZIF, Partner Site Hannover-Braunschweig, Braunschweig, Germany
e-mail: grassl.guntram@mh-hannover.de

14.1 The Intestinal Niche as a Reservoir of Microbiota and Enteric Pathogens

As we now know, the gastrointestinal tract of humans and other vertebrates is one of the most complex ecological niches imaginable. A multitude of different cells is combined to form the mucosal barrier and sample the molecules of various incoming compounds which can be both living organisms, such as bacteria and parasites, or other intake, such as viruses, food and environmental particles. The intestinal tract is the largest body surface and an important entry portal but also harbors a major defense system against pathogens. The intestinal cell lining, the mucosa, is characterized by a plethora of barrier functions and defensive factors (Fig. 14.1). Intestinal barrier functions can be of physical, chemical, or more complex, e.g., immunological nature. Physical factors consist of the tight connection between the cells of the mucosa and the overlying

D. Haller (ed.), *The Gut Microbiome in Health and Disease*,
https://doi.org/10.1007/978-3-319-90545-7_14

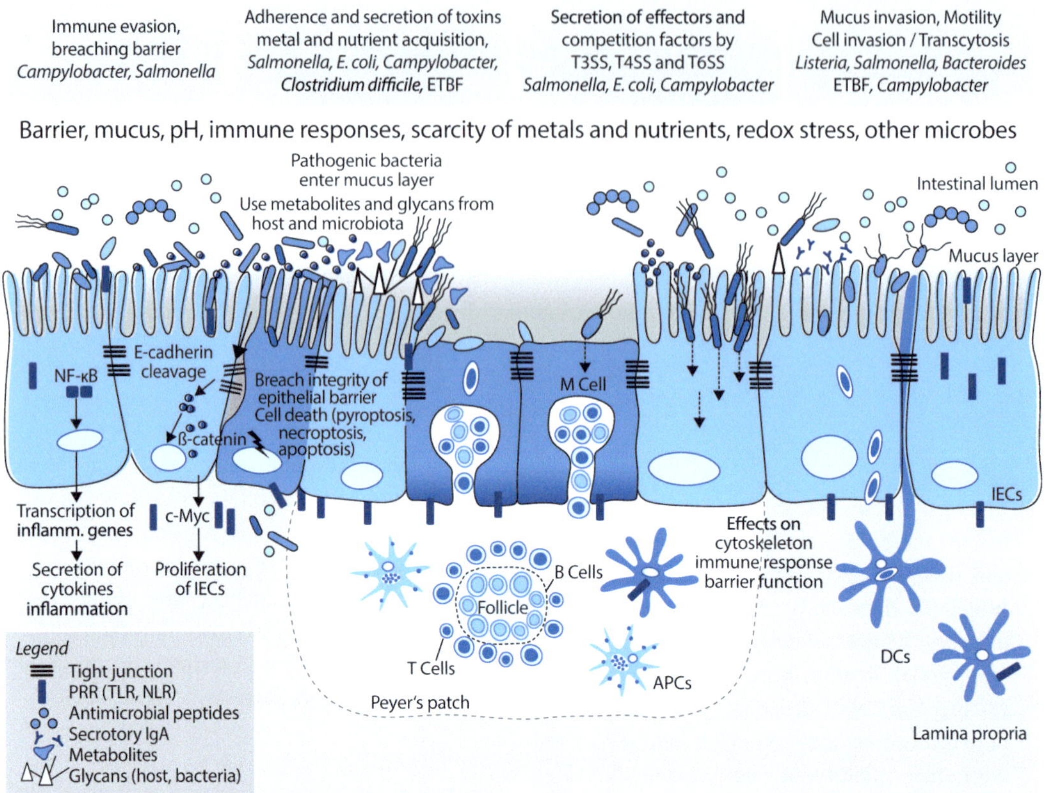

Fig. 14.1 Schematic overview of the intestinal niche, including its main components: the resident microbiota, immune factors, and pathogenic bacteria. The intestinal mucosa has numerous factors which shape the interaction with nonpathogenic and pathogenic agents and incoming components of food and the environment. Mucus and immune factors shape the microbiota and defend against pathogens. The stimulation of PRRs by pathogens and commensal bacteria leads to the activation of NF-κB and production of pro-inflammatory cytokines, which are necessary for a healthy balance of responses (homeostasis) but also for defense against pathogens. The hallmarks of detailed interactions of certain types of pathogens are included in the boxes above the schematic illustration

mucus layer. The mucus forms a netlike structure of various glycans, glycoproteins, and glycopeptides, the mucins (Bansil and Turner 2017; Arike and Hansson 2016), which are produced by specialized cells of the mucosa, the goblet cells (Birchenough et al. 2015; Johansson and Hansson 2016). The mucus consists of an inner layer, which is a very densely knit gel-like matrix and usually free of microbes, and an outer layer which is a gel of larger pore size and loosened structure (Johansson et al. 2013; Hansson 2012). The chemical defense system in the intestine includes a rather low pH, lytic enzymes (e.g., proteases, lysozyme), and secreted defensive factors produced by cells of the mucosa (Jäger et al. 2013; Sperandio et al. 2015; Johansson and Hansson 2016). The outer mucus layer of healthy mammals is easier to invade than the inner, dense layer and harbors a complex community of microbes, in particular bacteria (Johansson et al. 2013). These bacteria are termed the commensal microbiota. In recent years, significant advances have been made toward identifying the components of the commensal microbiota in detail. This major advance is primarily due to improved techniques such as deep sequencing which circumvent bacterial culture (Morgan and Huttenhower 2014; Hiergeist et al. 2016) but also to considerable success in cultivating numerous hitherto unculturable bacteria, ushering in a new era of microaerophilic and anaerobic "culturomics" (Lagier et al. 2012; Lagkouvardos et al. 2016; Browne 2016). The microbiota appears to be specific for any given host species (Ley et al. 2008; Chung et al. 2012; Eren et al. 2015; Goodrich et al. 2016a; Clavel et al. 2017) and are not naturally transferred between humans and other mammals, although the dominant phyla of the intestinal tract overlap between different vertebrate or mammalian hosts: *Bacteroidetes* and *Firmicutes* are the main phyla represented in the intestinal tract (Arumugam et al. 2011; Yatsunenko et al. 2012; Sankar et al. 2015; Hildebrand et al. 2013). In particular the *Clostridiales* order within the *Firmicutes* and its families *Lachnospiraceae*, *Ruminococcaceae*, and *Clostridiaceae* are important taxonomic groups, comprising many and diverse bacterial species in the intestinal niche. These taxa contain both pathogenic (*Clostridium difficile*) and numerous beneficial species (Buffie et al. 2015; Kim et al. 2017; Geva-Zatorsky et al. 2017). What is a healthy human microbiota, and how many species are contained in it? We cannot unambiguously answer this question yet; however, species estimates have become more precise and range from 100 to 500 species (Yatsunenko et al. 2012; Ursell et al. 2012; Thursby and Juge 2017). The individual variation in microbiota composition between persons is quite high. Microbiota can be inherited from family members, in particular parents, grandparents, and other caretakers in the community. Individual genetic outfit plays a role in maintaining a certain adapted and transmitted microbiota (Goodrich et al. 2014). This has been underscored in human twin studies and various animal experiments (Lim et al. 2014; Goodrich et al. 2016b; Nguyen et al. 2015). The microbiota matures naturally in a specific manner over the course of a lifetime. Drastic changes in the microbiota composition and diversity occur in particular in the newborn and infants after weaning (Koenig et al. 2011) and may be, transiently or permanently, influenced over time by environmental triggers (see last paragraphs of this chapter).

Many of the cultured intestinal bacteria (human, mouse) have been recently deposited in culture collections and databases, both as single organisms and also as communities (Lagkouvardos et al. 2016) (for details of the Human Microbiome Project, see https://hmpdacc.org/). Time will tell whether these artificial/synthetic communities remain stable in the autochthonous host or upon heterologous transfer in vivo and can be stably propagated outside of their original host or, considered more likely, whether these consortia will quickly degrade and evolve away from their original composition over time.

We are still missing information about certain intestinal microbiome components and their interplay, not only on the wealth of bacteria in it but also on fungi, parasites, and viruses (Ursell et al. 2012; Minot et al. 2013). Microbiome and viruses (the "virome") crosstalk with each other, and it is an important research goal to investigate in each case more closely, whether the interference is beneficial for the host or any intestinal pathogen and in which context (see recent review by Robinson and Pfeiffer 2014). Although not the focus of this chapter, it is worthwhile mentioning that frequent intestinal pathogenic viruses such as rotavirus or norovirus also influence the host's immune status and can partially recapitulate the beneficial impact of bacterial microbiota components on diverse immune functions in microbiota depletion models, an effect which is probably triggered via innate recognition and type I interferon signaling (Kernbauer et al. 2014).

Although the concept is still under debate, different "enterotypes," referring to the groupings of distinct microbiota community types of different host subgroups, have been described in humans and also in mice. The existence of enterotypes may be dependent on genetic differences between different human individuals and also on environmental factors (Arumugam et al. 2011; Hildebrand et al. 2013; Knights et al. 2014; Lim et al. 2014; Yin et al. 2017; Gibson et al. 2016; Costea et al. 2018).

Metagenomic approaches (16S rRNA gene amplicon sequencing and more comprehensive methods capturing the complete intestinal metagenome) on the basis of advanced high-throughput sequencing technologies have helped to dissect the functional and taxonomic diversity in the culturable and non-culturable intestinal microbiome (Qin et al. 2010; Sankar et al. 2015). Representation of functions and activities within an individual microbiome is highly redundant, which is independent of the presence of individual taxa. This might ensure that different complex microbiota types and their components are at least partially exchangeable for similar functionality (Geva-Zatorsky et al. 2017). Functions of microbiota are manifold and consist of catabolic and anabolic properties (degradation of proteins and complex glycans from food components, also associated with nutrient acquisition and, context-dependently, with induction of obesity), bile acid recycling and transformation, and provisioning of vitamins and essential amino acids (Blaut and Clavel 2007; Le Chatelier et al. 2013). In addition, one major function of the intestinal microbiota is the balancing of innate and adaptive intestinal immune functions [insight mainly gathered from germ-free rodent experiments; for recent comprehensive reviews, see Macpherson and McCoy (2015), Palm et al. (2015)]. This balance should lead to a long-term stable immune response and a homeostatic state of low inflammation in the uninfected intestinal tract, despite the presence of multiple microbial ligands able to stimulate the innate immune system by the ligation of pattern recognition receptors (PRR). All components of the immune system are influenced by the microbiota starting shortly after birth (Pabst et al. 2016; Tan et al. 2016): B-cells, T-cells, immune cell differentiation, and innateand adaptive lymphocytes. The adaptive immune system is stabilized via the microbiota by forming specific broadly reactive IgA species and IgA-producing memory B-cells (Kubinak and Round 2016; Planer et al. 2016; Pabst et al. 2016; McCoy et al. 2017; Okai et al. 2017) which have long-term beneficial effects on intestinal health, even after the induction of dysbiosis. Similar mechanisms hold true for the maturation and development of T-cell responses, which are likewise shaped by the microbiota even in the absence of pathogens (Lindner et al. 2015). The above-cited studies and others have shown that commensal microbiota, in addition to pathogens, can activate both innate and adaptive immune responses in the gut, for instance, by binding to PRR such as the Toll-like receptors (TLR) (Iwasaki and Medzhitov 2015; Thaiss et al. 2016).

Another striking trait of the intestinal niche is its richness and variable provisioning of various secondary metabolites, short-chain fatty acids (SCFA; Morrison and Preston 2016), glycans, peptides, and electron acceptors. This variability of resources for resident microbiota and pathogens alike is provided by variable "input," that is, incoming dietary components, and subsequently by varying metabolic properties of the resident microbiota, which lead to a variable "output" in residual metabolites. Moreover, inflammatory bouts or infections can also lead to a variation in these metabolic activities and output metabolites.

The intestinal niche characterized by the above-described, partially dynamic factors can be challenged by various niche-adapted, acute, and chronic pathogens, bacteria, viruses, eukaryotic parasites, and fungi. The microbiota forms a strong barrier against acute and chronic infections, termed colonization resistance (CR) (Buffie and Pamer 2013) (see also Sect. 14.3 for details and examples). The different factors forming resistance phenotypes include local metabolic resilience and innate and adaptive immune maturation and homeostasis (Rakoff-Nahoum et al. 2004; Macpherson and McCoy 2015; Ohno 2016), metabolic immunity (O'Neill et al. 2016; Próchnicki and Latz 2017; Buck et al. 2017), the balanced production of antimicrobial peptides by the host (Zasloff 2002; Günther et al. 2016; Perez-Lopez et al. 2016), and the release of antibiotic-like compounds by the intestinal microbiota (Donia et al. 2014; Donia and Fischbach 2015). This defensive role of the microbiota against invading pathogens such as *S. enterica* and *C. jejuni* has been impressively verified in various rodent infection models, where the microbiota has been artificially depleted, for instance, by antibiotic treatment (Kaiser et al. 2012; Ekmekciu et al. 2017), or in germ-free animals which have been compared to conventionally raised ones for infection susceptibility (Yrios and Balish 1986; Brugiroux et al. 2016). On the other hand, it has been demonstrated already but certainly merits to be further explored that specific intestinal pathogenic bacteria even use the innate host response in the presence of other microbes to enhance their colonization potential (Günther et al. 2016).

In the following paragraphs of this chapter, we will mainly focus on the well-studied pathogens *Salmonella* and *Campylobacter* as examples to illustrate the manifold ways how the microbiota and the host interact with intestinal pathogenic bacteria.

14.2 Specific Pathogens of the (Gastro)Intestinal Tract: Common Traits, Examples and Specific Properties in the Face of a Resident Microbiota

Intestinal bacterial pathogens, which are the focus of this chapter, cause for the most part acute rather than chronic infections. Acute pathogens comprise mostly strongly pro-inflammatory bacteria such as different *E. coli* pathotypes, *Salmonella enterica* serovars (Grassl and Finlay 2008), *Shigella*, *Yersinia* spp., other enterobacterial pathogens, *Clostridium difficile*, and *Vibrio cholerae*. Among the fewer, more chronic, human pathogens are enterotoxigenic *Bacteroides fragilis*, some *Campylobacter* sp., and *Listeria monocytogenes*. Under certain circumstances such as severe dysbiosis after antibiotic treatment, *Clostridium difficile* in the presence of its toxin can act as an acute severe pathogen, while it does not normally infect or multiply in the intestinal tract under microbiota-replete conditions.

14.2.1 Metabolic Competition of Pathogens with Commensal Bacteria (Fig. 14.2)

Salmonella enterica serovars can cause gastrointestinal disease (non-typhoidal *Salmonella* (NTS) serovars such as *S.* Typhimurium or *S.* Enteritidis) or invasive, systemic disease or typhoid fever (*S.* Typhi and *S.* Paratyphi). Some NTS strains (e.g., *S.* Typhimurium sequence type ST313) (Ramachandran et al. 2017; Okoro et al. 2012) can also cause invasive life-threatening disease (iNTS) especially when underlying comorbidities (malnutrition) or coinfections (HIV) are present.

Intestinal pathogens have to temporarily outcompete commensal bacteria in order to cause infection and disease. In addition to dedicated host interaction modules such as the pathogenicity island-encoded type III secretion systems (van der Heijden and Finlay 2012; Jennings et al. 2017), *Salmonella* possesses several mechanisms to overcome the CR provided by the complex microbiota, which are partially based on specific metabolic traits.

Early after ingestion of the bacteria with contaminated food and in the absence of inflammation, *Salmonella* Typhimurium expresses a nickel-iron hydrogenase (*hyb*) in the intestinal lumen which allows it to utilize molecular hydrogen (produced by commensals) and a fumarate reductase (*frdA*) to convert fumarate, a metabolite released by certain members of the microbiota, to succinate. Succinate is an excellent energy source for *Salmonella* which thus can grow to high numbers in the intestinal lumen (Maier et al. 2013; Sassone-Corsi and Raffatellu 2013). Free succinate in the intestine is also made as a fermentation product by commensal bacteria such as *Bacteroides thetaiotaomicron*. However, other commensals use succinate as an energy source, and therefore, in a gut colonized with a complex microbiota, succinate levels in the intestinal lumen are normally low. However, upon disturbance of the microbiota, for example, by antibiotic treatment or upon inflammation, succinate levels are elevated. It was recently shown that succinate is taken up by *Salmonella* and enters the tricarboxylic acid (TCA) cycle so that *Salmonella* expand and outcompete commensal strains (Spiga et al. 2017).

After expansion in the gut lumen, *Salmonella* travels through the mucus to the epithelium to invade the host tissue. Adhesion to and invasion into host cells result in the production of pro-inflammatory cytokines, chemoattraction of

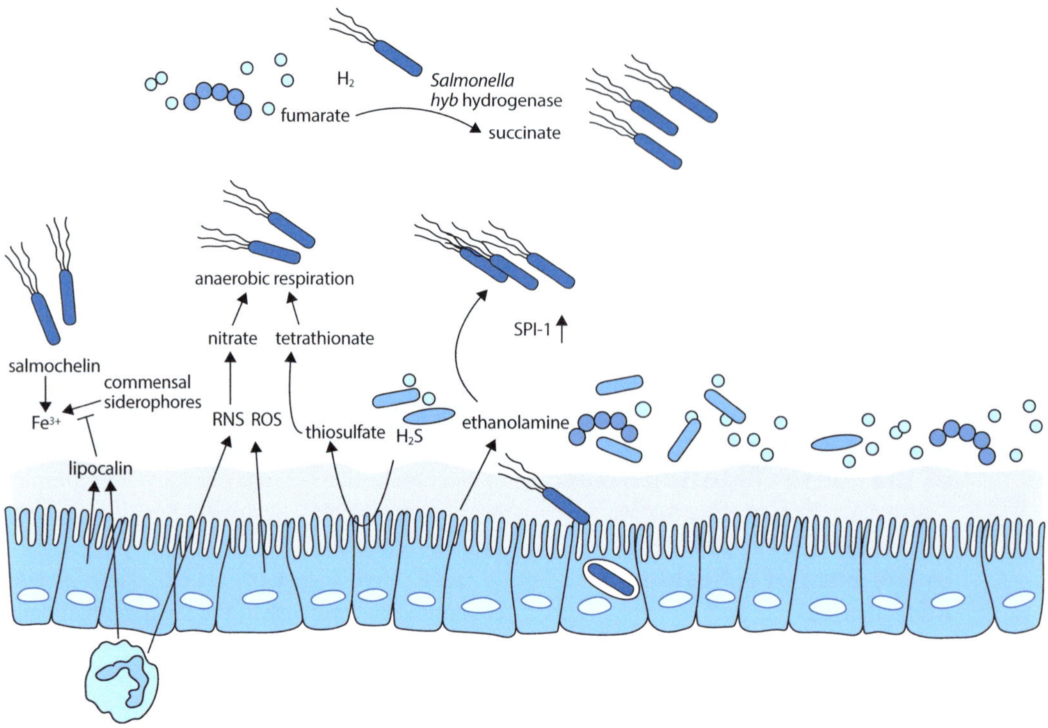

Fig. 14.2 This figure illustrates the metabolic interplay between commensal bacteria and the intestinal pathogen *Salmonella* as one example of interaction between the host, the microbiota, and a pathogen. Pathogens, as does the commensal microbiota, need nutrients, and their niche adaptation is geared toward the acquisition of those from all components of the niche. In the anaerobic environment of the intestinal lumen, *Salmonella* uses nickel-iron hydrogenase (*hyb*) to utilize molecular hydrogen, providing energy for *Salmonella* to multiply. The host cells produce lipocalin-2 which sequesters siderophores from commensal bacteria, thereby limiting their access to iron. However, *Salmonella*'s siderophore salmochelin cannot be bound by lipocalin-2 giving *Salmonella* priority access to iron. During inflammation, *Salmonella* also triggers the production of reactive oxygen species (ROS) and reactive nitrogen species (RNS) leading to the formation of alternative electron acceptors for *Salmonella*. SPI-1—*Salmonella* pathogenicity island-1

immune cells, and pathological tissue change. *Salmonella*-induced inflammation is characterized by the destruction of tight junctions between epithelial cells and the influx of leukocytes (neutrophils, macrophages, and dendritic cells) and lymphocytes and destruction of the normal gut wall architecture.

The metal ions iron and zinc are essential micronutrients for bacteria. During inflammation, host cells (mainly neutrophils) produce large amounts of lipocalin-2 which is able to sequester bacterial siderophores (iron carrier molecules) bound to iron. In addition, host cells secrete calprotectin to sequester free zinc (Behnsen et al. 2014). Thereby the host restricts access to these metal ions for bacteria. *Salmonella* circumvents these host defense mechanisms and outcompetes commensal bacteria by producing a modified siderophore termed salmochelin that cannot be bound by lipocalin-2 (Raffatellu et al. 2009) and by expressing a high-affinity Zn^{2+} transporter (Liu et al. 2012).

Another host defense against intracellular bacteria is the production of reactive nitrogen intermediates (RNI) and reactive oxygen intermediates (ROI). Yet again, *Salmonella* has developed countermeasures to take advantage of these defenses: A fraction of *Salmonella* bacteria reside intracellularly in the so-called *Salmonella*-containing vacuole (SCV) where they are protected from RNIs and ROIs. Furthermore, *Salmonella* can utilize metabolites made downstream of reactive intermediates for respiration or its own nutrition: for instance, reactive oxygen species can oxidize thiosulfate (a host metabolite derived from commensal-produced H_2S) to tetrathionate which is used by *Salmonella* as an electron acceptor for anaerobic respiration (Winter et al. 2010; Winter and Bäumler 2011). Under these circumstances, *Salmonella* can also use host-derived ethanolamine and microbiota-derived 1,2-propanediol as carbon sources (Price-Carter et al. 2001; Thiennimitr et al. 2011; Faber et al. 2017). Similarly, a by-product of RNI metabolism in the presence of microbiota is nitrate (NO_3) which can also be used by *Salmonella* for anaerobic respiration (Lopez et al. 2012).

14.2.2 The Role of Intestinal Glycans in the Interplay Between Microbiota, Host and Pathogens (Fig. 14.3)

Glycans are abundant in the intestinal tract and shape the interplay between resident microbiota, the host, and incoming pathogens. Both the epithelial surface and the mucus layers are heavily glycosylated. About 80% of the total mucus mass is due to (O-linked) glycosylation. In contrast, the epithelial glycocalyx consists of mostly N-linked glycoproteins and glycolipids. While there is only a single mucus layer in the small intestine, the large intestine is covered by an inner, dense mucus layer which is normally devoid of bacteria, followed by an outer loose mucus layer which is heavily colonized by commensals (Johansson et al. 2011; Ermund et al. 2013). Mucus serves as a barrier to protect the epithelium from luminal bacteria, pathogens, and commensals. However, pathogens have ways to penetrate the mucus, either by active movement through flagella or by targeting M cells (specialized epithelial cells that are located in the follicle-associated epithelium overlying Peyer's patches, which are not covered by a thick mucus layer). In addition, some pathogen-produced toxins can damage epithelial cells so that less mucus is produced (McGuckin et al. 2011). Apart from their barrier role, the glycosylated mucins that make up the mucus layers can serve as carbon and energy sources and also as adhesion factors for bacteria (Tailford et al. 2015; Ringot-Destrez et al. 2017; Juge 2012). In addition, the glycosylation patterns on intestinal epithelial cells and in the mucus can influence microbiota composition and thus modulate the host's susceptibility to infection and inflammation.

In order to invade and manipulate host cells, *Salmonella* has to get into close contact to the cells. For adhesion to mammalian cells or to the mucus, the bacteria use various extensions such as flagella, fimbriae, and pili, most of which bind to sugar residues on host glycoproteins or glycolipids. Enteric pathogens provide several excellent examples how bacteria can bind to terminal sugar residues on complex glycans. For example, *Salmonella* and other *Enterobacteriaceae* express type I fimbriae which bind to mannosylated glycans. In addition, *Salmonella* binds to fucose via StdA fimbriae (Chessa et al. 2009) and to GlcNAc and sialic acid via the adhesin SiiE (Gerlach et al. 2007; Wagner et al. 2014).

While some commensal bacteria can degrade complex glycans into shorter-chain sugars, many other commensal and most pathogenic bacteria can only utilize mono- or disaccharides. For example, the commensal *Bacteroides* sp. can hydrolyze complex glycans and take up and metabolize liberated sugars, while *Salmonella* utilizes microbiota-derived free fucose and sialic acid (Ng et al. 2013).

Histo-blood group antigens are widely expressed by the intestinal epithelium. Intestinal fucosylation (via fucosyltransferase (Fut)2) was shown to have a major impact on host susceptibility to various infections. For example, Fut2 deficiency protects from human *Norovirus* and *Helicobacter pylori* infections but increases the risk for developing Crohn's disease (Lindesmith et al. 2003; Azevedo et al. 2008; McGovern et al. 2010). While *fut2* is constitutively expressed in the large intestine, it is inducible by bacterial products in the ileum (Fig. 14.2) (Pickard et al. 2014): pathogens and commensals induce IL-23 production in dendritic cells (DCs) in the lamina propria, and IL-23 then induces IL-22 production

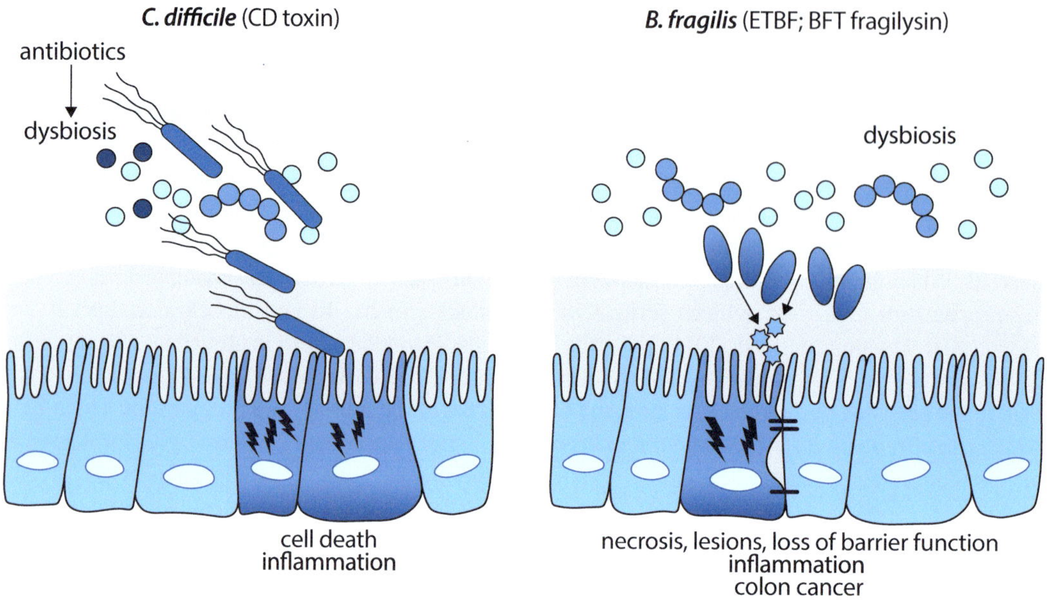

Fig. 14.3 One possibility to enhance pathogen survival in the gut are bacterial toxins, as illustrated in this figure. Toxin-mediated effects by the bacterial pathogens *C. difficile* and enterotoxigenic *B. fragilis* are given as examples. Intensive antibiotic treatment leads to the expansion of *C. difficile*, since the microbiota is severely depleted. Secretion of *Clostridium* toxins induces cell death and increased inflammation. *B. fragilis* toxin (BFT) cleaves E-cadherin and thereby destroys tight junctions, resulting in a leaky mucosal barrier and inflammation. Toxin effects by enteropathogens may ultimately contribute to tumorigenesis. ETBF—enterotoxigenic *B. fragilis*

in ILC-3 cells. Finally, ILC-3-derived IL-22 triggers *fut2* and antimicrobial peptide (AMP) expression by epithelial cells (Goto et al. 2014). Intestinal epithelial fucosylation can block *Salmonella* colonization (early after infection). However, how exactly epithelial fucosylation blocks *Salmonella* infection remains unclear (Goto et al. 2016). Others have shown that *Salmonella* possesses fimbriae that can bind to fucose (Chessa et al. 2009) and *Salmonella* can efficiently use fucose as a carbon and energy source (Ng et al. 2013). One mechanism how fucose can block a pathogen was demonstrated for another pathogen, enterohemorrhagic *E. coli* (EHEC): commensal-liberated L-fucose inhibits EHEC virulence gene expression (Pacheco et al. 2012).

Another intestinal blood group-related glycosyltransferase is *b4galnt2* which adds an N-acetylgalactosamine (GalNAc) as a terminal sugar onto glycoproteins and glycolipids (Dall'Olio et al. 2014). By modifying microbiota composition, *b4galnt2*-modified sugars increase the susceptibility to *Salmonella* infections (Rausch et al. 2015; Staubach et al. 2012). Other histo-blood group antigens serve as receptors for adhesion of other intestinal pathogens; e.g., Shiga toxin-producing *E. coli* binds to H type I and sulfated H type II blood group antigens via its F18 fimbriae (Moonens et al. 2012; Coddens et al. 2009). *Campylobacter* binds terminal fucose on H antigen (Ruiz-Palacios et al. 2003) (see also below).

Inflammation helps *Salmonella* to access glycans as energy sources. For example, during inflammation, motility allows *Salmonella* and other intestinal pathogens such as *Campylobacter* sp. to access inflammation-induced mucin components (Stecher et al. 2008). Antibiotic treatment leads to an increase in oxidized sugar products such as glucarate or galactarate. *Salmonella* can utilize these compounds and expand in their presence (Faber et al. 2016).

The enteric pathogens *Campylobacter* sp. of the *Epsilonproteobacteria*, represented by the species *C. jejuni* and *C. coli*, are less pro-inflammatory in humans but share a similar intestinal niche with *Salmonella* and seem to overlap in a few important characteristics with the enteric salmonellae as well, which will be illustrated in the next paragraph.

First of all, the resident microbiota is being strongly engaged to modulate the specific lifestyle of *Campylobacter*, as it is for *Salmonella*. Intestinal *Campylobacter* species colonize predominantly the intestinal mucus and intestinal crypts such as those of the chicken cecum (Beery et al. 1988). In the chicken gut, they cause no or rather mild symptoms (Humphrey et al. 2014); however, they become acutely pathogenic and can cause severe symptoms when they colonize the human bowel in the jejunum and colon (Young et al. 2007). Although *C. jejuni* are effectively separated from the bulk of the microbiota by mucus and their mucus-invading motility, their metabolism seems to be similarly well adjusted to the interplay between those pathogens, microbiota, and the host. Therefore, the microbiota is an important modulator of *Campylobacter* infection. Although we would like to focus here on *Campylobacter* as another important paradigm for pathogen-glycan interactions, it is worthwhile mentioning that *Campylobacter* are also able to source their metabolism and respiratory chain from various by-products/metabolites of a complex microbiota. One similar example as for the salmonellae (see above) includes the use of microbiota-derived tetrathionate as an electron acceptor for anaerobic respiration, together with the substrate ethanolamine (Liu et al. 2013). The interplay between host, microbiota, and specific pathogen is also illustrated by the fact that *C. jejuni* is metabolically very malleable, probably not only to adjust to different hosts but also to be able to adapt to different microbiota types, fluctuations, and compositions.

The importance of sugars for the *Campylobacter* intestinal lifestyle is substantial, both on the bacterial and on the host side (Day et al. 2012; Szymanski and Gaynor 2012). In addition to specific microbiota-derived glycans (mainly mono- and disaccharides) (Gripp et al. 2011; Stahl et al. 2011; Wagley et al. 2014; Vorwerk et al. 2015), some of the metabolites that are most likely sourced from the microbiota and used for *C. jejuni* metabolism in vivo are short-chain fatty acids (SCFA) such as propionate, butyrate, and acetate (Gripp et al. 2011; Molnár et al. 2015; Awad et al. 2016; Cresci et al. 2017), as well as dipeptides (breakdown products of proteins by the microbiota) (Gripp et al. 2011; Vorwerk et al. 2014). Intestinal *Campylobacter* species have been demonstrated to display various glycans on their surface and possess, which is unique among bacteria, both active N- and O-glycosylation gene clusters (Young et al. 2002; Szymanski and Gaynor 2012). Glycans of pathogens or chronic colonizers such as campylobacters are frequently mimicking human glycans as is the case for sialic acids (Guerry et al. 2000; Gilbert et al. 2002; Bax et al. 2011; Szymanski and Gaynor 2012). This kind of "molecular mimicry" is used by campylobacters to become rather invisible to the host immune system and may even act in a tolerogenic manner (Perdicchio et al. 2016). Some *C. jejuni* surface glycans have been shown to bind various human lectins (Kilcoyne et al. 2014; Stephenson et al. 2014; Lu et al. 2015; Phongsisay et al. 2015; Turonova et al. 2016). A few reports describe that *Campylobacter* glycans can functionally interact with human lectins, for instance, with Siglec-10 (Stephenson et al. 2014) or Siglec-7 (Avril et al. 2006; Heikema et al. 2013), which can lead to an increase in tolerogenic IL-10 (Stephenson et al. 2014). Sialoadhesin was also shown to bind to *C. jejuni* (Klaas et al. 2012). One may speculate that some lectin interactions influence immune polarization and, under certain circumstances, also induce immunological tolerance. However, more definitive data are required to strengthen this hypothesis. *Campylobacter* glycans can also bind to host glycans directly with a high affinity, without the need for lectin intermediaries on the host side or other protein adhesins on the bacterial side. This has been revealed using glycan arrays of *Campylobacter* surface sugars (Day et al. 2013, 2015).

Recently, two large transposon mutant studies which have explored the viability of *C. jejuni* mutants in different environments have expanded the range of established niche-adaptation factors of intestinal campylobacters (Gao et al. 2017; de

Vries et al. 2017), which also include some of those described above. Concerning the role of glycans, fucose uptake, and metabolism (Stahl et al. 2011; Muraoka and Zhang 2011), general glycosylation pathway and capsule synthesis were globally reported to be essential for mouse colonization (Gao et al. 2017). For chicken colonization, capsule biosynthesis, general glycosylation, and several lipooligosaccharide (LOS) glycosylation enzymes, but not flagellar glycosylation, were found to be required (de Vries et al. 2017). Interestingly, metabolome analysis (isotopologue profiling) in one of these recent studies showed that *C. jejuni* is able to recycle glycans (mainly glucose and galactose) of its outer shell derived from LOS and capsular polysaccharide (CPS) surface glycans, probably serving as a nutrient reservoir (Gao et al. 2017). In the chicken gut, one main niche of chronic intestinal *Campylobacter* sp. colonization, mucins seem to contain an abundance of alpha-1,2-fucose and sulfated O-glycans (Struwe et al. 2015), much more than human intestinal mucus. However, this composition in the human or chicken intestinal mucus may vary by the influence of dietary factors, microbiota, or other pathogens, which needs to be clarified. It is well known that specific intestinal terminal glycans such as alpha-1,2-fucose represent one main adherence factor for intestinal campylobacters (Day et al. 2009, 2013; Ruiz-Palacios et al. 2003). The differential expression of these glycans in different spatial regions of the intestine appears to inhibit the adherence and survival of *C. jejuni* much more in the lower versus the upper intestine (Day et al. 2009; Struwe et al. 2015). The overall number of glycans correlated inversely with the density of *C. jejuni* in vivo, and the presence of sulfated glycans may explicitly be required for inhibitory activities (Struwe et al. 2015). These environmentally abundant glycan cues may also explain the propensity of some strains to readily take up and metabolize host- and microbiota-derived fucose as a nutrient (Gripp et al. 2011; Stahl et al. 2011). *Campylobacter* glycans influence colonization and host immune responses by lectin interactions (Iovine et al. 2008; van Sorge et al. 2009; Jervis et al. 2012; Phongsisay 2016). The *Campylobacter* mimicry of human sugars is also likely to contribute to the detrimental paralytic disease Guillain-Barré syndrome in humans, by inducing antibody formation against human glycosphingolipids, for example, against the ganglioside GM1 (Godschalk et al. 2004; Bax et al. 2011; Phongsisay 2016; Lardone et al. 2016) or di-sialylated ganglioside mimics (Stephenson et al. 2013). Very recently, it has been shown that microbiota transplantation in humans may increase the formation of autoantibodies which induce such neurological problems in the presence of *Campylobacter* (Brooks et al. 2017). Interestingly, St Charles and colleagues have reported that antibiotic treatment leading to dysbiosis can also exacerbate Guillain-Barré-like symptoms in a *Campylobacter* mouse model (St Charles et al. 2017; Esan et al. 2017). In this respect, establishing markers for *Campylobacter* strains that are prone to expressing these glycans or inducing human autoantibodies in healthy or dysbiotic patients will be very important in the future to prevent these sequelae but has not been achieved yet. It will also be useful to better understand the impact of host glycan alterations during low- or high-level inflammation on the host interaction potential, metabolism, and pathogenesis of *Campylobacter* and other intestinal pathogenic bacteria.

We expect that metabolism- and glycan-dependent pathogen-microbiota crosstalk may be important for many other intestinal pathogens as well and deserves further intensive investigations. It is also anticipated that the discovery of novel intestinal pathogens or pathobionts will be facilitated by the ongoing extensive efforts to clarify microbiota composition and its contributions to health and disease.

14.2.3 Direct Competitive Behavior in Pathogen-Microbiota Interaction via Type VI Secretion Systems

Commensals and pathogens in the intestinal tract not only crosstalk with the host (Vogt et al. 2015) but also effectively communicate with each other

(Lustri et al. 2017). One additional, interesting aspect of microbe-host and microbe-pathogen interplay in the intestinal tract is the competitive behavior between different members of the gut microbiota. The competition for nutrients in the intestine is thought to be high, which raises the question of competitive behavior, in particular of closely related species. The mechanisms of competition can rely on different bacterial factors, the most important being metabolic competition (see above for examples), bacteriophages (Mills et al. 2013) and toxin-antitoxin systems. Type VI secretion systems (T6SS) are complex secretion systems in the outer membrane of Gram-negative bacteria (Cascales and Cambillau 2012) which have recently attracted substantial research activities. Apart from their effects on host cells, they are frequently being used to transport toxins or other bactericidal effectors into other, mostly closely related, *Campylobacter* spp. bacterial cells (Basler et al. 2013). Various intestinal bacterial species, pathogens and nonpathogens, including *Vibrio*, *Pseudomonas*, *Salmonella*, *Shigella*, and *Bacteroides* ssp., express T6SS and partly use them for competition against the microbiota (Pukatzki et al. 2007; Ho et al. 2014; Cianfanelli et al. 2016; Sana et al. 2016; Joshi et al. 2017; Anderson et al. 2017; Allsopp et al. 2017). Recently, a comprehensive overview study of the content of T6SS-bearing microbes in the human intestine and the influence of T6SS on human microbiome composition in children and adults was performed (Verster et al. 2017). Although far from delivering final conclusions, the article states that T6SS are enriched in the intestinal tract and have a substantial influence on the microbiota composition and that this influence seems particularly important during the intestinal microbiota maturation phase in children.

14.2.4 Toxin-Mediated Effects of Intestinal Pathogens on the Host (Fig. 14.4)

Some gut pathogens secrete toxins to manipulate host cells in order to destroy the epithelial barrier and improve their colonization and transmission. For instance, enterotoxigenic *Escherichia coli* (ETEC) (Madhavan and Sakellaris 2015) or some enterohemorrhagic-like *E. coli* (EHEC) produce various potent enterotoxins, including Shiga toxin (Stx). Interestingly, microbiota and diet-mediated microbiota modulation impacted on EHEC colonization levels, Stx tissue binding, and downstream toxin effects in an animal model (Zumbrun et al. 2013). *Vibrio cholerae* disease symptoms of watery diarrhea are largely caused by cholera toxin, which is encoded by a lysogenic bacteriophage (ctx Φphage). Likewise, the less-known pathogenic bacterium enterotoxigenic *Bacteroides fragilis* (ETBF) makes *B. fragilis* toxin (BFT), and *Clostridium difficile* produces clostridial toxins A and B. Some strains produce a binary toxin called *C. difficile* transferase (CDT) (Cowardin et al. 2016; Gerding et al. 2014) which appears to contribute independently to disease symptoms. *C. jejuni*, some intestinal pathogenic *E. coli* strains, and enterohepatic *Helicobacter* species produce a different toxin, cytolethal distending toxin (CDT), which possesses a DNA-degrading enzymatic activity (Faïs et al. 2016). These toxins damage epithelial cells and the tight junctions by yet largely unknown mechanisms and deplete certain cell types. Destroying goblet cells leads to a loss of mucus production and mucosal barrier dysfunction. Various toxin effects result in a leaky mucosa and make the underlying tissue accessible to the pathogens and commensal bacteria, thus increasing local inflammation and invasion. BFT destroys tight junctions between epithelial cells and cleaves E-cadherin, causing barrier dysfunction and colitis, and ultimately promotes tumorigenesis (Sears 2009; Wu et al. 2009). *C. difficile* toxins A and B glycosylate Rho proteins, thereby manipulating host cell cytoskeleton and compromising epithelial integrity. In many intestinal infections in conjunction with the resident microbiota, the full extent and mechanisms of these toxin effects are not yet understood and will require more detailed investigations. This set of open questions also concerns the potential effects of the toxins on components of the microbiota, which has not been studied so far.

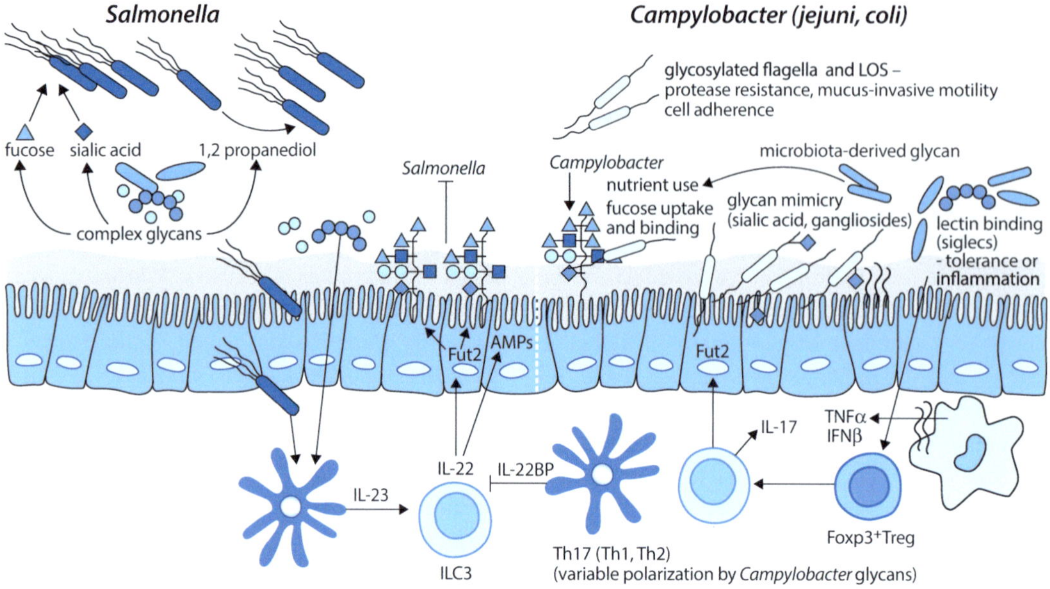

Fig. 14.4 Interaction between host glycans, the microbiota, and pathogenic bacteria is one major factor to shape the complex interplay in the intestinal niche and to provide nutrients to the well-adapted inhabitants. *Salmonella* and *Campylobacter* are given as prominent examples of pathogenic bacteria which make use of the glycan-rich environment for their own proliferation. For instance, both bacterial pathogens, *Salmonella* and *Campylobacter*, cannot degrade complex glycans present in the mucus on their own. However, after cleavage of complex glycans by commensal bacteria, various resulting simple sugars (monosaccharides and disaccharides) are readily taken up and metabolized by the pathogens, serving as energy sources. The pathogens and commensal bacteria can trigger cytokine production by host immune cells in a direct and indirect manner, which, in turn, can stimulate terminal fucosylation of epithelial glycans. Those and the production of antimicrobial peptides were, on one hand, shown to block *Salmonella* expansion and contribute to host defense mechanisms. On the other hand, terminal fucoses can also be utilized as nutrients by the pathogens, in particular by *Campylobacter*. Furthermore, *Campylobacter* uses variable glycosylation of its envelope as molecular mimicry, hiding from the host's immune system. For instance, lectin binding of *Campylobacter* can lead to the induction of immune tolerance by IL-10 production

14.3 CR Against Enteric Pathogens, Environmental Interference with CR and Intervention Strategies When CR Fails

The first reports showing that a complex microbiota confers colonization resistance (CR) against intestinal pathogens, including the abovementioned species, date back to the 1950s (Bohnhoff et al. 1954, 1964; Miller et al. 1954). CR can be provided by several direct and indirect mechanisms. Direct mechanisms include the production of antimicrobial substances, e.g., bacteriocins or microcins (Hegarty et al. 2016; Sassone-Corsi et al. 2016), or direct killing via T6SS (see above). Epithelial cells can be directly activated by commensal bacteria or their products such as flagellin and LPS to upregulate the production and release of AMPs and reactive oxygen and nitrogen species, which can kill invading bacteria.

Indirect mechanisms of CR include the competition for the same nutrient niche, the metabolic conversion of primary into secondary bile acids, as well as the stimulation of the host immune system and of the epithelium to produce mucus and antimicrobial peptides. Bacteriocins as an additional, direct mechanism are bacteria-produced active protein toxins with microbicidal activity, and most bacteriocins directly target a narrow set of related bacteria (Cotter et al. 2013). Harnessing the power of bacteriocins may lead to the development of a novel set of narrow-spectrum antibiotics. Additional metabolites such as the microbiota-produced SCFA can suppress virulence gene expression of pathogens, e.g., butyrate suppresses *Salmonella* SPI-1 gene expression (Gantois et al. 2006). In addition, butyrate is also a major energy source for epithelial cells and thus constitutes an important factor supporting a healthy epithelial barrier. Last but not least, these secondary metabolites shape the host immune status (Alvarez-Curto and Milligan 2016) which is important for tissue homeostasis and adequate defense against pathogens. These factors and their impact on pathogen CR may also be influenced by dietary components.

Antibiotic treatment is frequently a necessity and however provides one important, although unintended, environmental interference mechanism toward CR. Antibiotics, on one hand, serve to deplete pathogens but in all cases also alter the resident microbiota, even after only one or two courses. The changes inflicted by antibiotics on the microbiota may be pervasive and permanent (Buffie and Pamer 2013). In particular if administered in early childhood, several courses of antibiotics are assumed to lead to an irreversible loss of beneficial microbes (Cho et al. 2012; Blaser 2016; Kim et al. 2017; Olsan et al. 2017). This state which is accompanied by other general changes in the body (metabolic, immunological) is termed dysbiosis. This reduction in numbers and diversity of resident bacteria also reduces beneficial effects of CR and renders patients more susceptible on the short or even long term for infections such as common acute diarrheal pathogens (Kampmann et al. 2016), but also for *C. difficile* or *Enterococcus faecalis* infection. Therapeutically, some of these infections can be efficiently addressed by restoring the complex microbiota using fecal microbiota transplantation (FMT) (see Chap. 20). However, transferring a complex donor microbiota carries the risk for co-transferring other pathogens or non-infectious disease states such as metabolic syndrome or obesity (Baxter and Colville 2016). Therefore, there is an urgent need to define which set of harmless commensal strains are required to provide effective resistance to a specific infection and can be utilized for successful and long-lasting beneficial intervention strategies. Recently, several communities sufficient to confer resistance against intestinal infections with *S.* Typhimurium (Brugiroux et al. 2016) or *C. difficile* (Buffie et al. 2015; Lawley et al. 2012) have been defined. These studies are opening new ways to reconstitute the microbiome in several disease states with targeted interventions by restoring the "missing" commensal strains. Most likely, protective strains will vary depending on the patients' genetics and their endogenous microbiota. A couple of excellent recent reviews highlight the role of the resident microbiota for CR (Olsan et al. 2017; Lawley and Walker 2013; Stecher and Hardt 2011; Stecher et al. 2013; Yurist-Doutsch et al. 2014; Kim et al. 2017; Buffie and Pamer 2013).

Looking into the future, the term "precision engineering or editing of microbiome" has been coined as one arm of an individualized medicine approach and intervention strategy (Buffie et al. 2015), directly effective against pathogens or indirectly while modulating the immune response (Montalban-Arques et al. 2015). Two very recent examples do not directly add microbes but modulate microbiota indirectly by different metabolites or ions: One example is the unintended effect by the frequent food additive trehalose on microbiota and *C. difficile* (Collins et al. 2018), and the second one is the intentional reduction (precision editing) of *Salmonella* and other *Enterobacteriaceae* within the microbiota by tungstate to inhibit the bacterial molybdenum enzymes involved in inflammation-induced metabolism (Zhu et al. 2018). The approach of microbiota editing also includes the prospect to develop single bacterial species which can be protective or may even be used in a therapeutic manner against infectious and noninfectious, dysbiosis-triggered diseases.

▶ Controversy

1. To what extent do host genetics or the microbiome, respectively, influence susceptibility to intestinal infections?
2. What is the role of interaction between microbiome, virome, fungome, archaeome, and the host for bacterial/viral/fungal infections?

3. Can we define a universal core consortium of microbes providing CR against all or specific pathogens? Can we single out specific microbes/bacteria with particularly beneficial effects in certain disease states?
4. How can the host-specific contributions of immune response and microbiota be better defined for disease outcome of host-specific pathogens?
5. Are there direct effects of known bacterial (entero)toxins on commensals?
6. How does bacterial and inter-kingdom communication within the gut microbiota work and how effective is it?

History

1. Original description of intestinal microbiota and their specific functions (expanded in the 1980s); first descriptions of intestinal bacteria even originate from the end of the nineteenth century (Nuttal and Thierfelder 1895).
2. Description of the essential role of microbiota for intestinal and immune homeostasis (from 2000 to present).
3. The advent of "culturomics," which means the increase of culturable microbiota components/strains from human and animal intestine (from 2010 to present).
4. Modern revival of the microbiota transplant approaches in *Clostridium difficile* colitis and the strengthened concept, mainly underscored by germ-free animal models, that the microbiota in itself, in particular the intestinal resident microbiota, constitute an essential organ of the human body and are involved in CR (2010 to present).

Highlights

1. A complex microbiota confers (colonization) resistance to a variety of infectious agents, including pathogenic bacteria and viruses.
2. Restoring microbiota complexity can be used to treat infections and restore colonization resistance, providing novel strategies of disease intervention
3. Intestinal pathogens use the presence of microbiota, in particular of microbiota-derived metabolites, to improve their own nutrient acquisition and compete with residents
4. Sugars/glycans formed by the intestinal tissues and influenced by microbiota, diet, and inflammation are widely used by pathogens as adhesins and nutrient resources.

References

Allsopp, L. P., Wood, T. E., Howard, S. A., Maggiorelli, F., Nolan, L. M., Wettstadt, S., & Filloux, A. (2017). RsmA and AmrZ orchestrate the assembly of all three type VI secretion systems in Pseudomonas aeruginosa. *Proceedings of the National Academy of Sciences of the United States of America, 114*, 7707–7712.

Alvarez-Curto, E., & Milligan, G. (2016). Metabolism meets immunity: The role of free fatty acid receptors in the immune system. *Biochemical Pharmacology, 114*, 3–13.

Anderson, M. C., Vonaesch, P., Saffarian, A., Marteyn, B. S., & Sansonetti, P. J. (2017). Shigella sonnei encodes a functional T6SS used for interbacterial competition and niche occupancy. *Cell Host and Microbe, 21*, 769–776.e3.

Arike, L., & Hansson, G. C. (2016). The densely O-glycosylated MUC2 mucin protects the intestine and provides food for the commensal bacteria. *Journal of Molecular Biology, 428*, 3221–3229.

Arumugam, M., Raes, J., Pelletier, E., Le Paslier, D., Yamada, T., Mende, D. R., Fernandes, G. R., Tap, J., Bruls, T., Batto, J.-M., et al. (2011). Enterotypes of the human gut microbiome. *Nature, 473*, 174–180.

Avril, T., Wagner, E. R., Willison, H. J., & Crocker, P. R. (2006). Sialic acid-binding immunoglobulin-like lectin 7 mediates selective recognition of sialylated glycans expressed on Campylobacter jejuni lipooligosaccharides. *Infection and Immunity, 74*, 4133–4141.

Awad, W. A., Dublecz, F., Hess, C., Dublecz, K., Khayal, B., Aschenbach, J. R., & Hess, M. (2016). Campylobacter jejuni colonization promotes the translocation of

Escherichia coli to extra-intestinal organs and disturbs the short-chain fatty acids profiles in the chicken gut. *Poultry Science, 95*, 2259–2265.

Azevedo, M., Eriksson, S., Mendes, N., Serpa, J., Figueiredo, C., Resende, L. P., Ruvoën-Clouet, N., Haas, R., Borén, T., Le Pendu, J., et al. (2008). Infection by Helicobacter pylori expressing the BabA adhesin is influenced by the secretor phenotype. *The Journal of Pathology, 215*, 308–316.

Bansil, R., & Turner, B. S. (2017). The biology of mucus: Composition, synthesis and organization. *Advanced Drug Delivery Reviews, 124*, 3–15.

Basler, M., Ho, B. T., & Mekalanos, J. J. (2013). Tit-for-tat: Type VI secretion system counterattack during bacterial cell-cell interactions. *Cell, 152*, 884–894.

Bax, M., Kuijf, M. L., Heikema, A. P., van Rijs, W., Bruijns, S. C. M., García-Vallejo, J. J., Crocker, P. R., Jacobs, B. C., van Vliet, S. J., & van Kooyk, Y. (2011). Campylobacter jejuni lipooligosaccharides modulate dendritic cell-mediated T cell polarization in a sialic acid linkage-dependent manner. *Infection and Immunity, 79*, 2681–2689.

Baxter, M., & Colville, A. (2016). Adverse events in faecal microbiota transplant: A review of the literature. *The Journal of Hospital Infection, 92*, 117–127.

Beery, J. T., Hugdahl, M. B., & Doyle, M. P. (1988). Colonization of gastrointestinal tracts of chicks by Campylobacter jejuni. *Applied and Environmental Microbiology, 54*, 2365–2370.

Behnsen, J., Jellbauer, S., Wong, C. P., Edwards, R. A., George, M. D., Ouyang, W., & Raffatellu, M. (2014). The cytokine IL-22 promotes pathogen colonization by suppressing related commensal bacteria. *Immunity, 40*, 262–273.

Birchenough, G. M. H., Johansson, M. E. V., Gustafsson, J. K., Bergström, J. H., & Hansson, G. C. (2015). New developments in goblet cell mucus secretion and function. *Mucosal Immunology, 8*, 712–719.

Blaser, M. J. (2016). Antibiotic use and its consequences for the normal microbiome. *Science, 352*, 544–545.

Blaut, M., & Clavel, T. (2007). Metabolic diversity of the intestinal microbiota: Implications for health and disease. *The Journal of Nutrition, 137*, 751S–755S.

Bohnhoff, M., Drake, B. L., & Miller, C. P. (1954). Effect of streptomycin on susceptibility of intestinal tract to experimental Salmonella infection. *Proceedings of the Society for Experimental Biology and Medicine, 86*, 132–137.

Bohnhoff, M., Miller, C. P., & Martin, W. R. (1964). Resistance of the mouse's intestinal tract to experimental Salmonella infection. I. Factors which interfere with the initiation of infection by oral inoculation. *The Journal of Experimental Medicine, 120*, 805–816.

Brooks, P. T., Brakel, K. A., Bell, J. A., Bejcek, C. E., Gilpin, T., Brudvig, J. M., & Mansfield, L. S. (2017). Transplanted human fecal microbiota enhanced Guillain Barré syndrome autoantibody responses after Campylobacter jejuni infection in C57BL/6 mice. *Microbiome, 5*, 92.

Browne, H. (2016). Antibiotics, gut bugs and the young. *Nature Reviews. Microbiology, 14*, 336.

Brugiroux, S., Beutler, M., Pfann, C., Garzetti, D., Ruscheweyh, H.-J., Ring, D., Diehl, M., Herp, S., Lötscher, Y., Hussain, S., et al. (2016). Genome-guided design of a defined mouse microbiota that confers colonization resistance against Salmonella enterica serovar Typhimurium. *Nature Microbiology, 2*, 16215.

Buck, M. D., Sowell, R. T., Kaech, S. M., & Pearce, E. L. (2017). Metabolic instruction of immunity. *Cell, 169*, 570–586.

Buffie, C. G., & Pamer, E. G. (2013). Microbiota-mediated colonization resistance against intestinal pathogens. *Nature Reviews. Immunology, 13*, 790–801.

Buffie, C. G., Bucci, V., Stein, R. R., McKenney, P. T., Ling, L., Gobourne, A., No, D., Liu, H., Kinnebrew, M., Viale, A., et al. (2015). Precision microbiome reconstitution restores bile acid mediated resistance to Clostridium difficile. *Nature, 517*, 205–208.

Cascales, E., & Cambillau, C. (2012). Structural biology of type VI secretion systems. *Philosophical Transactions of the Royal Society of London. Series B, Biological Sciences, 367*, 1102–1111.

Chessa, D., Winter, M. G., Jakomin, M., & Baumler, A. J. (2009). Salmonella enterica serotype Typhimurium Std fimbriae bind terminal alpha(1,2)fucose residues in the cecal mucosa. *Molecular Microbiology, 71*, 864–875.

Cho, I., Yamanishi, S., Cox, L., Methé, B. A., Zavadil, J., Li, K., Gao, Z., Mahana, D., Raju, K., Teitler, I., et al. (2012). Antibiotics in early life alter the murine colonic microbiome and adiposity. *Nature, 488*, 621–626.

Chung, H., Pamp, S. J., Hill, J. A., Surana, N. K., Edelman, S. M., Troy, E. B., Reading, N. C., Villablanca, E. J., Wang, S., Mora, J. R., et al. (2012). Gut immune maturation depends on colonization with a host-specific microbiota. *Cell, 149*, 1578–1593.

Cianfanelli, F. R., Monlezun, L., & Coulthurst, S. J. (2016). Aim, load, fire: The type VI secretion system, a bacterial nanoweapon. *Trends in Microbiology, 24*, 51–62.

Clavel, T., Lagkouvardos, I., & Stecher, B. (2017). From complex gut communities to minimal microbiomes via cultivation. *Current Opinion in Microbiology, 38*, 148–155.

Coddens, A., Diswall, M., Angström, J., Breimer, M. E., Goddeeris, B., Cox, E., & Teneberg, S. (2009). Recognition of blood group ABH type 1 determinants by the FedF adhesin of F18-fimbriated Escherichia coli. *The Journal of Biological Chemistry, 284*, 9713–9726.

Collins, J., Robinson, C., Danhof, H., Knetsch, C. W., van Leeuwen, H. C., Lawley, T. D., Auchtung, J. M., & Britton, R. A. (2018). Dietary trehalose enhances virulence of epidemic Clostridium difficile. *Nature, 553*, 291–294.

Costea, P. I., Hildebrand, F., Manimozhiyan, A., Bäckhed, F., Blaser, M. J., Bushman, F. D., de Vos, W. M., Ehrlich, S. D., Fraser, C. M., Hattori, M., et al.

(2018). Enterotypes in the landscape of gut microbial community composition. *Nature Microbiology, 3*, 8–16.

Cotter, P. D., Ross, R. P., & Hill, C. (2013). Bacteriocins – a viable alternative to antibiotics? *Nature Reviews. Microbiology, 11*, 95–105.

Cowardin, C. A., Buonomo, E. L., Saleh, M. M., Wilson, M. G., Burgess, S. L., Kuehne, S. A., Schwan, C., Eichhoff, A. M., Koch-Nolte, F., Lyras, D., et al. (2016). The binary toxin CDT enhances Clostridium difficile virulence by suppressing protective colonic eosinophilia. *Nature Microbiology, 1*, 16108.

Cresci, G. A. M., Mayor, P. C., & Thompson, S. A. (2017). Effect of butyrate and Lactobacillus GG on a butyrate receptor and transporter during Campylobacter jejuni exposure. *FEMS Microbiology Letters, 364*. https://doi.org/10.1093/femsle/fnx046.

Dall'Olio, F., Malagolini, N., Chiricolo, M., Trinchera, M., & Harduin-Lepers, A. (2014). The expanding roles of the Sd(a)/Cad carbohydrate antigen and its cognate glycosyltransferase B4GALNT2. *Biochimica et Biophysica Acta, 1840*, 443–453.

Day, C. J., Tiralongo, J., Hartnell, R. D., Logue, C.-A., Wilson, J. C., von Itzstein, M., & Korolik, V. (2009). Differential carbohydrate recognition by Campylobacter jejuni strain 11168: Influences of temperature and growth conditions. *PLoS One, 4*, e4927.

Day, C. J., Semchenko, E. A., & Korolik, V. (2012). Glycoconjugates play a key role in Campylobacter jejuni infection: Interactions between host and pathogen. *Frontiers in Cellular and Infection Microbiology, 2*, 9.

Day, C. J., Tram, G., Hartley-Tassell, L. E., Tiralongo, J., & Korolik, V. (2013). Assessment of glycan interactions of clinical and avian isolates of Campylobacter jejuni. *BMC Microbiology, 13*, 228.

Day, C. J., Tran, E. N., Semchenko, E. A., Tram, G., Hartley-Tassell, L. E., Ng, P. S. K., King, R. M., Ulanovsky, R., McAtamney, S., Apicella, M. A., et al. (2015). Glycan:glycan interactions: High affinity biomolecular interactions that can mediate binding of pathogenic bacteria to host cells. *Proceedings of the National Academy of Sciences of the United States of America, 112*, E7266–E7275.

de Vries, S. P., Gupta, S., Baig, A., Wright, E., Wedley, A., Jensen, A. N., Lora, L. L., Humphrey, S., Skovgard, H., Macleod, K., et al. (2017). Genome-wide fitness analyses of the foodborne pathogen Campylobacter jejuni in in vitro and in vivo models. *Scientific Reports, 7*, 1251.

Donia, M. S., & Fischbach, M. A. (2015). HUMAN MICROBIOTA. Small molecules from the human microbiota. *Science, 349*, 1254766.

Donia, M. S., Cimermancic, P., Schulze, C. J., Wieland Brown, L. C., Martin, J., Mitreva, M., Clardy, J., Linington, R. G., & Fischbach, M. A. (2014). A systematic analysis of biosynthetic gene clusters in the human microbiome reveals a common family of antibiotics. *Cell, 158*, 1402–1414.

Ekmekciu, I., von Klitzing, E., Fiebiger, U., Escher, U., Neumann, C., Bacher, P., Scheffold, A., Kühl, A. A., Bereswill, S., & Heimesaat, M. M. (2017). Immune responses to broad-spectrum antibiotic treatment and fecal microbiota transplantation in mice. *Frontiers in Immunology, 8*, 397.

Eren, A. M., Morrison, H. G., Lescault, P. J., Reveillaud, J., Vineis, J. H., & Sogin, M. L. (2015). Minimum entropy decomposition: Unsupervised oligotyping for sensitive partitioning of high-throughput marker gene sequences. *The ISME Journal, 9*, 968–979.

Ermund, A., Schütte, A., Johansson, M. E. V., Gustafsson, J. K., & Hansson, G. C. (2013). Studies of mucus in mouse stomach, small intestine, and colon. I. Gastrointestinal mucus layers have different properties depending on location as well as over the Peyer's patches. *American Journal of Physiology. Gastrointestinal and Liver Physiology, 305*, G341–G347.

Esan, O. B., Pearce, M., van Hecke, O., Roberts, N., Collins, D. R. J., Violato, M., McCarthy, N., Perera, R., & Fanshawe, T. R. (2017). Factors associated with sequelae of Campylobacter and non-typhoidal Salmonella infections: A systematic review. *EBio Medicine, 15*, 100–111.

Faber, F., Tran, L., Byndloss, M. X., Lopez, C. A., Velazquez, E. M., Kerrinnes, T., Nuccio, S.-P., Wangdi, T., Fiehn, O., Tsolis, R. M., et al. (2016). Host-mediated sugar oxidation promotes post-antibiotic pathogen expansion. *Nature, 534*, 697–699.

Faber, F., Thiennimitr, P., Spiga, L., Byndloss, M. X., Litvak, Y., Lawhon, S., Andrews-Polymenis, H. L., Winter, S. E., & Bäumler, A. J. (2017). Respiration of microbiota-derived 1,2-propanediol drives Salmonella expansion during Colitis. *PLoS Pathogens, 13*, e1006129.

Faïs, T., Delmas, J., Serres, A., Bonnet, R., & Dalmasso, G. (2016). Impact of CDT toxin on human diseases. *Toxins, 8*. https://doi.org/10.3390/toxins8070220.

Gantois, I., Ducatelle, R., Pasmans, F., Haesebrouck, F., Hautefort, I., Thompson, A., Hinton, J. C., & Van Immerseel, F. (2006). Butyrate specifically down-regulates salmonella pathogenicity island 1 gene expression. *Applied and Environmental Microbiology, 72*, 946–949.

Gao, B., Vorwerk, H., Huber, C., Lara-Tejero, M., Mohr, J., Goodman, A. L., Eisenreich, W., Galán, J. E., & Hofreuter, D. (2017). Metabolic and fitness determinants for in vitro growth and intestinal colonization of the bacterial pathogen Campylobacter jejuni. *PLoS Biology, 15*, e2001390.

Gerding, D. N., Johnson, S., Rupnik, M., & Aktories, K. (2014). Clostridium difficile binary toxin CDT: Mechanism, epidemiology, and potential clinical importance. *Gut Microbes, 5*, 15–27.

Gerlach, R. G., Jackel, D., Stecher, B., Wagner, C., Lupas, A., Hardt, W. D., & Hensel, M. (2007). Salmonella Pathogenicity Island 4 encodes a giant non-fimbrial adhesin and the cognate type 1 secretion system. *Cellular Microbiology, 9*, 1834–1850.

Geva-Zatorsky, N., Sefik, E., Kua, L., Pasman, L., Tan, T. G., Ortiz-Lopez, A., Yanortsang, T. B., Yang, L., Jupp, R., Mathis, D., et al. (2017). Mining the human gut microbiota for immunomodulatory organisms. *Cell, 168*, 928–943.e11.

Gibson, T. E., Bashan, A., Cao, H.-T., Weiss, S. T., & Liu, Y.-Y. (2016). On the origins and control of community types in the human microbiome. *PLoS Computational Biology, 12*, e1004688.

Gilbert, M., Karwaski, M.-F., Bernatchez, S., Young, N. M., Taboada, E., Michniewicz, J., Cunningham, A.-M., & Wakarchuk, W. W. (2002). The genetic bases for the variation in the lipo-oligosaccharide of the mucosal pathogen, Campylobacter jejuni. Biosynthesis of sialylated ganglioside mimics in the core oligosaccharide. *The Journal of Biological Chemistry, 277*, 327–337.

Godschalk, P. C. R., Heikema, A. P., Gilbert, M., Komagamine, T., Ang, C. W., Glerum, J., Brochu, D., Li, J., Yuki, N., Jacobs, B. C., et al. (2004). The crucial role of Campylobacter jejuni genes in anti-ganglioside antibody induction in Guillain-Barre syndrome. *The Journal of Clinical Investigation, 114*, 1659–1665.

Goodrich, J. K., Waters, J. L., Poole, A. C., Sutter, J. L., Koren, O., Blekhman, R., Beaumont, M., Van Treuren, W., Knight, R., Bell, J. T., et al. (2014). Human genetics shape the gut microbiome. *Cell, 159*, 789–799.

Goodrich, J. K., Davenport, E. R., Waters, J. L., Clark, A. G., & Ley, R. E. (2016a). Cross-species comparisons of host genetic associations with the microbiome. *Science, 352*, 532–535.

Goodrich, J. K., Davenport, E. R., Beaumont, M., Jackson, M. A., Knight, R., Ober, C., Spector, T. D., Bell, J. T., Clark, A. G., & Ley, R. E. (2016b). Genetic determinants of the gut microbiome in UK twins. *Cell Host and Microbe, 19*, 731–743.

Goto, Y., Obata, T., Kunisawa, J., Sato, S., Ivanov, I. I., Lamichhane, A., Takeyama, N., Kamioka, M., Sakamoto, M., Matsuki, T., et al. (2014). Innate lymphoid cells regulate intestinal epithelial cell glycosylation. *Science, 345*, 1254009.

Goto, Y., Uematsu, S., & Kiyono, H. (2016). Epithelial glycosylation in gut homeostasis and inflammation. *Nature Immunology, 17*, 1244–1251.

Grassl, G. A., & Finlay, B. B. (2008). Pathogenesis of enteric Salmonella infections. *Current Opinion in Gastroenterology, 24*, 22–26.

Gripp, E., Hlahla, D., Didelot, X., Kops, F., Maurischat, S., Tedin, K., Alter, T., Ellerbroek, L., Schreiber, K., Schomburg, D., et al. (2011). Closely related Campylobacter jejuni strains from different sources reveal a generalist rather than a specialist lifestyle. *BMC Genomics, 12*, 584.

Guerry, P., Ewing, C. P., Hickey, T. E., Prendergast, M. M., & Moran, A. P. (2000). Sialylation of lipooligosaccharide cores affects immunogenicity and serum resistance of Campylobacter jejuni. *Infection and Immunity, 68*, 6656–6662.

Günther, C., Josenhans, C., & Wehkamp, J. (2016). Crosstalk between microbiota, pathogens and the innate immune responses. *International Journal of Medical Microbiology, 306*, 257–265.

Hansson, G. C. (2012). Role of mucus layers in gut infection and inflammation. *Current Opinion in Microbiology, 15*, 57–62.

Hegarty, J. W., Guinane, C. M., Ross, R. P., Hill, C., & Cotter, P. D. (2016). Bacteriocin production: A relatively unharnessed probiotic trait? *F1000Research, 5*, 2587.

Heikema, A. P., Jacobs, B. C., Horst-Kreft, D., Huizinga, R., Kuijf, M. L., Endtz, H. P., Samsom, J. N., & van Wamel, W. J. B. (2013). Siglec-7 specifically recognizes Campylobacter jejuni strains associated with oculomotor weakness in Guillain-Barré syndrome and Miller Fisher syndrome. *Clinical Microbiology and Infection, 19*, E106–E112.

Hiergeist, A., Reischl, U., Priority Program 1656 Intestinal Microbiota Consortium/quality assessment participants, & Gessner, A. (2016). Multicenter quality assessment of 16S ribosomal DNA-sequencing for microbiome analyses reveals high inter-center variability. *International Journal of Medical Microbiology, 306*, 334–342.

Hildebrand, F., Nguyen, T. L. A., Brinkman, B., Yunta, R. G., Cauwe, B., Vandenabeele, P., Liston, A., & Raes, J. (2013). Inflammation-associated enterotypes, host genotype, cage and inter-individual effects drive gut microbiota variation in common laboratory mice. *Genome Biology, 14*, R4.

Ho, B. T., Dong, T. G., & Mekalanos, J. J. (2014). A view to a kill: The bacterial type VI secretion system. *Cell Host and Microbe, 15*, 9–21.

Humphrey, S., Chaloner, G., Kemmett, K., Davidson, N., Williams, N., Kipar, A., Humphrey, T., & Wigley, P. (2014). Campylobacter jejuni is not merely a commensal in commercial broiler chickens and affects bird welfare. *MBio, 5*, e01364-01314.

Iovine, N. M., Pursnani, S., Voldman, A., Wasserman, G., Blaser, M. J., & Weinrauch, Y. (2008). Reactive nitrogen species contribute to innate host defense against Campylobacter jejuni. *Infection and Immunity, 76*, 986–993.

Iwasaki, A., & Medzhitov, R. (2015). Control of adaptive immunity by the innate immune system. *Nature Immunology, 16*, 343–353.

Jäger, S., Stange, E. F., & Wehkamp, J. (2013). Inflammatory bowel disease: An impaired barrier disease. *Langenbeck's Archives of Surgery, 398*, 1–12.

Jennings, E., Thurston, T. L. M., & Holden, D. W. (2017). Salmonella SPI-2 type III secretion system effectors: Molecular mechanisms and physiological consequences. *Cell Host and Microbe, 22*, 217–231.

Jervis, A. J., Butler, J. A., Lawson, A. J., Langdon, R., Wren, B. W., & Linton, D. (2012). Characterization of the structurally diverse N-linked glycans of Campylobacter species. *Journal of Bacteriology, 194*, 2355–2362.

Johansson, M. E. V., & Hansson, G. C. (2016). Immunological aspects of intestinal mucus and mucins. *Nature Reviews. Immunology, 16*, 639–649.

Johansson, M. E. V., Ambort, D., Pelaseyed, T., Schütte, A., Gustafsson, J. K., Ermund, A., Subramani, D. B., Holmén-Larsson, J. M., Thomsson, K. A., Bergström, J. H., et al. (2011). Composition and functional role of the mucus layers in the intestine. *Cellular and Molecular Life Sciences, 68*, 3635–3641.

Johansson, M. E. V., Sjövall, H., & Hansson, G. C. (2013). The gastrointestinal mucus system in health and disease. *Nature Reviews. Gastroenterology & Hepatology, 10*, 352–361.

Joshi, A., Kostiuk, B., Rogers, A., Teschler, J., Pukatzki, S., & Yildiz, F. H. (2017). Rules of engagement: The type VI secretion system in Vibrio cholerae. *Trends in Microbiology, 25*, 267–279.

Juge, N. (2012). Microbial adhesins to gastrointestinal mucus. *Trends in Microbiology, 20*, 30–39.

Kaiser, P., Diard, M., Stecher, B., & Hardt, W.-D. (2012). The streptomycin mouse model for Salmonella diarrhea: Functional analysis of the microbiota, the pathogen's virulence factors, and the host's mucosal immune response. *Immunological Reviews, 245*, 56–83.

Kampmann, C., Dicksved, J., Engstrand, L., & Rautelin, H. (2016). Composition of human faecal microbiota in resistance to Campylobacter infection. *Clinical Microbiology and Infection, 22*, 61–61.

Kernbauer, E., Ding, Y., & Cadwell, K. (2014). An enteric virus can replace the beneficial function of commensal bacteria. *Nature, 516*, 94–98.

Kilcoyne, M., Twomey, M. E., Gerlach, J. Q., Kane, M., Moran, A. P., & Joshi, L. (2014). Campylobacter jejuni strain discrimination and temperature-dependent glycome expression profiling by lectin microarray. *Carbohydrate Research, 389*, 123–133.

Kim, S., Covington, A., & Pamer, E. G. (2017). The intestinal microbiota: Antibiotics, colonization resistance, and enteric pathogens. *Immunological Reviews, 279*, 90–105.

Klaas, M., Oetke, C., Lewis, L. E., Erwig, L. P., Heikema, A. P., Easton, A., Willison, H. J., & Crocker, P. R. (2012). Sialoadhesin promotes rapid proinflammatory and type I IFN responses to a sialylated pathogen, Campylobacter jejuni. *Journal of Immunology (Baltimore, MD), 1950*(189), 2414–2422.

Knights, D., Ward, T. L., McKinlay, C. E., Miller, H., Gonzalez, A., McDonald, D., & Knight, R. (2014). Rethinking "enterotypes". *Cell Host and Microbe, 16*, 433–437.

Koenig, J. E., Spor, A., Scalfone, N., Fricker, A. D., Stombaugh, J., Knight, R., Angenent, L. T., & Ley, R. E. (2011). Succession of microbial consortia in the developing infant gut microbiome. *Proceedings of the National Academy of Sciences of the United States of America, 108*(Suppl 1), 4578–4585.

Kubinak, J. L., & Round, J. L. (2016). Do antibodies select a healthy microbiota? *Nature Reviews. Immunology, 16*, 767–774.

Lagier, J.-C., Armougom, F., Million, M., Hugon, P., Pagnier, I., Robert, C., Bittar, F., Fournous, G., Gimenez, G., Maraninchi, M., et al. (2012). Microbial culturomics: Paradigm shift in the human gut microbiome study. *Clinical Microbiology and Infection, 18*, 1185–1193.

Lagkouvardos, I., Pukall, R., Abt, B., Foesel, B. U., Meier-Kolthoff, J. P., Kumar, N., Bresciani, A., Martínez, I., Just, S., Ziegler, C., et al. (2016). The Mouse Intestinal Bacterial Collection (miBC) provides host-specific insight into cultured diversity and functional potential of the gut microbiota. *Nature Microbiology, 1*, 16131.

Lardone, R. D., Yuki, N., Irazoqui, F. J., & Nores, G. A. (2016). Individual restriction of fine specificity variability in anti-GM1 IgG antibodies associated with Guillain-Barré syndrome. *Scientific Reports, 6*, 19901.

Lawley, T. D., & Walker, A. W. (2013). Intestinal colonization resistance. *Immunology, 138*, 1–11.

Lawley, T. D., Clare, S., Walker, A. W., Stares, M. D., Connor, T. R., Raisen, C., Goulding, D., Rad, R., Schreiber, F., Brandt, C., et al. (2012). Targeted restoration of the intestinal microbiota with a simple, defined bacteriotherapy resolves relapsing Clostridium difficile disease in mice. *PLoS Pathogens, 8*, e1002995.

Le Chatelier, E., Nielsen, T., Qin, J., Prifti, E., Hildebrand, F., Falony, G., Almeida, M., Arumugam, M., Batto, J.-M., Kennedy, S., et al. (2013). Richness of human gut microbiome correlates with metabolic markers. *Nature, 500*, 541–546.

Ley, R. E., Lozupone, C. A., Hamady, M., Knight, R., & Gordon, J. I. (2008). Worlds within worlds: Evolution of the vertebrate gut microbiota. *Nature Reviews. Microbiology, 6*, 776–788.

Lim, M. Y., Rho, M., Song, Y.-M., Lee, K., Sung, J., & Ko, G. (2014). Stability of gut enterotypes in Korean monozygotic twins and their association with biomarkers and diet. *Scientific Reports, 4*, 7348.

Lindesmith, L., Moe, C., Marionneau, S., Ruvoen, N., Jiang, X., Lindblad, L., Stewart, P., LePendu, J., & Baric, R. (2003). Human susceptibility and resistance to Norwalk virus infection. *Nature Medicine, 9*, 548–553.

Lindner, C., Thomsen, I., Wahl, B., Ugur, M., Sethi, M. K., Friedrichsen, M., Smoczek, A., Ott, S., Baumann, U., Suerbaum, S., et al. (2015). Diversification of memory B cells drives the continuous adaptation of secretory antibodies to gut microbiota. *Nature Immunology, 16*, 880–888.

Liu, J. Z., Jellbauer, S., Poe, A. J., Ton, V., Pesciaroli, M., Kehl-Fie, T. E., Restrepo, N. A., Hosking, M. P., Edwards, R. A., Battistoni, A., et al. (2012). Zinc sequestration by the neutrophil protein calprotectin enhances Salmonella growth in the inflamed gut. *Cell Host and Microbe, 11*, 227–239.

Liu, Y. W., Denkmann, K., Kosciow, K., Dahl, C., & Kelly, D. J. (2013). Tetrathionate stimulated growth of Campylobacter jejuni identifies a new type of bi-functional tetrathionate reductase (TsdA) that is widely distributed in bacteria. *Molecular Microbiology, 88*, 173–188.

Lopez, C. A., Winter, S. E., Rivera-Chavez, F., Xavier, M. N., Poon, V., Nuccio, S. P., Tsolis, R. M., & Baumler, A. J. (2012). Phage-mediated acquisition of a type III secreted effector protein boosts growth of salmonella by nitrate respiration. *MBio, 3*. https://doi.org/10.1128/mBio.00143-12.

Lu, Q., Li, S., & Shao, F. (2015). Sweet talk: Protein glycosylation in bacterial interaction with the host. *Trends in Microbiology, 23*, 630–641.

Lustri, B. C., Sperandio, V., & Moreira, C. G. (2017). Bacterial chat: Intestinal metabolites and signals in host-microbiota-pathogen interactions. *Infection and Immunity, 85*. https://doi.org/10.1128/IAI.00476-17.

Macpherson, A. J., & McCoy, K. D. (2015). Standardised animal models of host microbial mutualism. *Mucosal Immunology, 8*, 476–486.

Madhavan, T. P., & Sakellaris, H. (2015). Colonization factors of enterotoxigenic Escherichia coli. *Advances in Applied Microbiology, 90*, 155–197.

Maier, L., Vyas, R., Cordova, C. D., Lindsay, H., Schmidt, T. S. B., Brugiroux, S., Periaswamy, B., Bauer, R., Sturm, A., Schreiber, F., et al. (2013). Microbiota-derived hydrogen fuels Salmonella typhimurium invasion of the gut ecosystem. *Cell Host and Microbe, 14*, 641–651.

McCoy, K. D., Ronchi, F., & Geuking, M. B. (2017). Host-microbiota interactions and adaptive immunity. *Immunological Reviews, 279*, 63–69.

McGovern, D. P., Jones, M. R., Taylor, K. D., Marciante, K., Yan, X., Dubinsky, M., Ippoliti, A., Vasiliauskas, E., Berel, D., Derkowski, C., et al. (2010). Fucosyltransferase 2 (FUT2) non-secretor status is associated with Crohn's disease. *Human Molecular Genetics, 19*, 3468–3476.

McGuckin, M. A., Lindén, S. K., Sutton, P., & Florin, T. H. (2011). Mucin dynamics and enteric pathogens. *Nature Reviews. Microbiology, 9*, 265–278.

Miller, C. P., Bohnhoff, M., & Drake, B. L. (1954). The effect of antibiotic therapy on susceptibility to an experimental enteric infection. *Transactions of the Association of American Physicians, 67*, 156–161.

Mills, S., Shanahan, F., Stanton, C., Hill, C., Coffey, A., & Ross, R. P. (2013). Movers and shakers: Influence of bacteriophages in shaping the mammalian gut microbiota. *Gut Microbes, 4*, 4–16.

Minot, S., Bryson, A., Chehoud, C., Wu, G. D., Lewis, J. D., & Bushman, F. D. (2013). Rapid evolution of the human gut virome. *Proceedings of the National Academy of Sciences of the United States of America, 110*, 12450–12455.

Molnár, A., Hess, C., Pál, L., Wágner, L., Awad, W. A., Husvéth, F., Hess, M., & Dublecz, K. (2015). Composition of diet modifies colonization dynamics of Campylobacter jejuni in broiler chickens. *Journal of Applied Microbiology, 118*, 245–254.

Montalban-Arques, A., De Schryver, P., Bossier, P., Gorkiewicz, G., Mulero, V., Gatlin, D. M., & Galindo-Villegas, J. (2015). Selective manipulation of the gut microbiota improves immune status in vertebrates. *Frontiers in Immunology, 6*, 512.

Moonens, K., Bouckaert, J., Coddens, A., Tran, T., Panjikar, S., De Kerpel, M., Cox, E., Remaut, H., & De Greve, H. (2012). Structural insight in histo-blood group binding by the F18 fimbrial adhesin FedF. *Molecular Microbiology, 86*, 82–95.

Morgan, X. C., & Huttenhower, C. (2014). Meta'omic analytic techniques for studying the intestinal microbiome. *Gastroenterology, 146*, 1437–1448.e1.

Morrison, D. J., & Preston, T. (2016). Formation of short chain fatty acids by the gut microbiota and their impact on human metabolism. *Gut Microbes, 7*, 189–200.

Muraoka, W. T., & Zhang, Q. (2011). Phenotypic and genotypic evidence for L-fucose utilization by Campylobacter jejuni. *Journal of Bacteriology, 193*, 1065–1075.

Ng, K. M., Ferreyra, J. A., Higginbottom, S. K., Lynch, J. B., Kashyap, P. C., Gopinath, S., Naidu, N., Choudhury, B., Weimer, B. C., Monack, D. M., et al. (2013). Microbiota-liberated host sugars facilitate post-antibiotic expansion of enteric pathogens. *Nature, 502*, 96–99.

Nguyen, T. L., Vieira-Silva, S., Liston, A., & Raes, J. (2015). How informative is the mouse for human gut microbiota research? *Disease Models and Mechanisms, 8*, 1–16.

Nuttal, G., & Thierfelder, H. (1895). Thierisches Leben ohne Bakterien im Verdauungskanal. *Hoppe-Seyler's Zeitschrift für Physiologische Chemie, 21*, 109–121.

O'Neill, L. A. J., Kishton, R. J., & Rathmell, J. (2016). A guide to immunometabolism for immunologists. *Nature Reviews. Immunology, 16*, 553–565.

Ohno, H. (2016). Intestinal M cells. *Journal of Biochemistry, 159*, 151–160.

Okai, S., Usui, F., Ohta, M., Mori, H., Kurokawa, K., Matsumoto, S., Kato, T., Miyauchi, E., Ohno, H., & Shinkura, R. (2017). Intestinal IgA as a modulator of the gut microbiota. *Gut Microbes, 8*, 486–492.

Okoro, C. K., Kingsley, R. A., Connor, T. R., Harris, S. R., Parry, C. M., Al-Mashhadani, M. N., Kariuki, S., Msefula, C. L., Gordon, M. A., de Pinna, E., et al. (2012). Intracontinental spread of human invasive Salmonella Typhimurium pathovariants in sub-Saharan Africa. *Nature Genetics, 44*, 1215–1221.

Olsan, E. E., Byndloss, M. X., Faber, F., Rivera-Chávez, F., Tsolis, R. M., & Bäumler, A. J. (2017). Colonization resistance: The deconvolution of a complex trait. *The Journal of Biological Chemistry, 292*, 8577–8581.

Pabst, O., Cerovic, V., & Hornef, M. (2016). Secretory IgA in the coordination of establishment and maintenance of the microbiota. *Trends in Immunology, 37*, 287–296.

Pacheco, A. R., Curtis, M. M., Ritchie, J. M., Munera, D., Waldor, M. K., Moreira, C. G., & Sperandio, V. (2012). Fucose sensing regulates bacterial intestinal colonization. *Nature, 492*, 113–117.

Palm, N. W., de Zoete, M. R., & Flavell, R. A. (2015). Immune-microbiota interactions in health and disease. *Clinical Immunology (Orlando, Florida), 159*, 122–127.

Perdicchio, M., Ilarregui, J. M., Verstege, M. I., Cornelissen, L. A. M., Schetters, S. T. T., Engels, S., Ambrosini, M., Kalay, H., Veninga, H., den Haan, J. M. M., et al. (2016). Sialic acid-modified antigens impose tolerance via inhibition of T-cell proliferation and de novo induction of regulatory T cells. *Proceedings of the National Academy of Sciences of the United States of America, 113*, 3329–3334.

Perez-Lopez, A., Behnsen, J., Nuccio, S.-P., & Raffatellu, M. (2016). Mucosal immunity to pathogenic intestinal bacteria. *Nature Reviews. Immunology, 16*, 135–148.

Phongsisay, V. (2016). The immunobiology of Campylobacter jejuni: Innate immunity and autoimmune diseases. *Immunobiology, 221*, 535–543.

Phongsisay, V., Hara, H., & Fujimoto, S. (2015). Toll-like receptors recognize distinct proteinase-resistant glycoconjugates in Campylobacter jejuni and Escherichia coli. *Molecular Immunology, 64*, 195–203.

Pickard, J. M., Maurice, C. F., Kinnebrew, M. A., Abt, M. C., Schenten, D., Golovkina, T. V., Bogatyrev, S. R., Ismagilov, R. F., Pamer, E. G., Turnbaugh, P. J., et al. (2014). Rapid fucosylation of intestinal epithelium sustains host-commensal symbiosis in sickness. *Nature, 514*, 638–641.

Planer, J. D., Peng, Y., Kau, A. L., Blanton, L. V., Ndao, I. M., Tarr, P. I., Warner, B. B., & Gordon, J. I. (2016). Development of the gut microbiota and mucosal IgA responses in twins and gnotobiotic mice. *Nature, 534*, 263–266.

Price-Carter, M., Tingey, J., Bobik, T. A., & Roth, J. R. (2001). The alternative electron acceptor tetrathionate supports B12-dependent anaerobic growth of Salmonella enterica serovar typhimurium on ethanolamine or 1,2-propanediol. *Journal of Bacteriology, 183*, 2463–2475.

Próchnicki, T., & Latz, E. (2017). Inflammasomes on the crossroads of innate immune recognition and metabolic control. *Cell Metabolism, 26*, 71–93.

Pukatzki, S., Ma, A. T., Revel, A. T., Sturtevant, D., & Mekalanos, J. J. (2007). Type VI secretion system translocates a phage tail spike-like protein into target cells where it cross-links actin. *Proceedings of the National Academy of Sciences of the United States of America, 104*, 15508–15513.

Qin, J., Li, R., Raes, J., Arumugam, M., Burgdorf, K. S., Manichanh, C., Nielsen, T., Pons, N., Levenez, F., Yamada, T., et al. (2010). A human gut microbial gene catalogue established by metagenomic sequencing. *Nature, 464*, 59–65.

Raffatellu, M., George, M. D., Akiyama, Y., Hornsby, M. J., Nuccio, S.-P., Paixao, T. A., Butler, B. P., Chu, H., Santos, R. L., Berger, T., et al. (2009). Lipocalin-2 resistance confers an advantage to Salmonella enterica serotype Typhimurium for growth and survival in the inflamed intestine. *Cell Host and Microbe, 5*, 476–486.

Rakoff-Nahoum, S., Paglino, J., Eslami-Varzaneh, F., Edberg, S., & Medzhitov, R. (2004). Recognition of commensal microflora by toll-like receptors is required for intestinal homeostasis. *Cell, 118*, 229–241.

Ramachandran, G., Panda, A., Higginson, E. E., Ateh, E., Lipsky, M. M., Sen, S., Matson, C. A., Permala-Booth, J., DeTolla, L. J., & Tennant, S. M. (2017). Virulence of invasive Salmonella Typhimurium ST313 in animal models of infection. *PLoS Neglected Tropical Diseases, 11*, e0005697.

Rausch, P., Steck, N., Suwandi, A., Seidel, J. A., Künzel, S., Bhullar, K., Basic, M., Bleich, A., Johnsen, J. M., Vallance, B. A., et al. (2015). Expression of the blood-group-related gene B4galnt2 alters susceptibility to Salmonella infection. *PLoS Pathogens, 11*, e1005008.

Ringot-Destrez, B., Kalach, N., Mihalache, A., Gosset, P., Michalski, J.-C., Léonard, R., & Robbe-Masselot, C. (2017). How do they stick together? Bacterial adhesins implicated in the binding of bacteria to the human gastrointestinal mucins. *Biochemical Society Transactions, 45*, 389–399.

Robinson, C. M., & Pfeiffer, J. K. (2014). Viruses and the microbiota. *Annual Review of Virology, 1*, 55–69.

Ruiz-Palacios, G. M., Cervantes, L. E., Ramos, P., Chavez-Munguia, B., & Newburg, D. S. (2003). Campylobacter jejuni binds intestinal H(O) antigen (Fuc alpha 1, 2Gal beta 1, 4GlcNAc), and fucosyloligosaccharides of human milk inhibit its binding and infection. *The Journal of Biological Chemistry, 278*, 14112–14120.

Sana, T. G., Flaugnatti, N., Lugo, K. A., Lam, L. H., Jacobson, A., Baylot, V., Durand, E., Journet, L., Cascales, E., & Monack, D. M. (2016). Salmonella Typhimurium utilizes a T6SS-mediated antibacterial weapon to establish in the host gut. *Proceedings of the National Academy of Sciences of the United States of America, 113*, E5044–E5051.

Sankar, S. A., Lagier, J.-C., Pontarotti, P., Raoult, D., & Fournier, P.-E. (2015). The human gut microbiome, a taxonomic conundrum. *Systematic and Applied Microbiology, 38*, 276–286.

Sassone-Corsi, M., & Raffatellu, M. (2013). A hydrogen boost for salmonella. *Cell Host and Microbe, 14*, 603–604.

Sassone-Corsi, M., Nuccio, S.-P., Liu, H., Hernandez, D., Vu, C. T., Takahashi, A. A., Edwards, R. A., & Raffatellu, M. (2016). Microcins mediate competition among Enterobacteriaceae in the inflamed gut. *Nature, 540*, 280–283.

Sears, C. L. (2009). Enterotoxigenic Bacteroides fragilis: A rogue among symbiotes. *Clinical Microbiology Reviews, 22*, 349–369 Table of Contents.

Sperandio, B., Fischer, N., & Sansonetti, P. J. (2015). Mucosal physical and chemical innate barriers:

Lessons from microbial evasion strategies. *Seminars in Immunology, 27*, 111–118.

Spiga, L., Winter, M. G., Furtado de Carvalho, T., Zhu, W., Hughes, E. R., Gillis, C. C., Behrendt, C. L., Kim, J., Chessa, D., Andrews-Polymenis, H. L., et al. (2017). An oxidative central metabolism enables Salmonella to utilize microbiota-derived succinate. *Cell Host and Microbe, 22*, 291–301.e6.

St Charles, J. L., Bell, J. A., Gadsden, B. J., Malik, A., Cooke, H., Van de Grift, L. K., Kim, H. Y., Smith, E. J., & Mansfield, L. S. (2017). Guillain Barré syndrome is induced in non-obese diabetic (NOD) mice following Campylobacter jejuni infection and is exacerbated by antibiotics. *Journal of Autoimmunity, 77*, 11–38.

Stahl, M., Friis, L. M., Nothaft, H., Liu, X., Li, J., Szymanski, C. M., & Stintzi, A. (2011). L-fucose utilization provides Campylobacter jejuni with a competitive advantage. *Proceedings of the National Academy of Sciences of the United States of America, 108*, 7194–7199.

Staubach, F., Künzel, S., Baines, A. C., Yee, A., McGee, B. M., Bäckhed, F., Baines, J. F., & Johnsen, J. M. (2012). Expression of the blood-group-related glycosyltransferase B4galnt2 influences the intestinal microbiota in mice. *The ISME Journal, 6*, 1345–1355.

Stecher, B., & Hardt, W. D. (2011). Mechanisms controlling pathogen colonization of the gut. *Current Opinion in Microbiology, 14*, 82–91.

Stecher, B., Barthel, M., Schlumberger, M. C., Haberli, L., Rabsch, W., Kremer, M., & Hardt, W. D. (2008). Motility allows S. Typhimurium to benefit from the mucosal defence. *Cellular Microbiology, 10*, 1166–1180.

Stecher, B., Berry, D., & Loy, A. (2013). Colonization resistance and microbial ecophysiology: Using gnotobiotic mouse models and single-cell technology to explore the intestinal jungle. *FEMS Microbiology Reviews, 37*, 793–829.

Stephenson, H. N., John, C. M., Naz, N., Gundogdu, O., Dorrell, N., Wren, B. W., Jarvis, G. A., & Bajaj-Elliott, M. (2013). Campylobacter jejuni lipooligosaccharide sialylation, phosphorylation, and amide/ester linkage modifications fine-tune human Toll-like receptor 4 activation. *The Journal of Biological Chemistry, 288*, 19661–19672.

Stephenson, H. N., Mills, D. C., Jones, H., Milioris, E., Copland, A., Dorrell, N., Wren, B. W., Crocker, P. R., Escors, D., & Bajaj-Elliott, M. (2014). Pseudaminic acid on Campylobacter jejuni flagella modulates dendritic cell IL-10 expression via Siglec-10 receptor: A novel flagellin-host interaction. *The Journal of Infectious Diseases, 210*, 1487–1498.

Struwe, W. B., Gough, R., Gallagher, M. E., Kenny, D. T., Carrington, S. D., Karlsson, N. G., & Rudd, P. M. (2015). Identification of O-glycan structures from chicken intestinal mucins provides insight into Campylobactor jejuni pathogenicity. *Molecular and Cellular Proteomics, 14*, 1464–1477.

Szymanski, C. M., & Gaynor, E. C. (2012). How a sugary bug gets through the day: Recent developments in understanding fundamental processes impacting Campylobacter jejuni pathogenesis. *Gut Microbes, 3*, 135–144.

Tailford, L. E., Crost, E. H., Kavanaugh, D., & Juge, N. (2015). Mucin glycan foraging in the human gut microbiome. *Frontiers in Genetics, 6*, 81.

Tan, T. G., Sefik, E., Geva-Zatorsky, N., Kua, L., Naskar, D., Teng, F., Pasman, L., Ortiz-Lopez, A., Jupp, R., Wu, H.-J. J., et al. (2016). Identifying species of symbiont bacteria from the human gut that, alone, can induce intestinal Th17 cells in mice. *Proceedings of the National Academy of Sciences of the United States of America, 113*, E8141–E8150.

Thaiss, C. A., Zmora, N., Levy, M., & Elinav, E. (2016). The microbiome and innate immunity. *Nature, 535*, 65–74.

Thiennimitr, P., Winter, S. E., Winter, M. G., Xavier, M. N., Tolstikov, V., Huseby, D. L., Sterzenbach, T., Tsolis, R. M., Roth, J. R., & Bäumler, A. J. (2011). Intestinal inflammation allows Salmonella to use ethanolamine to compete with the microbiota. *Proceedings of the National Academy of Sciences of the United States of America, 108*, 17480–17485.

Thursby, E., & Juge, N. (2017). Introduction to the human gut microbiota. *Biochemical Journal, 474*, 1823–1836.

Turonova, H., Neu, T. R., Ulbrich, P., Pazlarova, J., & Tresse, O. (2016). The biofilm matrix of Campylobacter jejuni determined by fluorescence lectin-binding analysis. *Biofouling, 32*, 597–608.

Ursell, L. K., Metcalf, J. L., Parfrey, L. W., & Knight, R. (2012). Defining the human microbiome. *Nutrition Reviews, 70*(Suppl 1), S38–S44.

van der Heijden, J., & Finlay, B. B. (2012). Type III effector-mediated processes in Salmonella infection. *Future Microbiology, 7*, 685–703.

van Sorge, N. M., Bleumink, N. M. C., van Vliet, S. J., Saeland, E., van der Pol, W.-L., van Kooyk, Y., & van Putten, J. P. M. (2009). N-glycosylated proteins and distinct lipooligosaccharide glycoforms of Campylobacter jejuni target the human C-type lectin receptor MGL. *Cellular Microbiology, 11*, 1768–1781.

Verster, A. J., Ross, B. D., Radey, M. C., Bao, Y., Goodman, A. L., Mougous, J. D., & Borenstein, E. (2017). The landscape of type VI secretion across human gut microbiomes reveals its role in community composition. *Cell Host and Microbe, 22*, 411–419.e4.

Vogt, S. L., Peña-Díaz, J., & Finlay, B. B. (2015). Chemical communication in the gut: Effects of microbiota-generated metabolites on gastrointestinal bacterial pathogens. *Anaerobe, 34*, 106–115.

Vorwerk, H., Mohr, J., Huber, C., Wensel, O., Schmidt-Hohagen, K., Gripp, E., Josenhans, C., Schomburg, D., Eisenreich, W., & Hofreuter, D. (2014). Utilization of host-derived cysteine-containing peptides overcomes the restricted sulphur metabolism of Campylobacter jejuni. *Molecular Microbiology, 93*, 1224–1245.

Vorwerk, H., Huber, C., Mohr, J., Bunk, B., Bhuju, S., Wensel, O., Spröer, C., Fruth, A., Flieger, A., Schmidt-Hohagen, K., et al. (2015). A transferable plasticity region in Campylobacter coli allows isolates of an otherwise non-glycolytic food-borne pathogen to catabolize glucose. *Molecular Microbiology, 98*, 809–830.

Wagley, S., Newcombe, J., Laing, E., Yusuf, E., Sambles, C. M., Studholme, D. J., La Ragione, R. M., Titball, R. W., & Champion, O. L. (2014). Differences in carbon source utilisation distinguish Campylobacter jejuni from Campylobacter coli. *BMC Microbiology, 14*, 262.

Wagner, C., Barlag, B., Gerlach, R. G., Deiwick, J., & Hensel, M. (2014). The Salmonella enterica giant adhesin SiiE binds to polarized epithelial cells in a lectin-like manner. *Cellular Microbiology, 16*, 962–975.

Winter, S. E., & Bäumler, A. J. (2011). A breathtaking feat: To compete with the gut microbiota, Salmonella drives its host to provide a respiratory electron acceptor. *Gut Microbes, 2*, 58–60.

Winter, S. E., Thiennimitr, P., Winter, M. G., Butler, B. P., Huseby, D. L., Crawford, R. W., Russell, J. M., Bevins, C. L., Adams, L. G., Tsolis, R. M., et al. (2010). Gut inflammation provides a respiratory electron acceptor for Salmonella. *Nature, 467*, 426–429.

Wu, S., Rhee, K.-J., Albesiano, E., Rabizadeh, S., Wu, X., Yen, H.-R., Huso, D. L., Brancati, F. L., Wick, E., McAllister, F., et al. (2009). A human colonic commensal promotes colon tumorigenesis via activation of T helper type 17 T cell responses. *Nature Medicine, 15*, 1016–1022.

Yatsunenko, T., Rey, F. E., Manary, M. J., Trehan, I., Dominguez-Bello, M. G., Contreras, M., Magris, M., Hidalgo, G., Baldassano, R. N., Anokhin, A. P., et al. (2012). Human gut microbiome viewed across age and geography. *Nature, 486*, 222–227.

Yin, Y., Fan, B., Liu, W., Ren, R., Chen, H., Bai, S., Zhu, L., Sun, G., Yang, Y., & Wang, X. (2017). Investigation into the stability and culturability of Chinese enterotypes. *Scientific Reports, 7*, 7947.

Young, N. M., Brisson, J. R., Kelly, J., Watson, D. C., Tessier, L., Lanthier, P. H., Jarrell, H. C., Cadotte, N., St Michael, F., Aberg, E., et al. (2002). Structure of the N-linked glycan present on multiple glycoproteins in the Gram-negative bacterium, Campylobacter jejuni. *The Journal of Biological Chemistry, 277*, 42530–42539.

Young, K. T., Davis, L. M., & Dirita, V. J. (2007). Campylobacter jejuni: Molecular biology and pathogenesis. *Nature Reviews. Microbiology, 5*, 665–679.

Yrios, J. W., & Balish, E. (1986). Pathogenesis of Campylobacter spp. in athymic and euthymic germfree mice. *Infection and Immunity, 53*, 384–392.

Yurist-Doutsch, S., Arrieta, M.-C., Vogt, S. L., & Finlay, B. B. (2014). Gastrointestinal microbiota-mediated control of enteric pathogens. *Annual Review of Genetics, 48*, 361–382.

Zasloff, M. (2002). Antimicrobial peptides in health and disease. *The New England Journal of Medicine, 347*, 1199–1200.

Zhu, W., Winter, M. G., Byndloss, M. X., Spiga, L., Duerkop, B. A., Hughes, E. R., Büttner, L., de Lima Romão, E., Behrendt, C. L., Lopez, C. A., et al. (2018). Precision editing of the gut microbiota ameliorates colitis. *Nature, 553*, 208–211.

Zumbrun, S. D., Melton-Celsa, A. R., Smith, M. A., Gilbreath, J. J., Merrell, D. S., & O'Brien, A. D. (2013). Dietary choice affects Shiga toxin-producing Escherichia coli (STEC) O157:H7 colonization and disease. *Proceedings of the National Academy of Sciences of the United States of America, 110*, E2126–E2133.

Microbiome and Diseases: Colorectal Cancer

15

Amina Iftekhar, Andreas Sperlich, Klaus-Peter Janssen, and Michael Sigal

Abstract

Cancers of the large intestine are among the most frequent malignomas worldwide and also rank among the most frequent causes for cancer-related mortality in developed countries, with an even increasing incidence in an aging population. Patient survival and treatment options in the metastatic form of this disease are still relatively poor. The cell-autonomous genetic and epigenetic changes associated with carcinogenesis, and the stepwise and consecutive progression along the adenoma-carcinoma sequence in the colorectum, have been studied intensively over the last decades. However, there is a growing interest in the impact of gut microbial communities on the initiation and progression of this cancer entity. Overwhelming evidence meanwhile suggests that the microbiota is an important and potentially causative factor for colorectal cancer (CRC). A disturbance in the microbial community may lead to impairment of epithelial barrier function, imbalance in epithelial self-renewal, DNA damage, and altered immune responses, thereby fostering tumor initiation and progression.

Authors "Klaus-Peter Janssen" and "Michael Sigal" share senior authorship for this chapter.

A. Iftekhar
Department of Molecular Biology, Max Planck Institute for Infection Biology, Berlin, Germany
e-mail: iftekhar@mpiib-berlin.mpg.de

A. Sperlich · K.-P. Janssen (✉)
Department of Surgery, Klinikum rechts der Isar, TUM, Munich, Germany
e-mail: andreas.sperlich@tum.de; klaus-peter.janssen@tum.de

M. Sigal
Department of Molecular Biology, Max Planck Institute for Infection Biology, Berlin, Germany

Department of Gastroenterology, Charité Universitätsmedizin, Berlin, Germany
e-mail: michael.sigal@charite.de

15.1 Introduction: Colorectal Cancer

Cancer is a multifactorial disease caused by genetic predispositions and environmental factors resulting in accumulation of genetic and epigenetic mutations leading to uncontrolled cell proliferation. According to the World Health Organization (WHO), CRC was the third most common cause of cancer related deaths worldwide in 2015. CRC is known to result from a well-established sequential cascade of mutations leading to loss of function of tumor suppressor genes or activation of oncogenes, where key genetic changes are associated with various stages of cancer progression (Fearon and Vogelstein 1990). The vast majority of CRCs harbor mutations in the canonical Wnt pathway leading to an increased cell proliferation. In addition, common mutations include the epidermal

D. Haller (ed.), *The Gut Microbiome in Health and Disease*,
https://doi.org/10.1007/978-3-319-90545-7_15

growth factor (EGF) pathway, p53 mutations, and mutations in the transforming growth factor-β (TGF-β) signaling pathway. Genetic lesions can be hereditary or acquired. Hereditary mutation in the adenomatous polyposis coli (APC) gene leads to a polyposis syndrome that is associated with development of CRC in all affected individuals. Thus, CRC is a genetic disease, but the vast majority of cases occur sporadically without a known genetic predisposition. Induction, accumulation, and persistence of mutations in the tissue are influenced by a variety of environmental factors, such as diet, lifestyle, comorbidities, and especially history of inflammatory bowel disease (IBD).

The human colon is a densely populated microbial ecosystem (Simon and Gorbach 1984). The gastrointestinal microbiota plays a crucial role in health and disease. The microbiota can be beneficial to the host by providing protection against pathogen, helping with digestive processes, contributing to metabolic pathways, and shaping gastrointestinal immune system. Absence of a healthy microbiota can be a result of an active disease or caused due to dietary habits. CRC is more prevalent in developed countries, believed to be associated with lower nutritional diversity in the food intake. An infection or nutritional imbalance leads to reduced colonization by beneficial microorganisms and an enrichment of pro-carcinogenic bacterial groups. Dysbiosis disturbs the immune system homeostasis causing inflammation, disrupting mucosal barrier, and increasing epithelial permeability, creating a microenvironment that further perpetuates pro-carcinogenic dysbiosis.

15.2 Alteration of Microbiota and Colorectal Carcinogenesis

Mouse models that develop colitis and promote neoplasia support the involvement of microbes in tumor development. Interleukin-10 (IL-10) is an anti-inflammatory cytokine, and IL-10 null (Il10−/−) mice are a recognized model for human inflammatory bowel disease. Il10−/− mice treated with a carcinogenic substance azoxymethane (AOM) develop chronic enterocolitis and colitis-associated adenocarcinomas, the severity of which is dependent on their microbiota. Germfree (GF) Il10−/− mice do not develop colitis (Sellon et al. 1998). An alternate model involves combining AOM and dextran sulfate sodium (DSS), to provide a two-step tumor model of colitis-associated cancer. DSS is an inflammatory agent that induces colonic epithelial damage and causes colitis in mice, mimicking features of human IBD. Transfer of microbiota from tumor-bearing AOM/DSS mice into germfree mice significantly increased the number and size of tumors compared to germfree mice inoculated with healthy microbiota (Zackular et al. 2013). In a later study, it was observed that stool from patients with CRC could promote colorectal carcinogenesis in mice. Stool from patients with CRC increases the numbers of polyps, intestinal dysplasia, proliferation, and inflammation in germfree mice and conventional mice injected with azoxymethane compared to mice fed with stool samples from healthy individuals (Wong et al. 2017). Thus, risk factors for development of CRC can be transferred with the pro-carcinogenic microbiota. Further, studies have proposed that the fecal microbiota changes as CRC progresses and can be potentially used for cancer screening and early diagnosis (Zeller et al. 2014). Metagenome-wide association studies (MGWAS) on stools from healthy subjects and advanced adenoma and carcinoma patients revealed a deficiency in lactic acid-producing commensals in the latter group. High intake of red meat relative to fruits and vegetables was shown to be a risk factor associated with this shift in microbiota, potentially promoting carcinogenesis (Feng et al. 2015).

In addition to these data linking the microbial composition to the risk for CRC development, specific bacteria that colonize the gut and their virulence factors have been associated with CRC. Mechanistic studies have revealed sophisticated bacterial tools that are mainly used by the bacteria for their own benefit and survival, but can result in disruption of genetic integrity and activation of aberrant pro-carcinogenic signaling in the colonic epithelium. Furthermore, the ability of bacteria to induce specific immune responses that promote

CRC development is currently under investigation, and major principles and advances are highlighted in the second part of this chapter.

15.3 Mechanisms by Which Bacteria Induce Aberrant Events Associated with Carcinogenesis

The finding showing that the gastric pathogen *Helicobacter pylori* is a causative agent for gastric cancer revolutionized our thinking of the pathophysiology of gastrointestinal disorders. These bacteria are now recognized as a type I carcinogen by the WHO. The association of *H. pylori* with gastric cancer was confirmed by many epidemiological studies (Parsonnet et al. 1991). In addition, several laboratories around the world were involved in discoveries of molecular mechanisms that drive the carcinogenesis (Amieva and El-Omar 2008). Specific virulence factors associated with cancer progression were identified. For example, a molecular syringe, the so-called type IV secretion system, was demonstrated to inject a CagA virulence factor into the host cells. The host cells do not recognize CagA as a foreign component and phosphorylate this protein. CagA phosphorylation induces many aberrant events in host cells including alteration in epithelial polarity (Amieva et al. 2003), proliferation, and migration (Bagnoli et al. 2005), and presence of CagA is associated with an increased cancer risk. In addition to CagA, other virulence factors have been identified that can alter the integrity of host DNA by causing DNA double-strand breaks (Koeppel et al. 2015), interfering with stemness (Sigal et al. 2015) and apoptosis and by blocking the responses of infected cells to inflammatory stimuli (Morey et al. 2018).

In vivo *H. pylori* has been demonstrated to persist either as free-swimming bacteria in the mucus or invade into gastric glands and attach to the apical junctions of epithelial cells (Sigal et al. 2015). *H. pylori* induces an expansion of long-lived stem cells and increases proliferation (Sigal et al. 2017), probably as a host cell response to the infection. The expanded stem cell compartment can be directly colonized by the bacteria leading to changes in signaling and potentially disrupting epithelial integrity of long-lived stem cells.

Studies about *H. pylori* and gastric cancer have elucidated several principles of how bacteria can manipulate epithelial cells. More recently, associations between other cancer types with the members of the microbiota were made, and specific molecular pathways of how epithelial integrity can be manipulated by these bacteria were described. We will summarize in this chaper the most recent findings about the members of the gut microbiota and gut pathogens that are linked to CRC and focus on described molecular pathways that were discovered to be linked with aberrant epithelial signaling and carcinogenesis.

15.3.1 *Salmonella enterica* Typhi/Paratyphi A

Typhoid fever is a life-threatening disease caused by the human-restricted *Salmonella enterica* serovars Typhi and Paratyphi A. Epidemiological studies have linked chronic carriage of *Salmonella enterica* Typhi/Paratyphi A with gallbladder cancer and the incidence of this adenocarcinoma is higher in countries where typhoid fever is endemic (Nagaraja and Eslick 2014). *Salmonella* resides in the gallbladders of chronic carriers intracellularly and extracellularly by forming biofilms on gallstones, which serve as a reservoir from where the bacteria are sporadically shed into the duodenum (Gonzalez-Escobedo and Gunn 2013). A key mechanism with which these human-specific *Salmonella enterica* serovars provoke genome instability is by the synthesis of the typhoid toxin that causes DNA damage (Song et al. 2013). Formation of actin stress fibers and cell distension has been observed in cells exposed to typhoid toxin-producing *Salmonella* (Haghjoo and Galan 2004). The genotoxicity of the typhoid toxin is dependent on the expression of the CdtB subunit, which is functionally and structurally homologous to the mammalian DNase I (Lara-Tejero and Galan 2000; Nesic et al. 2004) and is translocated into the nucleus of the intoxicated cell (Guidi et al. 2013b). This toxin is expressed by the bacterium only once it is internalized into the host cell and replicates within specialized vacuoles, known as *Salmonella*-containing vacuoles (Spano et al. 2008). The toxin is then synthesized and released within outer membrane vesicles, where it is protected from host proteases. The toxin can be retrogradely translocated to the nucleus through the trans-Golgi network where it causes DNA double-strand breaks (Guidi et al. 2013a). It has been shown that brefeldin A, an inhibitor of the transport of proteins from the endoplasmic reticulum to the Golgi, was able to inhibit the DNase activity of the toxin (Salcedo and Holden 2003). In addition, CdtB subunit can also be directly internalized into the host cells through the endocytic pathway (Guidi et al. 2013b).

Whether the described mechanisms of DNA damage induced by *Salmonella* species also occurs in the colon is not clear, but recent studies suggest an increased presence of *Salmonella* in premalignant or malignant colonic lesions (Mughini-Gras et al. 2018). Furthermore, typhoid toxin was shown to alter intestinal inflammation and promote persistence of *Salmonella* species in the host, suggesting that a chronic exposure of epithelial cells to the toxin could occur (Del Bel Belluz et al. 2016).

15.3.2 *Helicobacter hepaticus*

The CdtB subunit is common to another bacterial genotoxin, the cytolethal distending toxin (CDT). Multiple gram-negative bacterial species produce CDT, such as *Helicobacter hepaticus*. It has been directly linked to chronic hepatocellular and colon carcinoma in mice. *H. hepaticus* is a gram-negative, spiral-shaped, microaerophilic bacterium that is found predominantly in the cecum and colon of mice. CDT is a tripartite holotoxin consisting of three subunits, CdtA, CdtB, and CdtC (Lara-Tejero and Galan 2001). CdtB is the active domain and is delivered into the target cells with the aid of the accessory proteins, CdtA and CdtC (Haghjoo and Galan 2004). CDT has been shown to be essential for hepatic and intestinal persistence of *H. hepaticus* in infected mice, and these animals have an altered cytokine response (Ge et al. 2005; Pratt et al. 2006). CDT plays a key role in promoting dysplastic changes and increased hepatocyte proliferation in the livers of *H. hepaticus*-infected A/JCr mice (liver disease-susceptible mice). CDT-producing *H. hepaticus* can cause overproduction of antiapoptotic proteins, upregulate the NF-kB pathway, and cause subsequent inflammation-associated preneoplastic liver lesions (Ge et al. 2007).

CDT is essential for modulating host immune response by downregulating anti-inflammatory cytokine IL-10 in the colon of Swiss Webster mice to cause *H. hepaticus*-induced colitis and colon carcinoma (Ge et al. 2005). Recombinase-activating gene 2 (Rag-2)-deficient mice lack functional T and B lymphocytes because of an inability to initiate V(D)J rearrangement. *H. hepaticus* has been shown to induce colon tumorigenesis in Il10−/− mice and Rag-2-deficient mice (Erdman et al. 2003; Nagamine et al. 2008). CdtB-deficient *H. hepaticus* mutants have been reported to cause less severe typhlocolitis in Il-10−/− mice when compared to WT *H. hepaticus* (Young et al. 2004). CDT-producing *H. hepaticus* causes activation of the Tnfα/Il-6-Stat3 signaling pathway in the ceca of Rag-2-deficient mice (Ge et al. 2017). Expression of IL-10 by transferred CD4+ and CD25+ T regulatory cells in Rag-2-deficient mice inhibits the development of colitis following *H. hepaticus* infection (Erdman et al. 2003; Maloy et al. 2003). Chronic exposure to sublethal levels of recombinant CDTs from *H. hepaticus* has been shown to induce CdtB-dependent genomic instability and anchorage-independent growth (Guidi et al. 2013a). This study suggested that long-term exposure to CDT allows for the selection of cells that have overcome the tumorigenesis barrier, thus favoring tumor progression.

15.3.3 *pks+ Escherichia coli*

Another bacterial genotoxin, colibactin, was identified nearly 20 years after the discovery of CDT (Nougayrede et al. 2006). Colibactin is a secondary metabolite, a polyketide/nonribosomal peptide hybrid compound synthesized by a complex biosynthetic machinery. This toxin is encoded by a 54-kb *pks* pathogenicity island that is expressed by several *Enterobacteriaceae* members, such as *Escherichia coli* belonging to phylogenetic group B2. *E. coli* strains belonging to group B2 are less frequent in the environment but represent 30–50% of strains isolated from the feces of healthy humans in high-income countries (Escobar-Paramo et al. 2004). *E. coli* is a predominant gram-negative, facultative anaerobe that colonizes the human intestine few days after birth, and B2 phylogenetic strain has been found to be especially adapted for persisting in the gut microbiota throughout the lifetime of the host (Nowrouzian and Oswald 2012; Nowrouzian et al. 2005). The toxin is produced by the 19 genes (*clbA* to *clbS*) present on the *pks* genomic island. Similar to CDT, infection of epithelial cells with *pks+ E. coli* was shown to promote cell cycle arrest and a progressive enlargement of the cell body and nucleus and cause DNA double-strand breaks. The bacterial supernatant lacks toxicity, and the cytopathic effect of this toxin is contact-dependent of live bacteria with the host cell (Nougayrede et al. 2006). The mechanism by which colibactin is introduced into the host cell by the bacterium is unknown.

clbK gene of the *pks* island codes for thiazole-forming nonribosomal peptide synthases. Thiazole rings are present in pharmacophores, such as bleomycin and are known to intercalate DNA (Nougayrede et al. 2006; Vizcaino and Crawford 2015). The prodrug, precolibactin, is exported into the periplasm by efflux pump *clbM* and is cleaved by the periplasmic membrane-bound *clbP* peptidase to generate the active genotoxin. It was shown that precolibactin causes in vitro DNA alkylation and interstrand cross-links, demonstrating the DNA-damaging ability of this toxin (Vizcaino and Crawford 2015). It was later reported that *pks+ E. coli* were found in a significantly higher percentage in inflammatory bowel disease and CRC patients. Precancerous lesions exert a positive selection for the colonization and expansion of *pks+ E. coli* strains that were genetically diverse (Arthur et al. 2012; Buc et al. 2013; Sarshar et al. 2017). These bacteria also promoted invasive carcinoma in Il10−/− mice (Arthur et al. 2012). Progression of inflammation and development of CRC in Il10−/− mouse model was shown to induce nine genes in the *pks* island of *E. coli* NC101 (Muehlbauer et al. 2013). These studies indicate that colitis can alter microbial composition and microbial transcriptome response and cause expansion of genotoxic microorganism that can promote colon tumorigenesis. Inflammation is essential for *E. coli* to promote CRC in AOM/Il10−/− mice. Inflammation alters the natural barrier function of the colon epithelium, making it more readily assessable to the bacterium (Arthur et al. 2014). Short exposure of intestinal epithelial cells to *pks+ E. coli* has been shown to cause cells to divide despite incomplete DNA damage repair. Phenotypes observed include anaphase bridges, micronuclei, aneuploidy, ring chromosomes, and chromosome aberrations. Infected cells also have an increased gene mutation frequency and grow anchorage independently, demonstrating the mutagenic and transforming potential of the toxin (Cuevas-Ramos et al. 2010). DNA damage caused by *pks+ E. coli* has been shown to result in senescence and senescence-associated secretory phenotype (SASP). The prooxidant and pro-inflammatory mediators secreted by these senescent cells have been shown to cause DNA damage in bystander cells and promoted the growth of human colon carcinoma cells in a paracrine manner (Secher et al. 2013). Another study showed that colibactin induced SASP results in secretion of growth factors that promote proliferation of naïve recipient cells. Three-hour infection of tumor cells was sufficient to stimulate tumor growth of infected cell line in a xenograft model (Cougnoux et al. 2014). Further, early intestinal colonization by *pks+ E. coli* in neonatal rats has been shown to impair intestinal barrier function development, making them more vulnerable to develop intestinal immune-mediated diseases in adulthood (Secher et al. 2015).

15.3.4 *Fusobacterium nucleatum*

Fusobacterium nucleatum is an anaerobic gram-negative bacterium, which is part of the normal flora of the human oral and gut mucosa. Numerous studies have shown a predominant role of

F. nucleatum in the neoplastic process in colon cancer. Colon biopsies from adenomas showed a greater proportion of this bacterium, when compared to the adjacent normal mucosa. The same study showed that in *Apc*Min/+ mouse model (mice heterozygous for the tumor suppressor gene, APC) of intestinal tumorigenesis, *F. nucleatum* increased colorectal tumorigenesis by attracting tumor-infiltrating myeloid immune cells that can promote tumor progression. Infection of ApcMin/+ mice also resulted in the NF-κB pro-inflammatory signature known to facilitate colorectal tumorigenesis (Kostic et al. 2013; Yang et al. 2017). *F. nucleatum* adhesin, FadA helps the bacterium to attach and invade cancer cells. This adhesin has been shown to play an additional role of activating β-catenin and promoting tumor growth by modulating E-cadherin upon binding to it on epithelial cells. The study also showed that FadA gene levels are higher in the colon tissue from patients with adenomas compared to normal individuals. FadA stimulated the growth of human colon cancer cell lines that carry either an APC mutation or a β-catenin mutation and could not stimulate the growth of a noncancerous cell line, suggesting that it promotes carcinogenesis in cells with an existing tumor-initiating somatic mutation (Rubinstein et al. 2013). Another surface protein of the bacterium, Fap2 has been shown to bind to a host cell factor known to be overexpressed in human colorectal adenocarcinoma leading to fusobacterial enrichment (Abed et al. 2016). Fap2 has also been reported to facilitate impairment of host antitumor immunity. Fap2 binds and activates a receptor TIGIT on T cells and natural killer cells. Fap2 and TIGIT interaction leads to reduction in the cytotoxicity of infiltrating natural killer cells and lymphocytes toward tumor cells (Gur et al. 2015). Fap2 has also been shown to induce apoptosis in lymphocytes (Kaplan et al. 2005, 2010). These studies indicate that Fap2 is utilized by tumors to impair and evade the immune system. Further, studies have shown that *F. nucleatum* high status is associated with CpG island methylator phenotype and microsatellite instability in CRCs (Tahara et al. 2014). Recent findings also propose that *Fusobacterium* colonizes distant metastases from primary CRC. The bacterium and the associated microbiome also survive mouse xenografts of human primary colorectal adenocarcinomas and were retained through sequential xenograft passages. Treatment of these xenograft-bearing mice with metronidazole antibiotic decreased *Fusobacterium* load, cancer cell proliferation, and tumor growth (Bullman et al. 2017). In addition, *F. nucleatum* was shown to promote CRC chemoresistance by modulating autophagy (Yu et al. 2017).

15.3.5 *Bacteroides fragilis*

Bacteroides fragilis are gram-negative, obligate anaerobes and opportunistic pathogens that comprise 1–2% of the total colon microbial community (Dejea et al. 2013). Based on the absence or presence of the *bft* gene, they are classified into nontoxigenic *B. fragilis* (NTBF) and enterotoxigenic *B. fragilis* (ETBF), respectively. Enterotoxigenic *B. fragilis* (ETBF) strains are implicated in diarrheal disease, inflammatory bowel disease, and colorectal cancer (Basset et al. 2004; Sears 2009; Toprak et al. 2006; Wu et al. 2009). Their pathogenicity is due to a 20-kDa, heat-labile, zinc-dependent metalloprotease toxin called *B. fragilis* toxin (BFT). Binding of BFT to intestinal epithelial cells is temperature dependent and occurs only at 37 °C and is sensitive to cholesterol depletion. Metalloprotease activity of the toxin is essential for the binding, and this binding is resistant to acid washing, suggesting an irreversible interaction (Wu et al. 2006). Treatment of human colonic epithelial cell lines with BFT results in a time and concentration-dependent redistribution of actin microfilaments (F-actin), as well as an increase in cell volume. This study suggested that these changes in F-actin and cell volume by ETBF infection might lead to an alteration in tight-junction integrity in the intestinal epithelium, contributing to the pathogenesis of diarrhea (Koshy et al. 1996). BFT was also found to have greater biological activity on the basolateral membrane of polarized intestinal epithelial cells (Chambers et al. 1997; Obiso et al. 1997). Later studies revealed that *B. fragilis* alter tight junctional function of intestinal epithelial cells by cleaving tumor suppressor E-cadherin, the cellular substrate of BFT, thereby increasing mucosal permeability and activating the β-catenin signaling (Wu et al. 1998, 2003).

Fecal carriage of ETBF is 10–20% in the healthy population, as opposed to 40% for CRC patients (Toprak et al. 2006). The *bft* gene is more abundantly found in the colonic mucosa than in luminal samples of colorectal cancer (Boleij et al. 2015). Increased abundance of ETBF is seen in early-stage carcinogenic lesions, underlining its potential role in development of colorectal neoplasia (Purcell et al. 2017). ETBF colonization in C57BL/6J mice results in augmentation of dextran sodium sulfate (DSS)-induced colitis (Rabizadeh et al. 2007). BTF expression is essential for ETBF to induce persistent colitis in wild-type C57BL/6 mice (Rhee et al. 2009). ETBF triggers a Stat3- and T_{H17}-dependent pathway of inflammation-induced cancer in ApcMin/+ mice, and IL-17- and IL-23-blocking antibodies inhibited the ETBF-induced colon tumors (Wu et al. 2009). ETBF-induced inflammation is sufficient to produce microadenomas within 5 days in ApcMin/+ mice, but very early clearance of ETBF with antibiotics results in decreased mucosal IL-17A expression, allowing for the regression of the disease (DeStefano Shields et al. 2016). BFT-stimulated IL-8 production involves tyrosine kinase-dependent activation of nuclear factor-κB (NF-κB) and activation of the mitogen-activated protein kinases (MAPKs) (Wu et al. 2004). BFT-induced spermine oxidase (SMO) results in increased reactive oxygen species (ROS) production and consequent DNA damage in cell lines. SMO expression was upregulated in C57BL/6 mice with ETBF-induced colitis. Inhibitors of SMO significantly reduced ETBF-induced colon tumorigenesis in ApcMin/+ mice (Goodwin et al. 2011). Overall, research suggests that persistent ETBF infection may perturb apical junctional function and increase pro-inflammatory responses, resulting in premalignant transdifferentiation in the colon.

15.4 Immunological Aspects of Carcinogenesis in the Colorectum

The interactions between the immune system and solid tumors are highly complex; immune cells can have both tumor-promoting and tumor-inhibiting roles. As early as 1909, the German physician Paul Ehrlich proposed a cancer suppressive of the immune system (Strebhardt and Ullrich 2008), and deregulation of immunosurveillance is now considered a central hallmark of cancers (Hanahan and Weinberg 2011). Solid evidence indicates high intratumoral numbers of T lymphocytes, especially of CTLs and the T_{H1}-type subset, that predict good prognosis in colorectal cancer. Furthermore, there is ample preclinical and clinical evidence for tumor-enhancing effects of inflammatory processes, involving components of the immune system, as well as epithelial and mesenchymal cells. In this context, it is important to distinguish between two distinct phenomena, "inflammation-induced carcinogenesis" and "cancer-induced inflammation," which differ mechanistically, even though they may not occur in a mutually exclusive fashion. Inflammation-induced cancers of the large intestine are well-documented, though they represent only a minority of all clinically diagnosed cases of colorectal cancer (1–2%) (Breynaert et al. 2008). Of note, patients with inflammatory bowel diseases, such as ulcerative colitis or Crohn's disease, have a well-documented increased cancer risk (Bernstein et al. 2001; Eaden et al. 2001). However, even though the large majority of patients develop colorectal cancer in the absence of chronic inflammatory bowel disease, the immune system and inflammatory processes may still play a decisive role. The "cancer-induced inflammation" can be apparent in the formation of ectopic lymphoid structures which are frequently associated with colorectal cancer. The structures are heterogeneous from patient to patient, as well as within given tumors, but have been described as a positive prognostic factor (Coppola et al. 2011; Dieu-Nosjean et al. 2014; Pitzalis et al. 2014). Furthermore, pro-inflammatory signaling pathways are functional in many colorectal cancers, in addition to the well-established oncogenic growth factor receptor pathways (Fre et al. 2008), as evidenced clinically by the demonstrated efficacy of nonsteroidal anti-inflammatory drugs in the prevention of colorectal cancer (Lasry et al. 2016). Further, recent large-scale genome and transcriptome sequencing efforts have successfully defined molecular subgroups in colorectal cancer, revealing the existence of highly immunogenic tumors with good prognosis, associated with DNA mismatch repair defects, as well as mesenchymal-type tumors with signs of inflammation and bad prognosis (Dienstmann et al. 2017).

A central focus of current research is to unravel the still largely unresolved mechanisms underlying T-cell infiltration and activation in colorectal tumors, as well as the functional involvement of gut microbiota in these processes. In this section, we will address the major cell types and pathways from both the adaptive and the innate branch of the immune system that are currently known to be involved in the initiation and progression of colorectal cancer and discuss their contribution to the response toward antitumoral therapies.

15.4.1 Adaptive Immunity

As stated earlier, infiltration by immune cells is strongly linked with clinical outcome in patients with colorectal cancer (Fridman et al. 2012; Galon et al. 2006). High densities of cytotoxic effector T cells (Galon et al. 2006; Pages et al. 2010), as well as different subsets of CD4+ T-helper cells, such as T-helper type 1 cells expressing IFN-γ (Tosolini et al. 2011) and follicular-type T-helper cells expressing the chemokine receptor CXCR5 (Bindea et al. 2013), are associated with increased postoperative survival. In accordance with the positive role of T-helper cells, expression of HLA class II antigens was shown to be correlated with favorable clinical outcome (Sconocchia et al. 2014). However, the prognostic significance and contribution of interleukin (IL)-17-producing T-helper cells (T_{H17}) is being controversially discussed: intratumoral infiltration of T_{H17}-type cells was associated either with worse or improved prognosis (Amicarella et al. 2017; Tosolini et al. 2011). Somewhat surprisingly, and in contrast to observations for other solid tumor entities (Pere et al. 2012), immunosuppressive Foxp3+ regulatory T cells (T_{regs}) were associated with good prognosis in colorectal cancer (Frey et al. 2010; Salama et al. 2009). The group of J. Galon from Paris, France, has been able to establish infiltrating T-cell subsets as prognostic factors with high clinical relevance over the past decade (Galon et al. 2006, 2007; Pages et al. 2005, 2010). Of note, increased tumor infiltration by immune cells and high expression of specific T_{H1}-type transcripts were significantly correlated with the absence of histopathological parameters of early metastatic invasion (venous, lymphatic, and perineural invasion). Importantly, the density of tumor-infiltrating T lymphocytes (TILs), which can be quantified in a standardized way by immunohistochemical staining for the cell surface marker CD3 on tissue sections, had stronger predictive power for patients' survival than the well-established clinicopathological markers, such as the tumor-node-metastasis (TNM) classification system (Galon et al. 2006). The TNM system represents the current clinical standard in colorectal cancer and is accepted worldwide (UICC/AJCC), even though it has been shown to have significant limitations (Nitsche et al. 2011). Upon statistical multivariate regression analysis, the TIL density was retained as independent prognostic parameter, in contrast to the TNM stage, after adjustment to T-cell infiltration. The quantification of tumor-infiltrating CD3-positive cells further provided an impressive parameter for patient stratification. Patients with a high TIL density had a postoperative 5-year survival rate of 73%, whereas the subgroup with low T-cell density had a dismal outcome with a 5-year survival rate of only 30% (Galon et al. 2006). Among the tested T-cell subsets, CD8-positive memory T cells with the specific surface marker CD45RO had the highest prognostic impact, regarding both postoperative overall survival and disease-free survival.

Based on these impressive results, a scoring system based on histological evaluation has been proposed for colorectal as well as other solid cancers, called the "immunoscore" (Galon et al. 2014). This score takes into account the relative density of TILs in the center of the tumor, as well as in the invasive margin. As most important cell types for the score, CD45RO-/CD8-positive cells are retained. The subgroup of patients with a high density of memory CD8 T cells both in the center of the tumor and the invasive margin is associated with good prognosis, and the immunoscore is more strongly associated with disease relapse compared to the TNM system (Mlecnik et al. 2011). However, despite large efforts, the quantification of TILs in colorectal cancer has still not entered routine clinical practice, since the immune infiltrate is highly heterogeneous, with great interpatient as well as intratumoral differences, and the quantification of TILs is still far from being standardized.

15.4.2 Chemokine Networks Linking Gut Microbiota and Tumor-Infiltrating Immune Cells

Recently, research has focused on the chemotactic factors driving adaptive immune cell populations into colorectal cancer tissues, which are still largely undefined. Moreover, putative responding subsets within immune-infiltrating cell populations are still far from being fully understood. Expression of chemokines, secreted immune mediators that chemotactically recruit T cells, including CXCL9, CXCL10, CXCL16, and CX3CL, was reported to correlate with high densities of tumor-infiltrating lymphocytes (TILs) and predicted favorable clinical outcome (Hojo et al. 2007; Mlecnik et al. 2010). In accordance, we identified interferon-regulated CXC chemokines as excellent predictors of survival in colorectal cancer (Kistner et al. 2017). Increased intratumoral expression of CXCL9 and CXCL11 was significantly correlated with postoperative tumor-specific survival in patients with colon cancer; both parameters were retained as independent prognostic indicators upon multivariate analysis (Kistner et al. 2017). Of note, the three chemokines CXCL9, CXCL10, and CXCL11 share one common receptor, CXCR3, which is expressed on T_{H1} T cells and CTLs and induces their chemotactic recruitment (Groom and Luster 2011). Further chemokines of the CXC family were found to be strongly deregulated in colorectal cancer, CXCL3 (GRO3) being associated with metastasis formation and CXCL8 (Interleukin-8) with postoperative survival (Doll et al. 2010). Analysis of colorectal cancer cell lines and genetic mouse models for digestive tract cancer, as well as an ex vivo explant culture model based on resected human colon tumors, demonstrated that colorectal cancer cells produce significant amounts of CXCL11 and CXCL10 upon cytokine stimulation with IFN-γ or TNF. On tissue level, immunohistochemical analysis revealed that tumor cells are a major source of CXCL11 (Kistner et al. 2017). Moreover, high intratumoral chemokine co-expression of CXCL9, CXCL10, and CXCL11 was highly significantly correlated with the presence of CTLs and CD4+ T-helper cells within the tumor. However, no clear association with blood vessel density was found. In order to test the hypothesis that elevated CXC chemokines have a causal effect on tumorigenesis, an orthotopic mouse colorectal cancer model was established based on the isogenic murine colorectal carcinoma cells, stably expressing CXCL10. The in vivo model demonstrated that tumor growth was inhibited by intratumoral expression of CXCL10, largely mediated by the adaptive immune system (Kistner et al. 2017). Thus, CXC chemokines are promising potential targets for therapeutic intervention. Moreover, their expression predicts good response to neoadjuvant radiochemotherapy in rectal cancer, as shown by analysis of three independent patient collectives from Germany, Italy, and the USA (Agostini et al. 2015).

Of note, chemokine sources and microbial-derived stimuli leading to altered chemokine production within intestinal tissue remain largely unknown. Cell-autonomous genomic changes within colorectal cancer cells have been described, which may lead to amplification or conversely loss of function of chemokine genes (Bindea et al. 2013). In addition to genomic alterations, epigenetic silencing of T-helper type 1 chemokines through the polycomb repressive complex 2 has been described in colorectal cancer (Nagarsheth et al. 2016).

In addition to genetic and epigenetic effects, the gut microbiota play a pivotal role for chemokine secretion. There is ample evidence suggesting that gut commensal bacteria translocate across a dysfunctional epithelial barrier, stimulating the recruitment of immune cells in the lamina propria, which in turn secrete pro-inflammatory cytokines (Grivennikov et al. 2012). In a recent study, we evaluated whether gut flora-derived microbial stimuli induce the production of chemotactic factors, as well as the chemokine signaling network underlying and shaping T-cell infiltration into colorectal cancer (Cremonesi et al. 2018). Previous studies showed that gut commensal bacteria translocated across the neoplastic epithelium may interact with tumor cells and induce direct protumorigenic effects or by secretion of cancer-promoting cytokines (Arthur et al. 2012; Grivennikov et al. 2012; Rubinstein et al. 2013). Of note, upon stimulation by gut commensal bacteria in vitro and in vivo, colorectal cancer cell lines showed a strongly upregulate expression of multiple chemokines, recapitulating the profiles of chemokine gene expression of primary cancer cell isolates. Interestingly, exposure of tumor cells to gut bacteria ultimately results in higher T-cell recruitment in vivo in cancer xenografts, revealing a role of gut commensal bacteria in controlling extent of tumor infiltration by beneficial immune cells (Cremonesi et al. 2018). Consistent with findings obtained in mouse models, the extent of T-cell infiltration in human colorectal cancer was significantly associated with the presence of specific bacterial families and genera. Furthermore, the abundance of defined bacterial families was shown to be linked with the expression of specific chemokines, indicating that gut commensal bacteria induce the intratumoral production of specific immune cell-recruiting chemokines. Importantly, the composition of gut flora significantly predicted postoperative survival of patients with colorectal cancer. Even though much remains to be learned about the interactions between gut microbiota and the recruitment and activation state of intratumoral immune cell populations, it is conceivable that the individual gut microbiome in patients with colorectal cancer acts in concert with genetic alterations in cancer cells to determine the extent of immune cell infiltration, with pivotal effects on disease outcome. A central unresolved question remains whether specific individual bacterial species or strains or rather complex microbial communities contribute to altered chemokine expression patterns and immune cell infiltration in human colorectal cancer. This analysis is complicated by the well-established fact that a substantial fraction of human gut microbiota species remain to be described, are hitherto uncharacterized, and cannot be cultured. However, recent data actually show that different bacterial species may promote the expression of T-cell-recruiting chemokine genes. Ex vivo analysis of human samples showed that Firmicutes and in particular Lachnospiraceae and Ruminococcaceae were most significantly associated with the expression of T-cell-attracting factors. The abundance of *Bacteroides*, *Proteobacteria*, and, in particular, *Reyranella* was associated with the expression of most T-cell-recruiting chemokines, as well as with intratumoral densities of all T-cell subsets and furthermore with favorable disease outcome (Cremonesi et al. 2018). Moreover, defined bacteria types were associated with the concomitant expression of several key chemokines, indicating their capacity to promote parallel recruitment of different T-cell populations, like CTLs and helper cells (T_{H1}-type and CXCR5-positive) observed in cases with good prognosis. Of note, *Fusobacteria* species were recently reported to be associated with reduced survival rate for colorectal cancer patients (Mima et al. 2016), and were found in our study to be enriched in the subgroup of patients with low TIL density, and could evoke expression of T-cell-recruiting chemokines upon coculture with human colon cancer cell lines (Cremonesi et al. 2018). It has been described that *F. nucleatum* inhibits the functions of T cells and NK cells by the inhibitory receptor TIGIT (Gur et al. 2015). In contrast, a recent report demonstrated a link between the abundance of *Fusobacteria* species and increased expression of the cytokines IL-12 and TGF-β, shifting the T-cell population toward reduced Foxp3 expression, the hallmark transcription factor of immune-inhibitory T_{regs}, and was correlated with good prognosis (Saito et al. 2016). Therefore, it is currently unclear if intratumoral T-cell recruitment, as well as their activation or inhibition, is directly attributable to *Fusobacteria* and other individual bacterial species. Lastly, caution is required when extrapolating results from these studies, since different strains of a given bacterial species can be functionally different (Fig. 15.1).

15.4.3 Innate Immunity

In addition to T cells, cell types of the innate immune system were shown to affect prognosis and outcome of patients with colorectal cancer. Certain bacterial species might induce the recruitment of immune cells other than T cells, such as neutrophils or natural killer cells, which could also have a positive impact on prognosis. In fact, infiltration of natural killer cells (NK), as well as of natural killer T cells (NKT), was associated with good prognosis in colorectal cancer (Coca et al. 1997; Sandel et al. 2005; Tachibana et al. 2005). The role of macrophages in the context of cancer, however, is ambiguous (Balkwill and Mantovani 2012), and it is under debate whether tumor-associated macrophages actually constitute suitable targets for therapeutic intervention (Alahari et al. 2015). The intratumoral presence of CD16-positive myeloid cells expressing myeloperoxidase, i.e., activated

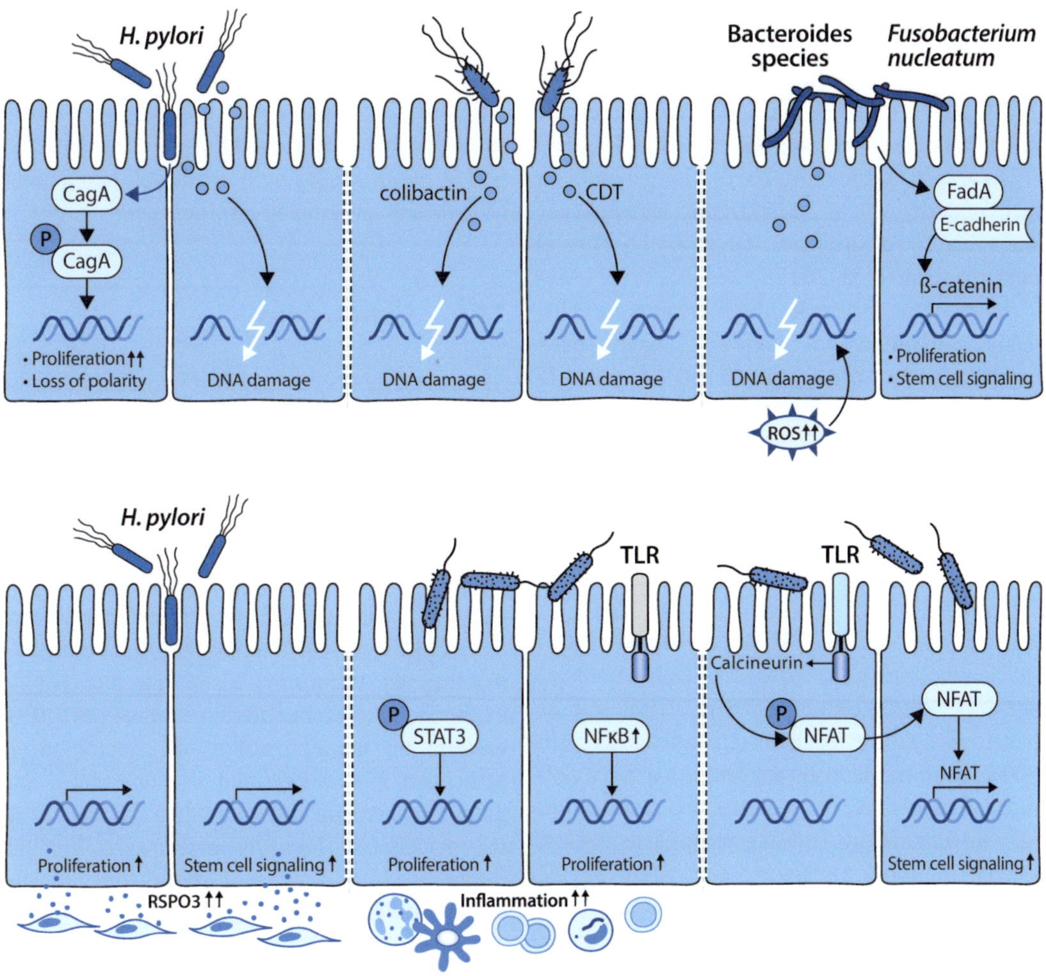

Fig. 15.1 Bacteria associated with colorectal cancer. Bacterial species associated with colorectal cancer covered in this article are visualized, and intracellular events resulting in an increased proliferation or DNA damage are highlighted

neutrophils, is an independent predictor of favorable prognosis (Droeser et al. 2013; Governa et al. 2017; Sconocchia et al. 2011). Further, an interplay between different innate cell types has been reported, since intratumoral neutrophils or natural killer cells were reported to increase the positive prognostic significance of cytotoxic T cells (Sconocchia et al. 2014). Since there is even greater functional plasticity in the myeloid cell lineage, it is not surprising that multiple reciprocal interactions between macrophages and T cells have been described (Biswas and Mantovani 2010). In a simplified view, tumor-associated macrophages (TAMs) can be assigned to two subgroups, defined by surface markers, functional aspects, and their contribution to tumorigenesis: where M1 TAMs are thought to play a more tumor-suppressive role by enhancing adaptive immunity and by producing pro-inflammatory cytokines like IL-6 and TNF, whereas M2 TAMs have a tumor-promoting effect and secrete immune-inhibitory factors like IL-10 and TGF-β (Biswas and Mantovani 2010). In general, an increased density of intratumoral macrophages was found to be correlated with favorable clinical outcome (Edin et al. 2013).

In addition to specialized innate immune cells, many inflammatory pathways have been described to be fully expressed and functional in intestinal epithelia, as well as in colorectal cancer cells (Abreu 2010). However, the molecular cross talk between colorectal cancer cells and gut microbiota remains to be elucidated. Enterocytes, as well as colonocytes, are capable of sensing gut bacteria through pattern recognition receptors, including Toll-like receptors (Abreu 2010). In addition to TLRs, the cytoplasmic NOD (nucleotide-binding and oligomerization domain) and NOD-like receptors have a critical role as pattern recognition receptors (Creagh and O'Neill 2006). Our recent data suggests that bacterial-induced expression of cytokines and chemokines may be initiated by Toll-like receptor-dependent signals on primary carcinoma cells from surgical explants (Kistner et al. 2017). Furthermore, in vitro stimulation with TLR agonists resulted in marked induction of chemokine expression in established colorectal cancer cell lines from mice and humans. However, further studies are required to reveal which TLRs, in concert with other pattern recognition receptors, are engaged by individual bacterial species associated with colorectal cancer. In conclusion, carcinoma cells can be regarded as a major chemokine source in colorectal cancer and likely in other solid tumor entities. Gut microbiota have a pivotal role in triggering chemokine production, which induces infiltration of T cells from blood vessels into the tumor tissue, ultimately resulting in the development of a stable antitumoral adaptive immune response and improved patient prognosis. These findings are summarized in Fig. 15.2. The recent advances in the field further our understanding of the complex interactions between gut microbiota, cancer cells, and the immune system and may enable the development of innovative treatment strategies that alter the gut flora, to cause infiltration of tumor tissues by immune cell populations of favorable prognostic significance.

15.5 Conclusions

Several members of the gut microbiota have the potential to interfere with the epithelial cell integrity. In addition to association between certain microbiota composition and carcinogenesis, molecular mechanisms of how bacteria induce DNA damage in epithelial cells or induce pro-carcinogenic signaling have been proposed and demonstrated. While these effects are observed in vitro, it is important to demonstrate that they also occur in vivo and indeed contribute to carcinogenesis. In addition to direct effects of bacteria on the epithelium, the microbiota is associated with differential immune states and this partly explain interindividual differences in the density of TILs, in antitumoral immune responses, and lastly in disease outcome and response to multimodal therapy.

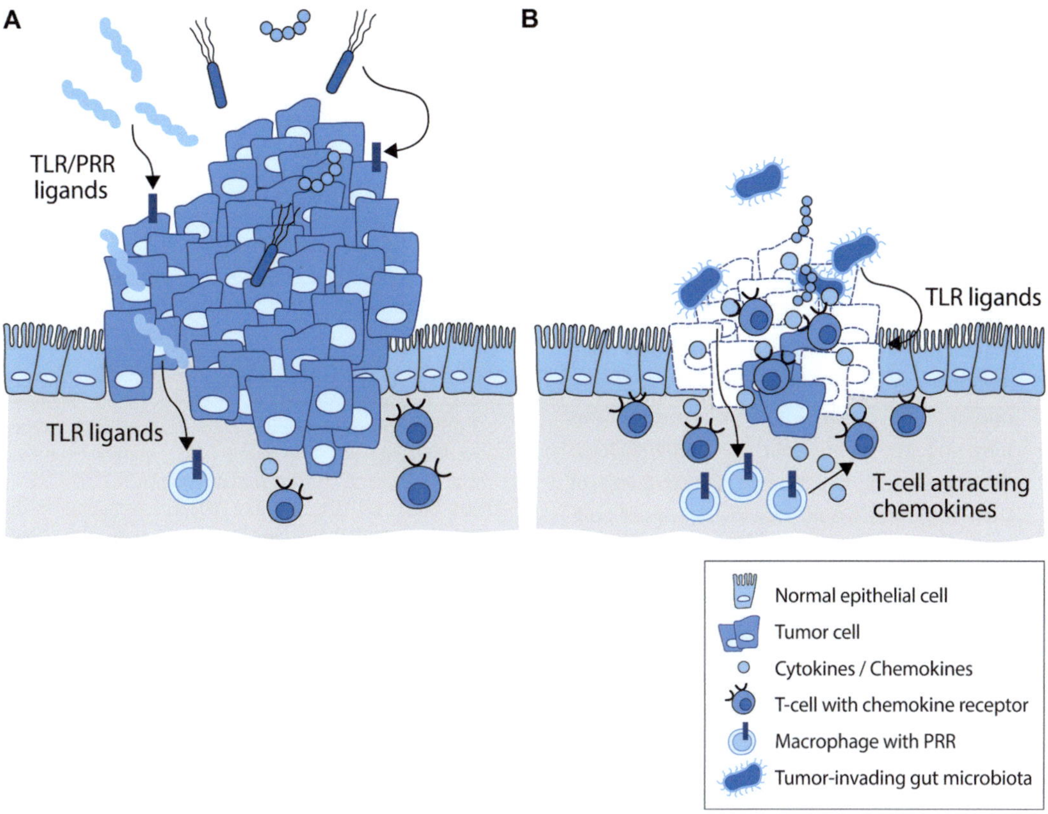

Fig. 15.2 (**a**) Specific gut microbiota are associated with bad prognosis and low T-cell infiltration. Microbial ligands engage TLRs and other pattern recognition receptors (PRRs) on carcinoma cells (enhancing cell survival) and immune cells, inhibiting their function. (**b**) In contrast, differing microbial communities are associated with favorable prognosis and high T-cell infiltration. Here, microbial signals trigger production of T-cell-attracting CXC chemokines by innate immune cells such as macrophages, leading to high densities of CTLs and T_{H1} cells, with effective antitumoral immune responses

▶ **Controversy**

Today, there is a growing interest in the impact of gut microbial communities on the initiation and progression of CRC. CRC was originally considered to be a strictly genetic disease, but recent studies reveal that the microbiota is an important contributing factor to this malignancy. A disturbance in the microbial community and expansion of pathogenic bacteria leads to inflammation, alternations in cellular microenvironment, contributing to precancerous and ultimately malignant tumors. Microbial dysbiosis observed in cancer can be a consequence of certain microorganisms gaining a competitive advantage in the tumor microenvironment or be an active contributor to the pathology. 16S rRNA sequencing and whole-genome shotgun sequencing have revealed presence of lower bacterial diversity and a higher abundance of certain protumorigenic bacteria in the intestinal microbiota of CRC patient (Ahn et al. 2013; Flemer et al. 2017; Vogtmann et al. 2016). Further, epidemiological evidence strongly suggests the involvement of specific microorganisms with certain gastrointestinal malignancies. Some of these bacteria have been shown to have the ability to directly damage the host cell DNA or interfere with signals that control epithelial proliferation and differentiation. While these new findings demonstrate that CRC could be driven by the microbiota, there is a lack of studies demonstrating a clear causative role in carcinogenesis, as in the case of infections with *H. pylori* in gastric and HPV in cervical cancer (de Martel et al. 2012). Interdisciplinary studies involving epidemiologists, molecular biologists and clinicians are required to demonstrate that the processes described in animal models indeed occur in humans and drive malignant transformation. Finally, the development of novel intervention strategies targeting the microbiota for prevention or therapy of CRC is required. Effectiveness of such strategies would provide the most convincing proof of a causative role of the microbiota in CRC development.

Highlights

- Colorectal cancer (CRC) is a genetic disease, but most cases occur sporadically. Characteristic changes in the microbiota of the patients with CRC suggest a causative role of the microbiota in CRC development.
- Transfer of microbiota from humans with CRC to mice increases the risk of colon tumor development compared to mice that receive microbiota from healthy individuals.
- Various bacteria that were considered commensals are now known to induce DNA damage and interfere with epithelial stem cell signaling and proliferation, indicating their potential to be pro-carcinogenic.
- Defined bacterial families are significantly associated with the intratumoral expression of specific immune cell-recruiting chemokines.
- The individual gut microbiome in patients with colorectal cancer acts in concert with genetic alterations in cancer cells to determine the extent of immune cells infiltration, with pivotal effects on disease outcome.

References

Abed, J., Emgard, J. E., Zamir, G., Faroja, M., Almogy, G., Grenov, A., et al. (2016). Fap2 mediates Fusobacterium nucleatum colorectal adenocarcinoma enrichment by binding to tumor-expressed Gal-GalNAc. *Cell Host and Microbe, 20*, 215–225.

Abreu, M. T. (2010). Toll-like receptor signalling in the intestinal epithelium: How bacterial recognition shapes intestinal function. *Nature Reviews Immunology, 10*, 131–144.

Agostini, M., Janssen, K. P., Kim, I. J., D'Angelo, E., Pizzini, S., Zangrando, A., et al. (2015). An integrative approach for the identification of prognostic and predictive biomarkers in rectal cancer. *Oncotarget, 6*, 32561–32574.

Ahn, J., Sinha, R., Pei, Z., Dominianni, C., Wu, J., Shi, J., et al. (2013). Human gut microbiome and risk for colorectal cancer. *Journal of the National Cancer Institute, 105*, 1907–1911.

Alahari, S. V., Dong, S., & Alahari, S. K. (2015). Are macrophages in tumors good targets for novel therapeutic approaches? *Molecules and Cells, 38*, 95–104.

Amicarella, F., Muraro, M. G., Hirt, C., Cremonesi, E., Padovan, E., Mele, V., et al. (2017). Dual role of tumour-infiltrating T helper 17 cells in human colorectal cancer. *Gut, 66*, 692–704.

Amieva, M. R., & El-Omar, E. M. (2008). Host-bacterial interactions in Helicobacter pylori infection. *Gastroenterology, 134*, 306–323.

Amieva, M. R., Vogelmann, R., Covacci, A., Tompkins, L. S., Nelson, W. J., & Falkow, S. (2003). Disruption of the epithelial apical-junctional complex by Helicobacter pylori CagA. *Science, 300*, 1430–1434.

Arthur, J. C., Perez-Chanona, E., Muhlbauer, M., Tomkovich, S., Uronis, J. M., Fan, T. J., et al. (2012). Intestinal inflammation targets cancer-inducing activity of the microbiota. *Science, 338*, 120–123.

Arthur, J. C., Gharaibeh, R. Z., Muhlbauer, M., Perez-Chanona, E., Uronis, J. M., McCafferty, J., et al. (2014). Microbial genomic analysis reveals the essential role of inflammation in bacteria-induced colorectal cancer. *Nature Communications, 5*, 4724.

Bagnoli, F., Buti, L., Tompkins, L., Covacci, A., & Amieva, M. R. (2005). Helicobacter pylori CagA induces a transition from polarized to invasive phenotypes in MDCK cells. *Proceedings of the National Academy of Sciences of the United States of America, 102*, 16339–16344.

Balkwill, F. R., & Mantovani, A. (2012). Cancer-related inflammation: Common themes and therapeutic opportunities. *Seminars in Cancer Biology, 22*, 33–40.

Basset, C., Holton, J., Bazeos, A., Vaira, D., & Bloom, S. (2004). Are Helicobacter species and enterotoxigenic Bacteroides fragilis involved in inflammatory bowel disease? *Digestive Diseases and Sciences, 49*, 1425–1432.

Bernstein, C. N., Blanchard, J. F., Kliewer, E., & Wajda, A. (2001). Cancer risk in patients with inflammatory bowel disease: A population-based study. *Cancer, 91*, 854–862.

Bindea, G., Mlecnik, B., Tosolini, M., Kirilovsky, A., Waldner, M., Obenauf, A. C., et al. (2013). Spatiotemporal dynamics of intratumoral immune cells reveal the immune landscape in human cancer. *Immunity, 39*, 782–795.

Biswas, S. K., & Mantovani, A. (2010). Macrophage plasticity and interaction with lymphocyte subsets: Cancer as a paradigm. *Nature Immunology, 11*, 889–896.

Boleij, A., Hechenbleikner, E. M., Goodwin, A. C., Badani, R., Stein, E. M., Lazarev, M. G., et al. (2015). The Bacteroides fragilis toxin gene is prevalent in the colon mucosa of colorectal cancer patients. *Clinical Infectious Diseases, 60*, 208–215.

Breynaert, C., Vermeire, S., Rutgeerts, P., & Van Assche, G. (2008). Dysplasia and colorectal cancer in inflammatory bowel disease: A result of inflammation or an intrinsic risk? *Acta Gastroenterologica Belgica, 71*, 367–372.

Buc, E., Dubois, D., Sauvanet, P., Raisch, J., Delmas, J., Darfeuille-Michaud, A., et al. (2013). High prevalence of mucosa-associated E. coli producing cyclomodulin and genotoxin in colon cancer. *PLoS One, 8*, e56964.

Bullman, S., Pedamallu, C. S., Sicinska, E., Clancy, T. E., Zhang, X., Cai, D., et al. (2017). Analysis of Fusobacterium persistence and antibiotic response in colorectal cancer. *Science, 358*, 1443–1448.

Chambers, F. G., Koshy, S. S., Saidi, R. F., Clark, D. P., Moore, R. D., & Sears, C. L. (1997). Bacteroides fragilis toxin exhibits polar activity on monolayers of human intestinal epithelial cells (T84 cells) in vitro. *Infection and Immunity, 65*, 3561–3570.

Coca, S., Perez-Piqueras, J., Martinez, D., Colmenarejo, A., Saez, M. A., Vallejo, C., et al. (1997). The prognostic significance of intratumoral natural killer cells in patients with colorectal carcinoma. *Cancer, 79*, 2320–2328.

Coppola, D., Nebozhyn, M., Khalil, F., Dai, H., Yeatman, T., Loboda, A., et al. (2011). Unique ectopic lymph node-like structures present in human primary colorectal carcinoma are identified by immune gene array profiling. *The American Journal of Pathology, 179*, 37–45.

Cougnoux, A., Dalmasso, G., Martinez, R., Buc, E., Delmas, J., Gibold, L., et al. (2014). Bacterial genotoxin colibactin promotes colon tumour growth by inducing a senescence-associated secretory phenotype. *Gut, 63*, 1932–1942.

Creagh, E. M., & O'Neill, L. A. (2006). TLRs, NLRs and RLRs: A trinity of pathogen sensors that co-operate in innate immunity. *Trends in Immunology, 27*, 352–357.

Cremonesi, E., Governa, V., Garzon, J. F. G., Mele, V., Amicarella, F., Muraro, M. G., Trella, E., Galati-Fournier, V., Oertli, D., Daster, S. R., et al. (2018, February 6). Gut microbiota modulate T cell trafficking into human colorectal cancer. *Gut*. pii: gutjnl-2016-313498. doi: 10.1136/gutjnl-2016-313498. [Epub ahead of print].

Cuevas-Ramos, G., Petit, C. R., Marcq, I., Boury, M., Oswald, E., & Nougayrede, J. P. (2010). Escherichia coli induces DNA damage in vivo and triggers genomic instability in mammalian cells. *Proceedings of the National Academy of Sciences of the United States of America, 107*, 11537–11542.

de Martel, C., Ferlay, J., Franceschi, S., Vignat, J., Bray, F., Forman, D., et al. (2012). Global burden of cancers attributable to infections in 2008: A review and synthetic analysis. *The Lancet Oncology, 13*, 607–615.

Dejea, C., Wick, E., & Sears, C. L. (2013). Bacterial oncogenesis in the colon. *Future Microbiology, 8*, 445–460.

Del Bel Belluz, L., Guidi, R., Pateras, I. S., Levi, L., Mihaljevic, B., Rouf, S. F., et al. (2016). The typhoid toxin promotes host survival and the establishment of a persistent asymptomatic infection. *PLoS Pathogens, 12*, e1005528.

DeStefano Shields, C. E., Van Meerbeke, S. W., Housseau, F., Wang, H., Huso, D. L., Casero, R. A., Jr., et al. (2016). Reduction of murine colon tumorigenesis driven by enterotoxigenic bacteroides fragilis using cefoxitin treatment. *The Journal of Infectious Diseases, 214*, 122–129.

Dienstmann, R., Vermeulen, L., Guinney, J., Kopetz, S., Tejpar, S., & Tabernero, J. (2017). Consensus molecular subtypes and the evolution of precision medicine in colorectal cancer. *Nature Reviews Cancer, 17*, 268.

Dieu-Nosjean, M. C., Goc, J., Giraldo, N. A., Sautes-Fridman, C., & Fridman, W. H. (2014). Tertiary lymphoid structures in cancer and beyond. *Trends in Immunology, 35*, 571–580.

Doll, D., Keller, L., Maak, M., Boulesteix, A. L., Siewert, J. R., Holzmann, B., et al. (2010). Differential expression of the chemokines GRO-2, GRO-3, and interleukin-8 in colon cancer and their impact on metastatic disease and survival. *International Journal of Colorectal Disease, 25*, 573–581.

Droeser, R. A., Hirt, C., Eppenberger-Castori, S., Zlobec, I., Viehl, C. T., Frey, D. M., et al. (2013). High myeloperoxidase positive cell infiltration in colorectal cancer is an independent favorable prognostic factor. *PLoS One, 8*, e64814.

Eaden, J. A., Abrams, K. R., & Mayberry, J. F. (2001). The risk of colorectal cancer in ulcerative colitis: A meta-analysis. *Gut, 48*, 526–535.

Edin, S., Wikberg, M. L., Rutegard, J., Oldenborg, P. A., & Palmqvist, R. (2013). Phenotypic skewing of macrophages in vitro by secreted factors from colorectal cancer cells. *PLoS One, 8*, e74982.

Erdman, S. E., Poutahidis, T., Tomczak, M., Rogers, A. B., Cormier, K., Plank, B., et al. (2003). CD4+ CD25+ regulatory T lymphocytes inhibit microbially induced colon cancer in Rag2-deficient mice. *The American Journal of Pathology, 162*, 691–702.

Escobar-Paramo, P., Grenet, K., Le Menac'h, A., Rode, L., Salgado, E., Amorin, C., et al. (2004). Large-scale population structure of human commensal Escherichia coli isolates. *Applied and Environmental Microbiology, 70*, 5698–5700.

Fearon, E. R., & Vogelstein, B. (1990). A genetic model for colorectal tumorigenesis. *Cell, 61*, 759–767.

Feng, Q., Liang, S., Jia, H., Stadlmayr, A., Tang, L., Lan, Z., et al. (2015). Gut microbiome development along the colorectal adenoma-carcinoma sequence. *Nature Communications, 6*, 6528.

Flemer, B., Lynch, D. B., Brown, J. M., Jeffery, I. B., Ryan, F. J., Claesson, M. J., et al. (2017). Tumour-associated and non-tumour-associated microbiota in colorectal cancer. *Gut, 66*, 633–643.

Fre, S., Vignjevic, D., Schoumacher, M., Duffy, S. L., Janssen, K.-P., Robine, S., et al. (2008). Epithelial morphogenesis and intestinal cancer: New insights in signaling mechanisms. In F. V. W. George & K. George (Eds.), *Advances in cancer research* (pp. 85–111). Academic Press.

Frey, D. M., Droeser, R. A., Viehl, C. T., Zlobec, I., Lugli, A., Zingg, U., et al. (2010). High frequency of tumor-infiltrating FOXP3(+) regulatory T cells predicts improved survival in mismatch repair-proficient colorectal cancer patients. *International Journal of Cancer, 126*, 2635–2643.

Fridman, W. H., Pages, F., Sautes-Fridman, C., & Galon, J. (2012). The immune contexture in human tumours: Impact on clinical outcome. *Nature Reviews Cancer, 12*, 298–306.

Galon, J., Costes, A., Sanchez-Cabo, F., Kirilovsky, A., Mlecnik, B., Lagorce-Pages, C., et al. (2006). Type, density, and location of immune cells within human colorectal tumors predict clinical outcome. *Science, 313*, 1960–1964.

Galon, J., Fridman, W. H., & Pages, F. (2007). The adaptive immunologic microenvironment in colorectal cancer: A novel perspective. *Cancer Research, 67*, 1883–1886.

Galon, J., Mlecnik, B., Bindea, G., Angell, H. K., Berger, A., Lagorce, C., et al. (2014). Towards the introduction of the 'Immunoscore' in the classification of malignant tumours. *The Journal of Pathology, 232*, 199–209.

Ge, Z., Feng, Y., Whary, M. T., Nambiar, P. R., Xu, S., Ng, V., et al. (2005). Cytolethal distending toxin is essential for Helicobacter hepaticus colonization in outbred Swiss Webster mice. *Infection and Immunity, 73*, 3559–3567.

Ge, Z., Rogers, A. B., Feng, Y., Lee, A., Xu, S., Taylor, N. S., et al. (2007). Bacterial cytolethal distending toxin promotes the development of dysplasia in a model of microbially induced hepatocarcinogenesis. *Cellular Microbiology, 9*, 2070–2080.

Ge, Z., Feng, Y., Ge, L., Parry, N., Muthupalani, S., & Fox, J. G. (2017). Helicobacter hepaticus cytolethal distending toxin promotes intestinal carcinogenesis in 129Rag2-deficient mice. *Cellular Microbiology, 19*.

Gonzalez-Escobedo, G., & Gunn, J. S. (2013). Gallbladder epithelium as a niche for chronic Salmonella carriage. *Infection and Immunity, 81*, 2920–2930.

Goodwin, A. C., Destefano Shields, C. E., Wu, S., Huso, D. L., Wu, X., Murray-Stewart, T. R., et al. (2011). Polyamine catabolism contributes to enterotoxigenic Bacteroides fragilis-induced colon tumorigenesis. *Proceedings of the National Academy of Sciences of the United States of America, 108*, 15354–15359.

Governa, V., Trella, E., Mele, V., Tornillo, L., Amicarella, F., Cremonesi, E., et al. (2017). The interplay between neutrophils and CD8(+) T cells improves survival in human colorectal cancer. *Clinical Cancer Research, 23*, 3847–3858.

Grivennikov, S. I., Wang, K., Mucida, D., Stewart, C. A., Schnabl, B., Jauch, D., et al. (2012). Adenoma-linked barrier defects and microbial products drive IL-23/IL-17-mediated tumour growth. *Nature, 491*, 254–258.

Groom, J. R., & Luster, A. D. (2011). CXCR3 in T cell function. *Experimental Cell Research, 317*, 620–631.

Guidi, R., Guerra, L., Levi, L., Stenerlow, B., Fox, J. G., Josenhans, C., et al. (2013a). Chronic exposure to the cytolethal distending toxins of Gram-negative bacteria promotes genomic instability and altered DNA damage response. *Cellular Microbiology, 15*, 98–113.

Guidi, R., Levi, L., Rouf, S. F., Puiac, S., Rhen, M., & Frisan, T. (2013b). Salmonella enterica delivers its genotoxin through outer membrane vesicles secreted from infected cells. *Cellular Microbiology, 15*, 2034–2050.

Gur, C., Ibrahim, Y., Isaacson, B., Yamin, R., Abed, J., Gamliel, M., et al. (2015). Binding of the Fap2 protein of Fusobacterium nucleatum to human inhibitory receptor TIGIT protects tumors from immune cell attack. *Immunity, 42*, 344–355.

Haghjoo, E., & Galan, J. E. (2004). Salmonella typhi encodes a functional cytolethal distending toxin that is delivered into host cells by a bacterial-internalization pathway. *Proceedings of the National Academy of Sciences of the United States of America, 101*, 4614–4619.

Hanahan, D., & Weinberg, R. A. (2011). Hallmarks of cancer: The next generation. *Cell, 144*, 646–674.

Hojo, S., Koizumi, K., Tsuneyama, K., Arita, Y., Cui, Z., Shinohara, K., et al. (2007). High-level expression of chemokine CXCL16 by tumor cells correlates with a good prognosis and increased tumor-infiltrating lymphocytes in colorectal cancer. *Cancer Research, 67*, 4725–4731.

Kaplan, C. W., Lux, R., Huynh, T., Jewett, A., Shi, W., & Haake, S. K. (2005). Fusobacterium nucleatum apoptosis-inducing outer membrane protein. *Journal of Dental Research, 84*, 700–704.

Kaplan, C. W., Ma, X., Paranjpe, A., Jewett, A., Lux, R., Kinder-Haake, S., et al. (2010). Fusobacterium nucleatum outer membrane proteins Fap2 and RadD induce cell death in human lymphocytes. *Infection and Immunity, 78*, 4773–4778.

Kistner, L., Doll, D., Holtorf, A., Nitsche, U., & Janssen, K. P. (2017). Interferon-inducible CXC-chemokines are crucial immune modulators and survival predictors in colorectal cancer. *Oncotarget, 8*, 89998–90012.

Koeppel, M., Garcia-Alcalde, F., Glowinski, F., Schlaermann, P., & Meyer, T. F. (2015). Helicobacter pylori infection causes characteristic DNA damage patterns in human cells. *Cell Reports, 11*, 1703–1713.

Koshy, S. S., Montrose, M. H., & Sears, C. L. (1996). Human intestinal epithelial cells swell and demonstrate actin rearrangement in response to the metalloprotease toxin of Bacteroides fragilis. *Infection and Immunity, 64*, 5022–5028.

Kostic, A. D., Chun, E., Robertson, L., Glickman, J. N., Gallini, C. A., Michaud, M., et al. (2013). Fusobacterium nucleatum potentiates intestinal tumorigenesis and modulates the tumor-immune microenvironment. *Cell Host and Microbe, 14*, 207–215.

Lara-Tejero, M., & Galan, J. E. (2000). A bacterial toxin that controls cell cycle progression as a deoxyribonuclease I-like protein. *Science, 290*, 354–357.

Lara-Tejero, M., & Galan, J. E. (2001). CdtA, CdtB, and CdtC form a tripartite complex that is required for cytolethal distending toxin activity. *Infection and Immunity, 69*, 4358–4365.

Lasry, A., Zinger, A., & Ben-Neriah, Y. (2016). Inflammatory networks underlying colorectal cancer. *Nature Immunology, 17*, 230–240.

Maloy, K. J., Salaun, L., Cahill, R., Dougan, G., Saunders, N. J., & Powrie, F. (2003). CD4+CD25+ T(R) cells suppress innate immune pathology through cytokine-dependent mechanisms. *The Journal of Experimental Medicine, 197*, 111–119.

Mima, K., Nishihara, R., Qian, Z. R., Cao, Y., Sukawa, Y., Nowak, J. A., et al. (2016). Fusobacterium nucleatum in colorectal carcinoma tissue and patient prognosis. *Gut, 65*, 1973–1980.

Mlecnik, B., Tosolini, M., Charoentong, P., Kirilovsky, A., Bindea, G., Berger, A., et al. (2010). Biomolecular network reconstruction identifies T-cell homing factors associated with survival in colorectal cancer. *Gastroenterology, 138*, 1429–1440.

Mlecnik, B., Tosolini, M., Kirilovsky, A., Berger, A., Bindea, G., Meatchi, T., et al. (2011). Histopathologic-based prognostic factors of colorectal cancers are associated with the state of the local immune reaction. *Journal of Clinical Oncology, 29*, 610–618.

Morey, P., Pfannkuch, L., Pang, E., Boccellato, F., Sigal, M., Imai-Matsushima, A., Dyer, V., Koch, M., Mollenkopf, H. J., Schlaermann, P., et al. (2018). Helicobacter pylori depletes cholesterol in gastric glands to prevent interferon gamma signaling and escape the inflammatory response. *Gastroenterology, 154*, 1391–1404 e1399.

Muehlbauer, M., Gharaibeh, R., Arthur, J. C., Fodor, A., & Jobin, C. (2013). 825 intestinal inflammation targets E. coli NC101 transcriptome response and promotes development of colorectal cancer (CRC). *Gastroenterology, 144*, S-144–S-145.

Mughini-Gras, L., Schaapveld, M., Kramers, J., Mooij, S., Neefjes-Borst, E. A., Pelt, W. V., et al. (2018). Increased colon cancer risk after severe Salmonella infection. *PLoS One, 13*, e0189721.

Nagamine, C. M., Rogers, A. B., Fox, J. G., & Schauer, D. B. (2008). Helicobacter hepaticus promotes azoxymethane-initiated colon tumorigenesis in BALB/c-IL10-deficient mice. *International Journal of Cancer, 122*, 832–838.

Nagaraja, V., & Eslick, G. D. (2014). Systematic review with meta-analysis: the relationship between chronic Salmonella typhi carrier status and gall-bladder cancer. *Alimentary Pharmacology & Therapeutics, 39*, 745–750.

Nagarsheth, N., Peng, D., Kryczek, I., Wu, K., Li, W., Zhao, E., et al. (2016). PRC2 epigenetically silences Th1-type chemokines to suppress effector T-cell trafficking in colon cancer. *Cancer Research, 76*, 275–282.

Nesic, D., Hsu, Y., & Stebbins, C. E. (2004). Assembly and function of a bacterial genotoxin. *Nature, 429*, 429–433.

Nitsche, U., Maak, M., Schuster, T., Kunzli, B., Langer, R., Slotta-Huspenina, J., et al. (2011). Prediction of prognosis is not improved by the seventh and latest edition of the TNM classification for colorectal cancer in a single-center collective. *Annals of Surgery, 254*, 793–800 discussion 800-791.

Nougayrede, J. P., Homburg, S., Taieb, F., Boury, M., Brzuszkiewicz, E., Gottschalk, G., et al. (2006). Escherichia coli induces DNA double-strand breaks in eukaryotic cells. *Science, 313*, 848–851.

Nowrouzian, F. L., & Oswald, E. (2012). Escherichia coli strains with the capacity for long-term persistence in the bowel microbiota carry the potentially genotoxic pks island. *Microbial Pathogenesis, 53*, 180–182.

Nowrouzian, F. L., Wold, A. E., & Adlerberth, I. (2005). Escherichia coli strains belonging to phylogenetic group B2 have superior capacity to persist in the intestinal microflora of infants. *The Journal of Infectious Diseases, 191*, 1078–1083.

Obiso, R. J., Jr., Azghani, A. O., & Wilkins, T. D. (1997). The Bacteroides fragilis toxin fragilysin disrupts the paracellular barrier of epithelial cells. *Infection and Immunity, 65*, 1431–1439.

Pages, F., Berger, A., Camus, M., Sanchez-Cabo, F., Costes, A., Molidor, R., et al. (2005). Effector memory T cells, early metastasis, and survival in colorectal cancer. *The New England Journal of Medicine, 353*, 2654–2666.

Pages, F., Galon, J., Dieu-Nosjean, M. C., Tartour, E., Sautes-Fridman, C., & Fridman, W. H. (2010).

Immune infiltration in human tumors: A prognostic factor that should not be ignored. *Oncogene, 29*, 1093–1102.

Parsonnet, J., Friedman, G. D., Vandersteen, D. P., Chang, Y., Vogelman, J. H., Orentreich, N., et al. (1991). Helicobacter pylori infection and the risk of gastric carcinoma. *The New England Journal of Medicine, 325*, 1127–1131.

Pere, H., Tanchot, C., Bayry, J., Terme, M., Taieb, J., Badoual, C., et al. (2012). Comprehensive analysis of current approaches to inhibit regulatory T cells in cancer. *Oncoimmunology, 1*, 326–333.

Pitzalis, C., Jones, G. W., Bombardieri, M., & Jones, S. A. (2014). Ectopic lymphoid-like structures in infection, cancer and autoimmunity. *Nature Reviews Immunology, 14*, 447–462.

Pratt, J. S., Sachen, K. L., Wood, H. D., Eaton, K. A., & Young, V. B. (2006). Modulation of host immune responses by the cytolethal distending toxin of Helicobacter hepaticus. *Infection and Immunity, 74*, 4496–4504.

Purcell, R. V., Pearson, J., Aitchison, A., Dixon, L., Frizelle, F. A., & Keenan, J. I. (2017). Colonization with enterotoxigenic Bacteroides fragilis is associated with early-stage colorectal neoplasia. *PLoS One, 12*, e0171602.

Rabizadeh, S., Rhee, K. J., Wu, S., Huso, D., Gan, C. M., Golub, J. E., et al. (2007). Enterotoxigenic bacteroides fragilis: A potential instigator of colitis. *Inflammatory Bowel Diseases, 13*, 1475–1483.

Rhee, K. J., Wu, S., Wu, X., Huso, D. L., Karim, B., Franco, A. A., et al. (2009). Induction of persistent colitis by a human commensal, enterotoxigenic Bacteroides fragilis, in wild-type C57BL/6 mice. *Infection and Immunity, 77*, 1708–1718.

Rubinstein, M. R., Wang, X., Liu, W., Hao, Y., Cai, G., & Han, Y. W. (2013). Fusobacterium nucleatum promotes colorectal carcinogenesis by modulating E-cadherin/beta-catenin signaling via its FadA adhesin. *Cell Host and Microbe, 14*, 195–206.

Saito, T., Nishikawa, H., Wada, H., Nagano, Y., Sugiyama, D., Atarashi, K., et al. (2016). Two FOXP3(+)CD4(+) T cell subpopulations distinctly control the prognosis of colorectal cancers. *Nature Medicine, 22*, 679–684.

Salama, P., Phillips, M., Grieu, F., Morris, M., Zeps, N., Joseph, D., et al. (2009). Tumor-infiltrating FOXP3+ T regulatory cells show strong prognostic significance in colorectal cancer. *Journal of Clinical Oncology, 27*, 186–192.

Salcedo, S. P., & Holden, D. W. (2003). SseG, a virulence protein that targets Salmonella to the Golgi network. *The EMBO Journal, 22*, 5003–5014.

Sandel, M. H., Speetjens, F. M., Menon, A. G., Albertsson, P. A., Basse, P. H., Hokland, M., et al. (2005). Natural killer cells infiltrating colorectal cancer and MHC class I expression. *Molecular Immunology, 42*, 541–546.

Sarshar, M., Scribano, D., Marazzato, M., Ambrosi, C., Aprea, M. R., Aleandri, M., et al. (2017). Genetic diversity, phylogroup distribution and virulence gene profile of pks positive Escherichia coli colonizing human intestinal polyps. *Microbial Pathogenesis, 112*, 274–278.

Sconocchia, G., Zlobec, I., Lugli, A., Calabrese, D., Iezzi, G., Karamitopoulou, E., et al. (2011). Tumor infiltration by FcgammaRIII (CD16)+ myeloid cells is associated with improved survival in patients with colorectal carcinoma. *International Journal of Cancer, 128*, 2663–2672.

Sconocchia, G., Eppenberger-Castori, S., Zlobec, I., Karamitopoulou, E., Arriga, R., Coppola, A., et al. (2014). HLA class II antigen expression in colorectal carcinoma tumors as a favorable prognostic marker. *Neoplasia, 16*, 31–42.

Sears, C. L. (2009). Enterotoxigenic bacteroides fragilis: A rogue among symbiotes. *Clinical Microbiology Reviews, 22*, 349–369 Table of Contents.

Secher, T., Samba-Louaka, A., Oswald, E., & Nougayrede, J. P. (2013). Escherichia coli producing colibactin triggers premature and transmissible senescence in mammalian cells. *PLoS One, 8*, e77157.

Secher, T., Payros, D., Brehin, C., Boury, M., Watrin, C., Gillet, M., et al. (2015). Oral tolerance failure upon neonatal gut colonization with Escherichia coli producing the genotoxin colibactin. *Infection and Immunity, 83*, 2420–2429.

Sellon, R. K., Tonkonogy, S., Schultz, M., Dieleman, L. A., Grenther, W., Balish, E., et al. (1998). Resident enteric bacteria are necessary for development of spontaneous colitis and immune system activation in interleukin-10-deficient mice. *Infection and Immunity, 66*, 5224–5231.

Sigal, M., Rothenberg, M. E., Logan, C. Y., Lee, J. Y., Honaker, R. W., Cooper, R. L., et al. (2015). Helicobacter pylori activates and expands Lgr5(+) stem cells through direct colonization of the gastric glands. *Gastroenterology, 148*, 1392–1404 e1321.

Sigal, M., Logan, C. Y., Kapalczynska, M., Mollenkopf, H. J., Berger, H., Wiedenmann, B., et al. (2017). Stromal R-spondin orchestrates gastric epithelial stem cells and gland homeostasis. *Nature, 548*, 451–455.

Simon, G. L., & Gorbach, S. L. (1984). Intestinal flora in health and disease. *Gastroenterology, 86*, 174–193.

Song, J., Gao, X., & Galan, J. E. (2013). Structure and function of the Salmonella Typhi chimaeric A(2)B (5) typhoid toxin. *Nature, 499*, 350–354.

Spano, S., Ugalde, J. E., & Galan, J. E. (2008). Delivery of a Salmonella Typhi exotoxin from a host intracellular compartment. *Cell Host & Microbe, 3*, 30–38.

Strebhardt, K., & Ullrich, A. (2008). Paul Ehrlich's magic bullet concept: 100 years of progress. *Nature Reviews Cancer, 8*, 473–480.

Tachibana, T., Onodera, H., Tsuruyama, T., Mori, A., Nagayama, S., Hiai, H., et al. (2005). Increased intratumor Valpha24-positive natural killer T cells: A prognostic factor for primary colorectal carcinomas. *Clinical Cancer Research, 11*, 7322–7327.

Tahara, T., Yamamoto, E., Suzuki, H., Maruyama, R., Chung, W., Garriga, J., et al. (2014). Fusobacterium in colonic flora and molecular features of colorectal carcinoma. *Cancer Research, 74*, 1311–1318.

Toprak, N. U., Yagci, A., Gulluoglu, B. M., Akin, M. L., Demirkalem, P., Celenk, T., et al. (2006). A possible role of Bacteroides fragilis enterotoxin in the aetiology of colorectal cancer. *Clinical Microbiology and Infection, 12*, 782–786.

Tosolini, M., Kirilovsky, A., Mlecnik, B., Fredriksen, T., Mauger, S., Bindea, G., et al. (2011). Clinical impact of different classes of infiltrating T cytotoxic and helper cells (Th1, th2, treg, th17) in patients with colorectal cancer. *Cancer Research, 71*, 1263–1271.

Vizcaino, M. I., & Crawford, J. M. (2015). The colibactin warhead crosslinks DNA. *Nature Chemistry, 7*, 411–417.

Vogtmann, E., Hua, X., Zeller, G., Sunagawa, S., Voigt, A. Y., Hercog, R., et al. (2016). Colorectal cancer and the human gut microbiome: Reproducibility with whole-genome shotgun sequencing. *PLoS One, 11*, e0155362.

Wong, S. H., Zhao, L., Zhang, X., Nakatsu, G., Han, J., Xu, W., et al. (2017). Gavage of fecal samples from patients with colorectal cancer promotes intestinal carcinogenesis in germ-free and conventional mice. *Gastroenterology, 153*, 1621–1633 e1626.

Wu, S., Lim, K. C., Huang, J., Saidi, R. F., & Sears, C. L. (1998). Bacteroides fragilis enterotoxin cleaves the zonula adherens protein, E-cadherin. *Proceedings of the National Academy of Sciences of the United States of America, 95*, 14979–14984.

Wu, S., Morin, P. J., Maouyo, D., & Sears, C. L. (2003). Bacteroides fragilis enterotoxin induces c-Myc expression and cellular proliferation. *Gastroenterology, 124*, 392–400.

Wu, S., Powell, J., Mathioudakis, N., Kane, S., Fernandez, E., & Sears, C. L. (2004). Bacteroides fragilis enterotoxin induces intestinal epithelial cell secretion of interleukin-8 through mitogen-activated protein kinases and a tyrosine kinase-regulated nuclear factor-kappaB pathway. *Infection and Immunity, 72*, 5832–5839.

Wu, S., Shin, J., Zhang, G., Cohen, M., Franco, A., & Sears, C. L. (2006). The Bacteroides fragilis toxin binds to a specific intestinal epithelial cell receptor. *Infection and Immunity, 74*, 5382–5390.

Wu, S., Rhee, K. J., Albesiano, E., Rabizadeh, S., Wu, X., Yen, H. R., et al. (2009). A human colonic commensal promotes colon tumorigenesis via activation of T helper type 17 T cell responses. *Nature Medicine, 15*, 1016–1022.

Yang, Y., Weng, W., Peng, J., Hong, L., Yang, L., Toiyama, Y., et al. (2017). Fusobacterium nucleatum Increases Proliferation of Colorectal Cancer Cells and Tumor Development in Mice by Activating Toll-Like Receptor 4 Signaling to Nuclear Factor-kappaB, and Up-regulating Expression of MicroRNA-21. *Gastroenterology, 152*, 851–866 e824.

Young, V. B., Knox, K. A., Pratt, J. S., Cortez, J. S., Mansfield, L. S., Rogers, A. B., et al. (2004). In vitro and in vivo characterization of helicobacter hepaticus cytolethal distending toxin mutants. *Infection and Immunity, 72*, 2521–2527.

Yu, T., Guo, F., Yu, Y., Sun, T., Ma, D., Han, J., et al. (2017). Fusobacterium nucleatum promotes chemoresistance to colorectal cancer by modulating autophagy. *Cell, 170*, 548–563 e516.

Zackular, J. P., Baxter, N. T., Iverson, K. D., Sadler, W. D., Petrosino, J. F., Chen, G. Y., et al. (2013). The gut microbiome modulates colon tumorigenesis. *MBio, 4*, e00692-00613.

Zeller, G., Tap, J., Voigt, A. Y., Sunagawa, S., Kultima, J. R., Costea, P. I., et al. (2014). Potential of fecal microbiota for early-stage detection of colorectal cancer. *Molecular Systems Biology, 10*, 766.

Microbiome and Diseases: Metabolic Disorders

16

Thomas Clavel and Josef Ecker

Abstract

Following the work of pioneers at the beginning of the twentieth century and later in the 1960s, there is nowadays a renewed interest for the impact of gut microbial communities on host metabolism. Nonetheless, we still ignore the real contribution of the intestinal microbiota to our energy homeostasis. The obesity pandemic is due to unbalanced energy intake (too high) vs. expenditure (too low) and not to the colonization of our intestine by unfavorable microbes. Nonetheless, gut microorganisms and in particular bacteria can metabolize a variety of dietary compounds, thereby releasing bioactive molecules that can influence metabolic functions of the host. Intestinal bacteria can also influence immune responses, which are known to play a role in the development of chronic metabolic disorders. Moreover, the approach of transferring gut microbial communities via fecal transplantation to germfree animals or human subjects has demonstrated the causal role of intestinal microbes in regulating host energy homeostasis and the development of metabolic pathologies. Experimental studies have even identified single cultured bacteria that influence metabolic responses, including underlying molecular mechanisms and bioactive molecules. All these findings speak in favor of the importance of the gut microbiota for host metabolism. In the present chapter, we summarize data in a critical manner and discuss key notions pertaining to the impact of gut microbial communities on host metabolism and ensuing implications for the development of metabolic disorders.

T. Clavel (✉)
Functional Microbiome Research Group, Institute of Medical Microbiology, University Hospital of RWTH, Aachen, Germany
e-mail: tclavel@ukaachen.de

J. Ecker
Department of Physiology of Human Nutrition, School of Life Sciences Weihenstephan, Technical University of Munich, Freising, Germany
e-mail: josef.ecker@tum.de

The intestine of mammals is colonized by a tremendous variety of microorganisms, especially bacteria, the density and diversity of which increase along the digestive tract from proximal to distal parts. Gut bacteria carry out a plethora of metabolic reactions and thus utilize a vast array of dietary compounds, thereby producing bioactive molecules that can regulate host physiology. This results in a mutualistic relationship that altogether benefits both the microbial ecosystem and the host. Major changes in lifestyles throughout industrialization, including important shifts in dietary habits, improved hygiene, and access to medication (e.g., antibiotics), have affected the gut microbial ecosystem and thus microbe-host interactions. To what extent these alterations

D. Haller (ed.), *The Gut Microbiome in Health and Disease*,
https://doi.org/10.1007/978-3-319-90545-7_16

contribute to the rising occurrence of metabolic diseases in industrialized countries, in particular obesity and associated comorbidities, has been a matter of intensive research. In the following sections, we touch upon the role of the gut microbiota in metabolic health and discuss important notions in the field, from dysbiosis and causal effects to single bacterial species and molecular crosstalk.

16.1 Disturbances of the Gut Microbiota in Metabolic Disorders

Dysbiosis is a term that characterizes a disturbed ecosystem in a specific pathological condition. It refers to shifts in the diversity and composition of the gut microbiota with functional implications for a given disease. Whereas many studies have reported disease-associated changes in the gut microbiota when compared to healthy state, it is very often unclear whether these shifts are directly relevant for the development of the given disease. Causal effects of the gut microbiota on metabolic health will be addressed below in Sect. 16.3. We here highlight major descriptive changes in the gut microbiota that are most often discussed in the context of metabolic disorders.

Dominant bacterial communities within the gut microbiota of individuals in Western countries are known to be relatively stable overtime, even over decades, in the absence of major dietary and lifestyle changes. The caloric content and composition of diets are linked to the development of overweight and obesity and have also been associated with major shifts in the gut microbiota, both in laboratory animals and in human beings (Daniel et al. 2014; David et al. 2014; Fava et al. 2013). However, on the short-term, the ecosystem is relatively resilient to changes induced by environmental stressors such as diet, i.e., it tends to rapidly return to original profiles upon stress cessation. In contrast, it has been observed that intake of a diet low in complex carbohydrates over four generations of mice leads to persistent alterations of the gut microbiota transmissible to the offspring (Sonnenburg et al. 2016). When extrapolated to the human situation, such a phenomenon, possibly amplified by additional dietary and environmental factors, may have contributed to the loss of specific gut bacterial species over time. This is corroborated by the observation that endemic populations maintaining a traditional lifestyle seem to harbor more microbial species in their intestine compared with individuals from Westernized countries (Martinez et al. 2015; Yatsunenko et al. 2012). It remains to be firmly proven that such changes have been really contributing to establishing a somewhat detrimental metabolic crosstalk between the gut microbiota and its host.

Hence, the first ecological parameter of interest pertaining to the gut microbial ecosystem in metabolic disorders is alpha-diversity, including simple enumeration of members of the ecosystem, referred to as richness. High diversity of the gut microbiota is usually considered to be a hallmark of ecosystem stability. Different species may possess similar properties resulting in so-called functional redundancy guaranteeing the robustness of complex metabolic and functional networks within the ecosystem. The loss of richness of the gut microbial ecosystem has been repeatedly associated with chronic intestinal inflammation (Lee et al. 2017; Pascal et al. 2017). In the context of metabolic disorders, the situation is a little more ambiguous. A metagenomic survey of 123 non-obese vs. 169 obese Danish individuals reported that, while gene counts (a proxy for diversity) in their fecal metagenomes ranged between 200,000 and 1,000,000, most subjects were characterized by either low (ca. 450,000) or high (ca. 750,000) gene counts (Le Chatelier et al. 2013). Interestingly, the group of individuals with low gene counts displayed detrimental metabolic features and responded less positively to a dietary intervention for weight loss and stabilization (Cotillard et al. 2013; Le Chatelier et al. 2013). A recent meta-analysis of ten studies looking at the association between obesity and gut microbiota profiles as obtained by 16S rRNA gene amplicon sequencing calculated that "obese individuals averaged 7.47% lower richness" and concluded that "obese individuals do have statistically significantly lower diversity than non-obese individuals; however, it is questionable whether the difference is biologically significant" (Sze and Schloss 2016).

Main additional features of disturbed intestinal ecosystems include an increased abundance of facultative anaerobes, mainly members of the family *Enterobacteriaceae*, and a leaky gut barrier—this latter aspect being presented in greater details in Sect. 16.4 (Cani et al. 2009; Ott et al. 2017; Qin et al. 2012; Teixeira et al. 2012; Xiao et al. 2014). In mouse models of diet-induced obesity, the intake of high-calorie diets is usually associated with a decrease in caecum volume and with a bloom in the family *Erysipelotrichaceae* within the phylum *Firmicutes* (Daniel et al. 2014; Fleissner et al. 2010; Turnbaugh et al. 2008). However, because of the heterogeneity of individual studies in terms of target pathology (e.g., obesity, diabetes, hepatic steatosis), experimental design (e.g., human study types or animal models), and methodologies (e.g., shotgun vs. targeted sequencing, different 16S rRNA gene regions amplified), it is to date still difficult to draw a uniform picture of changes in the gut microbiota associated with metabolic diseases. The particular case of the so-called *Firmicutes*-to-*Bacteroidetes* ratio is addressed in the "Controversy" box. One underlying reason for inconsistent intergroup comparisons is that gut microbiota diversity and composition is per definition individual-specific. Hence, approaches targeting overall differences between disease and healthy states will always be limited, and individualized strategies offer interesting perspectives (Zeevi et al. 2015).

The first recently published population-based study of the gut microbiome using 16S rRNA gene amplicon analysis of samples from thousands of individuals reported several confounders of microbiota analysis (e.g., stool consistency) that can influence sequence-based assessment of gut dysbiosis in metabolic disorders (Falony et al. 2016). This work also revealed that most data published in the past likely resulted from underpowered analysis and estimated that at least >500 human stool samples are required to reliably assess the association between gut microbial populations and body mass index. Another recent study brought to the forefront confounding effects of medication on the association between the gut microbiota and metabolic diseases: not considering the use of metformin, a drug commonly used for the treatment of diabetes, generated misleading findings in previous studies (Forslund et al. 2015). Issues on the reproducibility of findings on the gut microbiota in metabolic health will be further discussed below in Sect. 16.4 (especially Sects. 16.4.1 and 16.4.4) and in the "Controversy" box.

Altogether, additional work is required to replicate results and to provide a more detailed picture of disturbances of the gut microbiota in metabolic disorders. It is of particular importance to go beyond assessment of ecosystem changes, where half of the ecosystem members are still unknown, and identify key microbial species of particular relevance for host metabolism, including description of underlying molecular mechanisms.

16.2 From Complex Communities to Single Bacterial Species

Latest estimates of the cultured fraction of sequence-based diversity of gut microbiota range between 35% and 65% (Lagkouvardos et al. 2017). This means that a substantial proportion of microbial species in the gut remains to be described. Hence, it is essential to continue efforts in isolating and characterizing gut bacteria in order to define their possible role in the development or protection of chronic disorders, including metabolic diseases. In this section, we highlight some of the few bacterial species found to be implicated in metabolic health.

Because the focus of this chapter is on commensal dwellers of the intestine, it is not our purpose to address here the role of probiotics in detail, which would require an entire book chapter on its own. Their use for potentiating the growth of domestic animals was highlighted at the end of the 1980s (Fuller 1989). Since then, a multitude of studies on the effects of probiotics, especially lactobacilli, in the context of metabolic health have been published. Drawing clear conclusions from these studies is however a great challenge, due to the multitude of strains and experimental approaches used (Angelakis and Raoult 2010; Kumar et al. 2012; Lahtinen et al. 2012; Million et al. 2012; Nishizawa et al. 1988). The major limitation of all studies on probiotics in metabolic diseases is the lack of rationale for selecting strains. They usually correspond to bacteria that historically have a "generally regarded as safe" status but for which there is no clearly identified mechanisms that justify their use in specific metabolic pathologies.

Akkermansia muciniphila, the so far unique described species of the phylum *Verrucomicrobia* in the human gut, is currently the most studied bacterium in the context of metabolic health. This bacterium was first described in 2004 as a novel species capable of degrading the mucin that lines up the intestinal epithelial surface (Derrien et al. 2004). In 2008, the same authors showed that *A. muciniphila* is a dominant and prevalent bacterial species, which seems to colonize the human gut early in life (the earliest time point measured was 6 months of age) (Derrien et al. 2008). The first report of an association between the occurrence of *A. muciniphila* and host metabolism was published in 2012, when researchers observed that an antibiotic-driven bloom in the occurrence of the species was associated with lower incidence of diabetes in mice (Hansen et al. 2012). One year before, two human studies had revealed conflicting results on the relation between *A. muciniphila* occurrence in feces and body weight in pregnant women and infants (Collado et al. 2010; Santacruz et al. 2010). The first experimental proof of a causal effect of *A. muciniphila* on the improvement of high-fat diet-induced metabolic disorders was published shortly after in 2013 (Everard et al. 2013), but it took another 4 years until researchers identified a purified membrane protein that appears to be at least partly responsible for the positive metabolic effects of the species (Plovier et al. 2017). This is a major breakthrough in the field, because mechanisms underlying the positive effects of gut bacteria and most of all responsible molecules are very rarely described.

Other gut bacterial species shown to influence host metabolic health experimentally include *Christensenella minuta*, *Clostridium ramosum*, *Enterobacter cloacae*, and *Prevotella copri*. Using standardized mouse models colonized with a limited number of specific bacterial strains, Woting et al. showed that the presence of *C. ramosum* was associated with increased body weight and fat accompanied by higher food efficiency when fed with a high-fat diet (Woting et al. 2014). In 2013, Fei and Zhao reported that a strain of *Enterobacter cloacae* isolated from the intestine of a morbidly obese patient could induce obesity and insulin resistance upon colonization of germfree mice fed with a high-fat diet (Fei and Zhao 2013). More recently, Pedersen et al. found that the occurrence of *P. copri* in the stool of 277 nondiabetic Danish individuals was linked to serum levels of branched-chain fatty acids, most likely of microbial origin, in insulin-resistant subjects (Pedersen et al. 2016). The authors further demonstrated causal effects via treatment of conventional mice with the bacterium over 3 weeks, which led to aggravation of high-fat diet-induced glucose intolerance. Whereas the three bacterial species aforementioned were shown to have a rather negative impact on host metabolism experimentally, Goodrich et al. showed that colonization of gnotobiotic mice (previously inoculated with the fecal microbiota from an obese human donor) with *C. minuta* was able to reduce adiposity gain (Goodrich et al. 2014).

In conclusion, the plethora of reports in the literature on overall changes in the gut microbiota contrasts to the very few bacterial species identified as potentially playing an important role in metabolic diseases (Fig. 16.1). However, most of the evidence arises from experiments in mice. Moreover, findings must be extrapolated with caution, because different strains of a given species can be functionally different. Hence, the results presented above cannot be generalized to each bacterial species in general.

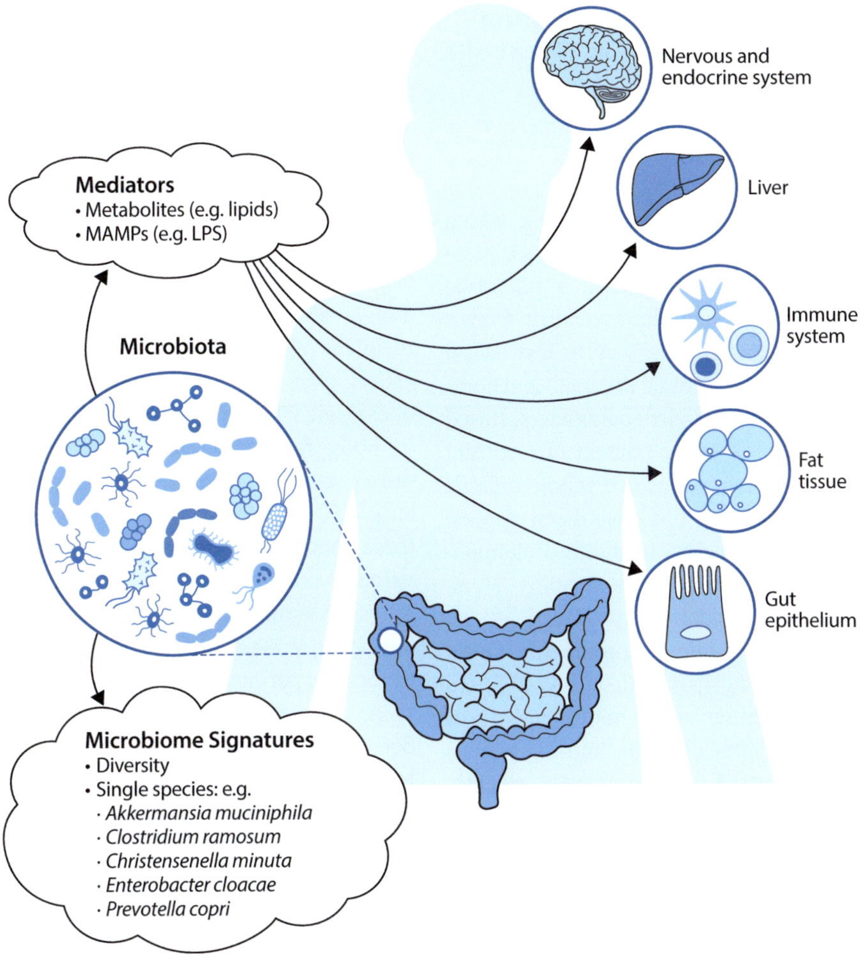

Fig. 16.1 Complex interactions between the gut microbiota and host metabolic responses. Microbe-host interactions in metabolic health result from intricate functional networks that we are just beginning to decipher. The gut microbiota consists of hundreds of microbial species in the colon, approximately half of which are still unknown, which substantially limits our understanding of its functional roles. Hallmarks of microbiome disturbances in metabolic diseases (microbiome signatures in the figure) are currently limited and rather vague (e.g., drop in diversity is also observed in chronic intestinal inflammation), partly due to the individual-specific nature of the gut microbiota (see Sect. 16.1). Functional interactions with the host are as complex as the ecosystem diversity, and currently known mediators originating from the microbes, including metabolites (e.g., amino acids and lipids) or microbe-associated molecular patterns (MAMPs), i.e., microbial molecules recognized by cells of the immune system such as lipopolysaccharides (LPS), are as scarce as the number of single species so far identified to regulate host metabolism. Species are highlighted in Sect. 16.2 and some of the mediators are detailed in Sects. 16.4 and 16.5 The complexity of microbe-host interactions in metabolic health also arises from the numerous host organs and systems, with which microbe-derived components can interact. The list is not exhaustive in the figure, and it is beyond the scope of this chapter to define in detail these interactions. The gut microbiota can, for instance, interact with: (1) The central nervous system and thereby regulate food intake and metabolic functions (Grasset et al. 2017; Torres-Fuentes et al. 2017), (2) The endocrine system and thereby influence growth (Schwarzer et al. 2016), (3) The liver via a plethora of metabolites and cellular products drained from the intestine (Abu-Shanab and Quigley 2010), (4) The immune system, which is known to play an important role in metabolic diseases (Kau et al. 2011), (5) The brown and white adipose tissue and thereby influence energy storage and expenditure (Zietak et al. 2016), (6) The gut epithelium and thereby regulate immune, endocrine, metabolic, and barrier functions (Blachier et al. 2017; Cherbuy et al. 1995; Plovier et al. 2017)

16.3 Evidence for the Implication of Gut Microbiota in Metabolic Pathologies

16.3.1 Transfer Experiments

Fecal microbiota transplantation (FMT), which consists in colonizing the intestine of a recipient person (or animal) by dispending a microbial suspension prepared using feces collected from a healthy donor, is an efficient way to test causal effects of changes in microbial diversity and composition on pathophysiological parameters. Based on very high efficacy in the context of recurrent infection by the pathogen *Clostridium difficile* (van Nood et al. 2013), FMT has been applied in a variety of pathologies, including inflammatory bowel diseases (IBD) (Paramsothy et al. 2017), hepatic encephalopathy (Bajaj et al. 2017), and the metabolic syndrome (de Groot et al. 2017). Following up on a first trial published in 2012 highlighting the improvement of insulin sensitivity after FMT in nine individuals (Vrieze et al. 2012), Kootte et al. very recently demonstrated an increase in peripheral insulin sensitivity 6 weeks after allogenic (microbiota from a different donor) but not autologous (own microbiota) FMT in 38 patients with metabolic syndrome (Kootte et al. 2017). Effects were transient, as no differences were seen 18 weeks after treatment. The authors also reported that lower diversity of the recipient microbiota at baseline was associated with positive response after treatment, stressing the need to determine fecal microbiota composition of patients at inclusion to aid in predicting efficacy of treatment. However, the biological relevance of a relatively low (<0.25) baseline difference in the Shannon index (one parameter of ecosystem diversity) is unclear, and treatment did not lead to an increase in diversity. These clinical studies clearly demonstrate causal effects of the gut microbiota in metabolic health and are very important experimental approaches to dissect microbe-host interactions. However, responses remain individual-specific, and the currently high risk-to-benefit ratio precludes routine clinical implementation.

Besides the clinical FMT studies aforementioned, the causal role of the gut microbiota in the development of metabolic diseases has also been highlighted by the fact that diet- or obesity-driven alterations of gut microbial communities in mice can lead to detrimental metabolic responses after transfer to germfree recipients. In 2008, Turnbaugh et al. compared the effects of transferring the gut microbiota from donor mice fed with a high-calorie (Western) or standard chow diet to recipient mice kept on the standard diet. Colonization with microbiota from the diet-induced obese mice led to significantly greater increase in body fat after 2 weeks without significant differences in feed consumption (Turnbaugh et al. 2008). Because of the short-term aspect of these mouse experiments and because absolute values were not provided (the paper reported a 43 vs. 25% increase in body fat as determined by dual-energy X-ray absorptiometry), extrapolation of these experimental data is somewhat difficult.

At the bridge between animal- and human-derived data, Ridaura et al. studied the impact of transferring the fecal microbiota from twin pairs discordant for obesity (i.e., one twin had a normal body mass index, while the other twin was obese) on mouse metabolism (Ridaura et al. 2013). Four twin pairs were used for colonizing mice kept on a standard diet, each sample being transferred into 3–4 mice in 1–5 independent experiments. On average, mice receiving the fecal microbiota from obese twins were characterized by a 10% to 15% increase in fat mass (as measured by quantitative magnetic resonance) after 4 to 5 weeks of colonization. Interestingly, the authors could repeat these findings by transferring cultured collections of fecal bacteria from the obese and lean donor of one twin pair instead of the whole undefined fecal communities, suggesting that it is possible to obtain single underlying species in the laboratory and reconstitute artificial communities to study molecular mechanisms. The authors also found that the increased fat mass phenotype after colonization with the obese microbiota could be reversed via cohousing with mice that had received the lean microbiota, which is interesting in terms of the permissive aspect of the obese microbiota in response to exogenous communities associated with favorable host conditions. These state-of-the-art experiments demonstrate again the transmissibility of metabolic phenotypes via the gut microbiota. Nonetheless, the complexity of assessing engraftment of complex fecal ecosystems (containing a substantial fraction of yet undescribed species) from human to mice (Arrieta et al. 2016) combined with the low-amplitude and short-term aspects of the fat mass changes as obtained with a limited selection of donor samples warrants confirmation prior to generalization of the results.

Experimental evidence of the causative role of the gut microbiota exists also in the context of metabolic pathologies other than obesity and type 2 diabetes. The liver is a central organ in host metabolism with an intricate link to the gut: the portal vein drains blood from the intestine, and the bile duct transports hepatic products into the intestinal lumen, which constitutes a flux of mediators referred to as enterohepatic circulation. Both organs form an efficient barrier to exogenous components and play key functions in nutrient absorption and metabolism and thus in host energy homeostasis. The rise of obesity worldwide is associated with the development of non-alcoholic fatty liver disease (NAFLD), characterized by abnormal accumulation of lipids in the liver with various complications ranging from chronic inflammation (steatohepatitis) to hepatocellular carcinoma. A study in mice previously showed that not all animals respond equally to a high caloric intake over 16 weeks: some of the mice (responders) were characterized by high levels of glycemia, systemic inflammation, and steatosis, whereas others (non-responders) did not develop any metabolic disorders despite obesity (Le Roy et al. 2013). These contrasting phenotypes could be recapitulated after colonization of germfree mice with the gut microbiota from responder and non-responder donors and following the same feeding protocol. These findings suggest that gut microbiota profiling could help predict the susceptibility to develop metabolic disorders and emphasize the possibility to modulate liver pathologies via the gut microbiota. Implementation of functional food (prebiotics and synbiotics) in the clinics for targeted manipulation of the gut microbiota may prove to be helpful in the context of NAFLD in the near future (Ferolla et al. 2016; Lambert et al. 2015).

16.3.2 Other Evidence

Colonization of the intestine of mammals following birth and throughout the first months of life is a very important physiological process for the protection against enteric infections and for the maturation of immune and metabolic responses. Hence, early life is an important window of opportunities where environmental factors (e.g., delivery mode, medication) or voluntary interventions (e.g., dietary supplements) may have long-lasting consequences for both the microbial ecosystem and health of the host. In particular, the link between obesity and repeated use of antibiotics at early age has been a subject of great interest (Cox and Blaser 2015). A study looking at prenatal exposure to antibiotics in the second or third trimester of pregnancy as recorded by questionnaires in a population of 436 mother-child dyads followed until 7 years of age reported a higher offspring risk of childhood obesity associated with antibiotic intake and also with caesarian section (Mueller et al. 2015). In contrast, a recent randomized intervention trial including 428 infants revealed that prolonged exposure to the antibiotics trimethoprim-sulfamethoxazole did not affect weight gain throughout the first 2 years of life (Edmonson and Eickhoff 2017). These and other findings (Ajslev et al. 2011; Azad et al. 2014; Kalliomaki et al. 2008; Saari et al. 2015) show that additional work will be needed prior to drawing firm conclusions on the impact of antibiotic treatment in early life on the development of obesity. Altogether, there is enough evidence suggesting that antibiotics can influence weight development later in life, but the effects remain individual-specific because they depend on many other parameters (e.g., time, type, and duration of treatment; combination with other factors such as delivery mode and mother weight). With respect to the use of antibiotics later in life, a recent randomized, placebo-controlled, double-blind study reported that 7-day treatment with either amoxicillin or vancomycin had no clinically relevant impact on metabolic health (insulin sensitivity, postprandial hormones, systemic inflammation, gut permeability, adipocyte size) in 57 obese, prediabetic men (Reijnders et al. 2016).

The implication of the host-microbiota crosstalk in metabolic diseases is also evident in the context of clinical interventions other than FMT. Bariatric surgery (also referred to as gastric bypass), which consists in reducing the size of the stomach, is one of the most efficient procedures for the treatment of morbid obesity. Gastric bypass was shown to be associated with marked overtime changes in the fecal microbiota (Furet et al. 2010), and later studies demonstrated the causal role of such changes by recapitulating favorable metabolic changes after colonizing germfree mice with the gut microbiota from animal (Liou et al. 2013) or human (Tremaroli et al. 2015) donors after surgery. With respect to nutritional factors, a very recent trial demonstrated efficacy of a 2-week intervention with an isocaloric, carbohydrate-restricted diet on hepatic steatosis in ten obese individuals (Mardinoglu et al. 2018). Rapid (as soon as 1 day after trial start) and marked reduction in liver fat and circulating triglycerides (>40% reduction on average at the end of the trial) was associated with rapid shifts in the fecal microbiota. Major changes included increased occurrence of *Streptococcus* and *Lactococcus* species linked to increased production of folate, which may have partly explained the improvement in liver fat metabolism observed in the participants.

The interplay between medication and the gut microbiota is also an area of great interest. A very recent randomized, double-blind, placebo-controlled study including 40 treatment-naïve type 2-diabetic patients showed an impact of metformin on the microbiota and demonstrated via transfer to germfree recipient mice that the positive effects of medication (e.g., reduced fat deposition) are partly conferred by these rearrangements in gut microbial communities (Wu et al. 2017).

Keeping in mind that the impact of microbe-host interactions on metabolic health shows a high degree of interindividual variability, it is important to complement our search of generalized notions valid for the majority of individuals to approaches that directly integrate this individual heterogeneity. Zeevi et al. recently followed such an approach and proposed an innovative workflow for predicting postprandial glycemic responses in human (Zeevi et al. 2015). They were able to propose meals adapted to individual characteristics and proved efficacy in the context of intervention trials. This innovative study paves the way for future implementation of personalized nutritional or interventional strategies that aims at potentiating the effects of a given treatment based on information about gut microbiota profiles.

In summary, there is solid experimental and clinical data that justify the implication of the gut microbiota in metabolic diseases, mostly based on interventional and microbiota transfer-based approaches. The main challenge for future research in the field is to define underlying molecular mechanisms.

16.4 Functional Consequences and Molecular Mechanisms

High diversity of the gut microbiota and the multiple organs involved in host metabolism contributes in forming a complex network of microbe-host interactions in metabolic health (Fig. 16.1). Knowledge of this network is still scant, although a few mechanisms have been repeatedly discussed in the field. They are presented below, albeit only briefly as interactions between the gut microbiota and each of the host systems and organs shown in Fig. 16.1 could fill an entire chapter.

16.4.1 Gut Barrier

The gut mucosa is the largest area of interface between cells of our own body and the environment, larger than the skin. One of the primary functions of this interface is to separate the self (body) from nonself (environment), which is referred to as barrier function. One important dogma is that the gut barrier is disturbed in metabolic diseases, leading to increased translocation of pro-inflammatory molecules of microbial origin such as lipopolysaccharides (LPS), a major component of the cell wall of Gram-negative bacteria. Experiments in genetically modified mice suffering from metabolic diseases (Brun et al. 2007) and in mouse models of diet-induced obesity (Cani et al. 2007) reported elevated endotoxemia, i.e., higher amounts of LPS in the blood. In 2009, Patrice Cani and colleagues (2009) proposed a mechanism in diet-induced obese mice by which decreased endotoxemia and improved gut barrier integrity after intake of the prebiotic inulin were linked to increased bifidobacteria numbers and elevated expression of GLP2 (glucagon-like peptide 2), a peptide secreted by enteroendocrine cells in the intestine which regulates energy absorption and mucosal morphology (Amato et al. 2016).

Results pertaining to gut barrier in animal models can however be affected by the type of high-calorie diets used, the animal facility, and the gut region considered (Hamilton et al. 2015; Kless et al. 2015; Muller et al. 2016). Moreover, the heterogeneity of LPS structures is substantial, and the immunogenic potential of various LPS can vary drastically (Steimle et al. 2016). Of note, not all LPS variants can be detected by most commonly used commercial assays, which are also sensitive to LPS contaminants from the environment. Furthermore, it is often not clear whether LPS is measured in portal or systemic blood. This is an important parameter that allows distinguishing between levels originating directly from the gut and those potentially having systemic functions. Hence, caution must be taken when interpreting results on gut barrier as measured by LPS translocation as they can be influenced by many technical parameters and are specific to the disease model used.

A breakthrough in the field was recently published by Cani and colleagues, who identified a bioactive molecule responsible for improvement in barrier function from *A. muciniphila* (for information on this bacterium, see Sect. 16.2). Amuc_1100*, a protein of 32 kDa encoded by a specific Type IV pili gene cluster, induced a lower body weight and fat mass gain and corrected hypercholesterolemia in mice fed with a high-calorie diet. This was accompanied by restored portal plasma concentrations of LPS to values observed in control animals on normal diet and by enhanced mRNA expression of tight junction proteins possibly via regulation of the endocannabinoid system and TLR-2-dependent pathways (Plovier et al. 2017). Moreover, there is also clinical evidence for the involvement of a leaky gut barrier in metabolic diseases. Several intervention studies reported that improvement of the metabolic health of participants was accompanied by improved barrier integrity (Damms-Machado et al. 2017; Karl et al. 2017; Ott et al. 2017; Russo et al. 2012; Xiao et al. 2014). Hence, it is altogether reasonable to assume that disturbances in gut physiology under conditions of metabolic stress are linked to distorted barrier function, but we are just starting to decipher how this works and what consequences for the host are.

16.4.2 At the Crossroad Between Immune and Metabolic Responses

One obvious functional consequence of increased fluxes of microbial molecules from the gut to the periphery via the liver is immune response activation. Adiposity is also associated with infiltration of immune cells into adipose tissues (Parseus et al. 2017), and it is nowadays acknowledged that the immune system plays an important role in metabolic disorders. Because interactions between the gut microbiota and immune responses are so complex, it is beyond the scope of this chapter to draw a comprehensive overview in the context of metabolic diseases, which has been touched upon by others (Kau et al. 2011; Zmora et al. 2017). We simply highlight some interesting findings.

Interleukin-22 (IL-22) was shown to alleviate metabolic disorders and restores mucosal immunity in diabetes, IL-22 knockout mice being prone to develop metabolic syndrome upon high-fat diet feeding and treatment with IL-22 restored this phenotype (Wang et al. 2014). Experimental studies demonstrated that the gut microbiota regulates T-cell differentiation in the gut via mechanisms that involve IL-22, with consequences on metabolic health (Garidou et al. 2015; Zelante et al. 2013). Dysregulation of innate immune pathways downstream of Toll-like receptor-5 and NLRP-6 and NLRP-3 inflammasomes was also shown to impact the development of metabolic diseases (Henao-Mejia et al. 2012; Singh et al. 2015; Vijay-Kumar et al. 2010). An important future task will be to disentangle the impact of local shifts in immune responses in the gut on systemic changes in metabolic responses.

With respect to changes affecting the adipose tissue, microbiota depletion by antibiotics or by using germfree mice was shown to promote browning of white adipose tissue and M2 macrophage polarization, thereby reducing obesity (Suarez-Zamorano et al. 2015). The translocation of pro-inflammatory molecules from the gut to adipose tissues and to the liver is an area of intensive research (Hersoug et al. 2016; Heymann and Tacke 2016). It is however unclear what the pathophysiological meaning of the so-called chronic, low-grade inflammation as proposed in overweight and obesity really is and what consequences for the development of metabolic comorbidities are. Underlying mechanisms may take years prior to the establishment of phenotypes and are thus difficult to test experimentally. A recent human study reported that interference with the gut microbiota by a 7-day antibiotic treatment had no clinically relevant impact on metabolic health and did not induce significant changes in blood concentrations of LBP (lipopolysaccharide-binding protein), IL-6, IL-8, and TNF used as proxy for systemic inflammation (Reijnders et al. 2016).

16.4.3 Dietary Factors and Hormones: Microbe-Host Interactions in Malnutrition

Because of a pandemic rise in Western countries, obesity is drawing most of the attention in the field of microbe-host interactions in metabolism. However, food shortage is also a major concern worldwide. It is thus worth addressing also issues about malnutrition in this chapter on the impact of gut microbiota on metabolic health, highlighting specific underlying molecular mechanisms.

Jeffrey Gordon and his group have performed pioneering work in this research area too. Via transfer of feces from infant Malawian donors to germfree mice, they showed that disturbances of the gut microbiota associated with malnutrition have causal effects on host metabolism (Blanton et al. 2016). In these experiments, they identified two species, *Ruminococcus gnavus* and *Clostridium symbiosum*, which ameliorated growth and metabolic abnormalities in recipient animals when added to the microbiota from malnourished children. In another study, they found that human milk oligosaccharides that are sialylated (i.e., bound to sialic acid) were significantly less abundant in the milk collected from 6-month-postpartum mothers ($n = 88$) of severely stunted (i.e., stopped abnormally in growth) compared with control Malawian infants (Charbonneau et al. 2016). In subsequent experiments in mice and pigs, the authors showed that feed supplementation with sialylated bovine milk oligosaccharides was associated with beneficial effects with respect to growth, i.e., greater ability to utilize nutrients for anabolism. Regarding the identification of specific gut bacterial players in undernutrition, Gordon and colleagues recently reported the occurrence of an enterotoxigenic strain of the species *Bacteroides fragilis* that was causally related to weight loss (Wagner et al. 2016). In contrast, the group of François Leulier found that *Lactobacillus plantarum* can promote systemic growth of the host during chronic undernutrition in a strain-dependent manner and proposed a molecular mechanism that involves the somatotropic hormone axis (Schwarzer et al. 2016). Germfree mice colonized with *L. plantarum* strain WJL grew better (25% in weight and 6% in length) than those colonized with a control *L. plantarum* strain when fed with a diet low in proteins, fat, and vitamins from weaning on until the age of 8 weeks. These effects were associated with increased serum levels of insulin-like growth factor-1 (a protein mainly secreted by the liver after stimulation by growth hormone and involved in many cellular processes and the growth of organs) and enhanced sensitivity to growth hormone treatment. Moreover, this work suggests mechanisms valid across phylogenetically very distant eukaryotic hosts, as strain WJL showing effects in mice was originally isolated from fruit flies (*Drosophila melanogaster*) and was a potent growth promoter in flies too.

16.4.4 Bacterial Meet Host Metabolism

Despite a renewed interest for the impact of gut microbiota on metabolic health and intensive research in the past decade, we are still not able to define precisely the caloric benefit for humans to host microbes in their gastrointestinal tract. In other words, what is the fraction of otherwise lost energy from food that is released by gut microbes and in which form, e.g., heat or metabolites usable as substrates by our own organs? Effects of microbial colonization on intestinal tissue maturation impacting absorption capacities are also unclear.

Comparisons between germfree and conventional animals and microbiota transfer experiments have been instrumental in revealing the role of gut microorganisms in host metabolism, but they did not deliver absolute quantitative values on the contribution of the gut microbiota to energy homeostasis. In 2004, Bäckhed and colleagues demonstrated that colonization of germfree mice increased adiposity by a factor of approx. 1.4 despite lower feed intake by the germfree mice (Backhed et al. 2004). However, early work in the 1960s had reported that growth differences between germfree and conventional animals depend on the reference microbiota that can vary across facilities (Pleasants 1968). Heterogeneity of models also concerns diet composition, which impacts the resistance of germfree mice to diet-induced obesity, as detailed in the "Controversy" box. In human, following the proposition that ratios in the relative abundance of the two major phyla *Firmicutes* and *Bacteroidetes* were related to obesity, Jumpertz et al. performed dietary intervention studies and estimated that a 20% increase in *Firmicutes* and corresponding decrease in *Bacteroidetes* were linked to an increase of approx. 150 kcal/d in net energy retained, as estimated by measurement of energy loss in feces (Jumpertz et al. 2011). However, the relevance of *Firmicutes*-to-*Bacteroidetes* ratios is debatable, as detailed in the "Controversy" box.

Increased energy harvest from the diet by the gut microbiome has been proposed to underline the development of obesity in a manner that is transmissible to offspring (Turnbaugh et al. 2006). One underlying hypothesis is that high-fat diet feeding is associated with increased production of short-chain fatty acids (SCFA), primarily butyrate used as a fuel for host cell metabolism. It has been reported that intestinal epithelial cells may derive up to 70% of their energy supply from butyrate oxidation (Roediger 1980). Moreover, studies in lambs showed that butyrate infusion in the abomasum (the fourth and final stomach compartment in ruminants) increased net nutrient flux to the portal vein and hepatic tissue (Foote and Freetly 2016). Under steady-state conditions, SCFA are produced primarily via the fermentation of complex carbohydrates by intricate bacterial trophic chains (Louis and Flint 2017). In diet-induced obesity where diets contain low amounts of carbohydrates, butyrate may arise from the metabolism of amino acids, although it is questionable whether this can sustain production of butyrate amounts higher than in steady-state conditions (Bui et al. 2015; Daniel et al. 2014). Harry Flint and colleagues showed that reduced dietary intake of carbohydrates by obese subjects resulted in decreased amounts of butyrate and butyrate-producing bacteria in feces (Duncan et al. 2007). Total SCFA amounts in the gut content of obese hosts may however increase in favor of SCFA other than butyrate. In 2010, Schwiertz and colleagues reported that SCFA concentrations were higher in obese vs. lean subjects and that the proportion of individual SCFA changed in favor of propionate in overweight and obese subjects (Schwiertz et al.

2010), propionate being usable as substrate for gluconeogenesis in the liver. Generally speaking, butyrate is a hallmark of normal fermentative processes within stable gut microbial communities, and disruption of this metabolic network has been associated with detrimental effects on host metabolism (Le Chatelier et al. 2013). In mouse studies, butyrate treatment by daily intragastric administration attenuated high-fat diet-induced steatohepatitis by improving gastrointestinal barrier (Zhou et al. 2017), even though butyrate application can have differential effects depending on the tissue under investigation (Matis et al. 2015). In summary, alterations of fermentation processes within gut microbiomes following the intake of high-calorie diets and in the context of metabolic diseases development are not clear. The dual functionality of butyrate, i.e., potentially contributing to obesity development under detrimental dietary conditions but otherwise beneficial for the host, remains to be elucidated.

Hence, an obvious way by which the gut microbiota can regulate metabolic responses is via the production of bioactive metabolites. Although the array of molecules produced by the microbiota is vast, the majority is still unknown, and only a few bacterial metabolites are yet known to influence host metabolism. The case of branched-chain amino acids and associated bacterial species was introduced above (see Sect. 16.2). Hydrogen sulfide produced by the activity of sulfate-reducing bacteria is also known to influence metabolic capacities of the intestinal epithelium (Blachier et al. 2017), although the impact on host metabolism overall is unclear. Lipids represent a major and very diverse class of compounds at the crossroad between diet, microbes, and the host. Whereas the conversion of long-chain fatty acids by microbial communities is known in environmental ecosystems (Alves et al. 2009), bacterial metabolism of dietary fatty acids in the intestine is unclear. Gut bacteria are so far only known to cleave triglycerides via lipase activity and to produce conjugated fatty acids, the latter compounds having well-documented effects on host metabolism (see Sect. 16.5.5) (McIntosh et al. 2009; Thorasin et al. 2015). Besides the production of SCFA (see paragraph above), gut bacteria can metabolize bile acids produced from cholesterol in the liver. This bacterial conversion of bile acids has major consequences on their bioavailability and bioactivities (Ridlon et al. 2006) and is of primary interest considering the clinical relevance of bile acids in metabolic health (Ma et al. 2017; Mudaliar et al. 2013; Thomas et al. 2008). In the following section, we provide further details on bioactive lipid molecules and the intricate interactions between the gut microbiota and host lipid metabolism.

16.5 Gut Microbiota and Host Lipid Metabolism

Nature is a source of an enormous diversity of lipid molecules. Fatty acids (FAs) are key modules for various lipids, including cell membrane lipids such as glycerophospholipids or triacylglycerols (TAG), which are the major components of lipid droplets (Ecker and Liebisch 2014; Wenk 2010). Excess lipids or defects in lipid storage are associated with diseases such as metabolic syndrome, which results from unbalanced energy intake and expenditure as well as complex interactions between genetic and environmental factors, including the gut microbiota (Ussar et al. 2015).

Lipids are either obtained from exogenous sources, i.e., diet, or from endogenous biosynthesis. They serve as precursors of signaling molecules such as eicosanoids with either pro- or anti-inflammatory effects (Schmitz and Ecker 2008). Numerous cellular processes such as cell growth or differentiation relay on de novo synthesis of FA primarily for cell membrane generation (Ecker et al. 2010a; van Meer et al. 2008).

16.5.1 Lipid Synthesis

The transcription factors sterol regulatory element-binding proteins (SREBP), which belong to the family of basic helix-loop-helix leucine zipper transcription factor, are major regulators of cellular lipid synthesis in mammalian cells (Horton et al. 2002). SREBP-2 mainly controls genes of the cholesterol pathway, and SREBP-1c preferentially regulates genes involved in FA biosynthesis. Fatty acid synthase is a well-characterized SREBP-1c target gene and a key enzyme required for endogenous FA synthesis in mammalian cells because it catalyzes the generation of palmitate (FA 16:0) from acetyl-CoA (Smith et al. 2003). Palmitate is either desaturated to palmitoleate (FA 16:1 $n-9$) by stearoyl-CoA desaturase (SCD) 1 or elongated to stearate (FA 18:0) by long-chain fatty acid elongase 6 (Matsuzaka et al. 2007; Miyazaki and Ntambi 2003). Stearate can be desaturated to oleate (FA 18:1 $n-9$) by SCD1, and palmitoleate can be elongated to vaccinate (FA 18:1 $n-7$) by long-chain fatty acid elongase 6.

SCD1 is a critical enzyme that has been linked to various diseases including diabetes, hypertriglyceridemia, cardiovascular disease, steatosis, bone health, and cancer (Hodson and Fielding 2013; Matsuzaka et al. 2007). Inhibition of SCD1 leads to severe atherosclerosis despite its effects on protecting from obesity, whereas increased SCD1 expression levels were shown to protect from endoplasmic reticulum (ER) stress and atherosclerosis (Brown et al. 2008; Erbay et al. 2009). Dendritic cells and macrophages, which originate from common myeloid precursor cells, significantly differ in their profiles of monounsaturated fatty acids (MUFA) and phospholipids due to different SCD1 activities (Ecker et al. 2010a). More recently, it was shown that SCD1 activity is absolutely required for autophagy, a major pathway for degradation of cytoplasmic components (Ogasawara et al. 2014).

Linoleic acid (FA 18:2 $n-6$) and α-linolenic acid (FA 18:3 $n-3$) are precursors of polyunsaturated fatty acids (PUFA) and thus important building blocks of both pro- and anti-inflammatory mediators (Buckley et al. 2014; Schmitz and Ecker 2008). They are essential fatty acids because, unlike plants, mammals do not express delta-12 and delta-15 desaturases, which are necessary to desaturate FA 18:1 and FA 18:2 (Pereira et al. 2003). A crucial organ for lipid synthesis is the liver, which plays critical roles in many metabolic diseases, especially type 2 diabetes (Moller 2001).

16.5.2 Gut Microbiota and Host Lipid Synthesis in Metabolic Diseases

To date, only a few studies have investigated effects of the gut microbiota on host lipid synthesis. One of the first studies that linked the gut microbiota to host lipid status was published in 2004 by Bäckhed and colleagues (2004). They compared germfree and conventionalized mice and concluded that the microbiota promotes the absorption of monosaccharides from the gut, leading to an induction of hepatic FASN mRNA expression via the transcription factors carbohydrate response element-binding protein and perhaps SREBP-1. They also proposed that the presence of gut microorganisms is associated with increased lipoprotein lipase (LPL) activity through suppression of intestinal expression fasting-induced adipocyte factor (FIAF), a circulating LPL inhibitor (Fig. 16.2). LPL hydrolyzes TAG to free FA from lipoproteins and leads to deposition of FFA in adipose tissues and TAG storage. In 2007, the same research group reported that germfree animals are protected from diet-induced obesity using a Western diet rich in simple sugars and beef tallow (Backhed et al. 2007). Germfree mice had higher levels of phosphorylated AMP-activated protein kinase (AMPK) in the liver and skeletal muscle and its downstream targets involved in β-oxidation.

The fact that germfree animals are protected from diet-induced obesity was, on the basis on this single study aforementioned, misleading. In 2010, it was reported that the absence of intestinal microbes does not protect mice from diet-induced obesity, when mice were fed with a different type of high-calorie diet (Fleissner et al. 2010). Moreover, germfree mice fed with a high-fat or Western diet showed no major changes in circulating FIAF levels compared to conventional animals in this study. These controversial findings show that the type of diets fed to mice, which can vary in caloric density, texture, amount, and composition

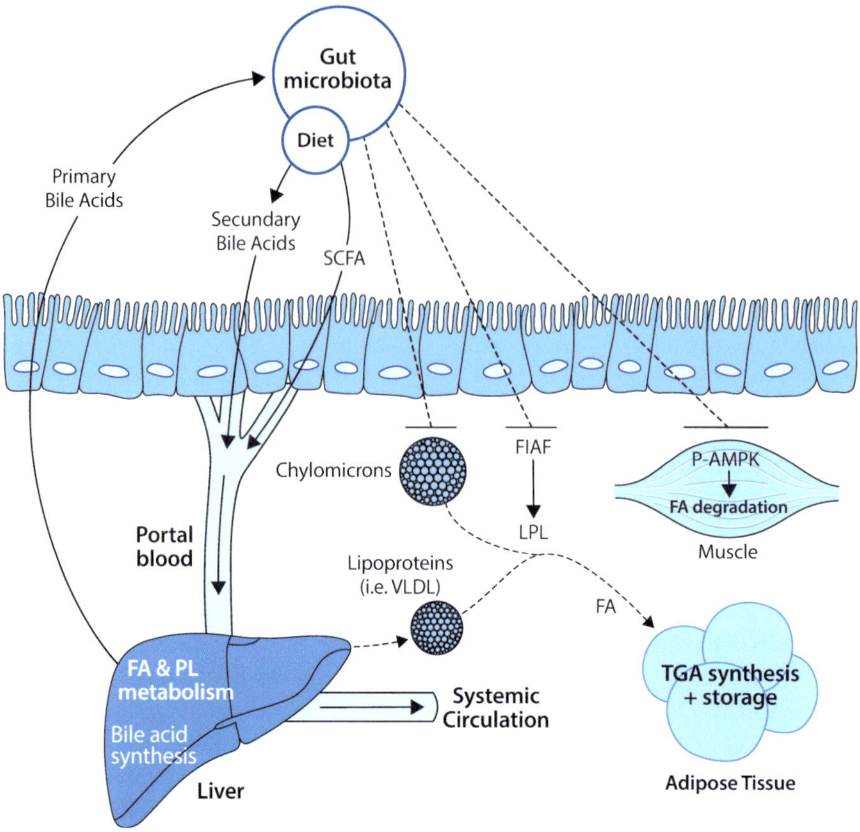

Fig. 16.2 The gut microbiota and host lipid metabolism. After fermentation of diet- and host-derived substrates (primarily carbohydrates but also proteins), the gut microbiota produces short-chain fatty acids (SCFA), such as acetate, propionate, and butyrate. Gut bacteria can also convert primary to secondary bile acids via deconjugation, dehydroxylation, and dehydrogenation. SCFA and secondary bile acids reach the liver and circulation via the portal vein and can alter host lipid metabolism and inflammatory responses substantially. The gut microbiota may also influence fatty acid (FA) synthesis and desaturation, phospholipid acyl-chain remodeling in the liver, and beta-oxidation of FA in muscles. It is still unclear whether circulating levels of fasting-induced adipocyte factor (FIAF), a lipase inhibitor, are influenced by microbial colonization. Reduced FIAF may enhance lipoprotein lipase (LPL) activity and thus lead to degradation of lipoproteins and deposition of free fatty acids in adipose tissue, which are precursors for triacylglycerols (TAG). Future studies should also verify whether the intestinal chylomicron output can be inhibited by the gut microbiota

of macronutrients and micronutrients, is very important (Clavel et al. 2014). Unfortunately, in most studies dealing with the microbiome in models of metabolic diseases, information about the diets used is not sufficiently detailed. Recently, it was demonstrated that fat source is very critical for the development of experimental obesity and for energy expenditure (Kubeck et al. 2016). Germfree mice, but not colonized mice, were resistant to diet-induced obesity when fed with a lard-based, cholesterol-rich high-fat diet. On a cholesterol-free, palm oil-based, high-fat diet, the development of obesity was independent of the gut microbiota.

Surprisingly, although SCFA (both acetate and propionate) can serve as precursors for FASN, a number of studies have shown that SCFA rather inhibit hepatic FA synthesis and fat storage in white adipose tissue while enhancing β-oxidation in multiple tissues (den Besten et al. 2013). Our understanding of underlying molecular mechanisms is still incomplete. It is discussed that SCFA activate AMPK in the liver and muscles either directly or indirectly via free fatty acid receptor 2 (also known as GPCR 43), which suppresses insulin-mediated fat accumulation and promotes energy expenditure (Kimura et al. 2013).

Gut microbes promote hepatic generation of monounsaturated fatty acids by SCD 1, leading to significant alterations of glycerophospholipid acyl chain profiles (Fig. 16.2). Further, Toll-like receptor 5-deficient mice, which are prone to develop microbiota-dependent metabolic syndrome, have increased hepatic SCD1 expression and oleic acid (FA18:1 $n-9$) levels (Singh et al. 2015). This finding might be of particular importance, since the phospholipid saturation degree contributes substantially to its flexibility. Membranes undergoing fast morphological changes contain high levels of polyunsaturated fatty acids (PUFA) (Ernst et al. 2016; Pinot et al. 2014). Further studies have to clarify whether these gut microbiota-driven alterations of host FA and phospholipid metabolism affect biophysical membrane properties or signaling processes like eicosanoid generation and thus impact general physiology as well as inflammatory and metabolic diseases.

16.5.3 Lipid Resorption

Resorption of dietary lipids is a three-step process comprising absorption of dietary FAs from the intestinal lumen, further trafficking from the apical to the basolateral site of enterocytes, and their secretion into the circulation.

Dietary lipids are hydrolyzed in the intestinal lumen by pancreatic lipases; thereby, TAG are converted to free FA and 2-monoacylglycerols (Barret et al. 2010; Hussain 2014). Lipids are in general relatively water insoluble; therefore, a key step is their emulsification with bile acids (BA) making lipids better substrates for hydrolysis by lipases (Barret et al. 2010). High levels of BA in the intestine (after contraction of the gallbladder) also induce the formation of micelles, which contain FA, cholesterol, and monoacylglycerols. Micelle formation solubilizes lipids and provides an effective mechanism for diffusion to the enterocytes. When FA concentrations are higher in the lumen than in enterocytes, the micelles move down their concentration gradient to the brush border of the mucosal cells, and the lipids may diffuse from micelles to enterocytes or are actively transported by membrane-associated proteins such as CD36, also known as FA translocase (Glatz et al. 2010). In human, the jejunum seems to be the major site of fat absorption in case of moderate food intake; the ileum becomes more important in case of a higher intake of dietary fat (Booth et al. 1961).

Once FAs have entered enterocytes (from the apical site), they are transported to various organelles by FA-binding proteins and shuttled to intracellular membranes, mainly to those of the ER, where they are esterified with monoacylglyerols by monoacyltransferases to form diacylglycerols (DAGs) (Hussain 2014; Hussain et al. 2013). DAG can be further converted to TAG by diacylglycerol acyltransferases. For lipid secretion into the circulation, lipids are packaged into chylomicrons (CMs) in the enterocyte and secreted to the lymph. During circulation, CMs are hydrolyzed by LPL so that their products, including free fatty acids, can enter peripheral tissues. The rest of the CM forms remnant particles, the so-called CM remnants, which are removed from the circulation by the liver. FA containing less than 10–12 carbons are enough water-soluble to pass the enterocyte unmodified and are actively transported as free FA into the portal blood (Barret et al. 2010).

16.5.4 Gut Microbiota and Lipid Resorption

It is hypothesized that the gut microbiota influences intestinal FA absorption, via, for instance, modulation of the luminal bile acid pool or the expression of transporters. While microbial colonization stimulated FA uptake and lipid droplet formation in the intestinal epithelium and liver of zebra fish as determined using in vivo imaging and fluorescent FA analogues (Semova et al. 2012), the situation in mammals is unclear. It is suggested that germfree mice have a significantly higher uptake of dietary lipids, since elevated levels of PUFA, i.e., DHA (FA22:6 $n-3$), which are of dietary origin, were detected in the plasma and liver. Additionally, 40% lower CM levels were measured in the serum of conventional mice vs. GF mice (Velagapudi et al. 2010). However, a pilot study testing administration of a lipid bolus to postprandial GF and conventionally raised mice did not demonstrate an altered lipid absorption in the absence of gut microbiota. Interestingly, studies done in the late 1960s showed that intestinal transit is faster in colonized vs. germfree mice (Abrams and Bishop 1967), but effects on lipid absorption are unknown.

16.5.5 Lipids as Bioactive Molecules Relevant for Microbe-Host Interactions

The intestinal epithelium is the largest and a very important interface to the external environment (Peterson and Artis 2014). Although the intestinal mucosa with its distinct morphology forms a rather tight barrier, it must also allow efficient absorption of nutrients and is in permanent contact with the luminal content, including microorganisms and their cellular components and metabolism products. The following lipids playing a role in microbe-host interactions are thought to be relevant for microbe-host communication:

1. *SCFA*
   Short-chain fatty acids are generated primarily by fermentation of complex carbohydrates by the gut microbiota, reach the circulation via the portal vein, and can then alter host metabolism and physiology substantially. Additionally to their diverse effects on the liver, muscle, and adipose tissue lipid metabolism (see Sect. 16.5.2), SCFA can activate gene expression of enzymes involved in intestinal gluconeogenesis either through a cAMP-dependent mechanism (in the case of butyrate) or via a gut-brain neuronal circuit involving the free fatty acid receptor (FFAR) 3 (in the case of propionate) (De Vadder et al. 2014).
2. *Bile acids*
   Bile acids are synthesized from cholesterol in the liver and can be metabolized to so-called secondary bile acids by bacteria in the gut. Besides their role in the formation of mixed micelles for lipid absorption, bile acids were shown to act as signaling molecules via membrane-bound G-protein-coupled receptors (GPCR) or via nuclear farnesoid X receptors (FXR) (Sayin et al. 2013). FXR control the expression of genes encoding enzymes involved in bile acid synthesis, metabolism, and transport of bile acids in liver and the sinusoid (Lefebvre et al. 2009).
3. *Phosphatidylcholine (PC)*
   A metabolomics study in a large clinical cohort originally identified trimethylamine-*N*-oxide (TMAO) as a predictive marker for cardiovascular disease and showed that dietary supplementation of TMAO can promote atherosclerosis in mice (Wang et al. 2011). Gut bacteria are involved in the production of trimethylamine from choline originating from dietary PC. Trimethylamine is further transformed by the liver to TMAO. Antibiotics-induced disturbances in the gut microbiota in atherosclerosis-prone mice inhibit atherosclerosis promoted by dietary choline.
4. *Conjugated fatty acids (CFA)*
   Conjugated FA is a collective term for positional and geometric isomers of FA with conjugated double bonds. CFA can be metabolized from linoleic acid by human intestinal bacteria including *Lactobacillus*, *Propionibacterium*, and *Bifidobacterium* species to different conjugated linoleic acid isomers such as *cis*-9, *trans*-11-CLA, *trans*-9, *trans*-11-CLA or *trans*-10, and *cis*-12-CLA. (Devillard et al. 2007; McIntosh et al. 2009). A major source for CLAs is also the diet, since ruminal bacteria such as *Butyrivibrio fibrisolvens* can metabolize LA to CLA (Wallace et al. 2007). CLA isomers can reach the circulation and peripheral tissues (Kuhnt et al. 2006; Wall et al. 2009). Several data from in vitro and animal studies show that

CLA isomers are beneficial and inhibit the progression of several diseases, including cardiovascular and inflammatory diseases as well as cancer (Bassaganya-Riera and Hontecillas 2006; Kelley et al. 2007; Mitchell and McLeod 2008). CFA are metabolized in colonic cells, where they can influence cellular lipid synthesis and degradation by modulating peroxisome proliferator-activated receptor (PPAR) and liver X receptor (LXR) activity and integrate into cell membrane and neutral lipids (Degen et al. 2011; Ecker et al. 2009, 2010b; Moya-Camarena et al. 1999).

5. *Hydroxy-fatty acids*
   Whereas fatty acid saturation does not occur in mammals, intestinal lactic acid bacteria such as *Lactobacillus plantarum* can saturate PUFA, thereby generating hydroxy-fatty acids, which can be detected in intestinal and plasma samples of mice (Kishino et al. 2013). In particular, higher levels of 10-hydroxy-fatty acids derived from linoleic and oleic acids can be detected in SPF vs. GF mice. Although in vivo functional studies have not yet been performed, it is clear that hydroxy-fatty acids act as ligands for PPARγ, a central transcription factor for lipid metabolism and storage involved in various metabolic diseases (Itoh et al. 2008).

16.5.6 Lipidomics and the Need to Improve Data Quality

The key technology for lipid analysis is mass spectrometry. Because lipidomic approaches are nowadays able to cover almost the full lipidome (Wenk 2010), the field is currently drawing much attention as it opens new ways to study lipid biology in numerous research areas, including gut microbiome research. However, this comes along with an increasing number of studies that report poor quality lipidomics data with false identification and inaccurate/inappropriate quantification of lipid molecules (Liebisch et al. 2017). Untargeted metabolomics approaches are particularly critical and often unsuitable for lipidomic investigations (Liebisch et al. 2015). Adequate controls of sample handling and processing (e.g., to assure analyte stability) are often missing, so are the use of appropriate internal standards, which are essential for proper quantification and identification of lipid species and for lipid annotation. The application of stable isotope-labelled tracers combined with reliable mass spectrometric analysis is a state-of-the-art approach to gain insight into synthesis, metabolism, and fluxes of lipid species (Ecker and Liebisch 2014).

16.6 Conclusions

The gut microbiota is of course not the primary cause of obesity but partly explains interindividual differences in body weight gain and the development of comorbidities. Gut microorganisms were also shown to play an important role in malnutrition. Needs for research in the field are depicted in Fig. 16.3. The functional importance of gut microbial communities has by now been demonstrated many times via microbiota transfer experiments, and we are just starting to identify individual microbial species with underlying mechanisms and bioactive molecules. It is important to remember that gut bacteria are per se not good or bad, that strain-dependent functional differences exist, and that the effects of specific strains depend on the microenvironment in which they proliferate. Hence, results must always be generalized with caution. The major task in the coming years will be to move away from purely descriptive work and intensify functional analysis, in order to come closer to defining the exact contribution of gut microbes to energy homeostasis and to learn how to manipulate this microbe-host crosstalk safely and with efficacy.

▶ Controversy

Research on the gut microbiome in metabolic health is still in its infancy but has reached already the broadcast media on several occasions and is thus under spotlights. This situation contributed to the establishment of a few misleading concepts in the field, of which we want to highlight two: (1) *Firmicutes*-to-*Bacteroidetes* ratio and

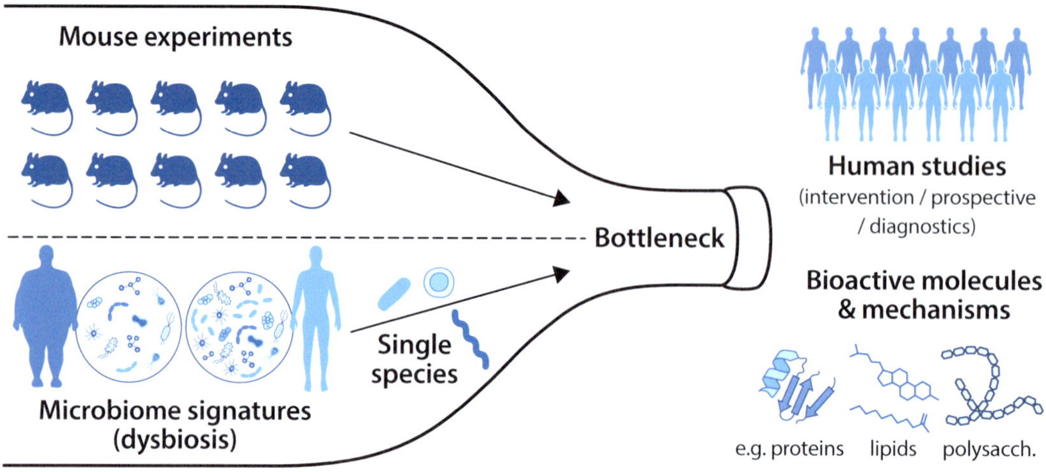

Fig. 16.3 Challenges and opportunities. Research in the field of microbe-host interactions in metabolic diseases delivered plentiful descriptive data in the past decade. The massive investments in sequencing projects contrast to the limited number of robust microbiome signatures available to data (see Sect. 16.1). Also, efforts in finding key microbial players (single species) with a functional role in host metabolism should be continued (see Sect. 16.3). Whereas the major asset of animal experiments is the ability to perform functional studies and to characterize molecular mechanisms, too many descriptive mouse studies have been published in the past. Even when functional implication of the gut microbiome was demonstrated, the meaning of results for human is often unclear. Hence, human and molecular studies are currently bottlenecks in the field. There is an urgent need to perform translational research and intensify the work in human populations. Large-scale, prospective studies (where individuals are followed overtime) will allow assessing the causal role of microbiome changes in the development of chronic diseases. In the clinics, it will be important to intensify work toward diagnostic approaches and intervention strategies. These activities should go hand-in-hand with the identification of more gut microbiota-derived molecules having an impact of host metabolism, with ensuing determination of their modes of action

(2) resistance of germfree mice to diet-induced obesity. We cannot present these topics in utmost details here in this info box. We also do not claim having definite answers or completely refute the notions proposed by others. We simply want to stress the fact that the entire research community must be careful before turning single findings into dogma that reach a standing that goes beyond scientific evidence. As noted by others (Perez-Munoz et al. 2017), scientific self-correction process is nowadays slower than the transfer of information to the public, and it takes quite some efforts to verify the validity of findings that attracted the spotlights.

1. *Firmicutes* and *Bacteroidetes* are the two major phyla of bacteria in the gut of human individuals and other mammals. Ley and colleagues reported in 2005 that genetically obese (ob/ob) mice were characterized by a 50% reduction in the abundance of *Bacteroidetes* and a proportional increase in *Firmicutes* (Ley et al. 2005). The same authors then showed in 2006 that the *Firmicutes*-to-*Bacteroidetes* ratio was increased in feces of 12 obese human subjects compared with seven control stool samples from lean individuals (Ley et al. 2006). The ratio decreased during therapy with low-calorie diets over a period of 1 year. Following this pioneering work, several other groups reported no or inverse changes in the *Firmicutes*-to-*Bacteroidetes* ratio (Duncan et al. 2008; Schwiertz et al. 2010). A recent population-based analysis of fecal microbiota using sequencing estimated that at least >500 samples are required to reliably assess the association between gut microbial populations and body mass index. Moreover, phyla correspond to very high taxonomic levels and

thus offer low resolution of analysis due to the tremendous diversity of species they contain. Low relative abundance of *Bacteroidetes* in obesity may make sense in view of the potential of *Prevotella* spp. to degrade fibers but less with respect to the possible association between *Bacteroides* spp. and dietary fat (De Filippis et al. 2016; Kovatcheva-Datchary et al. 2015; Wu et al. 2011). Ley and colleagues recently revisited their own view about the phylum *Bacteroidetes* in the context of metabolic diseases (Johnson et al. 2017). Hence, the *Firmicutes*-to-*Bacteroidetes* ratio is by no means a reference parameter with proven validity. Single studies on the gut microbiome in metabolic disorders vary a lot in terms of design (e.g., human or animal study; diet-induced obesity vs. genetic models; type of diets used), and a comparison of *Firmicutes*-to-*Bacteroidetes* ratios does not make sense unless done in a systematic manner.

2. In 2007, Bäckhed and colleagues published that germfree mice fed with a Western diet did not put on weight as their colonized counterparts (Backhed et al. 2007). They concluded that a hallmark of the absence of gut microbes is resistance against diet-induced obesity. In following years, Blaut and colleagues could not reproduce these findings and eventually published in 2010 that the resistance to obesity is driven by the type of diet fed to germfree mice: those fed with the Western diet as used by Bäckhed et al., characterized by higher amounts of simple sugars and fat from beef tallow, were indeed resistant to obesity, whereas those animals fed with a high-calorie diet containing wheat starch and fat from coconut oil were not (Fleissner et al. 2010). Last year, Klingenspor and colleagues went one step further and proposed that cholesterol is a major driver of the obesity-resistant phenotype in germfree mice (Kubeck et al. 2016). By using diets that differed only by their type of fat, the authors showed that germfree mice fed with a diet rich in palm oil became obese, whereas those fed with a lard-based high-fat diet stayed lean due to increased energy expenditure, preferential carbohydrate oxidation, and increased fecal fat and energy excretion. Altogether, these findings do not refute the importance of gut microbes in energy homeostasis but draw the attention to complex diet-gut microbiome interactions and their potential influence on experimental models and their representativeness.

History

Over millions of years of evolution, our body has developed efficient mechanisms to harvest and store energy from the diet. These mechanisms have somehow become deleterious in contemporary times of food abundance and high-calorie diets, at least in the Western world but also in blooming cities in developing countries. Nutrition has been a key factor of our evolution, and the microbes colonizing our gut can metabolize a vast array of nutrients and have coevolved with us, leading to the establishment of complex diet-microbe-host networks underlying energy regulation. Interactions between the diet and gut microbes and the consequence for human health were a subject of interest already at the beginning of the twentieth century (Rettger 1915). Ilya Metchnikov, the Nobel Prize Laureate for Physiology and Medicine in 1908, was very much interested in the impact of lactic acid bacteria and fermented milk products on health and proposed that "aging was the result of a continuous intoxication caused by intestinal bacteria." The idea of raising animal under sterile (germfree) conditions and observing changes in host physiology blossomed already at these early times but bloomed only later in the 1960s when the interplay between nutrition and gut colonization by commensals and pathogens was studied in greater details (Dubos et al. 1969; Dubos and Schaedler 1960; Wostmann 1975). Interestingly, it was reported that growth differences between germfree and conventional mice differed from one laboratory to the next, most likely due to the instability of conventional controls (Pleasants 1968). Facility-dependent colonization of the mouse gut was used decades later for the discovery of specific gut bacteria having distinct effects on the immune system (Ivanov et al. 2009). Also in the clinics, researchers in the 1980s were already interested in studying changes in the gut microbiota after gastric bypass, with the limitations of methods available back then (Bjorneklett et al. 1981). Hence, current research activities have precedents in the literature, which should not be forgotten. The renewed interest in the role of gut microbes in metabolic health, driven by new molecular methods and the common use of gnotobiology, is partly explained by the ever-rising prevalence of obesity and associated metabolic disorders, and generates hopes pertaining to modulation of metabolic health via the gut microbiome. However, research evidence falls short of this target because descriptive data have so far dominated mechanistic and applied research: there is no doubt that the gut microbiota regulates host metabolic responses, but there is an urgent need to define how underlying microbe-host interactions exactly work.

Highlights

- The gut microbiota is a very complex and dynamic ecosystem with a tremendous metabolic potential.
- Its causal role in malnutrition and the development of chronic metabolic disorders has been demonstrated via microbiota transfer experiments.
- Besides loss in ecosystem diversity, no consistent change in the gut microbiota can yet be associated with metabolic disorders.
- Many bacterial members of the ecosystem, including those species regulating metabolic responses, and the molecules involved remain to be identified.
- Lipid compounds (e.g., short-chain fatty acids, bile acids, lipopolysaccharides, conjugated fatty acids) and the crosstalk between lipid metabolism and the gut microbiota are important for host metabolic health.
- More human studies, including microbiome-based personalized approaches, are required to bridge the gap between theoretical knowledge and practical relevance.

References

Abrams, G. D., & Bishop, J. E. (1967). Effect of the normal microbial flora on gastrointestinal motility. *Proceedings of the Society for Experimental Biology and Medicine, 126*, 301–304.

Abu-Shanab, A., & Quigley, E. M. (2010). The role of the gut microbiota in nonalcoholic fatty liver disease. *Nature Reviews. Gastroenterology & Hepatology, 7*, 691–701.

Ajslev, T. A., Andersen, C. S., Gamborg, M., Sorensen, T. I., & Jess, T. (2011). Childhood overweight after establishment of the gut microbiota: The role of delivery mode, pre-pregnancy weight and early administration of antibiotics. *International Journal of Obesity, 35*, 522–529.

Alves, M. M., Pereira, M. A., Sousa, D. Z., Cavaleiro, A. J., Picavet, M., Smidt, H., et al. (2009). Waste lipids to energy: How to optimize methane production from long-chain fatty acids (LCFA). *Microbial Biotechnology, 2*, 538–550.

Amato, A., Baldassano, S., & Mule, F. (2016). GLP2: An underestimated signal for improving glycaemic control and insulin sensitivity. *The Journal of Endocrinology, 229*, R57–R66.

Angelakis, E., & Raoult, D. (2010). The increase of Lactobacillus species in the gut flora of newborn broiler chicks and ducks is associated with weight gain. *PLoS One, 5*, e10463.

Arrieta, M. C., Walter, J., & Finlay, B. B. (2016). Human microbiota-associated mice: A model with challenges. *Cell Host & Microbe, 19*, 575–578.

Azad, M. B., Bridgman, S. L., Becker, A. B., & Kozyrskyj, A. L. (2014). Infant antibiotic exposure and the development of childhood overweight and central adiposity. *International Journal of Obesity, 38*, 1290–1298.

Backhed, F., Ding, H., Wang, T., Hooper, L. V., Koh, G. Y., Nagy, A., et al. (2004). The gut microbiota as an environmental factor that regulates fat storage. *Proceedings of the National Academy of Sciences of the United States of America, 101*, 15718–15723.

Backhed, F., Manchester, J. K., Semenkovich, C. F., & Gordon, J. I. (2007). Mechanisms underlying the resistance to diet-induced obesity in germ-free mice. *Proceedings of the National Academy of Sciences of the United States of America, 104*, 979–984.

Bajaj, J. S., Kassam, Z., Fagan, A., Gavis, E. A., Liu, E., Cox, I. J., et al. (2017, December). Fecal microbiota transplant from a rational stool donor improves hepatic encephalopathy: A randomized clinical trial. *Hepatology, 66*(6), 1727–1738.

Barret, K., Brooks, H., Boitano, S., & Barman, S. (2010). Digestion, absorption, and nutritional principles (Chapter 27). In Ganong's review of medical physiology. In Lange medical book, pp. 451–467.

Bassaganya-Riera, J., & Hontecillas, R. (2006). CLA and n-3 PUFA differentially modulate clinical activity and colonic PPAR-responsive gene expression in a pig model of experimental IBD. *Clinical Nutrition, 25*, 454–465.

Bjorneklett, A., Viddal, K. O., Midtvedt, T., & Nygaard, K. (1981). Intestinal and gastric bypass. Changes in intestinal microecology after surgical treatment of morbid obesity in man. *Scand J Gastroenterol, 16*, 681–687.

Blachier, F., Beaumont, M., Andriamihaja, M., Davila, A. M., Lan, A., Grauso, M., et al. (2017). Changes in the luminal environment of the colonic epithelial cells and physiopathological consequences. *The American Journal of Pathology, 187*, 476–486.

Blanton, L. V., Charbonneau, M. R., Salih, T., Barratt, M. J., Venkatesh, S., Ilkaveya, O., et al. (2016). Gut bacteria that prevent growth impairments transmitted by microbiota from malnourished children. *Science, 351*.

Booth, C. C., Alldis, D., & Read, A. E. (1961). Studies on the site of fat absorption: 2 Fat balances after resection of varying amounts of the small intestine in man. *Gut, 2*, 168–174.

Brown, J. M., Chung, S., Sawyer, J. K., Degirolamo, C., Alger, H. M., Nguyen, T., et al. (2008). Inhibition of stearoyl-coenzyme A desaturase 1 dissociates insulin resistance and obesity from atherosclerosis. *Circulation, 118*, 1467–1475.

Brun, P., Castagliuolo, I., Di Leo, V., Buda, A., Pinzani, M., Palu, G., et al. (2007). Increased intestinal permeability in obese mice: New evidence in the pathogenesis of nonalcoholic steatohepatitis. *American Journal of Physiology. Gastrointestinal and Liver Physiology, 292*, G518–G525.

Buckley, C. D., Gilroy, D. W., & Serhan, C. N. (2014). Proresolving lipid mediators and mechanisms in the resolution of acute inflammation. *Immunity, 40*, 315–327.

Bui, T. P., Ritari, J., Boeren, S., de Waard, P., Plugge, C. M., & de Vos, W. M. (2015). Production of butyrate from lysine and the Amadori product fructoselysine by a human gut commensal. *Nature Communications, 6*, 10062.

Cani, P. D., Neyrinck, A. M., Fava, F., Knauf, C., Burcelin, R. G., Tuohy, K. M., et al. (2007). Selective increases of bifidobacteria in gut microflora improve high-fat-diet-induced diabetes in mice through a mechanism associated with endotoxaemia. *Diabetologia, 50*, 2374–2383.

Cani, P. D., Possemiers, S., Van de Wiele, T., Guiot, Y., Everard, A., Rottier, O., et al. (2009). Changes in gut microbiota control inflammation in obese mice through a mechanism involving GLP-2-driven improvement of gut permeability. *Gut, 58*, 1091–1103.

Charbonneau, M. R., O'Donnell, D., Blanton, L. V., Totten, S. M., Davis, J. C., Barratt, M. J., et al. (2016). Sialylated Milk Oligosaccharides Promote Microbiota-Dependent Growth in Models of Infant Undernutrition. *Cell, 164*, 859–871.

Cherbuy, C., Darcy-Vrillon, B., Morel, M. T., Pegorier, J. P., & Duee, P. H. (1995). Effect of germfree state on the capacities of isolated rat colonocytes to metabolize n-butyrate, glucose, and glutamine. *Gastroenterology, 109*, 1890–1899.

Clavel, T., Desmarchelier, C., Haller, D., Gerard, P., Rohn, S., Lepage, P., et al. (2014). Intestinal microbiota in metabolic diseases: From bacterial community structure and functions to species of pathophysiological relevance. *Gut Microbes, 5*, 544–551.

Collado, M. C., Isolauri, E., Laitinen, K., & Salminen, S. (2010). Effect of mother's weight on infant's microbiota acquisition, composition, and activity during early infancy: A prospective follow-up study initiated in early pregnancy. *The American Journal of Clinical Nutrition, 92*, 1023–1030.

Cotillard, A., Kennedy, S. P., Kong, L. C., Prifti, E., Pons, N., Le Chatelier, E., et al. (2013). Dietary intervention impact on gut microbial gene richness. *Nature, 500*, 585–588.

Cox, L. M., & Blaser, M. J. (2015). Antibiotics in early life and obesity. *Nature Reviews. Endocrinology, 11*, 182–190.

Damms-Machado, A., Louis, S., Schnitzer, A., Volynets, V., Rings, A., Basrai, M., et al. (2017). Gut permeability is related to body weight, fatty liver disease, and insulin resistance in obese individuals undergoing weight reduction. *The American Journal of Clinical Nutrition, 105*, 127–135.

Daniel, H., Gholami, A. M., Berry, D., Desmarchelier, C., Hahne, H., Loh, G., et al. (2014). High-fat diet alters gut microbiota physiology in mice. *The ISME Journal, 8*, 295–308.

David, L. A., Maurice, C. F., Carmody, R. N., Gootenberg, D. B., Button, J. E., Wolfe, B. E., et al. (2014). Diet rapidly and reproducibly alters the human gut microbiome. *Nature, 505*, 559–563.

De Filippis, F., Pellegrini, N., Laghi, L., Gobbetti, M., & Ercolini, D. (2016). Unusual sub-genus associations of faecal Prevotella and Bacteroides with specific dietary patterns. *Microbiome, 4*, 57.

de Groot, P. F., Frissen, M. N., de Clercq, N. C., & Nieuwdorp, M. (2017). Fecal microbiota transplantation in metabolic syndrome: History, present and future. *Gut Microbes, 8*, 253–267.

De Vadder, F., Kovatcheva-Datchary, P., Goncalves, D., Vinera, J., Zitoun, C., Duchampt, A., et al. (2014). Microbiota-generated metabolites promote metabolic benefits via gut-brain neural circuits. *Cell, 156*, 84–96.

Degen, C., Ecker, J., Piegholdt, S., Liebisch, G., Schmitz, G., & Jahreis, G. (2011). Metabolic and growth inhibitory effects of conjugated fatty acids in the cell line HT-29 with special regard to the conversion of t11,t13-CLA. *Biochimica et Biophysica Acta, 1811*, 1070–1080.

den Besten, G., van Eunen, K., Groen, A. K., Venema, K., Reijngoud, D. J., & Bakker, B. M. (2013). The role of short-chain fatty acids in the interplay between diet, gut microbiota, and host energy metabolism. *Journal of Lipid Research, 54*, 2325–2340.

Derrien, M., Vaughan, E. E., Plugge, C. M., & de Vos, W. M. (2004). Akkermansia muciniphila gen. nov., sp. nov., a human intestinal mucin-degrading bacterium. *International Journal of Systematic and Evolutionary Microbiology, 54*, 1469–1476.

Derrien, M., Collado, M. C., Ben-Amor, K., Salminen, S., & de Vos, W. M. (2008). The Mucin degrader Akkermansia muciniphila is an abundant resident of the human intestinal tract. *Applied and Environmental Microbiology, 74*, 1646–1648.

Devillard, E., McIntosh, F. M., Duncan, S. H., & Wallace, R. J. (2007). Metabolism of linoleic acid by human gut bacteria: Different routes for biosynthesis of conjugated linoleic acid. *Journal of Bacteriology, 189*, 2566–2570.

Dubos, R. J., & Schaedler, R. W. (1960). The effect of the intestinal flora on the growth rate of mice, and on their susceptibility to experimental infections. *The Journal of Experimental Medicine, 111*, 407–417.

Dubos, R., Lee, C. J., & Costello, R. (1969). Lasting biological effects of early environmental influences. V. Viability, growth, and longevity. *The Journal of Experimental Medicine, 130*, 963–977.

Duncan, S. H., Belenguer, A., Holtrop, G., Johnstone, A. M., Flint, H. J., & Lobley, G. E. (2007). Reduced dietary intake of carbohydrates by obese subjects results in decreased concentrations of butyrate and butyrate-producing bacteria in feces. *Applied and Environmental Microbiology, 73*, 1073–1078.

Duncan, S. H., Lobley, G. E., Holtrop, G., Ince, J., Johnstone, A. M., Louis, P., et al. (2008). Human colonic microbiota associated with diet, obesity and weight loss. *International Journal of Obesity, 32*, 1720–1724.

Ecker, J., & Liebisch, G. (2014). Application of stable isotopes to investigate the metabolism of fatty acids, glycerophospholipid and sphingolipid species. *Progress in Lipid Research, 54*, 14–31.

Ecker, J., Liebisch, G., Patsch, W., & Schmitz, G. (2009). The conjugated linoleic acid isomer trans-9,trans-11 is a dietary occurring agonist of liver X receptor alpha. *Biochemical and Biophysical Research Communications, 388*, 660–666.

Ecker, J., Liebisch, G., Englmaier, M., Grandl, M., Robenek, H., & Schmitz, G. (2010a). Induction of fatty acid synthesis is a key requirement for phagocytic differentiation of human monocytes. *Proceedings of the National Academy of Sciences of the United States of America, 107*, 7817–7822.

Ecker, J., Liebisch, G., Scherer, M., & Schmitz, G. (2010b). Differential effects of conjugated linoleic acid isomers on macrophage glycerophospholipid metabolism. *Journal of Lipid Research, 51*, 2686–2694.

Edmonson, M. B., & Eickhoff, J. C. (2017). Weight gain and obesity in infants and young children exposed to prolonged antibiotic prophylaxis. *JAMA Pediatrics, 171*, 150–156.

Erbay, E., Babaev, V. R., Mayers, J. R., Makowski, L., Charles, K. N., Snitow, M. E., et al. (2009). Reducing endoplasmic reticulum stress through a macrophage lipid chaperone alleviates atherosclerosis. *Nature Medicine, 15*, 1383–1391.

Ernst, R., Ejsing, C. S., & Antonny, B. (2016). Homeoviscous adaptation and the regulation of membrane lipids. *Journal of Molecular Biology, 428*, 4776–4791.

Everard, A., Belzer, C., Geurts, L., Ouwerkerk, J. P., Druart, C., Bindels, L. B., et al. (2013). Cross-talk between Akkermansia muciniphila and intestinal epithelium controls diet-induced obesity. *Proceedings of the National Academy of Sciences of the United States of America, 110*, 9066–9071.

Falony, G., Joossens, M., Vieira-Silva, S., Wang, J., Darzi, Y., Faust, K., et al. (2016). Population-level analysis of gut microbiome variation. *Science, 352*, 560–564.

Fava, F., Gitau, R., Griffin, B. A., Gibson, G. R., Tuohy, K. M., & Lovegrove, J. A. (2013). The type and quantity of dietary fat and carbohydrate alter faecal microbiome and short-chain fatty acid excretion in a metabolic syndrome 'at-risk' population. *International Journal of Obesity, 37*, 216–223.

Fei, N., & Zhao, L. (2013). An opportunistic pathogen isolated from the gut of an obese human causes obesity in germfree mice. *The ISME Journal, 7*, 880–884.

Ferolla, S. M., Couto, C. A., Costa-Silva, L., Armiliato, G. N., Pereira, C. A., Martins, F. S., et al. (2016). Beneficial effect of synbiotic supplementation on hepatic steatosis and anthropometric parameters, but not on gut permeability in a population with nonalcoholic steatohepatitis. *Nutrients, 8*.

Fleissner, C. K., Huebel, N., Abd El-Bary, M. M., Loh, G., Klaus, S., & Blaut, M. (2010). Absence of intestinal microbiota does not protect mice from diet-induced obesity. *The British Journal of Nutrition, 104*, 919–929.

Foote, A. P., & Freetly, H. C. (2016). Effect of abomasal butyrate infusion on net nutrient flux across the portal-drained viscera and liver of growing lambs. *Journal of Animal Science, 94*, 2962–2972.

Forslund, K., Hildebrand, F., Nielsen, T., Falony, G., Le Chatelier, E., Sunagawa, S., et al. (2015). Disentangling type 2 diabetes and metformin treatment signatures in the human gut microbiota. *Nature, 528*, 262–266.

Fuller, R. (1989). Probiotics in man and animals. *The Journal of Applied Bacteriology, 66*, 365–378.

Furet, J. P., Kong, L. C., Tap, J., Poitou, C., Basdevant, A., Bouillot, J. L., et al. (2010). Differential adaptation of human gut microbiota to bariatric surgery-induced weight loss: Links with metabolic and low-grade inflammation markers. *Diabetes, 59*, 3049–3057.

Garidou, L., Pomie, C., Klopp, P., Waget, A., Charpentier, J., Aloulou, M., et al. (2015). The gut microbiota regulates intestinal CD4 T cells expressing RORgammat and controls metabolic disease. *Cell Metabolism, 22*, 100–112.

Glatz, J. F., Luiken, J. J., & Bonen, A. (2010). Membrane fatty acid transporters as regulators of lipid metabolism: Implications for metabolic disease. *Physiological Reviews, 90*, 367–417.

Goodrich, J. K., Waters, J. L., Poole, A. C., Sutter, J. L., Koren, O., Blekhman, R., et al. (2014). Human genetics shape the gut microbiome. *Cell, 159*, 789–799.

Grasset, E., Puel, A., Charpentier, J., Collet, X., Christensen, J. E., Terce, F., et al. (2017). A specific gut microbiota dysbiosis of type 2 diabetic mice induces GLP-1 resistance through an enteric NO-dependent and gut-brain axis mechanism. *Cell Metabolism, 26*, 278.

Hamilton, M. K., Boudry, G., Lemay, D. G., & Raybould, H. E. (2015). Changes in intestinal barrier function and gut microbiota in high-fat diet-fed rats are dynamic and region dependent. *American Journal of Physiology. Gastrointestinal and Liver Physiology, 308*, G840–G851.

Hansen, C. H., Krych, L., Nielsen, D. S., Vogensen, F. K., Hansen, L. H., Sorensen, S. J., et al. (2012). Early life treatment with vancomycin propagates Akkermansia muciniphila and reduces diabetes incidence in the NOD mouse. *Diabetologia, 55*, 2285–2294.

Henao-Mejia, J., Elinav, E., Jin, C., Hao, L., Mehal, W. Z., Strowig, T., et al. (2012). Inflammasome-mediated dysbiosis regulates progression of NAFLD and obesity. *Nature, 482*, 179–185.

Hersoug, L. G., Moller, P., & Loft, S. (2016). Gut microbiota-derived lipopolysaccharide uptake and trafficking to adipose tissue: Implications for inflammation and obesity. *Obesity Reviews, 17*, 297–312.

Heymann, F., & Tacke, F. (2016). Immunology in the liver – From homeostasis to disease. *Nature Reviews. Gastroenterology and Hepatology, 13*, 88–110.

Hodson, L., & Fielding, B. A. (2013). Stearoyl-CoA desaturase: Rogue or innocent bystander? *Progress in Lipid Research, 52*, 15–42.

Horton, J. D., Goldstein, J. L., & Brown, M. S. (2002). SREBPs: Activators of the complete program of cholesterol and fatty acid synthesis in the liver. *The Journal of Clinical Investigation, 109*, 1125–1131.

Hussain, M. M. (2014). Intestinal lipid absorption and lipoprotein formation. *Current Opinion in Lipidology, 25*, 200–206.

Hussain, M. M., Leung, T. M., Zhou, L., & Abu-Merhi, S. (2013). Regulating intestinal function to reduce atherogenic lipoproteins. *Clinical Lipidology, 8*, 481–490.

Itoh, T., Fairall, L., Amin, K., Inaba, Y., Szanto, A., Balint, B. L., et al. (2008). Structural basis for the activation of PPARgamma by oxidized fatty acids. *Nature Structural and Molecular Biology, 15*, 924–931.

Ivanov, I. I., Atarashi, K., Manel, N., Brodie, E. L., Shima, T., Karaoz, U., et al. (2009). Induction of intestinal Th17 cells by segmented filamentous bacteria. *Cell, 139*, 485–498.

Johnson, E. L., Heaver, S. L., Walters, W. A., & Ley, R. E. (2017). Microbiome and metabolic disease: Revisiting the bacterial phylum Bacteroidetes. *The Journal of Molecular Medicine (Berlin), 95*, 1–8.

Jumpertz, R., Le, D. S., Turnbaugh, P. J., Trinidad, C., Bogardus, C., Gordon, J. I., et al. (2011). Energy-balance studies reveal associations between gut microbes, caloric load, and nutrient absorption in humans. *The American Journal of Clinical Nutrition, 94*, 58–65.

Kalliomaki, M., Collado, M. C., Salminen, S., & Isolauri, E. (2008). Early differences in fecal microbiota composition in children may predict overweight. *The American Journal of Clinical Nutrition, 87*, 534–538.

Karl, J. P., Margolis, L. M., Madslien, E. H., Murphy, N. E., Castellani, J. W., Gundersen, Y., et al. (2017). Changes in intestinal microbiota composition and metabolism coincide with increased intestinal permeability in young adults under prolonged physiological stress. *American Journal of Physiology. Gastrointestinal and Liver Physiology, 312*, G559–G571.

Kau, A. L., Ahern, P. P., Griffin, N. W., Goodman, A. L., & Gordon, J. I. (2011). Human nutrition, the gut microbiome and the immune system. *Nature, 474*, 327–336.

Kelley, N. S., Hubbard, N. E., & Erickson, K. L. (2007). Conjugated linoleic acid isomers and cancer. *The Journal of Nutrition, 137*, 2599–2607.

Kimura, I., Ozawa, K., Inoue, D., Imamura, T., Kimura, K., Maeda, T., et al. (2013). The gut microbiota suppresses insulin-mediated fat accumulation via the short-chain fatty acid receptor GPR43. *Nature Communications, 4*, 1829.

Kishino, S., Takeuchi, M., Park, S. B., Hirata, A., Kitamura, N., Kunisawa, J., et al. (2013). Polyunsaturated fatty acid saturation by gut lactic acid bacteria affecting host lipid composition. *Proceedings of the National Academy of Sciences of the United States of America, 110*, 17808–17813.

Kless, C., Muller, V. M., Schuppel, V. L., Lichtenegger, M., Rychlik, M., Daniel, H., et al. (2015). Diet-induced obesity causes metabolic impairment independent of alterations in gut barrier integrity. *Molecular Nutrition and Food Research, 59*, 968–978.

Kootte, R. S., Levin, E., Salojarvi, J., Smits, L. P., Hartstra, A. V., Udayappan, S. D., et al. (2017). Improvement of insulin sensitivity after lean donor feces in metabolic syndrome is driven by baseline intestinal microbiota composition. *Cell Metabolism, 26*, 611–619 e616.

Kovatcheva-Datchary, P., Nilsson, A., Akrami, R., Lee, Y. S., De Vadder, F., Arora, T., et al. (2015). Dietary fiber-induced improvement in glucose metabolism is associated with increased abundance of prevotella. *Cell Metabolism, 22*, 971–982.

Kubeck, R., Bonet-Ripoll, C., Hoffmann, C., Walker, A., Muller, V. M., Schuppel, V. L., et al. (2016). Dietary fat and gut microbiota interactions determine diet-induced obesity in mice. *Molecular Metabolism, 5*, 1162–1174.

Kuhnt, K., Wagner, A., Kraft, J., Basu, S., & Jahreis, G. (2006). Dietary supplementation with 11trans- and 12trans-18:1 and oxidative stress in humans. *The American Journal of Clinical Nutrition, 84*, 981–988.

Kumar, M., Nagpal, R., Kumar, R., Hemalatha, R., Verma, V., Kumar, A., et al. (2012). Cholesterol-lowering probiotics as potential biotherapeutics for metabolic diseases. *Experimental Diabetes Research, 2012*, 902917.

Lagkouvardos, I., Overmann, J., & Clavel, T. (2017). Cultured microbes represent a substantial fraction of the human and mouse gut microbiota. *Gut Microbes, 8* (5), 493–503.

Lahtinen, S. J., Davis, E., & Ouwehand, A. C. (2012). Lactobacillus species causing obesity in humans: Where is the evidence? *Beneficial Microbes, 3*, 171–174.

Lambert, J. E., Parnell, J. A., Eksteen, B., Raman, M., Bomhof, M. R., Rioux, K. P., et al. (2015). Gut microbiota manipulation with prebiotics in patients with non-alcoholic fatty liver disease: A randomized controlled trial protocol. *BMC Gastroenterology, 15*, 169.

Le Chatelier, E., Nielsen, T., Qin, J., Prifti, E., Hildebrand, F., Falony, G., et al. (2013). Richness of human gut microbiome correlates with metabolic markers. *Nature, 500*, 541–546.

Le Roy, T., Llopis, M., Lepage, P., Bruneau, A., Rabot, S., Bevilacqua, C., et al. (2013). Intestinal microbiota determines development of non-alcoholic fatty liver disease in mice. *Gut, 62*, 1787–1794.

Lee, T., Clavel, T., Smirnov, K., Schmidt, A., Lagkouvardos, I., Walker, A., et al. (2017). Oral versus intravenous iron replacement therapy distinctly alters the gut microbiota and metabolome in patients with IBD. *Gut, 66*, 863–871.

Lefebvre, P., Cariou, B., Lien, F., Kuipers, F., & Staels, B. (2009). Role of bile acids and bile acid receptors in metabolic regulation. *Physiological Reviews, 89*, 147–191.

Ley, R. E., Backhed, F., Turnbaugh, P., Lozupone, C. A., Knight, R. D., & Gordon, J. I. (2005). Obesity alters gut microbial ecology. *Proceedings of the National Academy of Sciences of the United States of America, 102*, 11070–11075.

Ley, R. E., Turnbaugh, P. J., Klein, S., & Gordon, J. I. (2006). Microbial ecology: Human gut microbes associated with obesity. *Nature, 444*, 1022–1023.

Liebisch, G., Ejsing, C. S., & Ekroos, K. (2015). Identification and annotation of lipid species in metabolomics studies need improvement. *Clinical Chemistry, 61*, 1542–1544.

Liebisch, G., Ekroos, K., Hermansson, M., & Ejsing, C. S. (2017). Reporting of lipidomics data should be standardized. *Biochimica et Biophysica Acta, 1862*, 747–751.

Liou, A. P., Paziuk, M., Luevano, J. M., Jr., Machineni, S., Turnbaugh, P. J., & Kaplan, L. M. (2013). Conserved shifts in the gut microbiota due to gastric bypass reduce host weight and adiposity. *Science Translational Medicine, 5*, 178ra141.

Louis, P., & Flint, H. J. (2017). Formation of propionate and butyrate by the human colonic microbiota. *Environmental Microbiology, 19*, 29–41.

Ma, H., Sales, V. M., Wolf, A. R., Subramanian, S., Matthews, T. J., Chen, M., et al. (2017). Attenuated effects of bile acids on glucose metabolism and insulin sensitivity in a male mouse model of prenatal undernutrition. *Endocrinology, 158*, 2441–2452.

Mardinoglu, A., Wu, H., Bjornson, E., Zhang, C., Hakkarainen, A., Rasanen, S. M., et al. (2018). An integrated understanding of the rapid metabolic benefits of a carbohydrate-restricted diet on hepatic steatosis in humans. *Cell Metabolism, 27*(3), 559–571.e5.

Martinez, I., Stegen, J. C., Maldonado-Gomez, M. X., Eren, A. M., Siba, P. M., Greenhill, A. R., et al. (2015). The gut microbiota of rural papua new guineans: Composition, diversity patterns, and ecological processes. *Cell Reports, 11*, 527–538.

Matis, G., Kulcsar, A., Turowski, V., Febel, H., Neogrady, Z., & Huber, K. (2015). Effects of oral butyrate application on insulin signaling in various tissues of chickens. *Domestic Animal Endocrinology, 50*, 26–31.

Matsuzaka, T., Shimano, H., Yahagi, N., Kato, T., Atsumi, A., Yamamoto, T., et al. (2007). Crucial role of a long-chain fatty acid elongase, Elovl6, in obesity-induced insulin resistance. *Nature Medicine, 13*, 1193–1202.

McIntosh, F. M., Shingfield, K. J., Devillard, E., Russell, W. R., & Wallace, R. J. (2009). Mechanism of conjugated linoleic acid and vaccenic acid formation in human faecal suspensions and pure cultures of intestinal bacteria. *Microbiology, 155*, 285–294.

Million, M., Maraninchi, M., Henry, M., Armougom, F., Richet, H., Carrieri, P., et al. (2012). Obesity-associated gut microbiota is enriched in Lactobacillus reuteri and depleted in Bifidobacterium animalis and

Methanobrevibacter smithii. *International Journal of Obesity, 36*, 817–825.

Mitchell, P. L., & McLeod, R. S. (2008). Conjugated linoleic acid and atherosclerosis: Studies in animal models. *Biochemistry and Cell Biology, 86*, 293–301.

Miyazaki, M., & Ntambi, J. M. (2003). Role of stearoyl-coenzyme A desaturase in lipid metabolism. *Prostaglandins, Leukotrienes, and Essential Fatty Acids, 68*, 113–121.

Moller, D. E. (2001). New drug targets for type 2 diabetes and the metabolic syndrome. *Nature, 414*, 821–827.

Moya-Camarena, S. Y., Vanden Heuvel, J. P., Blanchard, S. G., Leesnitzer, L. A., & Belury, M. A. (1999). Conjugated linoleic acid is a potent naturally occurring ligand and activator of PPARalpha. *Journal of Lipid Research, 40*, 1426–1433.

Mudaliar, S., Henry, R. R., Sanyal, A. J., Morrow, L., Marschall, H. U., Kipnes, M., et al. (2013). Efficacy and safety of the farnesoid X receptor agonist obeticholic acid in patients with type 2 diabetes and nonalcoholic fatty liver disease. *Gastroenterology, 145*, 574–582 e571.

Mueller, N. T., Whyatt, R., Hoepner, L., Oberfield, S., Dominguez-Bello, M. G., Widen, E. M., et al. (2015). Prenatal exposure to antibiotics, cesarean section and risk of childhood obesity. *International Journal of Obesity, 39*, 665–670.

Muller, V. M., Zietek, T., Rohm, F., Fiamoncini, J., Lagkouvardos, I., Haller, D., et al. (2016). Gut barrier impairment by high-fat diet in mice depends on housing conditions. *Molecular Nutrition and Food Research, 60*, 897–908.

Nishizawa, Y., Imaizumi, T., Tanishita, H., Yano, I., Kawai, Y., & Mormii, H. (1988). Relationship of fat deposition and intestinal microflora in VMH rats. *International Journal of Obesity, 12*, 103–110.

Ogasawara, Y., Itakura, E., Kono, N., Mizushima, N., Arai, H., Nara, A., et al. (2014). Stearoyl-CoA desaturase 1 activity is required for autophagosome formation. *The Journal of Biological Chemistry, 289*, 23938–23950.

Ott, B., Skurk, T., Hastreiter, L., Lagkouvardos, I., Fischer, S., Buttner, J., et al. (2017). Effect of caloric restriction on gut permeability, inflammation markers, and fecal microbiota in obese women. *Scientific Reports, 7*, 11955.

Paramsothy, S., Kamm, M. A., Kaakoush, N. O., Walsh, A. J., van den Bogaerde, J., Samuel, D., et al. (2017). Multidonor intensive faecal microbiota transplantation for active ulcerative colitis: A randomised placebo-controlled trial. *Lancet, 389*, 1218–1228.

Parseus, A., Sommer, N., Sommer, F., Caesar, R., Molinaro, A., Stahlman, M., et al. (2017). Microbiota-induced obesity requires farnesoid X receptor. *Gut, 66*, 429–437.

Pascal, V., Pozuelo, M., Borruel, N., Casellas, F., Campos, D., Santiago, A., et al. (2017). A microbial signature for Crohn's disease. *Gut, 66*, 813–822.

Pedersen, H. K., Gudmundsdottir, V., Nielsen, H. B., Hyotylainen, T., Nielsen, T., Jensen, B. A., et al. (2016). Human gut microbes impact host serum metabolome and insulin sensitivity. *Nature, 535*, 376–381.

Pereira, S. L., Leonard, A. E., & Mukerji, P. (2003). Recent advances in the study of fatty acid desaturases from animals and lower eukaryotes. *Prostaglandins, Leukotrienes, and Essential Fatty Acids, 68*, 97–106.

Perez-Munoz, M. E., Arrieta, M. C., Ramer-Tait, A. E., & Walter, J. (2017). A critical assessment of the "sterile womb" and "in utero colonization" hypotheses: Implications for research on the pioneer infant microbiome. *Microbiome, 5*, 48.

Peterson, L. W., & Artis, D. (2014). Intestinal epithelial cells: Regulators of barrier function and immune homeostasis. *Nature Reviews. Immunology, 14*, 141–153.

Pinot, M., Vanni, S., Pagnotta, S., Lacas-Gervais, S., Payet, L. A., Ferreira, T., et al. (2014). Lipid cell biology. Polyunsaturated phospholipids facilitate membrane deformation and fission by endocytic proteins. *Science, 345*, 693–697.

Pleasants, J. R. (1968). Characteristics of the germ-free animal. In M. E. Coates, H. A. Gordon, & B. S. Wostmann (Eds.), *The germ-free animal in research*. London and New York: Academic Press.

Plovier, H., Everard, A., Druart, C., Depommier, C., Van Hul, M., Geurts, L., et al. (2017). A purified membrane protein from Akkermansia muciniphila or the pasteurized bacterium improves metabolism in obese and diabetic mice. *Nature Medicine, 23*, 107–113.

Qin, J., Li, Y., Cai, Z., Li, S., Zhu, J., Zhang, F., et al. (2012). A metagenome-wide association study of gut microbiota in type 2 diabetes. *Nature, 490*, 55–60.

Reijnders, D., Goossens, G. H., Hermes, G. D., Neis, E. P., van der Beek, C. M., Most, J., et al. (2016). Effects of gut microbiota manipulation by antibiotics on host metabolism in obese humans: A randomized double-blind placebo-controlled trial. *Cell Metabolism, 24*, 63–74.

Rettger, L. F. (1915). The influence of milk feeding on mortality and growth, and on the character of the intestinal flora. *The Journal of Experimental Medicine, 21*, 365–388.

Ridaura, V. K., Faith, J. J., Rey, F. E., Cheng, J., Duncan, A. E., Kau, A. L., et al. (2013). Gut microbiota from twins discordant for obesity modulate metabolism in mice. *Science, 341*, 1241214.

Ridlon, J. M., Kang, D. J., & Hylemon, P. B. (2006). Bile salt biotransformations by human intestinal bacteria. *Journal of Lipid Research, 47*, 241–259.

Roediger, W. E. (1980). Role of anaerobic bacteria in the metabolic welfare of the colonic mucosa in man. *Gut, 21*, 793–798.

Russo, F., Linsalata, M., Clemente, C., Chiloiro, M., Orlando, A., Marconi, E., et al. (2012). Inulin-enriched pasta improves intestinal permeability and modifies the circulating levels of zonulin and glucagon-like peptide 2 in healthy young volunteers. *Nutrition Research, 32*, 940–946.

Saari, A., Virta, L. J., Sankilampi, U., Dunkel, L., & Saxen, H. (2015). Antibiotic exposure in infancy and risk of being overweight in the first 24 months of life. *Pediatrics, 135*, 617–626.

Santacruz, A., Collado, M. C., Garcia-Valdes, L., Segura, M. T., Martin-Lagos, J. A., Anjos, T., et al. (2010). Gut microbiota composition is associated with body weight, weight gain and biochemical parameters in pregnant women. *The British Journal of Nutrition, 104*, 83–92.

Sayin, S. I., Wahlstrom, A., Felin, J., Jantti, S., Marschall, H. U., Bamberg, K., et al. (2013). Gut microbiota regulates bile acid metabolism by reducing the levels of tauro-beta-muricholic acid, a naturally occurring FXR antagonist. *Cell Metabolism, 17*, 225–235.

Schmitz, G., & Ecker, J. (2008). The opposing effects of n-3 and n-6 fatty acids. *Progress in Lipid Research, 47*, 147–155.

Schwarzer, M., Makki, K., Storelli, G., Machuca-Gayet, I., Srutkova, D., Hermanova, P., et al. (2016). Lactobacillus plantarum strain maintains growth of infant mice during chronic undernutrition. *Science, 351*, 854–857.

Schwiertz, A., Taras, D., Schafer, K., Beijer, S., Bos, N. A., Donus, C., et al. (2010). Microbiota and SCFA in lean and overweight healthy subjects. *Obesity (Silver Spring), 18*, 190–195.

Semova, I., Carten, J. D., Stombaugh, J., Mackey, L. C., Knight, R., Farber, S. A., et al. (2012). Microbiota regulate intestinal absorption and metabolism of fatty acids in the zebrafish. *Cell Host and Microbe, 12*, 277–288.

Singh, V., Chassaing, B., Zhang, L., San Yeoh, B., Xiao, X., Kumar, M., et al. (2015). Microbiota-dependent hepatic lipogenesis mediated by stearoyl CoA desaturase 1 (SCD1) promotes metabolic syndrome in TLR5-deficient mice. *Cell Metabolism, 22*, 983–996.

Smith, S., Witkowski, A., & Joshi, A. K. (2003). Structural and functional organization of the animal fatty acid synthase. *Progress in Lipid Research, 42*, 289–317.

Sonnenburg, E. D., Smits, S. A., Tikhonov, M., Higginbottom, S. K., Wingreen, N. S., & Sonnenburg, J. L. (2016). Diet-induced extinctions in the gut microbiota compound over generations. *Nature, 529*, 212–215.

Steimle, A., Autenrieth, I. B., & Frick, J. S. (2016). Structure and function: Lipid A modifications in commensals and pathogens. *International Journal of Medical Microbiology, 306*, 290–301.

Suarez-Zamorano, N., Fabbiano, S., Chevalier, C., Stojanovic, O., Colin, D. J., Stevanovic, A., et al. (2015). Microbiota depletion promotes browning of white adipose tissue and reduces obesity. *Nature Medicine, 21*, 1497–1501.

Sze, M. A., & Schloss, P. D. (2016). Looking for a signal in the noise: Revisiting obesity and the microbiome. *MBio, 7*, e01018–e01016.

Teixeira, T. F., Souza, N. C., Chiarello, P. G., Franceschini, S. C., Bressan, J., Ferreira, C. L., et al. (2012). Intestinal permeability parameters in obese patients are correlated with metabolic syndrome risk factors. *Clinical Nutrition, 31*, 735–740.

Thomas, C., Pellicciari, R., Pruzanski, M., Auwerx, J., & Schoonjans, K. (2008). Targeting bile-acid signalling for metabolic diseases. *Nature Reviews. Drug Discovery, 7*, 678–693.

Thorasin, T., Hoyles, L., & McCartney, A. L. (2015). Dynamics and diversity of the 'Atopobium cluster' in the human faecal microbiota, and phenotypic characterization of 'Atopobium cluster' isolates. *Microbiology, 161*, 565–579.

Torres-Fuentes, C., Schellekens, H., Dinan, T. G., & Cryan, J. F. (2017). The microbiota-gut-brain axis in obesity. *Lancet Gastroenterology and Hepatology, 2*, 747–756.

Tremaroli, V., Karlsson, F., Werling, M., Stahlman, M., Kovatcheva-Datchary, P., Olbers, T., et al. (2015). Roux-en-Y gastric bypass and vertical banded gastroplasty induce long-term changes on the human gut microbiome contributing to fat mass regulation. *Cell Metabolism, 22*, 228–238.

Turnbaugh, P. J., Ley, R. E., Mahowald, M. A., Magrini, V., Mardis, E. R., & Gordon, J. I. (2006). An obesity-associated gut microbiome with increased capacity for energy harvest. *Nature, 444*, 1027–1031.

Turnbaugh, P. J., Backhed, F., Fulton, L., & Gordon, J. I. (2008). Diet-induced obesity is linked to marked but reversible alterations in the mouse distal gut microbiome. *Cell Host and Microbe, 3*, 213–223.

Ussar, S., Griffin, N. W., Bezy, O., Fujisaka, S., Vienberg, S., Softic, S., et al. (2015). Interactions between gut microbiota, host genetics and diet modulate the predisposition to obesity and metabolic syndrome. *Cell Metabolism, 22*, 516–530.

van Meer, G., Voelker, D. R., & Feigenson, G. W. (2008). Membrane lipids: Where they are and how they behave. *Nature Reviews. Molecular Cell Biology, 9*, 112–124.

van Nood, E., Vrieze, A., Nieuwdorp, M., Fuentes, S., Zoetendal, E. G., de Vos, W. M., et al. (2013). Duodenal infusion of donor feces for recurrent Clostridium difficile. *The New England Journal of Medicine, 368*, 407–415.

Velagapudi, V. R., Hezaveh, R., Reigstad, C. S., Gopalacharyulu, P., Yetukuri, L., Islam, S., et al. (2010). The gut microbiota modulates host energy and lipid metabolism in mice. *Journal of Lipid Research, 51*, 1101–1112.

Vijay-Kumar, M., Aitken, J. D., Carvalho, F. A., Cullender, T. C., Mwangi, S., Srinivasan, S., et al. (2010). Metabolic syndrome and altered gut microbiota in mice lacking Toll-like receptor 5. *Science, 328*, 228–231.

Vrieze, A., Van Nood, E., Holleman, F., Salojarvi, J., Kootte, R. S., Bartelsman, J. F., et al. (2012). Transfer of intestinal microbiota from lean donors increases insulin sensitivity in individuals with metabolic syndrome. *Gastroenterology, 143*, 913–916 e917.

Wagner, V. E., Dey, N., Guruge, J., Hsiao, A., Ahern, P. P., Semenkovich, N. P., et al. (2016). Effects of a gut pathobiont in a gnotobiotic mouse model of childhood undernutrition. *Science Translational Medicine, 8*, 366ra164.

Wall, R., Ross, R. P., Shanahan, F., O'Mahony, L., O'Mahony, C., Coakley, M., et al. (2009). Metabolic activity of the enteric microbiota influences the fatty acid composition of murine and porcine liver and adipose tissues. *The American Journal of Clinical Nutrition, 89*, 1393–1401.

Wallace, R. J., McKain, N., Shingfield, K. J., & Devillard, E. (2007). Isomers of conjugated linoleic acids are synthesized via different mechanisms in ruminal digesta and bacteria. *Journal of Lipid Research, 48*, 2247–2254.

Wang, Z., Klipfell, E., Bennett, B. J., Koeth, R., Levison, B. S., Dugar, B., et al. (2011). Gut flora metabolism of phosphatidylcholine promotes cardiovascular disease. *Nature, 472*, 57–63.

Wang, X., Ota, N., Manzanillo, P., Kates, L., Zavala-Solorio, J., Eidenschenk, C., et al. (2014). Interleukin-22 alleviates metabolic disorders and restores mucosal immunity in diabetes. *Nature, 514*, 237–241.

Wenk, M. R. (2010). Lipidomics: New tools and applications. *Cell, 143*, 888–895.

Wostmann, B. S. (1975). Nutrition and metabolism of the germfree mammal. *World Review of Nutrition and Dietetics, 22*, 40–92.

Woting, A., Pfeiffer, N., Loh, G., Klaus, S., & Blaut, M. (2014). Clostridium ramosum promotes high-fat diet-induced obesity in gnotobiotic mouse models. *MBio, 5*, e01530-01514.

Wu, G. D., Chen, J., Hoffmann, C., Bittinger, K., Chen, Y. Y., Keilbaugh, S. A., et al. (2011). Linking long-term dietary patterns with gut microbial enterotypes. *Science, 334*, 105–108.

Wu, H., Esteve, E., Tremaroli, V., Khan, M. T., Caesar, R., Manneras-Holm, L., et al. (2017). Metformin alters the gut microbiome of individuals with treatment-naive type 2 diabetes, contributing to the therapeutic effects of the drug. *Nature Medicine, 23*, 850–858.

Xiao, S., Fei, N., Pang, X., Shen, J., Wang, L., Zhang, B., et al. (2014). A gut microbiota-targeted dietary intervention for amelioration of chronic inflammation underlying metabolic syndrome. *FEMS Microbiology Ecology, 87*, 357–367.

Yatsunenko, T., Rey, F. E., Manary, M. J., Trehan, I., Dominguez-Bello, M. G., Contreras, M., et al. (2012). Human gut microbiome viewed across age and geography. *Nature, 486*, 222–227.

Zeevi, D., Korem, T., Zmora, N., Israeli, D., Rothschild, D., Weinberger, A., et al. (2015). Personalized nutrition by prediction of glycemic responses. *Cell, 163*, 1079–1094.

Zelante, T., Iannitti, R. G., Cunha, C., De Luca, A., Giovannini, G., Pieraccini, G., et al. (2013). Tryptophan catabolites from microbiota engage aryl hydrocarbon receptor and balance mucosal reactivity via interleukin-22. *Immunity, 39*, 372–385.

Zhou, D., Pan, Q., Xin, F. Z., Zhang, R. N., He, C. X., Chen, G. Y., et al. (2017). Sodium butyrate attenuates high-fat diet-induced steatohepatitis in mice by improving gut microbiota and gastrointestinal barrier. *World Journal of Gastroenterology, 23*, 60–75.

Zietak, M., Kovatcheva-Datchary, P., Markiewicz, L. H., Stahlman, M., Kozak, L. P., & Backhed, F. (2016). Altered microbiota contributes to reduced diet-induced obesity upon cold exposure. *Cell Metabolism, 23*, 1216–1223.

Zmora, N., Bashiardes, S., Levy, M., & Elinav, E. (2017). The role of the immune system in metabolic health and disease. *Cell Metabolism, 25*, 506–521.

17 Microbiome and Diseases: Hepatic Disorders

Ina Bergheim and Detlef Schuppan

Abstract

Intensive research efforts aim to understand the multifaceted molecular mechanisms underlying disease onset and progression of nonalcoholic fatty liver disease (NAFLD) and alcohol-induced liver disease (ALD). Taken together, NAFLD and ALD are the most common liver diseases worldwide, and universally accepted therapies other than lifestyle interventions either focusing on weight reduction and physical exercise or alcohol abstinence are lacking. During the last decade, alterations of intestinal microbiota composition and intestinal barrier function leading to an increased translocation of bacterial endotoxin and of metabolites originating from an altered intestinal microbiome are emerging as key pathogenic factors in both diseases. In this book chapter, present knowledge and understanding of the interplay of intestinal microbiota, intestinal barrier function, and the development of nonalcoholic and alcoholic liver diseases, respectively, are summarized.

17.1 Introduction

As the liver receives ~70% of its blood supply from the intestine through the portal vein, an interaction between the liver and the gut via the intestinal microbiome has been discussed for several decades. Indeed, already in the 1950s, studies were published showing that germfree animals were protected from the development of liver necrosis (György 1954; Luckey et al. 1954). Furthermore, treating rats with diet-induced steatohepatitis with antibiotics like aureomycin protected them from the development of fibrosis and cirrhosis (Rutenburg et al. 1957). Interestingly, this study showed that nonabsorbable antibiotics were markedly more efficient in delaying the progression to advanced fibrosis when compared to absorbable ones (Rutenburg et al. 1957). The first reports of a novel disease entity named nonalcoholic steatohepatitis (NASH), mainly found in overweight women with type 2 diabetes, appeared in 1980 (Ludwig et al. 1980). An involvement of the intestinal microbiome in the development of NASH was suggested by a study in 1989 when a patient suffering from small bowel diverticulosis and bacterial overgrowth in the small intestine, the latter being defined as excessive bacteria in the small intestine (Dukowicz et al. 2007), developed NASH (Nazim

I. Bergheim (✉)
Lehrstuhl Ernährungsphysiologie – Department für Ernährungswissenschaften, Universität Wien, Wien, Germany
e-mail: ina.bergheim@univie.ac.at

D. Schuppan
Institut für Translationale Immunolgie (TIM), Universitätsmedizin Mainz – Johannes Gutenberg-Universität Mainz, Mainz, Germany
e-mail: detlef.schuppan@unimedizin-mainz.de

D. Haller (ed.), *The Gut Microbiome in Health and Disease*,
https://doi.org/10.1007/978-3-319-90545-7_17

et al. 1989). Over the next 30 years, an exponentially increasing number of publications suggested that alterations at the level of intestinal microbiota composition usually coupled with intestinal barrier functional defects are related to liver health or to the severity of liver diseases of various etiologies. The liver receives venous blood from the intestine being rich in nutrients but also other substances ingested with diet and derived from intestinal microbiota that have crossed the intestinal barrier. Indeed, recent studies have proposed that the liver forms a first vascular and metabolic firewall capturing xenobiotics and endobiotics in settings of intestinal barrier dysfunction, before they may enter the blood circulation (Balmer et al. 2014). Finally, the liver-gut axis dramatically affects extraintestinal health and disease.

This chapter focuses on the present knowledge and understanding of the interplay of intestinal microbiota, intestinal barrier function, and the development of metabolic liver diseases, i.e., NAFLD, specifically inflammatory NASH, and ALD and their extraintestinal and extrahepatic comorbidities.

17.2 Nonalcoholic Fatty Liver Disease

NAFLD, frequently regarded as the hepatic manifestation of the metabolic syndrome (Kim and Younossi 2008), is strongly correlated with overweight/obesity and insulin resistance or overt type 2 diabetes (Townsend and Newsome 2016). With still increasing prevalence, NAFLD is now regarded as the most common liver disease worldwide affecting between 20 and 30% of most developed and developing populations (Younossi et al. 2016). Its prevalence is highest in the Middle East, the USA, and South America and lowest in Africa (Younossi et al. 2016). Furthermore, liver-specific and overall mortality of patients with NAFLD were estimated to be 1.94 (range, 1.28–2.92) and 1.05 (range, 0.70–1.56) (Younossi et al. 2016). NAFLD comprises a wide spectrum of diseases ranging from simple steatosis to steatohepatitis (nonalcoholic steatohepatitis, NASH) regarded as the inflammatory, aggressive form of NAFLD and to fibrosis and finally cirrhosis and primary hepatocellular carcinoma (HCC) (annual progression rate for NASH ~40.76% according to Younossi et al. 2016) (Neuschwander-Tetri and Caldwell 2003) (see Fig. 17.1). Notably, once patients have developed cirrhosis, their relative risk of liver-related mortality increases more than 50-fold when compared to non-fibrotic NAFLD (Dulai et al. 2017). However, the latter are rare while still serious complications of NAFLD (e.g., 0.44/1000 person-years; range, 0.29–0.66) (Starley et al. 2010). In the USA (cirrhotic) NASH has become the second leading cause of liver transplantation in adults (Satapathy and Sanyal 2015).

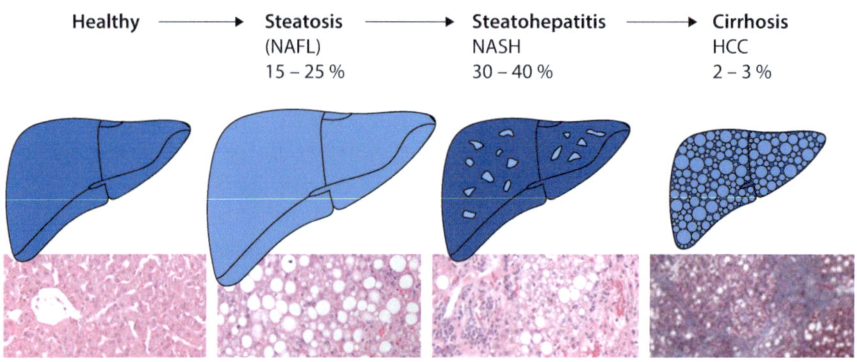

Fig. 17.1 Stages and probability of progression of nonalcoholic fatty liver disease. *NAFL* nonalcoholic fatty liver, *NASH* nonalcoholic steatohepatitis, *HCC* hepatocellular carcinoma

17.2.1 Nonalcoholic Fatty Liver Disease: Intestinal Microbiota and Intestinal Barrier Function

Despite these alarming numbers, molecular mechanisms involved in disease development and progression are not yet fully understood, and universally accepted prevention and therapeutic strategies except for weight loss and physical exercise are lacking (Schuppan and Schattenberg 2013). Polymorphisms of genes involved in the regulation of lipid and glucose metabolism and the renin-angiotensin system have been proposed to be critical in the onset of NAFLD but also in its progression to later stages (e.g., inflammation, fibrosis, and even HCC) (Miyaaki and Nakao 2017). Indeed, numerous studies and meta-analyses have established that a genetic polymorphism of position 148 (rs738409 C/G) in the gene encoding for patatin-like phospholipase domain-containing protein (PNPLA3) is a strong genetic predisposition not only associated with hepatic fat accumulation but also the development of inflammation, fibrosis, and HCC in humans (Chen et al. 2015). Yet, studies also suggest that the impact of this genetic modification but also those of others on NAFLD may differ among genetic groups (Chen et al. 2015). Besides genetics, physical inactivity and general overnutrition, especially overnutrition with the so-called Western-style diet, i.e., a diet rich in saturated fatty acids and mono- and disaccharides like fructose (especially as high-fructose corn syrup) and sucrose as well as cholesterol, repeatedly have been identified as a critical factor in the development and progression of NAFLD (Barrea et al. 2017; Romero-Gomez et al. 2017). More recently, intestinal microbiota composition and barrier dysfunction have been proposed to be also critical factors in the development of NAFLD (see Fig. 17.2). Indeed, patients with different stages of NAFLD/NASH suffer from endotoxemia associated with an increased intestinal permeability and a loss of tight junctional proteins in the duodenum (Miele et al. 2009; Thuy et al. 2008; Volynets et al. 2012). These patients also frequently have elevated plasma levels of the lipopolysaccharide (LPS)-binding protein (LBP) as well as enhanced expression of the endotoxin receptor TLR (Toll-like receptor)-4 and tumor necrosis factor (TNF)-α in liver tissue (Kanuri et al. 2011; Ruiz et al. 2007; Wigg et al. 2001). Furthermore, not only expression of TLR-4 but also other TLRs, e.g., TLR-1–5 (but not TLR-6–10), are induced in livers of patients with simple steatosis, to increase in NASH and NASH with beginning fibrosis (Kanuri et al. 2015). This suggests that not only the translocation of LPS derived from Gram-negative bacteria might be increased in settings of NAFLD and NASH but also that of other bacterial toxins as well as viral and bacterial proteins, modified proteins, DNA, and RNA (see Fig. 17.2). Results of rodent models using mutants of leptin signaling like ob/ob and db/db, or applying various feeding schemes, such as diets rich in fructose, sucrose, and/or fat and cholesterol, and especially combinations of these macronutrients to induce NAFLD, lend further support to the hypothesis that an increased translocation of bacterial endotoxins from the gut and the subsequent activation of TLR-4-dependent signaling in the liver might be critical factors in the development of NAFLD and especially NASH (Brun et al. 2007; Cani et al. 2008; Engstler et al. 2016, 2017; Jin et al. 2016; Sellmann et al. 2015; Spruss et al. 2009. 2011; Sutter et al. 2015; Wagnerberger et al. 2012). Thus endotoxin reduction by treatment with non-resorbable antibiotics exerting their effect primarily in the gut rather than systemic can protect rodents from the development of diet-induced NAFLD (Bergheim et al. 2008), being associated with a mitigated TLR response and downstream signaling cascades in the liver (Wagnerberger et al. 2012). Also, increasing orocecal transit time in NAFLD patients being defined as the time from lactulose ingestion to a sustained increase of over 5 ppm above fasting levels in the end-expiratory hydrogen concentration or the interval to that of over 10 ppm (Hirakawa et al. 1988) through treating them with a prokinetic drug like cisapride has been shown to ameliorate endotoxemia (Fu and Jiang 2006). Furthermore, short-term and long-term treatment of patients with NASH and with minimal hepatic encephalopathy due to liver disease in

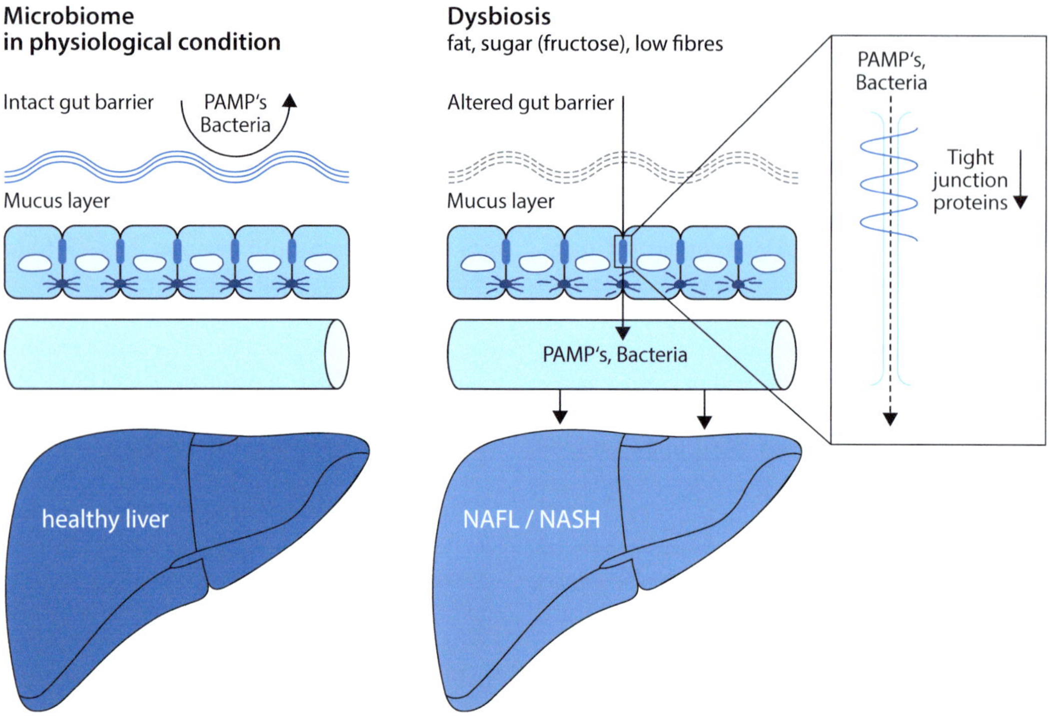

Fig. 17.2 Interaction of microbiome and host is critical in the development of NAFLD. Diet can trigger dysbiosis in the small and/ or large intestine triggering a loss of tight junction proteins and intestinal barrier dysfunction. A subsequent translocation of bacterial products into the portal vein can induce hepatic inflammation and cause disease progression. *PAMPs* pathogen-associated molecular patterns, *NAFL* nonalcoholic fatty liver, *NASH* nonalcoholic steatohepatitis

general with the non-resorbable antibiotic rifaximin decreased bacterial endotoxin levels in blood and laboratory markers of liver disease and improved hepatic encephalopathy (Bajaj et al. 2013; Gangarapu et al. 2015). Genetic modifications and pharmacological interventions that prevent the loss of tight junction proteins in intestinal tissue or the activation of TLR-4 and downstream signaling cascades in the liver have repeatedly been shown to be associated with a marked protection against the development of NAFLD and NASH (Jin et al. 2017; Spruss et al. 2009, 2012). Older, more descriptive studies already demonstrated that these pathologies were frequently associated with altered orocecal transit time and a higher prevalence of small intestinal bacterial overgrowth (Hirakawa et al. 1988; Soza et al. 2005; Wigg et al. 2001). In line with these findings, more recent analyses of fecal microbiota composition revealed marked differences between intestinal microbiota composition of patients with NAFLD/NASH when compared with controls. Indeed, Bajaj et al. (2014) suggest that progressive changes in gut microbiota composition are associated with developing cirrhosis and that these changes become more severe in hepatic functional decompensation (Bajaj et al. 2014). Notwithstanding, in this study a patient population with various etiologies of cirrhosis was studied. Boursier et al. (2016) suggested that the severity of NAFLD/NASH in adults is associated with alterations of fecal microbiota composition and the bacterial metabolome and that *Bacteroides* is independently associated with NASH, whereas *Ruminococcus* was more abundant in patients with hepatic fibrosis (Boursier et al. 2016; Loomba et al. 2017); however, in this study no disease-free controls were included. In line with these findings, pediatric patients with simple steatosis or NASH had lower levels of *Oscillospira* which were associated with higher abundance of *Dorea* and *Ruminococcus* than controls (Del Chierico et al. 2017). These alterations were associated with higher levels of 2-butanone and 4-methyl-2-pentanone in fecal samples of NAFLD patients compared to controls (Del Chierico et al. 2017). In another study, in which NAFLD patients were not stratified by disease stage, it was shown that relative abundances of *Escherichia*, *Anaerobacter*, *Lactobacillus*, and *Streptococcus* were higher in patients, while that of *Ruminococcaceae* and herein especially the two genera *Oscillibacter* and *Flavonifractor* were higher in stool samples of healthy controls (Jiang et al. 2015). In the latter study, patients with NAFLD/NASH also showed lower protein levels of the tight junction proteins occludin and increased levels of proinflammatory cytokines in the duodenal mucosa, especially TNF-α and IL-6, in this part of the small intestine when compared to healthy controls. The differences between these two studies might have resulted from differences in study design (only patients with NAFLD vs. a comparison with healthy controls and NAFLD/NASH patients without a histologically defined stage), differences in age (e.g., pediatric patients vs. adult patients), but also the lack of information on dietary intake. Indeed, one study is from China, while the other one enrolled patients in France and Italy. It has been suggested previously that mammalian gut bacterial communities are strongly dependent on dietary pattern (Muegge et al. 2011) and that, for instance, the ruminococci population can be altered through dietary intake (David et al. 2014).

17.2.2 NAFLD: Diet Pattern, Intestinal Microbiota, and the Barrier

Several studies suggest that diet strongly affects human gut microbiome composition (David et al. 2014; Muegge et al. 2011; Wu et al. 2011). For instance, David et al. (2014) demonstrated that microbial activity is dramatically different between herbivorous and carnivorous mammals, thereby reciprocally calibrating between carbohydrate and protein fermentation. Furthermore, it has been shown that foodborne microbes including bacteria, fungi, and even viruses ingested with different diets colonize the gut (David et al. 2014).

Results of recent rodent-based studies suggest that not only dietary content of fiber, sugars like fructose and sucrose, or saturated fats (Beilharz et al. 2016; Jena et al. 2016; Lam et al. 2015; Patterson et al. 2014) may impact intestinal microbiota composition but also different sources of protein such as red meat (e.g., beef and pork), white meat (chicken and fish), and other sources of protein like casein and soy which distinctly affect the intestinal microbiota (Zhu et al. 2015). In these studies, white meat consumption was associated with a markedly higher relative abundance of *Lactobacillus* in the cecum, while relative abundance of *Ruminococcus* was highest in soy protein-fed animals (Zhu et al. 2017). Interestingly, in rodents fed meat protein or casein, levels of circulating lipopolysaccharide-binding protein (LBP) as well as hepatic expression of CD14, both markers of innate immune activation, were significantly lower than in those animals fed soy protein (Zhu et al. 2017).

Also in humans, dietary protein affects fecal microbiota composition. In a smaller study with NAFLD patients ingesting a hypocaloric high-protein diet, the observed decrease in body weight and liver fat was associated with a reduced relative abundance of *Lachnospira*, whereas abundance of *Blautia* and *Butyricicoccus* was increased in feces of study participants (Pataky et al. 2016). The observed decrease in liver fat was negatively correlated with bacteria belonging to the *Firmicutes* and *Bacteroidetes* phyla (Pataky et al. 2016). Intervention studies in healthy subjects further support the concept that nutritional imbalances, such as a diet low in choline, may induce fatty liver associated with the growth of genera like *Gammaproteobacteria* that are associated with intestinal inflammation and concurrent metabolic dysfunction (Spencer et al. 2011). Yet, results of this study also suggest that host factors like genetic predisposition may also affect response to different diets and that even under identical dietary conditions the gut microbiota remains distinct between individuals over an extended time (Spencer et al. 2011).

Notably, earlier and recent rodent studies clearly showed that obese animals fed a high-fat, high-carbohydrate diet harbor intestinal microbiota that more thoroughly metabolize carbohydrates and generate more caloric value from a given meal (Backhed et al. 2004; Ley et al. 2005; Sonnenburg and Backhed 2016). Transplantation of these microbiota into lean mice makes these fatter, and vice versa colonization of obese mice with the lean microbiota leads to weight loss subsequent to a less efficient energy production. Moreover, the microbiota from obese mice with fatty liver has been shown to activate the intestinal and liver inflammasome, thereby promoting inflammatory NASH (Henao-Mejia et al. 2012). These results could be reproduced in a small but well-performed clinical study in which obese patients with fatty liver received a stool transplantation either with an obese or lean microbiota (Vrieze et al. 2012). Finally, the beneficial effects of certain drugs like metformin on insulin resistance are in part due to the drug changing the intestinal microbiota (Wu et al. 2017).

Taken together, results stemming from human studies in different regions of the world, and rodent studies including active interventions with microbiota transplantation, suggest that the intestinal microbiota composition differs markedly between patients with NAFLD/NASH and healthy individuals. Studies also suggest that lifestyle and herein especially dietary patterns may markedly impact intestinal microbiota composition and barrier function. Indeed, recently it has been proposed that personalized microbiome-based approaches including nutrition-based interventions or the treatment with specific pro- or prebiotics might be an option for a preventive or supportive treatment of NAFLD/NASH and the associated metabolic syndrome [for overview see Shapiro et al. (2017), Wiest et al. (2017), Zmora et al. (2016)]. However, as knowledge and understanding underlying the molecular mechanisms of the interplay of intestinal microbiota and barrier and the liver is still limited in humans but also in rodent models of the disease and long-term effects of altering intestinal microbiota composition have not yet been fully understood, more research is needed before these individualized approaches to precision medicine can be taken.

17.3 Alcoholic Liver Disease, Intestinal Microbiota, and Intestinal Barrier Function: Current Knowledge

Despite intense public information and many public campaigns informing about the risks of high and especially chronic alcohol ingestion, alcohol consumption is still among the leading causes of liver damage in many countries worldwide (World Health Organization 2014; Organization and Unit 2014). Similar to NAFLD alcoholic liver disease (ALD) comprises a spectrum of conditions ranging from simple fatty liver to alcoholic hepatitis, cirrhosis, and even hepatocellular carcinoma (Yeh and Brunt 2014). Indeed, it is estimated that ~10–15% of alcoholics will develop cirrhosis (Mann et al. 2003). Yet, postmortem studies suggest even higher numbers in men exceeding a daily alcohol intake of 80 g/d (Savolainen et al. 1993). Interestingly, results of a meta-analysis suggest that the same average consumption of ethanol is related to a higher risk of liver cirrhosis in women than in men (Rehm et al. 2010). Furthermore, it has been suggested by several studies that alcohol consumption and herein especially high intake are also associated with adverse outcomes in patients with liver diseases other than alcoholic liver disease [for overview see Hagstrom (2017)]. In spite of intense research efforts, it is not clear why only a minor fraction of alcoholics develops cirrhosis, and molecular mechanisms underlying the development of ALD are still not fully understood, and therapies primarily focus on abstinence being fraught with high relapse rate.

17.3.1 ALD: Intestinal Barrier Function, Bacterial Endotoxemia, and Intestinal Microbiota

Already more than 30 years ago, the group of Bode et al. and others demonstrated that ALD is associated with elevated bacterial endotoxin levels (Bode et al. 1987; Fukui et al. 1991; Gaeta et al. 1982). Furthermore, bacterial endotoxin levels correlated with enhanced intestinal permeability in patients with ALD (Parlesak et al. 2000) as well as with disease severity (Lin et al. 1995); however, the latter study did include different etiologies of liver disease. By now numerous animal and human studies confirmed these findings [for overview see Starkel and Schnabl (2016)] but also suggest that even acute high-dose intake (a single dose of >20g pure ethanol) is sufficient to increase bacterial endotoxin levels in peripheral blood in humans (Bala et al. 2014) (also see Fig. 17.3). In line with these findings, results of mouse and rat ethanol binge drinking models (5–6 g ethanol given intragastrically once) have shown that persistence of the marked hepatic fat accumulation found within 6–12 h after the acute alcohol ingestion is highly dependent upon the translocation of endotoxin (Enomoto et al. 2001; Wagnerberger et al. 2013). Chronic and acute high-dose intake of ethanol can lead to a loss of epithelial cells from the villous tips and to hemorrhagic erosions in the lamina propria [for overview also see Elamin et al. (2013)]. Furthermore, animal studies using chronic feeding models suggest that the mucus layer and the expression of certain mucins but also the expression antimicrobial peptides are altered (Kirpich et al. 2013). In the liver, gut-derived bacterial endotoxins have been suggested to add to the hypermetabolic state, e.g., increased hepatic oxygen uptake and

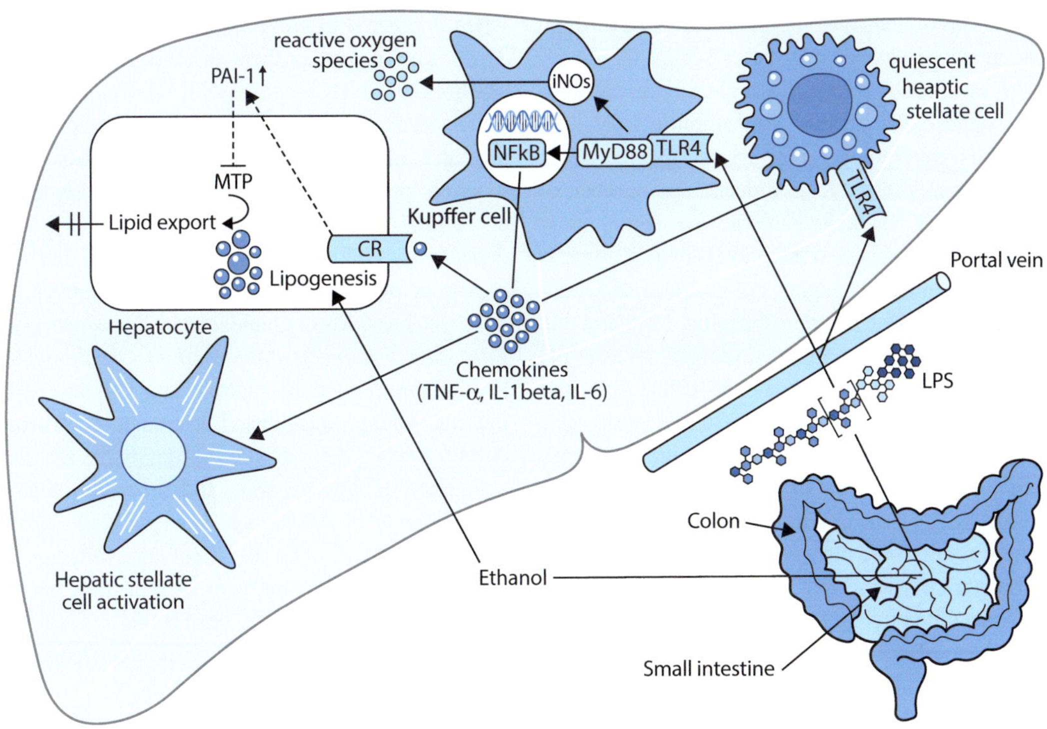

Fig. 17.3 Consequences of intestinal dysbiosis and barrier dysfunction induced by alcohol consumption on the liver. Alcohol ingestion leads to intestinal bacterial overgrowth and dysbiosis as well as barrier dysfunction. Lipopolysaccharides reach the liver through the portal vein resulting in an induction of TLR-4-dependent signaling cascades and subsequently the formation of reactive oxygen species and cytokines which in turn alter hepatic lipid export and further add to the activation of hepatic stellate cells. *CR* cytokine receptors, *NFκB* nuclear factor kappa B, *TLR-4* Toll-like receptor-4, *LPS* lipopolysaccharide, *PAI-1* plasminogen activator inhibitor 1, *iNOS* inducible nitric oxide synthase, *TNF-α* tumor necrosis factor alpha, *MyD88* myeloid differentiation primary response protein 88, *MTP* microsomal triglyceride transfer protein

accumulation of lipids after acute alcohol exposure (Rivera et al. 1998; Yuki and Thurman 1980) (for mechanisms involved also see Fig. 17.3). Indeed, treatment with both, non-resorbable antibiotics (that largely eliminate the microbiota) and GdCl3 (that eliminates proinflammatory Kupffer cells) almost completely abolished these hypermetabolic alterations after acute high alcohol ingestion in rodents (Rivera et al. 1998). Similar results were also reported in settings of chronic alcohol exposure (Adachi et al. 1994, 1995; Koop et al. 1997; Uesugi et al. 2001). Indeed, interventions, be they genetic or pharmacological, aiming to disrupt the recognition of endotoxin by liver cells in settings of chronic alcohol ingestion are associated with a marked protection from the development of ALD [for overview also see Szabo (2015)].

In support that gut-derived bacterial toxins and alterations of intestinal microbiota induced through the intake of alcohol may be critical in the development of ALD, some older and more recent studies suggest that ethanol consumption may alter intestinal microbiota composition (Mutlu et al. 2012) (see Table 17.1). Indeed, the number of anaerobic and aerobic bacteria was significantly higher in jejunal juice of alcoholics than in controls with a significantly higher colonization of Gram-negative and endospore-forming bacteria in alcoholics and that correlated closely with the pH found in gastric juice (Bode et al. 1984). By using a hydrogen breath test, the same group also showed that the prevalence of small intestinal bacterial overgrowth was significantly more frequent in alcoholics than in controls (Bode et al. 1993). These results were also supported by several other groups all showing that small intestinal bacterial overgrowth is frequently associated with the presence of cirrhosis regardless of the cause of liver damage (Casafont Morencos et al. 1996; Lakshmi et al. 2010; Morencos et al. 1995; Pande et al. 2009; Yang et al. 1998). Interestingly, Jun et al. (2010) showed that the prevalence of cirrhosis was related to the presence of bacterial DNA in peripheral blood (Jun et al. 2010), further supporting the hypothesis that dysbiotic alterations of intestinal microbiota are a risk factor of increased intestinal permeability. However, in this study no details were provided regarding the etiology of cirrhosis. In line with these findings, Mutlu et al. (2012) reported in alcoholics with and without liver disease that chronic alcohol intake was associated with changes of mucosa-associated colonic bacterial composition, e.g., a lower abundance of *Bacteroidetes* and a higher abundance of *Proteobacteria* in a subset of alcoholics when compared to healthy controls, and that this dysbiosis also correlated with endotoxemia in the alcoholics (Mutlu et al. 2012). Furthermore, it has been shown that patients with alcoholic cirrhosis may have a distinctly altered functional composition of the fecal microbiome, e.g., a depletion of functional genes involved in nutrient metabolism including amino acids and lipid and nucleotide metabolism (Chen et al. 2014).

In patients with alcoholic steatohepatitis, the abundance of fecal *A. muciniphila* is decreased when compared to controls (Grander et al. 2018). The same study further showed that treating alcohol-fed mice orally with *A. muciniphila* protected animals from alcohol-induced gut leakiness and enhanced mucus thickness as well as tight junction protein expression. Similarly, Leclercq et al. (2014a) showed that patients with alcohol dependence and a high intestinal permeability, as measured by ^{51}Cr-EDTA, but without fibrosis or cirrhosis, had a lower abundance of *Ruminococcus*, *Faecalibacterium*, *Subdoligranulum*, *Oscillibacter*, and *Anaerofilum* than controls and patients with low intestinal permeability (Leclercq et al. 2014b). After a 3-week detoxification program, the relative abundance of *Ruminococcus* and *Subdoligranulum* and the level of *Bifidobacterium* spp. and *Lactobacillus* spp. were increased in subjects with a prior high intestinal permeability. In another study bacterial endotoxin and peptidoglycan levels and expression of TLRs in peripheral blood mononuclear cells decreased during detoxification in noncirrhotic alcohol-dependent subjects (Leclercq et al. 2014a). Notably, even a short-term 5-day treatment of patients with alcoholic psychosis and mild alcohol-induced liver injury with *Bifidobacterium bifidum* and *Lactobacillus plantarum 8PA3* in combination with the standard therapy (abstinence and vitamin supplementation) resulted in a significant increase in the number of bifidobacteria and

Table 17.1 Summary of studies assessing microbiome and intestinal barrier function in patients with alcoholic liver disease

Subjects	*n*	Methods	Key findings	References
AD Controls	27 13	Jejunal fluid aspirates Anaerobic incubation	↑ Gram-negative anaerobic bacteria; coliform bacteria with AD Correlation between gastric juice pH and NOM	Bode et al. (1984)
Cirrhosis AIC Controls	18 12 12	Fecal samples Functional gene array	Dysbiosis with cirrhosis and AIC ↑ *Proteobacteria*; *Fusobacteria* ↓ *Bacteroidetes* Correlation between enrichment of functional genes (e.g., xenobiotic metabolism, virulence) in microbiome and alcohol consumption	Chen et al. (2014)
AD AIC Controls	29 19 18	Colonic biopsy Length heterogeneity PCR Multitag pyrosequencing	Increased dysbiosis with AD and AIC ↑ *Proteobacteria*, *Bacilli* ↓ *Bacteroidetes*; *Clostridia* No correlation between duration of sobriety and dysbiosis	Mutlu et al. (2012)
AD Controls	60 15	Fecal samples Cr-EDTA method qPCR/pyrosequencing of 16S rDNA	Correlation between intestinal permeability and dysbiosis with AD ↑ *Lachnospiraceae* ↓ *Ruminocococcaceae; Bifidobacterium; Lactobacillus; Clostridia* Incomplete microbiota recovery after period of sobriety	Leclercq et al. (2014b)
Cirrhosis Controls	45 28	Glucose hydrogen breath test Methane breath test	Increased SIBO with cirrhosis Correlation between SIBO and severity of liver disease	Yang et al. (1998)
AD Controls	45 60	Hydrogen breath test	Increased SIBO with AD No difference in alcoholics with/without cirrhosis	Bode et al. (1993)
Cirrhosis EHPVO Controls	174 28 51	Glucose hydrogen breath test	Increased SIBO with cirrhosis No correlation between SIBO and: cirrhosis etiology; degree of liver dysfunction	Lakshmi et al. (2010)
Cirrhosis Controls	53 42	Lactulose hydrogen/methane breath test Multiplex PCR	Increased SIBO with cirrhosis Correlation between SIBO and bacterial translocation	Jun et al. (2010)
AIC Controls	89 40	Glucose hydrogen breath test	Increased SIBO with AIC Correlation between increased NOM and development of spontaneous bacterial peritonitis	Morencos et al. (1995)
Cirrhosis AIC Controls	62 16 15	Glucose hydrogen breath test	Increased SIBO with cirrhosis Correlation between SIBO and severity of liver disease	Pande et al. (2009)

AD alcohol dependence, *AIC* alcohol-induced cirrhosis, *EHPVO* extrahepatic portal venous obstruction, *NOM* number of microorganisms, *SIBO* small intestinal bacterial overgrowth

lactobacilli and a more rapid fall of elevated transaminases in blood when compared with standard therapy alone (abstinence plus vitamins).

Taken together, similar to the findings for NAFLD, animal and human studies suggest that alterations of the intestinal microbiota composition are critical in the development of ALD (also see Figs. 17.2 and 17.3). Studies further suggest that abstinence as the therapy of choice for ALD is associated with a rapid change of intestinal microbiota composition and intestinal barrier function; finally, probiotics aid in reconstitution of the protective mucosal barrier and alleviate endotoxin-induced liver damage. Several ongoing clinical trials target the intestinal microbiota in settings of chronic alcohol-induced liver disease to enhance recovery of patients with early stages of ALD and to assist with stabilization of cirrhosis (Wiest et al. 2017). Results of these trials are expected to yield valuable information to direct future probiotic therapies for ALD and possibly also NAFLD/NASH.

17.4 Conclusion

Both NAFLD and ALD are multifactorial diseases associated with altered nutritional patterns, e.g., over- and malnutrition, and elevated alcohol intake, respectively. Accordingly, lifestyle interventions targeting nutritional patterns through caloric restriction and encouraging abstinence are still the primary therapeutic approaches. Nonetheless, results of animal and human studies strongly suggest that both clinical pictures are associated with alterations at the level of intestinal microbiota and barrier function. Furthermore, results of intervention studies targeting intestinal microbiota and/or barrier may improve or even prevent disease progression further suggesting that altering intestinal microbiota composition and subsequently microbial metabolite profile might provide excellent targets for the therapy and may be also the prevention of both NAFLD and ALD. Despite promising clinical evidence, a better characterization of the intestinal microbiota and its metabolome using larger and better characterized patient cohorts as well as unified sample collection and detection methods is needed to develop universally acceptable treatment recommendations. Using these more generalized methodological approaches, subsets of patients suffering from ALD or NAFLD may be identified that benefit from therapeutic interventions targeting the intestinal microbiome and barrier function, thereby adding to the development of a personalized microbiome-focused therapy that may serve as adjunctive treatment and that may help to prevent NAFLD and ALD.

Controversy

Despite being among the most common liver diseases worldwide, molecular mechanisms underlying the development of alcoholic liver disease and nonalcoholic fatty liver disease are not fully understood. While an interaction of intestinal microbiota, intestinal barrier integrity, and the development of ALD has been proposed for more than 30 years, current therapies and prevention strategies of ALD still primarily focus on lifestyle interventions, i.e., strict abstinence. However, while abstinence would be wishful in settings of a chronic and excessive alcohol intake, this is often not achieved, and patients reach irreversible stages of the disease with liver transplantation being the only remaining cure of choice. Therefore, novel (adjunctive) therapies preventing the progression of ALD are at need. Lifestyle interventions focusing on normalizing body weight are the therapy but also the prevention strategy of choice for NAFLD and NASH. However, similar to ALD, compliance is low and relapse rates are high. First studies targeting the intestinal microbiota and barrier bear promising results; still, as data are limited, more research is needed to unravel mechanisms underlying the interaction of host and microbiota before a personalized microbiome-focused therapy or even prevention of NAFLD and ALD can be a general recommendation.

History

NAFLD, NASH, and alcoholic liver disease (ALD) are by now among the most common liver diseases worldwide. NAFLD/NASH is strongly correlated with the presence of overweight/obesity and insulin resistance, while ALD results from the chronic elevated intake of ethanol. Despite intense research efforts, molecular mechanisms involved in the development of ALD and even more so of NAFLD/NASH are not yet been fully understood, as exemplified by an often unpredictable disease severity or progression, despite similar major predisposing factors. Indeed, for both diseases lifestyle interventions, either focusing on weight reduction and increase of physical activity in the case of NAFLD/NASH or the avoidance of alcohol intake in the case of ALD, are still the first lines and most effective treatments and preventive measures. However, lifestyle interventions are frequently fraught with low compliance and high relapse rates. Results of human and animal studies have shown that genetic predisposition, physical inactivity, and general overnutrition but also a diet rich in saturated fat and/or sugars as well as excess cholesterol are associated with the development of NAFLD/NASH. In recent years, similar to the findings in ALD, alterations of the intestinal microbiota composition and barrier function have been proposed to be key factors for the development of NAFLD and NASH. Intervention studies focusing on the gut and especially targeting intestinal microbiota as well as intestinal barrier function are promising; however, long-term results are still missing and knowledge on molecular mechanism and particular microbial species and metabolites is just evolving.

Highlights

- Nonalcoholic fatty liver disease (NAFLD) is by now regarded the most common liver disease worldwide.
- Alterations of intestinal microbiota and increased bacterial endotoxin levels are associated with both alcoholic liver disease and nonalcoholic fatty liver disease.
- Nutrition and dietary patterns are critical modulators of intestinal microbiome composition and the development of alcoholic liver disease, NAFLD, and NASH.
- A better understanding of the interaction of the intestinal microbiota, intestinal barrier dysfunction, and the liver will help to improve prevention and therapy of NAFLD and alcoholic liver disease.

References

Adachi, Y., Bradford, B. U., Gao, W., Bojes, H. K., & Thurman, R. G. (1994). Inactivation of Kupffer cells prevents early alcohol-induced liver injury. *Hepatology, 20*, 453–460.

Adachi, Y., Moore, L. E., Bradford, B. U., Gao, W., & Thurman, R. G. (1995). Antibiotics prevent liver injury in rats following long-term exposure to ethanol. *Gastroenterology, 108*, 218–224.

Backhed, F., Ding, H., Wang, T., Hooper, L. V., Koh, G. Y., Nagy, A., et al. (2004). The gut microbiota as an environmental factor that regulates fat storage. *Proceedings of the National Academy of Sciences of the United States of America, 101*, 15718–15723.

Bajaj, J. S., Heuman, D. M., Sanyal, A. J., Hylemon, P. B., Sterling, R. K., Stravitz, R. T., et al. (2013). Modulation of the metabiome by rifaximin in patients with cirrhosis and minimal hepatic encephalopathy. *PLoS One, 8*, e60042.

Bajaj, J. S., Heuman, D. M., Hylemon, P. B., Sanyal, A. J., White, M. B., Monteith, P., et al. (2014). Altered profile of human gut microbiome is associated with cirrhosis and its complications. *Journal of Hepatology, 60*, 940–947.

Bala, S., Marcos, M., Gattu, A., Catalano, D., & Szabo, G. (2014). Acute binge drinking increases serum endotoxin and bacterial DNA levels in healthy individuals. *PLoS One, 9*, e96864.

Balmer, M. L., Slack, E., de Gottardi, A., Lawson, M. A., Hapfelmeier, S., Miele, L., et al. (2014). The liver may act as a firewall mediating mutualism between the host and its gut commensal microbiota. *Science Translational Medicine, 6*, 237ra266.

Barrea, L., Di Somma, C., Muscogiuri, G., Tarantino, G., Tenore, G. C., Orio, F., et al. (2017). Nutrition, inflammation and liver-spleen axis. *Critical Reviews in Food Science and Nutrition, 57*(16), 3472–3488.

Beilharz, J. E., Kaakoush, N. O., Maniam, J., & Morris, M. J. (2016). The effect of short-term exposure to energy-matched diets enriched in fat or sugar on memory, gut microbiota and markers of brain inflammation and plasticity. *Brain, Behavior, and Immunity, 57*, 304–313.

Bergheim, I., Weber, S., Vos, M., Kramer, S., Volynets, V., Kaserouni, S., et al. (2008). Antibiotics protect against fructose-induced hepatic lipid accumulation in mice: role of endotoxin. *Journal of Hepatology, 48*, 983–992.

Bode, J. C., Bode, C., Heidelbach, R., Durr, H. K., & Martini, G. A. (1984). Jejunal microflora in patients with chronic alcohol abuse. *Hepato-Gastroenterology, 31*, 30–34.

Bode, C., Kugler, V., & Bode, J. C. (1987). Endotoxemia in patients with alcoholic and non-alcoholic cirrhosis and in subjects with no evidence of chronic liver disease following acute alcohol excess. *Journal of Hepatology, 4*, 8–14.

Bode, C., Kolepke, R., Schafer, K., & Bode, J. C. (1993). Breath hydrogen excretion in patients with alcoholic liver disease – evidence of small intestinal bacterial overgrowth. *Zeitschrift für Gastroenterologie, 31*, 3–7.

Boursier, J., Mueller, O., Barret, M., Machado, M., Fizanne, L., Araujo-Perez, F., et al. (2016). The severity of nonalcoholic fatty liver disease is associated with gut dysbiosis and shift in the metabolic function of the gut microbiota. *Hepatology, 63*, 764–775.

Brun, P., Castagliuolo, I., Di Leo, V., Buda, A., Pinzani, M., Palu, G., et al. (2007). Increased intestinal permeability in obese mice: new evidence in the pathogenesis of nonalcoholic steatohepatitis. *American Journal of Physiology. Gastrointestinal and Liver Physiology, 292*, G518–G525.

Cani, P. D., Bibiloni, R., Knauf, C., Waget, A., Neyrinck, A. M., Delzenne, N. M., et al. (2008). Changes in gut microbiota control metabolic endotoxemia-induced inflammation in high-fat diet-induced obesity and diabetes in mice. *Diabetes, 57*, 1470–1481.

Casafont Morencos, F., de las Heras Castano, G., Martin Ramos, L., Lopez Arias, M. J., Ledesma, F., & Pons Romero, F. (1996). Small bowel bacterial overgrowth in patients with alcoholic cirrhosis. *Digestive Diseases and Sciences, 41*, 552–556.

Chen, Y., Qin, N., Guo, J., Qian, G., Fang, D., Shi, D., et al. (2014). Functional gene arrays-based analysis of fecal microbiomes in patients with liver cirrhosis. *BMC Genomics, 15*, 753.

Chen, L. Z., Xin, Y. N., Geng, N., Jiang, M., Zhang, D. D., & Xuan, S. Y. (2015). PNPLA3 I148M variant in nonalcoholic fatty liver disease: demographic and ethnic characteristics and the role of the variant in nonalcoholic fatty liver fibrosis. *World Journal of Gastroenterology, 21*, 794–802.

David, L. A., Maurice, C. F., Carmody, R. N., Gootenberg, D. B., Button, J. E., Wolfe, B. E., et al. (2014). Diet rapidly and reproducibly alters the human gut microbiome. *Nature, 505*, 559–563.

Del Chierico, F., Nobili, V., Vernocchi, P., Russo, A., Stefanis, C., Gnani, D., et al. (2017). Gut microbiota profiling of pediatric nonalcoholic fatty liver disease and obese patients unveiled by an integrated meta-omics-based approach. *Hepatology, 65*, 451–464.

Dukowicz, A. C., Lacy, B. E., & Levine, G. M. (2007). Small intestinal bacterial overgrowth: A comprehensive review. *Gastroenterology and Hepatology (New York), 3*, 112–122.

Dulai, P. S., Singh, S., Patel, J., Soni, M., Prokop, L. J., Younossi, Z., et al. (2017). Increased risk of mortality by fibrosis stage in nonalcoholic fatty liver disease: Systematic review and meta-analysis. *Hepatology, 65*, 1557–1565.

Elamin, E. E., Masclee, A. A., Dekker, J., & Jonkers, D. M. (2013). Ethanol metabolism and its effects on the intestinal epithelial barrier. *Nutrition Reviews, 71*, 483–499.

Engstler, A. J., Aumiller, T., Degen, C., Durr, M., Weiss, E., Maier, I. B., et al. (2016). Insulin resistance alters hepatic ethanol metabolism: Studies in mice and children with non-alcoholic fatty liver disease. *Gut, 65*, 1564–1571.

Engstler, A. J., Sellmann, C., Jin, C. J., Brandt, A., Herz, K., Priebs, J., et al. (2017). Treatment with alpha-galactosylceramide protects mice from early onset of nonalcoholic steatohepatitis: Role of intestinal barrier function. *Molecular Nutrition and Food Research, 61* (5). https://doi.org/10.1002/mnfr.201600985

Enomoto, N., Ikejima, K., Yamashina, S., Hirose, M., Shimizu, H., Kitamura, T., et al. (2001). Kupffer cell sensitization by alcohol involves increased permeability to gut-derived endotoxin. *Alcoholism: Clinical and Experimental Research, 25*, 51S–54S.

Fu, X. S., & Jiang, F. (2006). Cisapride decreasing orocecal transit time in patients with nonalcoholic steatohepatitis. *Hepatobiliary and Pancreatic Diseases International: HBPD INT, 5*, 534–537.

Fukui, H., Brauner, B., Bode, J. C., & Bode, C. (1991). Plasma endotoxin concentrations in patients with alcoholic and non-alcoholic liver disease: reevaluation with an improved chromogenic assay. *Journal of Hepatology, 12*, 162–169.

Gaeta, G. B., Perna, P., Adinolfi, L. E., Utili, R., & Ruggiero, G. (1982). Endotoxemia in a series of 104 patients with chronic liver diseases: Prevalence and significance. *Digestion, 23*, 239–244.

Gangarapu, V., Ince, A. T., Baysal, B., Kayar, Y., Kilic, U., Gok, O., et al. (2015). Efficacy of rifaximin on circulating endotoxins and cytokines in patients with nonalcoholic fatty liver disease. *European Journal of Gastroenterology and Hepatology, 27*, 840–845.

Grander, C., Adolph, T. E., Wieser, V., Lowe, P., Wrzosek, L., Gyongyosi, B., et al. (2018). Recovery of ethanol-induced Akkermansia muciniphila depletion ameliorates alcoholic liver disease. *Gut, 67*(5), 891–901. https://doi.org/10.1136/gutjnl-2016-313432

György, P. (1954). Antibiotics and liver injury. *Annals of the New York Academy of Sciences, 57*, 925–931.

Hagstrom, H. (2017). Alcohol consumption in concomitant liver disease: How much is too much? *Current Hepatology Reports, 16*, 152–157.

Henao-Mejia, J., Elinav, E., Jin, C., Hao, L., Mehal, W. Z., Strowig, T., et al. (2012). Inflammasome-mediated dysbiosis regulates progression of NAFLD and obesity. *Nature, 482*, 179–185.

Hirakawa, M., Iida, M., Kohrogi, N., & Fujishima, M. (1988). Hydrogen breath test assessment of orocecal transit time: comparison with barium meal study. *The American Journal of Gastroenterology, 83*, 1361–1363.

Jena, P. K., Prajapati, B., Mishra, P. K., & Seshadri, S. (2016). Influence of gut microbiota on inflammation and pathogenesis of sugar rich diet induced diabetes. *Immunome Research, 12*(1), 109.

Jiang, W., Wu, N., Wang, X., Chi, Y., Zhang, Y., Qiu, X., et al. (2015). Dysbiosis gut microbiota associated with inflammation and impaired mucosal immune function in intestine of humans with non-alcoholic fatty liver disease. *Scientific Reports, 5*, 8096.

Jin, C. J., Engstler, A. J., Sellmann, C., Ziegenhardt, D., Landmann, M., Kanuri, G., et al. (2016). Sodium butyrate protects mice from the development of the early signs of nonalcoholic fatty liver disease: Role of melatonin and lipid peroxidation. *The British Journal of Nutrition, 23*, 1–12.

Jin, C. J., Engstler, A. J., Ziegenhardt, D., Bischoff, S. C., Trautwein, C., & Bergheim, I. (2017). Loss of lipopolysaccharide-binding protein attenuates the development of diet-induced non-alcoholic fatty liver disease in mice. *Journal of Gastroenterology and Hepatology, 32*, 708–715.

Jun, D. W., Kim, K. T., Lee, O. Y., Chae, J. D., Son, B. K., Kim, S. H., et al. (2010). Association between small intestinal bacterial overgrowth and peripheral bacterial DNA in cirrhotic patients. *Digestive Diseases and Sciences, 55*, 1465–1471.

Kanuri, G., Spruss, A., Wagnerberger, S., Bischoff, S. C., & Bergheim, I. (2011). Role of tumor necrosis factor α (TNFα) in the onset of fructose-induced nonalcoholic fatty liver disease in mice. *The Journal of Nutritional Biochemistry, 22*, 527–534.

Kanuri, G., Ladurner, R., Skibovskaya, J., Spruss, A., Konigsrainer, A., Bischoff, S. C., et al. (2015). Expression of toll-like receptors 1-5 but not TLR 6-10 is elevated in livers of patients with non-alcoholic fatty liver disease. *Liver International, 35*, 562–568.

Kim, C. H., & Younossi, Z. M. (2008). Nonalcoholic fatty liver disease: A manifestation of the metabolic syndrome. *Cleveland Clinic Journal of Medicine, 75*, 721–728.

Kirpich, I. A., Feng, W., Wang, Y., Liu, Y., Beier, J. I., Arteel, G. E., et al. (2013). Ethanol and dietary unsaturated fat (corn oil/linoleic acid enriched) cause intestinal inflammation and impaired intestinal barrier defense in mice chronically fed alcohol. *Alcohol (Fayetteville, New York), 47*, 257–264.

Koop, D. R., Klopfenstein, B., Iimuro, Y., & Thurman, R. G. (1997). Gadolinium chloride blocks alcohol-dependent liver toxicity in rats treated chronically with intragastric alcohol despite the induction of CYP2E1. *Molecular Pharmacology, 51*, 944–950.

Lakshmi, C. P., Ghoshal, U. C., Kumar, S., Goel, A., Misra, A., Mohindra, S., et al. (2010). Frequency and factors associated with small intestinal bacterial overgrowth in patients with cirrhosis of the liver and extra hepatic portal venous obstruction. *Digestive Diseases and Sciences, 55*, 1142–1148.

Lam, Y. Y., Ha, C. W., Hoffmann, J., Oscarsson, J., Dinudom, A., Mather, T. J., et al. (2015). Effects of dietary fat profile on gut permeability and microbiota and their relationships with metabolic changes in mice. *Obesity, 23*, 1429–1439.

Leclercq, S., De Saeger, C., Delzenne, N., de Timary, P., & Stärkel, P. (2014a). Role of inflammatory pathways, blood mononuclear cells, and gut-derived bacterial products in alcohol dependence. *Biological Psychiatry, 76*, 725–733.

Leclercq, S., Matamoros, S., Cani, P. D., Neyrinck, A. M., Jamar, F., Stärkel, P., et al. (2014b). Intestinal permeability, gut-bacterial dysbiosis, and behavioral markers of alcohol-dependence severity. *Proceedings of the National Academy of Sciences of the United States of America, 111*, E4485–E4493.

Ley, R. E., Backhed, F., Turnbaugh, P., Lozupone, C. A., Knight, R. D., & Gordon, J. I. (2005). Obesity alters gut microbial ecology. *Proceedings of the National Academy of Sciences of the United States of America, 102*, 11070–11075.

Lin, R.-S., Lee, F.-Y., Lee, S.-D., Tsai, Y.-T., Lin, H. C., Rei-Hwa, L., et al. (1995). Endotoxemia in patients with chronic liver diseases: Relationship to severity of liver diseases, presence of esophaegeal varices, and hyperdynamic circulation. *Journal of Hepatology, 22*, 165–172.

Loomba, R., Seguritan, V., Li, W., Long, T., Klitgord, N., Bhatt, A., et al. (2017). Gut microbiome-based metagenomic signature for non-invasive detection of advanced fibrosis in human nonalcoholic fatty liver disease. *Cell Metabology, 25*, 1054–1062 e1055.

Luckey, T., Reyniers, J., György, P., & Forbes, M. (1954). Germfree animals and liver necrosis. *Annals of the New York Academy of Sciences, 57*, 932–935.

Ludwig, J., Viggiano, T. R., McGill, D. B., & Oh, B. J. (1980). Nonalcoholic steatohepatitis: Mayo Clinic experiences with a hitherto unnamed disease. *Mayo Clinic Proceedings, 55*, 434–438.

Mann, R. E., Smart, R. G., & Govoni, R. (2003). The epidemiology of alcoholic liver disease. *Alcohol Research and Health, 27*, 209–219.

Miele, L., Valenza, V., La Torre, G., Montalto, M., Cammarota, G., Ricci, R., et al. (2009). Increased intestinal permeability and tight junction alterations in nonalcoholic fatty liver disease. *Hepatology, 49*, 1877–1887.

Miyaaki, H., & Nakao, K. (2017). Significance of genetic polymorphisms in patients with nonalcoholic fatty liver disease. *Clinical Journal of Gastroenterology, 10*, 201–207.

Morencos, F. C., de las Heras Castano, G., Martin Ramos, L., Lopez Arias, M. J., Ledesma, F., & Pons Romero, F. (1995). Small bowel bacterial overgrowth in patients with alcoholic cirrhosis. *Digestive Diseases and Sciences, 40*, 1252–1256.

Muegge, B. D., Kuczynski, J., Knights, D., Clemente, J. C., González, A., Fontana, L., et al. (2011). Diet drives convergence in gut microbiome functions across mammalian phylogeny and within humans. *Science (New York), 332*, 970–974.

Mutlu, E. A., Gillevet, P. M., Rangwala, H., Sikaroodi, M., Naqvi, A., Engen, P. A., et al. (2012). Colonic microbiome is altered in alcoholism. *American Journal of Physiology – Gastrointestinal and Liver Physiology, 302*, G966–G978.

Nazim, M., Stamp, G., & Hodgson, H. J. (1989). Non-alcoholic steatohepatitis associated with small intestinal diverticulosis and bacterial overgrowth. *Hepato-Gastroenterology, 36*, 349–351.

Neuschwander-Tetri, B. A., & Caldwell, S. H. (2003). Nonalcoholic steatohepatitis: Summary of an AASLD single topic conference. *Hepatology, 37*, 1202–1219.

Pande, C., Kumar, A., & Sarin, S. K. (2009). Small-intestinal bacterial overgrowth in cirrhosis is related to the severity of liver disease. *Alimentary Pharmacology and Therapeutics, 29*, 1273–1281.

Parlesak, A., Schafer, C., Schutz, T., Bode, J. C., & Bode, C. (2000). Increased intestinal permeability to macromolecules and endotoxemia in patients with chronic alcohol abuse in different stages of alcohol-induced liver disease. *Journal of Hepatology, 32*, 742–747.

Pataky, Z., Genton, L., Spahr, L., Lazarevic, V., Terraz, S., Gaia, N., et al. (2016). Impact of hypocaloric hyperproteic diet on gut microbiota in overweight or obese patients with nonalcoholic fatty liver disease: A pilot study. *Digestive Diseases and Sciences, 61*, 2721–2731.

Patterson, E., O'Doherty, R. M., Murphy, E. F., Wall, R., O'Sullivan, O., Nilaweera, K., et al. (2014). Impact of dietary fatty acids on metabolic activity and host intestinal microbiota composition in C57BL/6J mice. *British Journal of Nutrition, 111*, 1905–1917.

Rehm, J., Taylor, B., Mohapatra, S., Irving, H., Baliunas, D., Patra, J., et al. (2010). Alcohol as a risk factor for liver cirrhosis: A systematic review and meta-analysis. *Drug and Alcohol Review, 29*, 437–445.

Rivera, C. A., Bradford, B. U., Seabra, V., & Thurman, R. G. (1998). Role of endotoxin in the hypermetabolic state after acute ethanol exposure. *The American Journal of Physiology, 275*, G1252–G1258.

Romero-Gomez, M., Zelber-Sagi, S., & Trenell, M. (2017). Treatment of NAFLD with diet, physical activity and exercise. *Journal of Hepatology, 67*(4), 829–846.

Ruiz, A. G., Casafont, F., Crespo, J., Cayón, A., Mayorga, M., Estebanez, A., et al. (2007). Lipopolysaccharide-binding protein plasma levels and liver TNF-alpha gene expression in obese patients: Evidence for the potential role of endotoxin in the pathogenesis of non-alcoholic steatohepatitis. *Obesity Surgery, 17*, 1374.

Rutenburg, A. M., Sonnenblick, E., Koven, I., Aprahamian, H. A., Reiner, L., & Fine, J. (1957). The role of intestinal bacteria in the development of dietary cirrhosis in rats. *The Journal of Experimental Medicine, 106*, 1–14.

Satapathy, S. K., & Sanyal, A. J. (2015). Epidemiology and natural history of nonalcoholic fatty liver disease. *Seminars in Liver Disease, 35*, 221–235.

Savolainen, V., Liesto, K., Männikkö, A., Penttilä, A., & Karhunen, P. (1993). Alcohol consumption and alcoholic liver disease: evidence of a threshold level of effects of ethanol. *Alcoholism: Clinical and Experimental Research, 17*, 1112–1117.

Schuppan, D., & Schattenberg, J. M. (2013). Non-alcoholic steatohepatitis: pathogenesis and novel therapeutic approaches. *Journal of Gastroenterology and Hepatology, 28*(*Suppl 1*), 68–76.

Sellmann, C., Priebs, J., Landmann, M., Degen, C., Engstler, A. J., Jin, C. J., et al. (2015). Diets rich in fructose, fat or fructose and fat alter intestinal barrier function and lead to the development of nonalcoholic fatty liver disease over time. *The Journal of Nutritional Biochemistry, 26*, 1183–1192.

Shapiro, H., Suez, J., & Elinav, E. (2017). Personalized microbiome-based approaches to metabolic syndrome management and prevention. *Journal of Diabetes, 9*, 226–236.

Sonnenburg, J. L., & Backhed, F. (2016). Diet-microbiota interactions as moderators of human metabolism. *Nature, 535*, 56–64.

Soza, A., Riquelme, A., Gonzalez, R., Alvarez, M., Perez-Ayuso, R. M., Glasinovic, J. C., et al. (2005). Increased orocecal transit time in patients with nonalcoholic fatty liver disease. *Digestive Diseases and Sciences, 50*, 1136–1140.

Spencer, M. D., Hamp, T. J., Reid, R. W., Fischer, L. M., Zeisel, S. H., & Fodor, A. A. (2011). Association between composition of the human gastrointestinal microbiome and development of fatty liver with choline deficiency. *Gastroenterology, 140*, 976–986.

Spruss, A., Kanuri, G., Wagnerberger, S., Haub, S., Bischoff, S. C., & Bergheim, I. (2009). Toll-like receptor 4 is involved in the development of fructose-induced hepatic steatosis in mice. *Hepatology, 50*, 1094–1104.

Spruss, A., Kanuri, G., Uebel, K., Bischoff, S. C., & Bergheim, I. (2011). Role of the inducible nitric oxide synthase in the onset of fructose-induced steatosis in mice. *Antioxidants and Redox Signaling, 14*, 2121–2135.

Spruss, A., Kanuri, G., Stahl, C., Bischoff, S. C., & Bergheim, I. (2012). Metformin protects against the development of fructose-induced steatosis in mice: role of the intestinal barrier function. *Laboratory Investigation, 92*, 1020–1032.

Starkel, P., & Schnabl, B. (2016). Bidirectional communication between liver and gut during alcoholic liver disease. *Seminars in Liver Disease, 36*, 331–339.

Starley, B. Q., Calcagno, C. J., & Harrison, S. A. (2010). Nonalcoholic fatty liver disease and hepatocellular carcinoma: A weighty connection. *Hepatology, 51*, 1820–1832.

Sutter, A. G., Palanisamy, A. P., Lench, J. H., Jessmore, A. P., & Chavin, K. D. (2015). Development of steatohepatitis in Ob/Ob mice is dependent on Toll-like receptor 4. *Annals of Hepatology, 14*, 735–743.

Szabo, G. (2015). Gut-liver axis in alcoholic liver disease. *Gastroenterology, 148*, 30–36.

Thuy, S., Ladurner, R., Volynets, V., Wagner, S., Strahl, S., Konigsrainer, A., et al. (2008). Nonalcoholic fatty liver disease in humans is associated with increased plasma endotoxin and plasminogen activator inhibitor 1 concentrations and with fructose intake. *The Journal of Nutrition, 138*, 1452–1455.

Townsend, S. A., & Newsome, P. N. (2016). Non-alcoholic fatty liver disease in 2016. *British Medical Bulletin, 119*, 143–156.

Uesugi, T., Froh, M., Arteel, G. E., Bradford, B. U., & Thurman, R. G. (2001). Toll-like receptor 4 is involved in the mechanism of early alcohol-induced liver injury in mice. *Hepatology, 34*, 101–108.

Volynets, V., Kuper, M. A., Strahl, S., Maier, I. B., Spruss, A., Wagnerberger, S., et al. (2012). Nutrition, intestinal permeability, and blood ethanol levels are altered in patients with nonalcoholic fatty liver disease (NAFLD). *Digestive Diseases and Sciences, 57*, 1932–1941.

Vrieze, A., Van Nood, E., Holleman, F., Salojarvi, J., Kootte, R. S., Bartelsman, J. F., et al. (2012). Transfer of intestinal microbiota from lean donors increases insulin sensitivity in individuals with metabolic syndrome. *Gastroenterology, 143*, 913–916 e917.

Wagnerberger, S., Spruss, A., Kanuri, G., Volynets, V., Stahl, C., Bischoff, S. C., et al. (2012). Toll-like receptors 1-9 are elevated in livers with fructose-induced hepatic steatosis. *The British Journal of Nutrition, 107*, 1727–1738.

Wagnerberger, S., Fiederlein, L., Kanuri, G., Stahl, C., Millonig, G., Mueller, S., et al. (2013). Sex-specific differences in the development of acute alcohol-induced liver steatosis in mice. *Alcohol and Alcoholism (Oxford, Oxfordshire), 48*, 648–656.

Wiest, R., Albillos, A., Trauner, M., Bajaj, J. S., & Jalan, R. (2017). Targeting the gut-liver axis in liver disease. *Journal of Hepatology, 67*(5), 1084–1103.

Wigg, A., Roberts-Thomson, I., Dymock, R., McCarthy, P., Grose, R., & Cummins, A. (2001). The role of small intestinal bacterial overgrowth, intestinal permeability, endotoxaemia, and tumour necrosis factor α in the pathogenesis of non-alcoholic steatohepatitis. *Gut, 48*, 206–211.

World Health Organization. (2014). *Age-standardized death rates of liver cirrhosis*. World Health Organization.

World Health Organization, and Unit, W.H.O.M.o.S.A. (2014). Global status report on alcohol and health, 2014. World Health Organization.

Wu, G. D., Chen, J., Hoffmann, C., Bittinger, K., Chen, Y. Y., Keilbaugh, S. A., et al. (2011). Linking long-term dietary patterns with gut microbial enterotypes. *Science, 334*, 105–108.

Wu, H., Esteve, E., Tremaroli, V., Khan, M. T., Caesar, R., Manneras-Holm, L., et al. (2017). Metformin alters the gut microbiome of individuals with treatment-naive type 2 diabetes, contributing to the therapeutic effects of the drug. *Nature Medicine, 23*, 850–858.

Yang, C. Y., Chang, C. S., & Chen, G. H. (1998). Small-intestinal bacterial overgrowth in patients with liver cirrhosis, diagnosed with glucose H2 or CH4 breath tests. *Scandinavian Journal of Gastroenterology, 33*, 867–871.

Yeh, M. M., & Brunt, E. M. (2014). Pathological features of fatty liver disease. *Gastroenterology, 147*, 754–764.

Younossi, Z. M., Koenig, A. B., Abdelatif, D., Fazel, Y., Henry, L., & Wymer, M. (2016). Global epidemiology of nonalcoholic fatty liver disease-meta-analytic assessment of prevalence, incidence, and outcomes. *Hepatology, 64*(1), 73–84. https://doi.org/10.1002/hep.28431

Yuki, T., & Thurman, R. G. (1980). The swift increase in alcohol metabolism. Time course for the increase in hepatic oxygen uptake and the involvement of glycolysis. *Biochemical Journal, 186*, 119–126.

Zhu, Y., Lin, X., Zhao, F., Shi, X., Li, H., Li, Y., et al. (2015). Meat, dairy and plant proteins alter bacterial composition of rat gut bacteria. *Scientific Reports, 5*, 15220.

Zhu, Y., Shi, X., Lin, X., Ye, K., Xu, X., Li, C., et al. (2017). Beef, chicken, and soy proteins in diets induce different gut microbiota and metabolites in rats. *Frontiers in Microbiology, 8*, 1395.

Zmora, N., Zeevi, D., Korem, T., Segal, E., & Elinav, E. (2016). Taking it personally: Personalized utilization of the human microbiome in health and disease. *Cell Host and Microbe, 19*, 12–20.

Microbiome and Diseases: Neurological Disorders

18

Anne E. Slingerland and Christoph K. Stein-Thoeringer

Autointoxication genera is a historical concept from the early era of gut-brain axis research used to explain mental illness:
"It is far from our mind to conceive that all mental conditions have the same etiological factor, but we feel justified in recognizing the existence of cases of mental disorders which have as a basic etiological factor a toxic condition arising in the gastrointestinal tract."
Armando Ferraro and Joseph E. Kilman;
The New York Psychiatric Institute, in Psychiatric Quarterly *(Ferraro and Kilman 1933)*

Abstract

An increasing amount of evidence implicates that the gastrointestinal microbiota affects a vast range of neuronal functions from neurodevelopment and synaptic signaling to behavior. This microbe-host interplay occurs at the level of the peripheral and central nervous system and is commonly referred to as "microbiome-gut-brain axis." Preclinical and clinical data also highlight the significant association of gut microbiome dysbiosis and the development and progression of psychiatric and neurological disorders. The present chapter will outline how the gut microbiota signals to the brain and how it affects brain development and function describing representative examples. Finally, it will discuss the complex interaction of intestinal microbiota and neuropsychiatric disorders.

A. E. Slingerland
Immunology Program, Sloan Kettering Institute, Memorial Sloan Kettering Cancer Center, New York, NY, USA
e-mail: steinthc@mskcc.org

C. K. Stein-Thoeringer (✉)
Immunology Program, Sloan Kettering Institute, Memorial Sloan Kettering Cancer Center, New York, NY, USA

Klinik und Poliklinik für Innere Medizin II, Klinikum rechts der Isar, Technische Universität München, München, Germany

18.1 Introduction

Infectious disease research has a long tradition of investigation into how pathogenic microbes can affect brain function and induce mental and neurological disorders. Infections by human immunodeficiency virus (HIV), *Treponema pallidum*, *Borrelia burgdorferi*, or infectious diseases like measles or malaria can lead to severe neurological and psychiatric disorders in late stages and if not adequately treated. Latent brain toxoplasmosis is another example of a parasitic infection that has been linked with psychiatric disorders or suicidal behaviors (Coccaro et al. 2016; Sutterland et al. 2015). These neurological manifestations of systemic infectious diseases are predominantly

D. Haller (ed.), *The Gut Microbiome in Health and Disease*,
https://doi.org/10.1007/978-3-319-90545-7_18

caused by pathogen spread into the central nervous system (CNS) and subsequent activation of local inflammatory processes or direct interference of the pathogen with neurotransmitter signaling, e.g., disruption of dopamine signaling in brain toxoplasmosis (Vyas 2015).

The advent of novel techniques to explore and characterize the microbiota has revolutionized our understanding of gastrointestinal and metabolic processes. We now understand that all bacteria, not merely pathogenic interlopers, play a fundamental role in homeostasis and immunity, and the influence of this paradigm shift has more recently been seen in neuroscience and neuropsychiatry research. Over the past few years, a large amount of evidence has been collected implicating the intestinal microbiome—a unique assemblage of commensal microorganisms (i.e., bacteria, archaea, fungi, and viruses) residing in various niches in our gastrointestinal tract—in the development and progression of psychiatric and neurological disorders. The biological intersection between the mammalian gut microbiome and the CNS is commonly referred to as "microbiome-gut-brain axis," a bidirectional system that enables intestinal microbes to signal to the spinal cord and brain and the nervous system to communicate back to the gut (Fung et al. 2017). These microbe-brain communications are carried out directly via afferent and efferent nerves or through indirect mechanisms that include endocrine and metabolic pathways. Commensal bacteria also have the ability to influence the status of the immune system, modulating how immune cells subsequently interact with the CNS.

This chapter explains how the gut microbiota signals to the brain and how it affects brain development and function, concentrating on those gut-brain interactions that operate directly via neural pathways, rather than via immune signaling (microbe-immune system interactions are covered in detail in other chapters). It also outlines the contribution of the intestinal microbiota to the development and severity of neurological and psychiatric disorders.

18.2 Microbe Effects on Neural Function

18.2.1 The Microbiota-Gut-Brain Axis: Neural and Humoral Microbe-to-Brain Signaling

Interactions between a host and its intestinal microbiota are intricate, and the influence of gut microbes on adjacent and distant host organs occurs through multiple pathways. Figure 18.1 gives an overview of these different connections.

The gastrointestinal (GI) tract harbors multiple distinct cell types including specialized epithelial and mesenchymal cells (e.g., microfold [M] cells and subepithelial dome (SED) cells in Peyer's patches), endocrine cells, neurons, and immune cells, all of which facilitate homeostasis and symbiosis between microbes and the host (Belkaid and Harrison 2017; Bellono et al. 2017; Nagashima et al. 2017; Rios et al. 2016; Yano et al. 2015). In the GI tract, cells from the CNS, the autonomic nervous system (PNS), and the enteric nervous system (ENS) form a dense neural network capable of sensing and responding to intrinsic, extrinsic, and environmental cues. This network serves as a principle branch of the gut-brain axis (Yano et al. 2015; Yoo and Mazmanian 2017).

Enteric Nervous System (ENS) The ENS, an intricate network of neurons and glia along the intestinal tract, works autonomously to regulate, upon many other functions, GI microcirculation, smooth muscle motility, and enteroendocrine cell secretion (Furness et al. 2014). Positioned adjacent to the mucosa, submucosal and myenteric plexuses of the ENS serve as direct portals for microbial signals. The first evidence for the impact of the commensal microflora on ENS physiology and functionality was the observation that germ-free and antibiotic-treated mice in comparison with control have fewer neurons in the ENS, as well as less neuronal excitability, and reduced GI motility (Anitha et al. 2012; Collins et al. 2014; McVey Neufeld et al. 2013). In addition to neuronal development and function,

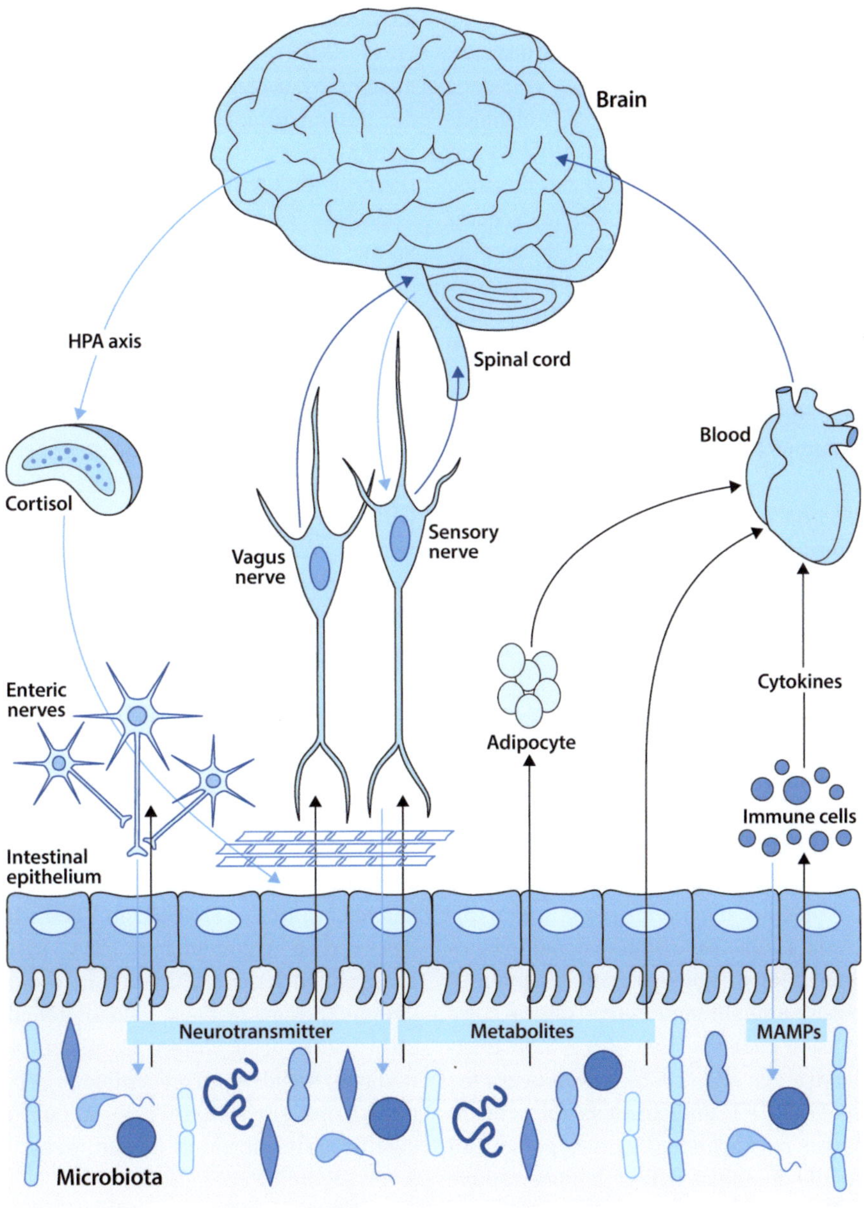

Fig. 18.1 Main principal mechanisms of bidirectional interactions in the microbiota-gut-brain axis

indigenous gut microbiota also regulate early postnatal colonization, as well as renewal of the gut lamina propria through enteric glia (Kabouridis et al. 2015). Mechanistically, a complex interaction of dietary factors and production of unconjugated bile acids by specific commensal bacterial species has been shown to modulate gut motility in a gnotobiotic mouse model (Dey et al. 2015). Further, gut indigenous, spore-forming bacterial species such as *Clostridia* can regulate serotonin synthesis in enterochromaffin cells, which can directly stimulate myenteric neurons and regulate gut motility (Yano et al. 2015). Consistent with this finding, bacterial metabolites derived from fiber fermentation in the colon such as short-chain fatty acids (SCFAs) also stimulate intestinal serotonin synthesis (Reigstad et al. 2015). Butyrate, a major SCFA, for instance, can also directly change the excitability of enteric neurons through modulations of K^+ channels (Hamodeh et al. 2004). The gut microbiota has also been demonstrated to regulate signaling between myenteric plexus neurons and muscularis macrophages, facilitating proper GI motility (Muller et al. 2014). Conversely, ENS-to-microbe signaling is essential for the maintenance of intestinal microbial homeostasis. For instance, an ENS knockout in zebrafish was accompanied by a dysbiotic gut microbiota that induced a gut inflammatory phenotype (Rolig et al. 2017).

Extrinsic Neural Connections In addition to the ENS, extrinsic sympathetic (from spinal cord) and parasympathetic nerves (primarily N. vagus) of the autonomic nervous system connect the gut and the CNS via afferent and efferent fibers. Efferent endings synapse directly on smooth muscles, submucosa, and enteric ganglia (Browning and Travagli 2014). Afferent neuronal fibers are found in the GI mucosa, muscle, and subserous layers and convey sensory information to the CNS. Evidence for the involvement of commensal bacteria in afferent nerve signaling comes from preclinical experiments in which different bacterial strains were administered to laboratory animals. Rats displayed less activation of lumbar sensory afferent nerve fibers during colorectal distension after *Lactobacillus reuteri* treatment (Kamiya et al. 2006). Moreover the inhibition of TRPV1 ion channels on primary sensory nerves by *Lactobacillus* is hypothesized to play a role in the apparent antinociceptive activity of the microbe (Perez-Burgos et al. 2015).

It is known that the vagus nerve conveys sensory information from the viscera directly to the brain. In mouse experiments, it has been observed that administration of *Lactobacillus rhamnosus* significantly increased nerve firing in the vagus nerve upon intestinal distension (Perez-Burgos et al. 2013). The treatment with either *Bifidobacterium longum* or *Lactobacillus rhamnosus* decreased anxiety-related and nociceptive behavior. This action was abolished in vagotomized mice (Bercik et al. 2011b; Bravo et al. 2011). Activation of the vagus nerve nucleus by acetate, another SCFA produced by the gut microbiota, is likely to be involved in insulin-glucose metabolism and obesity in mice (Perry et al. 2016).

SCFAs, such as those produced by *Clostridia* in the colon, are also able to directly regulate peripheral nerve function. It has been demonstrated that GPR41, a Gi/o protein-coupled receptor for SCFAs, is expressed on both autonomic nervous system ganglia and dorsal root ganglia of the somatic nervous system (Nøhr et al. 2015). As such, SCFAs can directly regulate excitability of sympathetic visceral nerves via GPR41 (Kimura et al. 2011).

Hypothalamic-Pituitary-Adrenal (HPA) Axis and Other Factors The HPA axis is a neuroendocrine system that governs stress regulation and other autonomic processes. Neuroendocrine neurons in the hypothalamus synthesize and secrete vasopressin and corticotropin-releasing hormone, which stimulate the secretion of adrenocorticotropic hormone by the anterior pituitary gland. This hormone, in turn, regulates production of glucocorticoid hormones by the adrenal cortex. Studies in germ-free mice observed a dysregulated HPA system with exaggerated serum corticosterone responses after stress exposure in these animals (Neufeld et al. 2011; Sudo et al. 2004). Consistent with these findings, broad-spectrum antibiotic decontamination of the gut flora of conventional mice also significantly increased serum stress hormone levels (Fröhlich et al. 2016).

Recently, a novel microbe-to-brain humoral signaling pathway has been found involving bacterial peptidoglycan, a major component of the bacterial cell wall. It can translocate into the brain through an intact blood-brain barrier (BBB) and is further sensed by pattern recognition receptors and nucleotide-binding oligomerization domain (NOD)-like receptors expressed on neurons during brain development and later in life on innate immune cells within the CNS (Arentsen et al. 2017). Genetic deletion of a major bacterial peptidoglycan sensor in mice not only affects brain development but also changes social and anxiety-related behaviors in these animals (Arentsen et al. 2018).

18.2.2 Gut Microbiota Influences on CNS Physiology

Brain Morphology Studies investigating the influence of gut microbiota on brain morphology have largely been performed in germ-free mice. These animals show alterations in the ultrastructure of dendrites and spines of neurons within the amygdala and hippocampus and changes in axon fiber myelination in the prefrontal cortex (Hoban et al. 2016; Luczynski et al. 2016). Furthermore, neurogenesis is increased in the hippocampus of adult germ-free animals, with increased survival of newly born neurons (Ogbonnaya et al. 2015). In contrast, long-term antibiotic treatment of adult mice resulting in gut decontamination decreased hippocampal neurogenesis, which could be reversed by probiotic treatment or allowing access to a running wheel (Möhle et al. 2016). Gut microbiota has been shown to be also crucially involved in maintaining the BBB, as germ-free mice have a more permeable BBB, a phenotype that was reversed when these mice were colonized with SCFA-producing microbes (Braniste et al. 2014). SCFAs and the gut microbiota have also been linked with brain microglia homeostasis, as germ-free mice or animals lacking conventional intestinal flora display structural microglia abnormalities and an immature immune phenotype (Erny et al. 2015). These reports suggest that development, maturation, and renewal of several cell types of the CNS are regulated by signals from the gut microbiota, although more research is needed to understand the mechanisms of this cross talk.

Neurophysiology The existing evidence for the involvement of the gut microbiota in the hardwiring of the brain introduced the possibility of a similar impact on neurophysiology. In a germ-free mouse model, it was observed that the absence of a conventional gut flora is associated with lower levels of serotonin receptors (e.g., 5HT1AR) or brain-derived neurotrophic factor (BDNF) and glutamate receptors (NMDA subunits) in the limbic system (Bercik et al. 2011a; Diaz Heijtz et al. 2011; Neufeld et al. 2011; Sudo et al. 2004). It has also been found that these mice have an increased turnover of monoamine neurotransmitters (e.g., serotonin, dopamine, noradrenaline) and a deficient metabolism of tryptophan, resulting in lower availability of this essential precursor for serotonin synthesis (Clarke et al. 2013). Administering *Lactobacillus rhamnosus* to mice also changed the expression of the receptor for gamma-aminobutyric acid (GABA), the most important inhibitory neurotransmitter, in cortical and limbic regions of the mouse brain (Bravo et al. 2011). Along these lines, there is preliminary evidence that diet-induced gut microbial dysbiosis also impacts long-term synaptic potentiation in neurons, the cellular correlate of synaptic plasticity and memory (Buffington et al. 2016).

These findings reflect the profound impact of the gut microbiota on neural function and, as a result, the influence on various host behaviors (Fig. 18.2): a vast amount of data, mostly from animal studies, has already been published linking various gut microbes or defined bacterial ecologies with social behaviors, motor function, pain, anxiety, stress-coping behaviors, learning and memory, or regulations of the circadian rhythm, feeding, and appetite (Burokas et al. 2017; De Vadder et al. 2014; Hsiao et al. 2013; Kelly et al. 2017; Luczynski et al. 2017; Thaiss et al. 2014).

18.3 Gut Microbiota and Brain Disorders

The examples above represent microbe-host relationships under steady state and healthy conditions. However, it is now recognized that the perturbation of gut indigenous microbiota can contribute to many aspects of neurophysiological

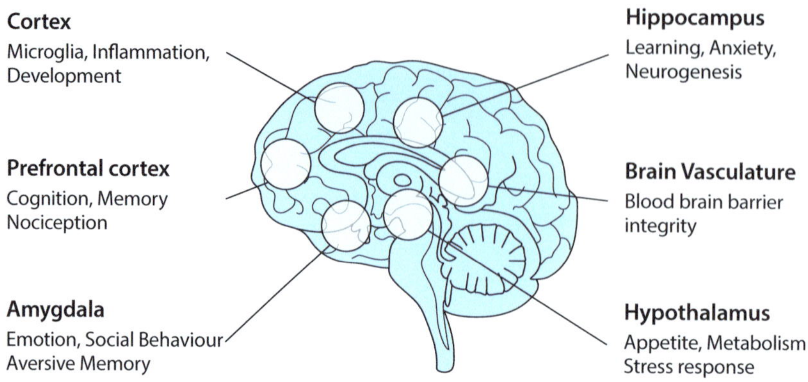

Fig. 18.2 Potential impact of the gut microbiota on several neuronal and neurobehavioral functions in distinct brain centers

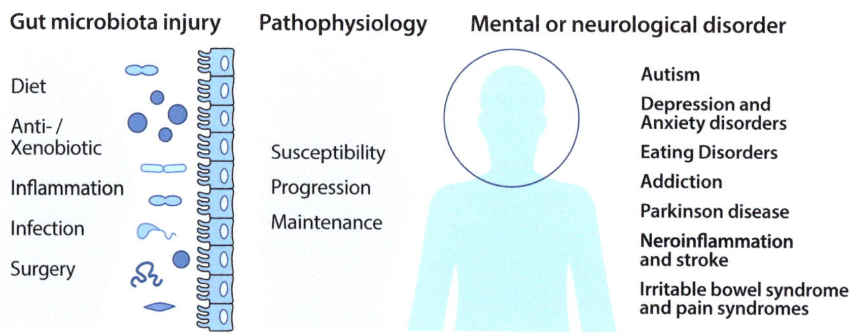

Fig. 18.3 Concept of gut microbiota injury influencing neurological disorders

dysfunction, leading to different neurological diseases and mental disorders (for an overview see Fig. 18.3). This is also reflected in the prevalence of comorbid GI pathologies and psychiatric and/or neurological illnesses (Shah et al. 2014). In the following sections, we selected several brain diseases and mental health disorders with strong evidence for a pathophysiological role of the intestinal microbiota.

18.3.1 Neurological Disorders

Multiple Sclerosis Multiple sclerosis (MS) is an autoimmune disease of the CNS that is characterized by demyelination, axonal damage, and progressive neurological disability and triggered by aberrant T-cell-mediated immune responses against myelin antigens (Ota et al. 1990). Only recently, evidence for an influence of the gut microbiota on CNS autoimmunity was found in germ-free mouse models. In a model of spontaneously developing experimental autoimmune encephalomyelitis (EAE), germ-free mice develop a significantly attenuated pathology. Further, it was observed that the commensal microbiota in laboratory mice stimulates myelin-specific CD4+ T cells and also B cells producing autoantibodies against myelin oligodendrocyte glycoprotein (Berer et al. 2011). Furthermore, segmented filamentous bacteria have been shown to trigger pro-inflammatory Th17 immune responses in the brain which was accompanied by spontaneous EAE in these mice (Lee et al. 2011).

Several groups have sequenced the fecal gut microbiota of MS patients and observed an altered flora composition when compared to healthy controls (Shahi et al. 2017). Most of these studies analyzed fecal samples only at a single time point and in rather heterogeneous patient populations. Nevertheless, several bacterial genera like *Akkermansia*, *Acinetobacter*, *Prevotella*, or *Parabacteroides* were found to be differently abundant in the microbiota of MS patients compared to healthy control floras across different studies (Berer et al. 2017; Cekanaviciute et al. 2017; Chen et al. 2016; Jangi et al. 2016). In addition to intestinal microbiota analyses, one study also sequenced brain biopsies for bacterial traces and found a predominance of *Proteobacteria* in white matter lesions of progressive MS patients compared to non-MS controls (Branton et al. 2016).

Functionally, it was observed that colonizing laboratory mice with gut microbiota from MS patients increased the severity of EAE. This flora also failed to induce anti-inflammatory IL10+ FoxP3+ regulatory T cells. Stimulating peripheral blood mononuclear cells (PBMCs) with *Akkermansia* and *Acinetobacter*, two MS-associated intestinal bacterial taxa, induced significant pro-inflammatory Th1 immune responses (Cekanaviciute et al. 2017). As an example of the significance of anti-inflammatory bacteria, *Prevotella histicola*, isolated from human celiac disease patients and possessing known immunomodulatory capacities, can suppress disease in EAE models via induction of FoxP3+ regulatory T cells and tolerogenic dendritic cells, and a decrease in pro-inflammatory Th1 and Th17 responses (Mangalam et al. 2017). Together, these results provide functional evidence that human GI microbiome has large effects on CNS-specific autoimmunity.

Ischemic Stroke Increasing knowledge about the role of gut microbiota has also been collected in cerebral ischemia using again animal models. The interaction between resident brain immune cells and peripheral immune cells that infiltrate the CNS after ischemia contributes to tissue damage and repair (Macrez et al. 2011). Diet-microbiota pathways also contribute to stroke pathophysiology because of the strong impact of the gut flora on atherosclerosis and metabolic syndrome (Schroeder and Bäckhed 2016), both major risk factors for stroke. These interactions have been discussed in a previous chapter of the book (Chap. 16).

In mouse models of focal cerebral ischemia, antibiotic-induced depletion of gut microbiota significantly decreased survival after stroke, an effect that was reversed by colonization with a conventional murine microflora (Winek et al. 2016). Another preclinical study also demonstrated the impact of an antibiotic-induced gut flora dysbiosis on stroke outcome and hypothesized that it was because of poststroke bacterial priming of gut dendritic cells with enhancement of effector T cells traveling to the brain. Here, these Th17 T cells localize in the meninges and enhance ischemic neuroinflammation and tissue damage (Benakis et al. 2016). Interestingly, experimental stroke itself can induce microbial dysbiosis in mice via changes in GI motility. Mice who were colonized with this dysbiotic flora demonstrated worse deterioration in an experimental stroke model due to induction of pro-inflammatory Th1 and Th17 T cells in the intestines that are able to migrate to the poststroke brain. In turn, restoration of a healthy microbiota in mice undergoing experimental brain ischemia via fecal matter transfers (FMTs) was observed to be neuroprotective (Singh et al. 2016).

18.3.2 Neurodegenerative Disorders

Parkinson's Disease (PD) This debilitating neurological disorder is caused by a gradual degeneration of neurons in the substantia nigra, a major motor area in the midbrain, and a consecutive loss of dopaminergic neurotransmission (Hornykiewicz 2002). It primarily comes with motor symptoms such as slow movements, resting tremor, rigidity, and postural instability. Aggregation of the protein alpha-synuclein was proven to be central in the development of the neurodegenerative process (Ingelsson 2016).

The gut microbiome has only recently been investigated in PD patients. Nevertheless, several studies reported an increased intestinal abundance of *Akkermansia* and *Lactobacillus* and decreased fecal levels of *Prevotella* in PD patients vs. matched controls (Bedarf et al. 2017; Hasegawa et al. 2015; Hill-Burns et al. 2017; Hopfner et al. 2017; Scheperjans et al. 2015). Some of these studies could also link microbe signature with PD drug treatment, dietary factors (Hill-Burns et al. 2017), and severity of PD symptoms (Scheperjans et al. 2015). Further evidence for microbial effects on PD comes from a recent study showing that gut microbiota changes precede PD motor symptoms. Here, a reduced abundance of *Prevotella* in patient's stool was found to be associated with both motor prodromi and full PD (Heintz-Buschart et al. 2018).

Given that constipation is a major confounder in PD, the validity of these finding is somewhat controversial. To address this question, Sampson et al. demonstrated that when mice genetically overexpressing alpha-synuclein and, thus, showing PD-related symptoms are raised germ-free or received antibiotic treatment, the development of motor abnormalities and alpha-synuclein-dependent activation of microglia are significantly reduced (Sampson et al. 2016). Treating these mice with SCFAs, in turn, promoted motor deficits. Finally, colonizing germ-free alpha-synuclein mutant mice with gut microbiota from PD patients vs. healthy controls also increased PD-related motor symptoms in "PD-humanized" mice providing causal evidence for a functional contribution by the intestinal microbiota to synucleinopathies.

Other Neurodegenerative Diseases Changes in the intestinal microbiota have also been reported in patients with amyotrophic lateral sclerosis (Brenner et al. 2017) and in patients with multiple system atrophy (Tan et al. 2018) in comparison to matched healthy controls. These neurodegenerative disorders, however, affect several organs secondarily, and that implies frequent medical interventions, e.g., parenteral nutrition, antibiotic treatments, and surgical interventions. As these factors mutually interact with the status of the commensal flora, an investigation of the possible pathogenetic role of the indigenous microbiota in these neurodegenerative disorders is extremely complex.

18.3.3 Psychiatric Disorders

As described above, increasing evidence from both rodent and human studies points to a major role of the commensal gut flora in shaping neurochemical networks in the brain and its profound influence on social, cognitive, and emotional functions. In this context, changes in the gut microbiota have also been associated with the onset and manifestation of autism spectrum disorders, schizophrenia, and depression. This interaction is further supported by the fact that gastrointestinal comorbidities and food allergies are common in neurodevelopmental disorders (de Theije et al. 2014).

Autism Spectrum Disorders (ASD) This neuropsychiatric disorder manifests primarily during early childhood, and in most cases, symptoms persist throughout adulthood. It comprises a set of several neurodevelopmental disabilities with repetitive/stereotypic behaviors and deficits in communication and social interaction (Lord et al. 2000). The gut microbiome of ASD children has been analyzed in cross-sectional studies with multiple different cohorts (Kelly et al. 2017). A recent meta-analysis of 15 microbiome studies found differences in the abundance of bacteria in the phyla *Firmicutes*, *Bacteroidetes*, and *Proteobacteria* in ASD patients vs. controls (Cao et al. 2013). However, due to substantial heterogeneity in patient populations and methodology (e.g., lack of data regarding the use of medications, dietary differences, and supplement administration), no definite conclusions on the significance of these different ecologies in ASD could be drawn.

Another line of evidence for the impact of the gut microbiota on ASD comes from a small, interventional clinical study where children with ASD received oral, non-absorbed, vancomycin for 12 weeks (Sandler et al. 2000). During treatment, eight out of ten children showed significant improvement of behavioral symptoms, but these gains largely waned after discontinuation of treatment. Another study investigated the effects of probiotics in ASD patients using a formulation of

Lactobacillus, *Bifidobacterium*, and *Streptococcus* but observed only effects on the patient's microbiome and fecal cytokine levels and no significant change in behavior (Tomova et al. 2015). FMT has also been studied as potential microbiota intervention strategy in ASD: 18 children received repeated oral or rectal application of standardized human gut microbiota formulations from healthy, unrelated, adult donors. It was observed that this treatment significantly improved behavioral ASD symptoms and concomitant GI symptoms (Kang et al. 2017). FMT increased bacterial diversity and the abundance of *Bifidobacterium*, *Prevotella*, and *Desulfovibrio* bacteria, which were initially found to be absent in ASD patients' gut microfloras (Kang et al. 2013).

The maternal immune activation (MIA) mouse model is known to induce ASD-like features in the offspring and has been reported to also induce significant microbiome alterations and GI barrier defects in these mice. MIA also leads to an altered serum metabolome, with 4-ethylphenyl sulfate (4EPS) as a major microbial metabolite linked to anxiety-related behaviors in this model (Hsiao et al. 2013). Treatment of MIA offspring with a single bacterium, *Bacteroides fragilis*, was able to reverse GI barrier dysfunction and to rescue some of the MIA-induced behavioral deficits including stereotypic behaviors, disturbed vocalizations, and anxiety behaviors.

Another frequently used mouse model for ASD research is the BTBR T+ Itpr3tf/J (BTBR) strain in which autism-like behavior is driven by multiple genetic alterations (Meyza and Blanchard 2017). These mice display an altered gut microbiota with reduced metabolism of bile acids, tryptophan, and SCFAs. Further, GI dysfunction was reported on the ENS and epithelium level, and several ASD-related behavioral features such as reduced sociability, increased compulsive behaviors, and anxiety-related behaviors were observed (Golubeva et al. 2017). Together, these preclinical and clinical data support the concept of a dysfunctional microbiota-gut-brain axis in ASD.

Schizophrenia So far, only a limited number of clinical studies have investigated the role of gut microbiota in schizophrenia. A recent study conducted in patients with first-episode psychosis revealed an elevated abundance of bacteria from the *Lactobacillus* family vs. controls that also correlated with the severity of symptoms (Schwarz et al. 2018). A subgroup of these patients that displayed the strongest microbiota differences during initial assessment also showed poorer response to treatment after 12 months of antipsychotic administration. Interestingly, *Lactobacilli* and *Bifidobacteria* were also found in another study to be overrepresented in oropharyngeal samples from schizophrenic patients vs. controls (Castro-Nallar et al. 2015). Another clinical study analyzed a blood-specific microbiome in schizophrenic patients and observed an increased alpha and beta diversity compared to controls and other patients with neuropsychiatric disorders (Olde Loohuis 2018).

Given the potential of probiotics to restore a disturbed gut microbiota, Dickerson et al. carried out a controlled 14-week probiotic intervention study in schizophrenic patients using a combination of *Lactobacillus rhamnosus* and *Bifidobacterium animalis* strains (Dickerson et al. 2014). This intervention only alleviated bowel movement difficulties associated with schizophrenia and/or medication in this cohort and did not improve psychiatric symptoms.

In these studies, it is noteworthy that the vast majority of patients in these studies received antipsychotic medication, which can impact gut microbiota composition and, therefore, can profoundly bias disease-microbiome associations or microbiota intervention trials (Bahr et al. 2015; Bahra et al. 2015; Davey et al. 2013).

Major Depressive Disorder (MDD) It has been reported that antibiotic treatment during the first year of life is associated with symptoms of depression and impaired neurocognitive abilities in childhood (Slykerman et al. 2017).

Several cross-sectional studies have compared gut microbiota signatures of MDD patients to healthy controls and have observed reductions in diversity and microbial shifts with overrepresentations of *Proteobacteria* and *Acinetobacteria* (Jiang et al. 2015; Kelly et al. 2016; Zheng et al. 2016). On the metagenome level, Stevens et al. observed that the gut microbiome of patients with depression is characterized by bacteria with increased lipopolysaccharide (LPS) biosynthesis and altered pathways of neurotransmitter metabolism and mucin production. Intriguingly, these patients also showed enhanced plasma LPS (likely derived from the skewed distribution of gram-negative taxa in this cohort) and also increased markers of perturbed GI epithelial barrier integrity (Stevens et al. 2017). These data substantiate prior human and preclinical studies linking gut dysbiosis with intestinal barrier dysfunction and mental illness (Lasselin et al. 2016).

Studying cause vs. consequence of microbiota changes in depressive disorders, two independent research groups transplanted fecal material from patients with depressive disorders and controls into germ-free mice or rats. Notably, both of them found that laboratory animals colonized with stool material from MDD patients exhibit depressive-like phenotypes and depression-associated neurochemical changes (Kelly et al. 2016; Zheng et al. 2016).

In healthy humans, a recent study found that consuming a fermented milk product containing different probiotic bacteria can directly impact brain activity during emotional and attentional tasks in healthy subjects as measured by brain fMRI (Tillisch et al. 2013). In MDD patients, a placebo-controlled, randomized, blinded administration of a cocktail of different probiotics was even able to significantly reduce depressive symptoms in this cohort (Akkasheh et al. 2016). Similarly, randomized, placebo-controlled treatment with *Bifidobacterium longum* of patients with irritable bowel syndrome, characterized by frequently co-occurring depressive symptoms, also significantly reduced depressive features (Pinto-Sanchez et al. 2017); notably, an fMRI brain scan showed that probiotic administration also reduced responses to negative emotional stimuli in multiple brain areas, including the amygdala and fronto-limbic region.

In summary, these observations in humans exemplify the mounting evidence for a functional association between gut dysbiosis and features of depression. However, the complex mechanisms and pathways that contribute to the intersection of gut microbiota and mental health are still poorly explored and demand huge, future efforts before novel strategies of microbiota interventions in neuropsychiatry are ready for clinical implementation.

▶ Controversy

Gastrointestinal comorbidities are significantly more prevalent in patients with neurological and psychiatric disorders. Further, medication, lifestyle changes (i.e., smoking or drug use), different diets, and other environmental factors need to be taken into account. These comorbidities and additional factors can contribute to the observed differences in the gut microbiome in these patient populations (de Theije et al. 2014; Son et al. 2015; Williams et al. 2011, 2012).

Preclinical studies using rodent models indicate that certain domains of a psychiatric, neurologic or neurodevelopmental diseases are influenced by the gut microbiota. However, preclinical models have many limitations, especially in neuropsychiatry, and translating preclinical data into the patient population is particularly challenging for these complex disorders. Another bias in microbiota research using mouse models comes from a recent observation that standard laboratory mice have an altered gut microbiome with low diversity compared to the rich microbiota of wild-living mice (Rosshart et al. 2017). Importantly, such a diverse natural microbiota was associated with reduced colonic inflammation, protection from influenza virus, and improved resistance against mutagen-/inflammation-induced colorectal tumorigenesis (Rosshart et al. 2017).

In addition to diet and other environmental contributions, human genetic variation has emerged as another important factor that profoundly shapes the gut microbiome of an individual (Goodrich et al. 2017). For instance, a strong interplay between host genetics and gut microbiota signatures has been observed in inflammatory bowel disease (Imhann et al. 2016). However, this gene-microbiome intersection has not yet been studied in neuropsychiatric disorders, but it would be of tremendous relevance given the intrinsic genetic component of such pathologies (Birnbaum and Weinberger 2017; Network and Pathway Analysis Subgroup of Psychiatric Genomics Consortium 2015).

History

The idea that gut bacteria may influence brain function and mental health dates back to the nineteenth and early twentieth centuries. "Autointoxication," "intestinal stasis," or "intestinal intoxication" described disease processes of toxins coming from the gut and affecting the central nervous system, specifically mental health (Bested et al. 2013).

Daniel R. Brower published in 1898 in JAMA on "autointoxication" and melancholia: he reported that gut bacteria produce toxin compounds like indoles and lactic acid that are "detoxified" by the liver and kidneys in healthy individuals. However, these detoxification process would fail in patients with melancholia (Brower 1898). To "detoxify the gut and body," physicians used colon irrigations or even surgical extractions of tonsils or teeth.

Later, probiotic formulas, mostly containing *Lactobacillus* strains, were used to "detoxify" and rebalance the gut. In this context, Ilja Metschnikow, microbiologist and Nobel laureate of 1908, wrote in 1912 (Cosmopolitan): "In effect, we fight microbe with microbe... there seems hope that we shall in time be able to transform the entire intestinal flora from a harmful to an innocuous one... the beneficent effect of this transformation must be enormous...."

With this statement he arguably predicted the trajectory of modern microbiota-gut-brain axis research directed to understand the role of the gut microflora in neuropsychiatric disorders with anti-, pre-, and post-biotics or even FMT.

Highlights

- The intestinal microbiota is a key factor in the early development of the enteric, peripheral, and central nervous system for both neurons and glia and in neurogenesis in the adult brain.
- Microbe-to-brain communication via the gut-brain axis influences neurochemical signaling and synaptic physiology in the brain with profound impact on cognition, learning and memory, emotional behaviors, and neuroinflammatory processes.
- Animal and human studies revealed a major role of the gut commensal microbiota in the pathogenesis of neurological and neurodegenerative disorders, e.g., multiple sclerosis, ischemic stroke, or Parkinson's disease, predominantly through microbial priming of immune responses or metabolites interfering with brain protein biochemistry.
- Increasing preclinical and clinical evidence suggests that the indigenous gut microbiota is strongly involved in the development of psychiatric disorders such as autism spectrum disorders, schizophrenia, or major depressive episodes.

References

Akkasheh, G., Kashani-Poor, Z., Tajabadi-Ebrahimi, M., Jafari, P., Akbari, H., Taghizadeh, M., et al. (2016). Clinical and metabolic response to probiotic administration in patients with major depressive disorder: A randomized, double-blind, placebo-controlled trial. *Nutrition, 32*, 315–320.

Anitha, M., Vijay-Kumar, M., Sitaraman, S. V., Gewirtz, A. T., & Srinivasan, S. (2012). Gut microbial products regulate murine gastrointestinal motility via Toll-like receptor 4 signaling. *Gastroenterology, 143*, 1006–1016.e1004.

Arentsen, T., Qian, Y., Gkotzis, S., Femenia, T., Wang, T., Udekwu, K., et al. (2017). The bacterial peptidoglycan-sensing molecule Pglyrp2 modulates brain development and behavior. *Molecular Psychiatry, 22*, 257–266.

Arentsen, T., Khalid, R., Qian, Y., & Diaz Heijtz, R. (2018). Sex-dependent alterations in motor and anxiety-like behavior of aged bacterial peptidoglycan sensing molecule 2 knockout mice. *Brain, Behavior, and Immunity, 67*, 345–354.

Bahr, S. M., Tyler, B. C., Wooldridge, N., Butcher, B. D., Burns, T. L., Teesch, L. M., et al. (2015). Use of the second-generation antipsychotic, risperidone, and secondary weight gain are associated with an altered gut microbiota in children. *Translational Psychiatry, 5*, e652.

Bahra, S. M., Weidemann, B. J., Castro, A. N., Walsh, J. W., deLeon, O., Burnett, C. M., et al. (2015). Risperidone-induced weight gain is mediated through shifts in the gut microbiome and suppression of energy expenditure. *eBioMedicine, 2*, 1725–1734.

Bedarf, J. R., Hildebrand, F., Coelho, L. P., Sunagawa, S., Bahram, M., Goeser, F., et al. (2017). Functional implications of microbial and viral gut metagenome changes in early stage L-DOPA-naïve Parkinson's disease patients. *Genome Medicine, 9*, 39.

Belkaid, Y., & Harrison, O. J. (2017). Homeostatic Immunity and the Microbiota. *Immunity, 46*, 562–576.

Bellono, N. W., Bayrer, J. R., Leitch, D. B., Castro, J., Zhang, C., O'Donnell, T. A., et al. (2017). Enterochromaffin cells are gut chemosensors that couple to sensory neural pathways. *Cell, 170*, 185–198.e116.

Benakis, C., Brea, D., Caballero, S., Faraco, G., Moore, J., Murphy, M., et al. (2016). Commensal microbiota affects ischemic stroke outcome by regulating intestinal γδ T cells. *Nature Medicine, 22*, 516–523.

Bercik, P., Denou, E., Collins, J., Jackson, W., Lu, J., Jury, J., et al. (2011a). The intestinal microbiota affect central levels of brain-derived neurotropic factor and behavior in mice. Gastroenterology *141*, 599–609, 609.e591–593.

Bercik, P., Park, A. J., Sinclair, D., Khoshdel, A., Lu, J., Huang, X., et al. (2011b). The anxiolytic effect of Bifidobacterium longum NCC3001 involves vagal pathways for gut-brain communication. *Neurogastroenterology and Motility, 23*, 1132–1139.

Berer, K., Mues, M., Koutrolos, M., Rasbi, Z. A., Boziki, M., Johner, C., et al. (2011). Commensal microbiota and myelin autoantigen cooperate to trigger autoimmune demyelination. *Nature, 479*, 538–541.

Berer, K., Gerdes, L. A., Cekanaviciute, E., Jia, X., Xiao, L., Xia, Z., et al. (2017). Gut microbiota from multiple sclerosis patients enables spontaneous autoimmune encephalomyelitis in mice. *Proceedings of the National Academy of Sciences of the United States of America, 114*, 10719–10724.

Bested, A. C., Logan, A. C., & Selhub, E. M. (2013). Intestinal microbiota, probiotics and mental health: From Metchnikoff to modern advances: Part I – autointoxication revisited. *Gut Pathogens, 5*, 5.

Birnbaum, R., & Weinberger, D. R. (2017). Genetic insights into the neurodevelopmental origins of schizophrenia. *Nature Reviews. Neuroscience, 18*(12), 727–740.

Braniste, V., Al-Asmakh, M., Kowal, C., Anuar, F., Abbaspour, A., Tóth, M., et al. (2014). The gut microbiota influences blood-brain barrier permeability in mice. *Science Translational Medicine, 6*, 263ra158.

Branton, W. G., Lu, J. Q., Surette, M. G., Holt, R. A., Lind, J., Laman, J. D., et al. (2016). Brain microbiota disruption within inflammatory demyelinating lesions in multiple sclerosis. *Scientific Reports, 6*, 37344.

Bravo, J. A., Forsythe, P., Chew, M. V., Escaravage, E., Savignac, H. M., Dinan, T. G., et al. (2011). Ingestion of Lactobacillus strain regulates emotional behavior and central GABA receptor expression in a mouse via the vagus nerve. *Proceedings of the National Academy of Sciences of the United States of America, 108*, 16050–16055.

Brenner, D., Hiergeist, A., Adis, C., Mayer, B., Gessner, A., Ludolph, A. C., et al. (2017). The fecal microbiome of ALS patients. *Neurobiology of Aging, 61*, 132–137.

Brower, D. R. (1898). Auto-intoxication in its relations to the diseases of the nervous system. *JAMA, 30*, 575–577.

Browning, K. N., & Travagli, R. A. (2014). Central nervous system control of gastrointestinal motility and secretion and modulation of gastrointestinal functions. *Comprehensive Physiology, 4*, 1339–1368.

Buffington, S. A., Di Prisco, G. V., Auchtung, T. A., Ajami, N. J., Petrosino, J. F., & Costa-Mattioli, M. (2016). Microbial reconstitution reverses maternal diet-induced social and synaptic deficits in offspring. *Cell, 165*, 1762–1775.

Burokas, A., Arboleya, S., Moloney, R. D., Peterson, V. L., Murphy, K., Clarke, G., et al. (2017). Targeting the microbiota-gut-brain axis: Prebiotics have anxiolytic and antidepressant-like effects and reverse the impact of chronic stress in mice. *Biological Psychiatry, 82*, 472–487.

Cao, X., Lin, P., Jiang, P., & Li, C. (2013). Characteristics of the gastrointestinal microbiome in children with autism spectrum disorder: A systematic review. *Shanghai Archives of Psychiatry, 25*, 342–353.

Castro-Nallar, E., Bendall, M. L., Pérez-Losada, M., Sabuncyan, S., Severance, E. G., Dickerson, F. B., et al. (2015). Composition, taxonomy and functional diversity of the oropharynx microbiome in individuals with schizophrenia and controls. *PeerJ, 3*, e1140.

Cekanaviciute, E., Yoo, B. B., Runia, T. F., Debelius, J. W., Singh, S., Nelson, C. A., et al. (2017). Gut bacteria from multiple sclerosis patients modulate human T cells and exacerbate symptoms in mouse models. *Proceedings of the National Academy of Sciences of the United States of America, 114*, 10713–10718.

Chen, J., Chia, N., Kalari, K. R., Yao, J. Z., Novotna, M., Soldan, M. M., et al. (2016). Multiple sclerosis patients have a distinct gut microbiota compared to healthy controls. *Scientific Reports, 6*, 28484.

Clarke, G., Grenham, S., Scully, P., Fitzgerald, P., Moloney, R. D., Shanahan, F., et al. (2013). The microbiome-gut-brain axis during early life regulates the hippocampal serotonergic system in a sex-dependent manner. *Molecular Psychiatry, 18*, 666–673.

Coccaro, E. F., Lee, R., Groer, M. W., Can, A., Coussons-Read, M., & Postolache, T. T. (2016). Toxoplasma gondii infection: Relationship with aggression in psychiatric subjects. *The Journal of Clinical Psychiatry, 77*, 334–341.

Collins, J., Borojevic, R., Verdu, E. F., Huizinga, J. D., & Ratcliffe, E. M. (2014). Intestinal microbiota influence the early postnatal development of the enteric nervous system. *Neurogastroenterology and Motility, 26*, 98–107.

Davey, K. J., Cotter, P. D., O'Sullivan, O., Crispie, F., Dinan, T. G., Cryan, J. F., et al. (2013). Antipsychotics and the gut microbiome: Olanzapine-induced metabolic dysfunction is attenuated by antibiotic administration in the rat. *Translational Psychiatry, 3*, e309.

de Theije, C. G., Bavelaar, B. M., Lopes da Silva, S., Korte, S. M., Olivier, B., Garssen, J., et al. (2014). Food allergy and food-based therapies in neurodevelopmental disorders. *Pediatric Allergy and Immunology, 25*, 218–226.

De Vadder, F., Kovatcheva-Datchary, P., Goncalves, D., Vinera, J., Zitoun, C., Duchampt, A., et al. (2014). Microbiota-generated metabolites promote metabolic benefits via gut-brain neural circuits. *Cell, 156*, 84–96.

Dey, N., Wagner, V. E., Blanton, L. V., Cheng, J., Fontana, L., Haque, R., et al. (2015). Regulators of gut motility revealed by a gnotobiotic model of diet-microbiome interactions related to travel. *Cell, 163*, 95–107.

Diaz Heijtz, R., Wang, S., Anuar, F., Qian, Y., Björkholm, B., Samuelsson, A., et al. (2011). Normal gut microbiota modulates brain development and behavior. *Proceedings of the National Academy of Sciences of the United States of America, 108*, 3047–3052.

Dickerson, F. B., Stallings, C., Origoni, A., Katsafanas, E., Savage, C. L., Schweinfurth, L. A., et al. (2014). Effect of probiotic supplementation on schizophrenia symptoms and association with gastrointestinal functioning: A randomized, placebo-controlled trial. *The Primary Care Companion for CNS Disorders, 16*.

Erny, D., Hrabě de Angelis, A. L., Jaitin, D., Wieghofer, P., Staszewski, O., David, E., et al. (2015). Host microbiota constantly control maturation and function of microglia in the CNS. *Nature Neuroscience, 18*, 965–977.

Ferraro, A., & Kilman, J. (1933). Experimental toxic approach to mental illness. *Psychiatric Quarterly, 7*, 115–153.

Fröhlich, E. E., Farzi, A., Mayerhofer, R., Reichmann, F., Jačan, A., Wagner, B., et al. (2016). Cognitive impairment by antibiotic-induced gut dysbiosis: Analysis of gut microbiota-brain communication. *Brain, Behavior, and Immunity, 56*, 140–155.

Fung, T. C., Olson, C. A., & Hsiao, E. Y. (2017). Interactions between the microbiota, immune and nervous systems in health and disease. *Nature Neuroscience, 20*, 145–155.

Furness, J. B., Callaghan, B. P., Rivera, L. R., & Cho, H. J. (2014). The enteric nervous system and gastrointestinal innervation: Integrated local and central control. *Advances in Experimental Medicine and Biology, 817*, 39–71.

Golubeva, A. V., Joyce, S. A., Moloney, G., Burokas, A., Sherwin, E., Arboleya, S., et al. (2017). Microbiota-related changes in bile acid and tryptophan metabolism are associated with gastrointestinal dysfunction in a mouse model of autism. *eBioMedicine, 24*, 166–178.

Goodrich, J. K., Davenport, E. R., Clark, A. G., & Ley, R. E. (2017). The relationship between the human genome and microbiome comes into view. *Annual Review of Genetics, 51*, 413–433.

Hamodeh, S. A., Rehn, M., Haschke, G., & Diener, M. (2004). Mechanism of butyrate-induced hyperpolarization of cultured rat myenteric neurones. *Neurogastroenterology and Motility, 16*, 597–604.

Hasegawa, S., Goto, S., Tsuji, H., Okuno, T., Asahara, T., Nomoto, K., et al. (2015). Intestinal dysbiosis and lowered serum lipopolysaccharide-binding protein in Parkinson's disease. *PLoS One, 10*, e0142164.

Heintz-Buschart, A., Pandey, U., Wicke, T., Sixel-Döring, F., Janzen, A., Sittig-Wiegand, E., et al. (2018). The nasal and gut microbiome in Parkinson's disease and idiopathic rapid eye movement sleep behavior disorder. *Movement Disorders, 33*(1), 88–98.

Hill-Burns, E. M., Debelius, J. W., Morton, J. T., Wissemann, W. T., Lewis, M. R., Wallen, Z. D., et al. (2017). Parkinson's disease and Parkinson's disease medications have distinct signatures of the gut microbiome. *Movement Disorders, 32*, 739–749.

Hoban, A. E., Stilling, R. M., Ryan, F. J., Shanahan, F., Dinan, T. G., Claesson, M. J., et al. (2016). Regulation of prefrontal cortex myelination by the microbiota. *Translational Psychiatry, 6*, e774.

Hopfner, F., Künstner, A., Müller, S. H., Künzel, S., Zeuner, K. E., Margraf, N. G., et al. (2017). Gut microbiota in Parkinson disease in a northern German cohort. *Brain Research, 1667*, 41–45.

Hornykiewicz, O. (2002). Dopamine miracle: From brain homogenate to dopamine replacement. *Movement Disorders, 17*, 501–508.

Hsiao, E. Y., McBride, S. W., Hsien, S., Sharon, G., Hyde, E. R., McCue, T., et al. (2013). Microbiota modulate behavioral and physiological abnormalities associated with neurodevelopmental disorders. *Cell, 155*, 1451–1463.

Imhann, F., Vich Vila, A., Bonder, M. J., Fu, J., Gevers, D., Visschedijk, M. C., et al. (2016). Interplay of host genetics and gut microbiota underlying the onset and clinical presentation of inflammatory bowel disease. *Gut, 67*(1), 108–119.

Ingelsson, M. (2016). Alpha-synuclein oligomers-neurotoxic molecules in Parkinson's disease and other lewy body disorders. *Frontiers in Neuroscience, 10*, 408.

Jangi, S., Gandhi, R., Cox, L. M., Li, N., von Glehn, F., Yan, R., et al. (2016). Alterations of the human gut microbiome in multiple sclerosis. *Nature Communications, 7*, 12015.

Jiang, H., Ling, Z., Zhang, Y., Mao, H., Ma, Z., Yin, Y., et al. (2015). Altered fecal microbiota composition in patients with major depressive disorder. *Brain, Behavior, and Immunity, 48*, 186–194.

Kabouridis, P. S., Lasrado, R., McCallum, S., Chng, S. H., Snippert, H. J., Clevers, H., et al. (2015). Microbiota controls the homeostasis of glial cells in the gut lamina propria. *Neuron, 85*, 289–295.

Kamiya, T., Wang, L., Forsythe, P., Goettsche, G., Mao, Y., Wang, Y., et al. (2006). Inhibitory effects of Lactobacillus reuteri on visceral pain induced by colorectal distension in Sprague-Dawley rats. *Gut, 55*, 191–196.

Kang, D. W., Park, J. G., Ilhan, Z. E., Wallstrom, G., Labaer, J., Adams, J. B., et al. (2013). Reduced incidence of Prevotella and other fermenters in intestinal microflora of autistic children. *PLoS One, 8*, e68322.

Kang, D. W., Adams, J. B., Gregory, A. C., Borody, T., Chittick, L., Fasano, A., et al. (2017). Microbiota transfer therapy alters gut ecosystem and improves gastrointestinal and autism symptoms: An open-label study. *Microbiome, 5*, 10.

Kelly, J. R., Borre, Y., O'Brien, C., Patterson, E., El Aidy, S., Deane, J., et al. (2016). Transferring the blues: Depression-associated gut microbiota induces neurobehavioural changes in the rat. *Journal of Psychiatric Research, 82*, 109–118.

Kelly, J. R., Minuto, C., Cryan, J. F., Clarke, G., & Dinan, T. G. (2017). Cross talk: The microbiota and neurodevelopmental disorders. *Frontiers in Neuroscience, 11*, 490.

Kimura, I., Inoue, D., Maeda, T., Hara, T., Ichimura, A., Miyauchi, S., et al. (2011). Short-chain fatty acids and ketones directly regulate sympathetic nervous system via G protein-coupled receptor 41 (GPR41). *Proceedings of the National Academy of Sciences of the United States of America, 108*, 8030–8035.

Lasselin, J., Elsenbruch, S., Lekander, M., Axelsson, J., Karshikoff, B., Grigoleit, J. S., et al. (2016). Mood disturbance during experimental endotoxemia: Predictors of state anxiety as a psychological component of sickness behavior. *Brain, Behavior, and Immunity, 57*, 30–37.

Lee, Y. K., Menezes, J. S., Umesaki, Y., & Mazmanian, S. K. (2011). Proinflammatory T-cell responses to gut microbiota promote experimental autoimmune encephalomyelitis. *Proceedings of the National Academy of Sciences of the United States of America, 108(Suppl 1)*, 4615–4622.

Lord, C., Cook, E. H., Leventhal, B. L., & Amaral, D. G. (2000). Autism spectrum disorders. *Neuron, 28*, 355–363.

Luczynski, P., Whelan, S. O., O'Sullivan, C., Clarke, G., Shanahan, F., Dinan, T. G., et al. (2016). Adult microbiota-deficient mice have distinct dendritic morphological changes: Differential effects in the amygdala and hippocampus. *The European Journal of Neuroscience, 44*, 2654–2666.

Luczynski, P., Tramullas, M., Viola, M., Shanahan, F., Clarke, G., O'Mahony, S., et al. (2017). Microbiota regulates visceral pain in the mouse. *eLife, 6*, e25887.

Macrez, R., Ali, C., Toutirais, O., Le Mauff, B., Defer, G., Dirnagl, U., et al. (2011). Stroke and the immune system: From pathophysiology to new therapeutic strategies. *Lancet Neurology, 10*, 471–480.

Mangalam, A., Shahi, S. K., Luckey, D., Karau, M., Marietta, E., Luo, N., et al. (2017). Human gut-derived commensal bacteria suppress CNS inflammatory and demyelinating disease. *Cell Reports, 20*, 1269–1277.

McVey Neufeld, K. A., Mao, Y. K., Bienenstock, J., Foster, J. A., & Kunze, W. A. (2013). The microbiome is essential for normal gut intrinsic primary afferent neuron excitability in the mouse. *Neurogastroenterology and Motility, 25*, 183–e188.

Meyza, K. Z., & Blanchard, D. C. (2017). The BTBR mouse model of idiopathic autism – Current view on mechanisms. *Neuroscience and Biobehavioral Reviews, 76*, 99–110.

Möhle, L., Mattei, D., Heimesaat, M. M., Bereswill, S., Fischer, A., Alutis, M., et al. (2016). Ly6C (hi) monocytes provide a link between antibiotic-induced changes in gut microbiota and adult hippocampal neurogenesis. *Cell Reports, 15*, 1945–1956.

Muller, P. A., Koscsó, B., Rajani, G. M., Stevanovic, K., Berres, M. L., Hashimoto, D., et al. (2014). Crosstalk between muscularis macrophages and enteric neurons regulates gastrointestinal motility. *Cell, 158*, 300–313.

Nagashima, K., Sawa, S., Nitta, T., Tsutsumi, M., Okamura, T., Penninger, J. M., et al. (2017). Identification of subepithelial mesenchymal cells that induce IgA and diversify gut microbiota. *Nature Immunology, 18*, 675–682.

Network and Pathway Analysis Subgroup of Psychiatric Genomics Consortium. (2015). Psychiatric genome-wide association study analyses implicate neuronal, immune and histone pathways. *Nature Neuroscience, 18*, 199–209.

Neufeld, K. M., Kang, N., Bienenstock, J., & Foster, J. A. (2011). Reduced anxiety-like behavior and central neurochemical change in germ-free mice. *Neurogastroenterology and Motility, 23*, 255–264, e119.

Nøhr, M. K., Egerod, K. L., Christiansen, S. H., Gille, A., Offermanns, S., Schwartz, T. W., et al. (2015). Expression of the short chain fatty acid receptor GPR41/FFAR3 in autonomic and somatic sensory ganglia. *Neuroscience, 290*, 126–137.

Ogbonnaya, E. S., Clarke, G., Shanahan, F., Dinan, T. G., Cryan, J. F., & O'Leary, O. F. (2015). Adult hippocampal neurogenesis is regulated by the microbiome. *Biological Psychiatry, 78*, e7–e9.

Olde Loohuis, L. M. (2018). Transcriptome analysis in whole blood reveals increased microbial diversity in schizophrenia. *Translational Psychiatry, 8*, 96. https://doi.org/10.1038/s41398-018-0107-9

Ota, K., Matsui, M., Milford, E. L., Mackin, G. A., Weiner, H. L., & Hafler, D. A. (1990). T-cell recognition of an immunodominant myelin basic protein epitope in multiple sclerosis. *Nature, 346*, 183–187.

Perez-Burgos, A., Wang, B., Mao, Y. K., Mistry, B., McVey Neufeld, K. A., Bienenstock, J., et al. (2013). Psychoactive bacteria Lactobacillus rhamnosus (JB-1) elicits rapid frequency facilitation in vagal afferents. *American Journal of Physiology. Gastrointestinal and Liver Physiology, 304*, G211–G220.

Perez-Burgos, A., Wang, L., McVey Neufeld, K. A., Mao, Y. K., Ahmadzai, M., Janssen, L. J., et al. (2015). The TRPV1 channel in rodents is a major target for antinociceptive effect of the probiotic Lactobacillus reuteri DSM 17938. *The Journal of Physiology, 593*, 3943–3957.

Perry, R. J., Peng, L., Barry, N. A., Cline, G. W., Zhang, D., Cardone, R. L., et al. (2016). Acetate mediates a microbiome-brain-β-cell axis to promote metabolic syndrome. *Nature, 534*, 213–217.

Pinto-Sanchez, M. I., Hall, G. B., Ghajar, K., Nardelli, A., Bolino, C., Lau, J. T., et al. (2017). Probiotic bifidobacterium longum NCC3001 reduces depression scores and alters brain activity: A pilot study in patients with irritable bowel syndrome. *Gastroenterology, 153*, 448–459.e448.

Reigstad, C. S., Salmonson, C. E., Rainey, J. F., Szurszewski, J. H., Linden, D. R., Sonnenburg, J. L., et al. (2015). Gut microbes promote colonic serotonin production through an effect of short-chain fatty acids on enterochromaffin cells. *The FASEB Journal, 29*, 1395–1403.

Rios, D., Wood, M. B., Li, J., Chassaing, B., Gewirtz, A. T., & Williams, I. R. (2016). Antigen sampling by intestinal M cells is the principal pathway initiating mucosal IgA production to commensal enteric bacteria. *Mucosal Immunology, 9*, 907–916.

Rolig, A. S., Mittge, E. K., Ganz, J., Troll, J. V., Melancon, E., Wiles, T. J., et al. (2017). The enteric nervous system promotes intestinal health by constraining microbiota composition. *PLoS Biology, 15*, e2000689.

Rosshart, S. P., Vassallo, B. G., Angeletti, D., Hutchinson, D. S., Morgan, A. P., Takeda, K., et al. (2017). Wild mouse gut microbiota promotes host fitness and improves disease resistance. *Cell, 171*, 1015.e13–1028.e13.

Sampson, T. R., Debelius, J. W., Thron, T., Janssen, S., Shastri, G. G., Ilhan, Z. E., et al. (2016). Gut microbiota regulate motor deficits and neuroinflammation in a model of Parkinson's disease. *Cell, 167*, 1469–1480.e1412.

Sandler, R. H., Finegold, S. M., Bolte, E. R., Buchanan, C. P., Maxwell, A. P., & Väisänen, M. L. (2000). Short-term benefit from oral vancomycin treatment of regressive-onset autism. *Journal of Child Neurology, 15*, 429–435.

Scheperjans, F., Aho, V., Pereira, P. A., Koskinen, K., Paulin, L., Pekkonen, E., et al. (2015). Gut microbiota are related to Parkinson's disease and clinical phenotype. *Movement Disorders, 30*, 350–358.

Schroeder, B. O., & Bäckhed, F. (2016). Signals from the gut microbiota to distant organs in physiology and disease. *Nature Medicine, 22*, 1079–1089.

Schwarz, E., Maukonen, J., Hyytiäinen, T., Kieseppä, T., Orešič, M., Sabunciyan, S., et al. (2018). Analysis of microbiota in first episode psychosis identifies preliminary associations with symptom severity and treatment response. *Schizophrenia Research, 192*, 398–403.

Shah, E., Rezaie, A., Riddle, M., & Pimentel, M. (2014). Psychological disorders in gastrointestinal disease: Epiphenomenon, cause or consequence? *Annals of Gastroenterology, 27*, 224–230.

Shahi, S. K., Freedman, S. N., & Mangalam, A. K. (2017). Gut microbiome in multiple sclerosis: The players involved and the roles they play. *Gut Microbes, 6*, 1–9.

Singh, V., Roth, S., Llovera, G., Sadler, R., Garzetti, D., Stecher, B., et al. (2016). Microbiota dysbiosis controls the neuroinflammatory response after stroke. *The Journal of Neuroscience, 36*, 7428–7440.

Slykerman, R. F., Thompson, J., Waldie, K. E., Murphy, R., Wall, C., & Mitchell, E. A. (2017). Antibiotics in the first year of life and subsequent neurocognitive outcomes. *Acta Paediatrica, 106*, 87–94.

Son, J. S., Zheng, L. J., Rowehl, L. M., Tian, X., Zhang, Y., Zhu, W., et al. (2015). Comparison of fecal microbiota in children with autism spectrum disorders and neurotypical siblings in the simons simplex collection. *PLoS One, 10*, e0137725.

Stevens, B. R., Goel, R., Seungbum, K., Richards, E. M., Holbert, R. C., Pepine, C. J., et al. (2017). Increased human intestinal barrier permeability plasma biomarkers zonulin and FABP2 correlated with plasma LPS and altered gut microbiome in anxiety or depression. *Gut*. in press, https://doi.org/10.1136/gutjnl-2017-314759

Sudo, N., Chida, Y., Aiba, Y., Sonoda, J., Oyama, N., Yu, X. N., et al. (2004). Postnatal microbial colonization programs the hypothalamic-pituitary-adrenal system for stress response in mice. *The Journal of Physiology, 558*, 263–275.

Sutterland, A. L., Fond, G., Kuin, A., Koeter, M. W., Lutter, R., van Gool, T., et al. (2015). Beyond the association. Toxoplasma gondii in schizophrenia, bipolar disorder, and addiction: Systematic review and meta-analysis. *Acta Psychiatrica Scandinavica, 132*, 161–179.

Tan, A. H., Chong, C. W., Song, S. L., Teh, C. S. J., Yap, I. K. S., Loke, M. F., et al. (2018). Altered gut microbiome and metabolome in patients with multiple system atrophy. *Movement Disorders, 33*(1), 174–176.

Thaiss, C. A., Zeevi, D., Levy, M., Zilberman-Schapira, G., Suez, J., Tengeler, A. C., et al. (2014). Transkingdom control of microbiota diurnal oscillations promotes metabolic homeostasis. *Cell, 159*, 514–529.

Tillisch, K., Labus, J., Kilpatrick, L., Jiang, Z., Stains, J., Ebrat, B., et al. (2013). Consumption of fermented milk product with probiotic modulates brain activity. *Gastroenterology, 144*, 1394–1401 1401.e1391–1394.

Tomova, A., Husarova, V., Lakatosova, S., Bakos, J., Vlkova, B., Babinska, K., et al. (2015). Gastrointestinal microbiota in children with autism in Slovakia. *Physiology and Behavior, 138*, 179–187.

Vyas, A. (2015). Mechanisms of host behavioral change in Toxoplasma gondii rodent association. *PLoS Pathogens, 11*, e1004935.

Williams, B. L., Hornig, M., Buie, T., Bauman, M. L., Cho Paik, M., Wick, I., et al. (2011). Impaired carbohydrate digestion and transport and mucosal dysbiosis in the intestines of children with autism and gastrointestinal disturbances. *PLoS One, 6*, e24585.

Williams, B. L., Hornig, M., Parekh, T., & Lipkin, W. I. (2012). Application of novel PCR-based methods for detection, quantitation, and phylogenetic characterization of Sutterella species in intestinal biopsy samples from children with autism and gastrointestinal disturbances. *MBio, 3*, e00261–e00211.

Winek, K., Engel, O., Koduah, P., Heimesaat, M. M., Fischer, A., Bereswill, S., et al. (2016). Depletion of cultivatable gut microbiota by broad-spectrum antibiotic pretreatment worsens outcome after murine stroke. *Stroke, 47*, 1354–1363.

Yano, J. M., Yu, K., Donaldson, G. P., Shastri, G. G., Ann, P., Ma, L., et al. (2015). Indigenous bacteria from the gut microbiota regulate host serotonin biosynthesis. *Cell, 161*, 264–276.

Yoo, B. B., & Mazmanian, S. K. (2017). The enteric network: Interactions between the immune and nervous systems of the gut. *Immunity, 46*, 910–926.

Zheng, P., Zeng, B., Zhou, C., Liu, M., Fang, Z., Xu, X., et al. (2016). Gut microbiome remodeling induces depressive-like behaviors through a pathway mediated by the host's metabolism. *Molecular Psychiatry, 21*, 786–796.

Clinical Implementation of High-Throughput Sequencing

19

Andreas Hiergeist and André Gessner

Abstract

Rapid advances in high-throughput sequencing-based technologies and computational tools have opened up entirely new strategies for extensively characterizing the microbial ecology of human body habitats, independent of laboratory cultivation. Several large-scale seminal studies have revealed that various human diseases are closely associated with compositional changes in the intestinal microbiota. However, the causal connection between these microbial imbalances and clinical symptomology and the underlying pathophysiological mechanisms of microbial-host interactions are still essentially unknown for many pathologies. The transfer of findings from basic biomedical research into clinical application is one of the major challenges in microbiome research and is impeded by large interindividual variations and the lack of knowledge about potential confounding factors such as diet or host and environmental influences. Clinical application of microbiome analyses requires a diligent implementation of quality-controlled standardized wet lab and bioinformatic protocols, as well as continuous quality monitoring and accreditation in addition to well-controlled cohort studies. Furthermore, additional tools for the functional analysis of microbiome signatures are needed. Only if these conditions are met can high-throughput sequencing-based quantitative metagenomics be successfully applied as a prognostic tool in clinical practice or for improving the development of individualized therapies based on microbiota profiles.

The commercial launch of the first next-generation sequencing (NGS) platforms more than a decade ago has provided insight into various microbial communities of different human body habitats, without the need for applying sophisticated large-scale cultivation techniques. Due to their substantial advances over the Sanger method in terms of cost and enabling the highly parallel sequencing of millions of DNA molecules in one run, high-throughput sequencing (HTS) techniques have become established in both research and medical diagnostic laboratories. Nowadays, such platforms have become the standard practice in the fields of human genetics and clinical oncology for the discovery of genetic mutations associated with various diseases (Gagan and Van Allen 2015). The detection of such sequence variations has allowed for the development of individual-specific preventive and therapeutic strategies. In microbiological diagnostics, the main applications of HTS include sequencing of bacterial genomes for outbreak

A. Hiergeist · A. Gessner (✉)
Institute of Clinical Microbiology and Hygiene, University Hospital Regensburg, Regensburg, Germany
e-mail: andreas.hiergeist@ukr.de; andre.gessner@ukr.de

D. Haller (ed.), *The Gut Microbiome in Health and Disease*,
https://doi.org/10.1007/978-3-319-90545-7_19

investigations, identification of genome-wide single nucleotide polymorphisms (SNPs), and the detection of genetically encoded resistance mechanisms. More recently, whole shotgun metagenomic sequencing has been successfully used for the detection of unknown pathogens from clinical specimen (Wilson et al. 2017). Particularly, the metagenomic identification of novel viruses or viral variants has offered promising approaches for the identification of currently unknown pathogenic agents in patients with unclear etiology or for monitoring the emergence and reemergence of viral diseases.

For various reasons, HTS-based methods for the characterization of human microbiota have not yet reached the application stage in routine clinical diagnostics. Current strategies are mainly based on the (targeted) amplification of phylogenetically conserved eubacterial, archaeal, or fungal marker genes like 16S and 18S ribosomal RNA (rRNA) genes or internal transcribed spacer (ITS) sequences. In addition, shotgun metagenome-based sequencing enables the assessment of the functional diversity of complex ecosystems and allows for the (untargeted) detection and concurrent identification of prokaryotic, eukaryotic, and viral signatures. Both approaches, as well as other "omics"-based techniques, share in their analytical and methodological complexity, which is comprised of multistage downstream processes. However, this intricacy makes standardization of these methods challenging. Their application requires an accurate knowledge of inherent methodological pitfalls and their disruptive impact on microbiome profiles.

Beyond this, metagenome-wide association studies (MWAS) including large cohorts have identified key taxa as well as functional genes of the intestinal microbiota, which were potentially associated with diseases such as type 2 diabetes, atherosclerosis, obesity, liver cirrhosis, colorectal cancer (CRC), rheumatoid arthritis, and other pathologies (Wang and Jia 2016). Different species can potentially fill the same ecological or functional niche. Thus, changes in microbial patterns in a dynamically changing ecosystem as a response to external factors coming from the environment or the host may vary individually. This aspect strengthens the need for an individualized interpretation of microbiome data and personalized adjustment of therapeutic interventions. Nonetheless, clinical application has not yet been approved for most associations.

Meta-analyses of microbiome studies have shown that a common characteristic of most microbiota-associated diseases is a general loss of diversity when compared to healthy cohorts. In addition, specific patterns associated with certain diseases have been described. Dysbiotic states can be reflected either by the increase in pathogens or the loss of beneficial microorganisms. For instance, in colorectal cancer an increase in pathogen-associated genera *Fusobacterium*, *Porphyromonas*, *Peptostreptococcus*, *Parvimonas*, and *Enterobacter* has been found across several studies, while in IBD the depletion of several butyrate-producing genera of the order *Clostridiales* has been observed relative to healthy controls (Duvallet et al. 2017). Identification of shared and disease-specific patterns plus additional biomarkers will be necessary to use HTS-based microbiota profiling as a diagnostic tool, especially when it involves derivation of a diagnosis from single patient microbiome signatures. Additionally, a wide variety of numerous environmental or host-derived confounding factors have to be considered. In most instances, their changing effect on the composition of the human microbiota has yet to be investigated.

To date, fecal microbiota transfer (FMT) has often been applied for the eradication of recurrent *Clostridium difficile* infections with a high rate of success (Taur and Pamer 2014). FMT is utilized in the restoration of dysbiotic states of the intestinal microbiota in cases involving Crohn's disease, ulcerative colitis, or the therapy of a post-antibiotic dysbiosis accompanied by an overgrowth of antibiotic-resistant bacteria such as vancomycin-resistant enterococci or multiresistant Gram-negative bacteria (Davido et al. 2017). FMT will also undoubtedly play a major role in the therapy of many other pathologies in the near future, before more selective strategies such as defined probiotic consortia, prebiotics, or phage therapies become available.

HTS-based microbiota profiling is also a suitable method for the comprehensive screening of donors and longitudinal monitoring of microbiota stability in the donor after transplantation.

Actually, whole shotgun metagenome sequencing enables the tracking of the colonization after FMT on a genome-scale (Lee et al. 2017). In general, HTS-based techniques allow for the individualized monitoring of other therapeutic interventions such as dietary modulation of the microbiota by probiotics and prebiotics.

Further simplification, as well as progressing standardization and optimization of these techniques, will most certainly open new horizons in medical research, disease diagnostics, and therapy management. Therefore, we intend to address the many requirements and challenges involved in the implementation of HTS-based microbiome analysis in clinical practice. Furthermore, we will focus primarily on the clinical implementation of 16S rRNA gene-based analysis of the human microbiota since this is the method most commonly being used at present. It is important to note, however, that many points also apply to other HTS-based approaches, such as whole shotgun metagenomics.

19.1 Challenges for the Clinical Implementation of Microbiome Sequencing

Population-scale analyses of healthy cohorts, such as the human microbiome project (HMP), have shown certain congruencies in microbial patterns on the taxonomic level. However, results from such datasets and other population-scale studies have not yet led to a clear definition of a "healthy" microbiome. This is mainly due to the fact that the way in which the multitude of environmental and host factors shape the human microbiome is still poorly understood. However, in a clinical setting these confounding factors must be considered in the diagnosis and derivation of appropriate therapeutic interventions for a specific dysbiotic microbiota pattern. Factors such as age, sex, body weight, geography, diet, lifestyle, inflammation, host genetics, domestic environments, host-to-host transmission, hygiene, birth mode, or physical activity can all potentially have a significant influence on microbial patterns.

Clearly, antibiotics have an enormous impact on the human microbiota. The effects of duration and dose after administration, depending on the antibiotic class and their spectrum of activity, are not yet well understood for the majority of commensal microorganisms. For instance, the administration of antibiotics early in life may have lasting effects, including behavioral changes of the host. (Leclercq et al. 2017). Furthermore, disruption of microbiota profiles resulting from the intake of host-targeted drugs during the course of a disease can also lead to inaccurate conclusions (Maurice et al. 2013). For example, differentially abundant microbiota in former type 2 diabetes studies were identified as a result of metformin administration, and therefore suspected causal relationships arose to a great extent from antidiabetic medication and not the underlying disease (Forslund et al. 2015). Fortunately, it was possible to separate gut microbiota signatures from those of the medication, but this clearly demonstrates that confounding factors have to be taken into consideration.

Microbiome-adapted questionnaires for patient interviews will also be necessary to identify known and currently unknown influencing factors. Xenobiotics, or substances foreign to the human body, such as dietary compounds, chemicals, or pharmaceuticals, are partially modified by intestinal microorganisms (Koppel et al. 2017), although the majority of biochemical transformations cannot yet be linked to specific enzymatic reactions or microorganisms. Xenobiotics may shape the structure or function of intestinal microbiota by the creation of new ecological niches, or the products of biotransformation may interact with the host. One well-studied example is the modification of dietary choline to trimethylamine by the intestinal microbiota, which is then hepatically converted to trimethylamine N-oxide, a metabolite associated with atherogenesis (plaque formation in larger blood vessels). Further examples of bioconversions by the gastrointestinal microbiota include the modification of drugs like irinotecan used in anticancer therapy or digoxin, a Na+/K+ ATPase inhibitor used for therapy of congestive heart failure. The latter compound was found to be inactivated by conversion to the derivative dihydrodigoxin by species of the gut commensal *Eggerthella lenta* (Koppel et al. 2017).

Apart from these interindividual differences, intestinal microbiota have been found to exhibit a diurnal rhythm which influences transcriptional oscillations of the host (Thaiss et al. 2016). Testing of multiple samples during the day, or the definition of sampling protocols in longitudinal studies, is necessary to correct for daytime-dependent fluctuations. Also, microhabitats within the gut ecosystem lead to local differences in the composition of the gut microbiota. Diluting effects toward distal parts of the gastrointestinal tract may lead to blurred signatures, impeding explicit detection of disease-associated microbiota shifts (Donaldson et al. 2015). The mucosa represents a separate individual ecosystem within the gastrointestinal tract. Therefore, mucosa-associated profiles differ significantly from fecal signatures (Zoetendal et al. 2002). Particularly for the study of inflammatory processes, the mucosal microbiota and its metabolic capacity play an important role due to the close proximity of the mucosa to immune cells. Samples should therefore be taken at the presumed place of effect in the gastrointestinal tract.

Considering the multitude and wide variety of potentially confounding factors, linking microbial patterns from a single patient to specific diseases still appears to be very idealistic. Based on the numerous interindividual differences, prospective cohort studies require a large number of cases to reach sufficient statistical power. However, studying a homogenous phenotype will support association studies. Conducting longitudinal studies instead of single-point measured values should also enable the correction for microbiota fluctuations.

19.2 Methodological Bias: Steps Toward the Standardization of Metagenome-Based Microbiome Analyses

An accurate, reliable, and reproducible methodology is essential for the solid implementation of microbiome sequencing into a diagnostic workflow in order to deduce therapeutic options based on characteristic microbiota profiles or to enable the comparability of microbiome data between laboratories and studies. Even though new sequencing technologies have been applied for microbiome analysis in numerous large-scale and seminal studies for over a decade, initiatives to establish harmonized and quality-controlled protocols are still in their infancy. However, two large, international research group consortia, the International Human Microbiome Standards (IHMS) project (Costea et al. 2017) and the Microbiome Quality Control (MBQC) project (Sinha et al. 2017), are working toward large-scale benchmarking of protocols and encouraging the development of standard operating procedures.

Over the past decade, considerable efforts have been made to identify major methodological confounders throughout the entire process of metagenomics, starting with sampling through DNA extraction, PCR amplification, and DNA sequencing to bioinformatic analysis (Fig. 19.1). The impact of each intermediate step on the resulting microbiota profiles should be carefully examined when comparing data from different laboratories and studies. Furthermore, the subsequent deposition of methodological metadata, along with clinical information and sequencing data, into public meta-databases is necessary to increase, or even allow, the overall comparability of microbiome data. The availability and usage of such databases are at best rudimentary at the moment. Large sequence read databases, such as the Sequence Read Archive (SRA) (Leinonen et al. 2011), do not provide adequate and easy-to-use solutions for the deposition of metadata along with metagenomic reads. However, new microbiome-focused databases such as QIITA (https://qiita.ucsd.edu), MetaMetaDB (Yang and Iwasaki 2014), or IMNGS (Lagkouvardos et al. 2016) are pursuing more functional approaches.

19.3 Considerations for the Development of Standard Operating Procedures for Microbiome Analyses

One particularly critical but frequently disregarded factor is the preservation of the original microbial patterns immediately after sample

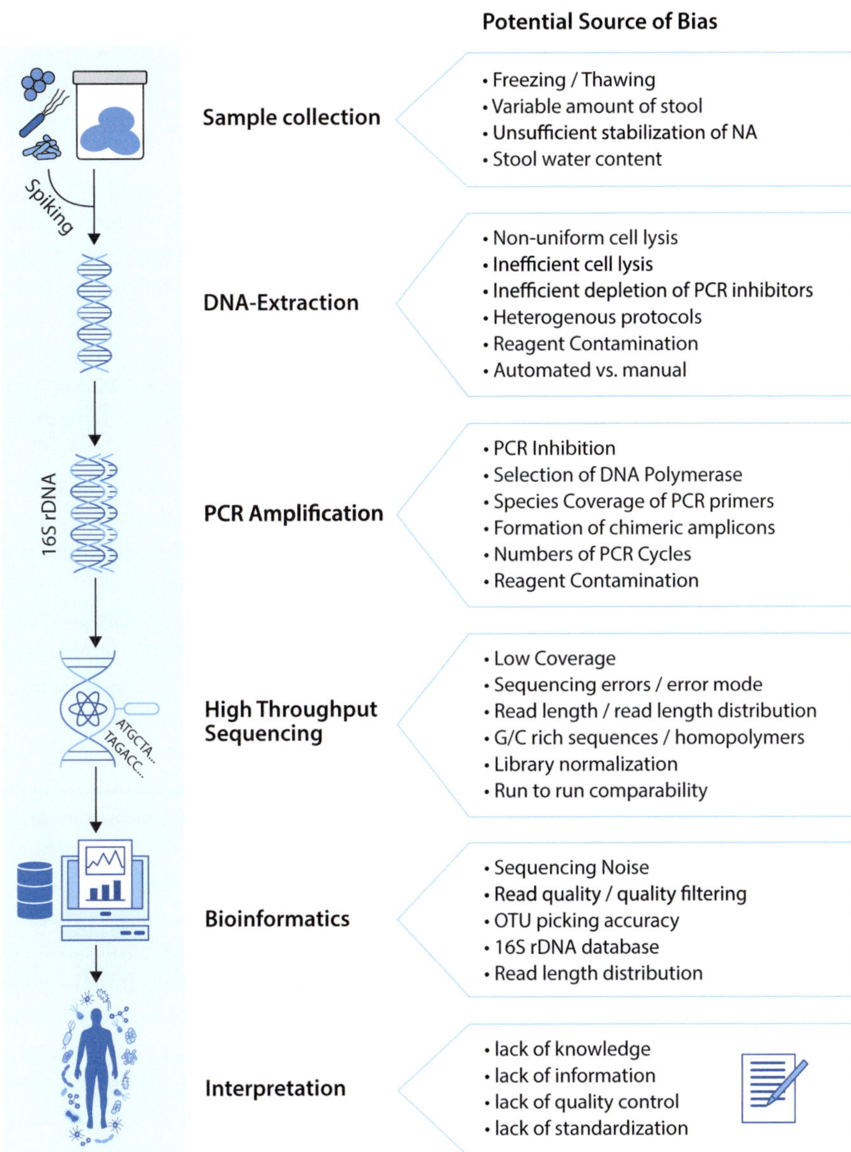

Fig. 19.1 Methodological pitfalls in marker gene-based microbiome analyses

collection throughout the entire period of storage. While the water content of stool samples or homogenization of fecal samples prior to DNA extraction only has minor effects on the bacterial composition (Santiago et al. 2014), other very significant effects of sampling and storage conditions have been observed. In general, sample handling is more manageable in animal studies, but the application in clinical practice requires tailored and easy-to-use solutions to enable the collection of a fixed quantity of sample material, as well as the stabilization of microbial nucleic acids at room temperature. Overgrowth of a large majority of obligate anaerobic bacteria in stool samples by rapidly growing facultative anaerobic bacteria, due to the availability of oxygen or the degradation of nucleic acids by nuclease activity or autolytic processes, is the major factor destroying the original microbial structure in specimens being examined. A wide range of methods have been applied to overcome these issues. However, differences in microbiota profiles applying different storage temperatures (Choo et al. 2015) or using preservation substances like 95 % ethanol or RNAlater® have been reported (Franzosa et al. 2014). In a comprehensive study by Song and coworkers, the effects of storage conditions in a total of 1200 human and dog fecal specimens were compared using five different methods. Samples were stored for 8 weeks, and additional variations like freeze-thaw cycles or temperature shifts were applied. The use of 95% ethanol, FTA cards, and a commercial gut microbiome sampling kit proved to sufficiently preserve microbiome profiles (Song et al. 2016). In general, liquid preservation solutions allow for better standardization in terms of constant sample amounts by using well-developed collection devices and maintaining temperature control during sampling by enabling storage at ambient temperature for several days.

The human intestinal microbiota is comprised of a large fraction of Gram-positive as well as endospore-forming and lysis-resistant bacteria (Kearney et al. 2017). One of the most difficult challenges for the methodological standardization of metagenomic studies is the uniform and effective extraction of nucleic acids from complex microbial mixtures, with the aim of quantitatively preserving the original microbial composition and effectively purifying nucleic acids from complex sample matrices and PCR inhibitors like bile, polyphenols, urate, or hemoglobin. In addition to these inherent difficulties, the wide variety of different protocols available, ranging from manual to automatic purification, the numerous options of pretreatment steps from enzymatic to mechanical cell lysis, and the difficulties in accurately assessing homogeneity and performance of protocols all aggravate the development of standardized practices. The selection of a particular DNA extraction protocol impacts the extent of DNA fragmentation, the yield, the purity, as well as the ratio between Gram-positive and Gram-negative bacteria and alpha diversity. The International Human Microbiome Standards (IHMS) project (Costea et al. 2017) is perhaps the most comprehensive evaluation of DNA extraction protocols to date. They systematically compared the reproducibility as well as interlaboratory variations of the most commonly used DNA extraction protocols. A total of 21 protocols were tested after applying different experimental modifications, and shotgun metagenomic sequencing was performed to obtain the diversity and distance measures. Substantial technical variation was observed between the various protocols, and factors primarily influencing the outcome were DNA fragmentation, purity, yield, and ratio of Gram-positive to Gram-negative bacteria and alpha diversity.

Consistent with previous results (Santiago et al. 2014), mechanical pretreatment using repeated bead beating was necessary to efficiently lyse Gram-positive bacteria. Furthermore, the higher abundance of Gram-positives was correlated with an overall higher diversity as measured by calculating the Shannon diversity index. The suggested protocol for DNA extraction along with standard operating procedures for sampling and storage procedures can be obtained from http://www.microbiome-standards.org.

In marker gene-based approaches, the PCR amplification step is often considered to be the primary source of error when compared to whole shotgun metagenomics protocols due to the amplification of alterations introduced in previous steps. The choice of DNA polymerase with regard to proofreading activity and processivity, template concentration, and the number of PCR cycles has an effect on the error rate and formation of chimeric DNA molecules during amplification, which both artificially inflate alpha diversity (Gohl et al. 2016). Analyzing bacterial mock communities, Bahl et al. (2017) found that genomic GC% content was negatively correlated with read abundances of "high GC" *Proteobacteria*, in contrast to "low GC" *Firmicutes*, which were mostly overestimated. Increasing the initial denaturation elevated the read abundances of bacteria with high genomic GC% content without influencing the overall species evenness (Bahl et al. 2017).

Beyond this, the choice of conserved 16S rRNA gene-specific primers was found to have a notable influence on the microbiota profiles, which was mainly due to the varying species coverage of universal primers (Hiergeist et al. 2016). It is well known that published universal PCR primers differ considerably regarding coverage of bacterial and archaeal species and specificity (Klindworth et al. 2013). As an example, the highly published and widely used universal 16S rRNA gene primer 27f is missing several *Bifidobacterium* and other bacterial species from the phylum *Actinobacteria* (Walker et al. 2015). Therefore, the performance characteristics of PCR primers as well as PCR conditions should be carefully evaluated in silico and by analyzing bacterial mock communities (Fouhy et al. 2016). This also applies to various strategies for the generation of 16S rRNA amplicon sequencing libraries, since different indexing strategies using the Illumina MiSeq platform were found to have an influence on the final microbiota profiles (Jones et al. 2015; Raju et al. 2018).

In terms of sequencing technology, the choice of HTS platform, which is closely interlinked with and interdependent of read length, sequencing depth, and error profiles, is the determining factor for the resulting microbiome profiles. With the Illumina MiSeq and the IonTorrent PGM and S5 platforms, the field of 16S rRNA gene-based microbiome sequencing is currently dominated primarily by short-read technologies. Depending on the composition of the analyzed samples and the choice of universal 16S rRNA gene primers, these technologies, in most instances, fail to identify bacteria on a species level. Read length and the coverage of hypervariable regions of the 16S rRNA gene significantly influence the reliable taxonomic identification of operational taxonomic units to the species level (Hiergeist et al. 2015). Strain- or even species-specific differences mainly remain undetected using marker gene-based approaches, and differentially changed profiles of microorganisms identified in microbiome studies often reflect mixed effects on a genus or family level. Sequencing of alternative targets like recombinase or gyrase genes may be helpful in achieving more discriminatory power for desired species, but the existence of well-curated sequence databases is limited. On the other hand, whole shotgun metagenome-based approaches have the potential to identify microorganisms at the strain level by the assembly of whole genomes from complex metagenomics samples (Nielsen et al. 2014). However, the hurdles yet to overcome in bioinformatic analysis are enormous, and the existence of reference databases containing well-curated non-draft whole genome sequences is limited.

Alternative protocols have been developed to help overcome read length limitations in 16S rRNA gene-based approaches. Near full-length 16S rRNA gene reads can be generated on the MiSeq platform using randomly barcoded primers and a tagmentation-based library construction using the Nextera® Tn5 transposase. After sequencing the tagmented DNA fragments, reads are grouped according to their random tags and assembled (Burke and Darling 2016). Simultaneous sequencing of various separate amplicons covering multiple variable regions may increase the chance of allocating an ID for species, which were indistinguishable using a single amplicon approach (Barb et al. 2016). Such multiple amplicon approaches are done at the expense of sequencing depth, and their suitability for analyzing large numbers of complex samples still needs to be tested.

Moreover, the applicability of new long-read technologies for generating near full-length 16S rRNA gene reads has been examined. Single-molecule real-time (SMRT) sequencing (Wagner et al. 2016) and nanopore sequencing technologies (Benitez-Paez et al. 2015) have already successfully been used to sequence full-length 16S rRNA gene reads or even near full-length ribosomal operons (Benitez-Paez et al. 2015) using bacterial mock communities. However, high error rates and low sequencing depth are still major obstacles to successfully applying these strategies for species-level characterization of complex microbial communities, although full-length reads are required for adequate discriminatory power on the species level. In addition, currently existing 16S rRNA reference databases are biased since they contain a large fraction of partial sequences, which are truncated at the 3′ or 5′ ends. This impedes the precise evaluation of primers amplifying full-length 16S rRNA genes. Furthermore, taxonomic misclassification or inconsistencies between databases also complicate the comparison of microbiota profiles.

While reagent contamination (e.g., "kitome") does not significantly interfere with samples containing high microbial loads, such as fecal samples (Velásquez-Mejía et al. 2018), low microbial biomass samples can be prone to environmental contamination of reagents and sample material (Kim et al. 2017), which can affect biological conclusions. Sources of contamination can originate from DNA contamination of kits and reagents, as well as from the mouth and skin of the investigators. Therefore, the integration of contamination controls into the analysis workflow for every sequencing run, together with batch release testing of kits and reagents, is strongly recommended. The establishment of a three-room or modified two-room concept of a molecular diagnostics lab design can greatly reduce the risk of contamination.

The choice of bioinformatic solutions for the analysis of DNA sequences generated by HTS platforms also represents a major source of variation. Today, a wide variety of open-source tools and algorithms exist for quality control of sequencing data, OTU clustering, taxonomic assignment, sequence alignment, and building of phylogenetic trees. Furthermore, statistical characterization of species distributions, differential abundances, or co-abundances of microbial networks are diverse, and general procedures like the use of exact sequence variants instead of OTUs (Callahan et al. 2017) are in discussion. Although effective new tools for handling compositional datasets and the implementation of stable algorithms have yet to be developed, it is also critical that they be carefully selected based on the exact nature of the datasets to be analyzed.

With respect to implementing an efficient quality management system, updates of 16S rRNA databases with taxonomic renaming or reclassification of bacteria, or the introduction of newly discovered species, critically affect the comparability of datasets. Changes have to be carefully validated by reanalyzing and comparing selected reference samples, or via longitudinal studies, with up-to-date databases. This also applies to software package updates and tools for bioinformatic analyses, which have to be monitored and validated, and can be error-prone and time-consuming. Integration of software solutions, which includes an automated reanalysis of datasets with up-to-date databases and taxonomies, is critical in accredited laboratories to enable direct comparability of datasets before making clinical or therapeutic decisions.

19.4 The Use of Spike-In Controls as Internal Process and Quality Controls

International standards and management system requirements for accreditation of clinical and medical laboratories are defined in ISO 15189 (medical laboratories: particular requirements for quality and competence), with the goal of maintaining standardized outcomes from testing biological samples which are suitable for clinical decision-making with respect to diagnosis, treatment, and prevention of diseases. The implementation of an accredited quality management system ensures the analytical validity of the entire process from sample arrival to clinical interpretation and includes validation and periodical verification of laboratory protocols as well as external quality assessment and testing of reference materials.

As discussed in the previous section, the analysis of the human microbiota using high-throughput sequencing technologies also contains potential sources of bias throughout the intermediate steps. Standardization of these processes is challenging, since methodological deviations have to be tracked to ensure sustained quality and unaffected microbiota profiles. Thus, the implementation of appropriate quality controls is a prerequisite. Internal controls consisting of whole bacterial or fungal cells should be favored over genomic DNA or the use of synthetic spike-in controls (Schirmer et al. 2015) to allow the examination of bias introduced by cell lysis and DNA extraction protocols. Cells are quantitatively spiked into defined amounts of the original

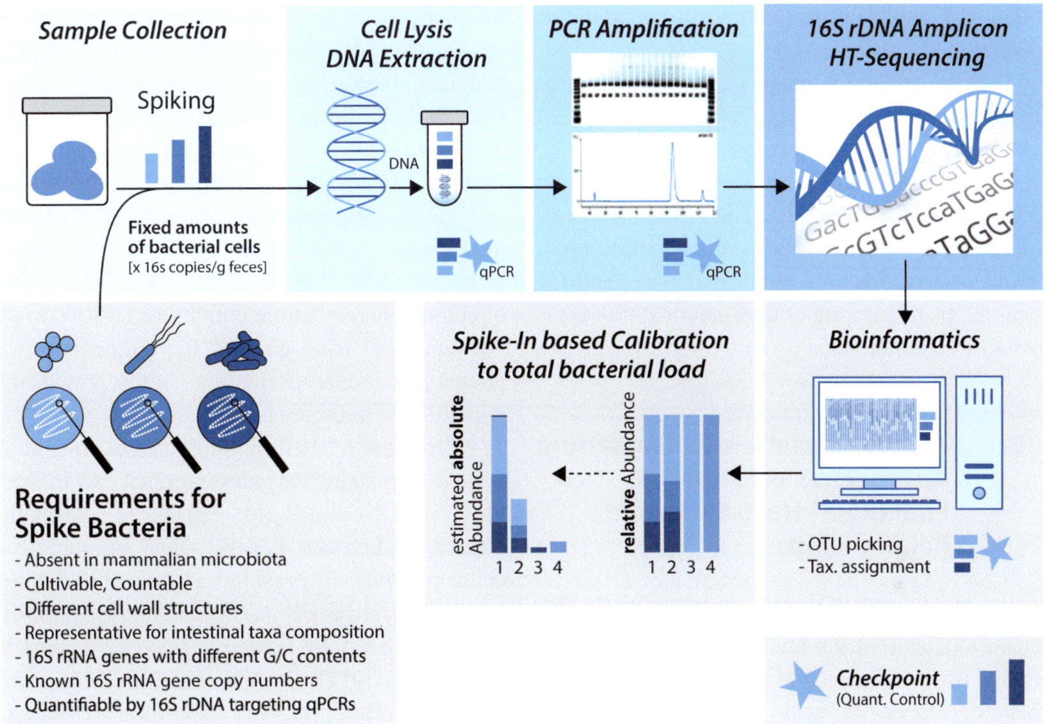

Fig. 19.2 Use of spike-in controls for 16S rRNA gene-based microbiome analyses

sample matrix and are passed through the entire process from DNA extraction, PCR amplification, and high-throughput sequencing to bioinformatic sequence analysis. This facilitates the tracking of potential method-specific errors after every single protocol step and can be easily assayed by validated specific qPCR protocols (Fig. 19.2). The selection of bacterial species suitable for process control of microbiome analyses should meet several requirements. They have to be absent and clearly distinguishable from the mammalian microbiota and should be cultivable in selected media under laboratory conditions. Cell counts should be measurable by microscopy or automated cell counters to be able to spike quantified amounts of cells. It is also advisable to select bacteria with different cell wall structures including Gram-positive, Gram-negative, or endospore-forming bacteria to map different lysis efficiencies of DNA extraction protocols. Taxonomic classification of selected bacteria should be representative of the taxonomic distribution of the analyzed body habitat and genomic GC% content, plus the genomic 16S rRNA gene copy numbers should be known. These copy numbers should be quantifiable by species-specific qPCR assays, which can then be applied to quantify genome equivalents and detect deviations from required specifications after DNA extraction and PCR amplification. After DNA sequencing of samples, taxonomic assignment of clustered operational taxonomic units (OTUs) can then be used to compare their relative read abundances to relative qPCR quantities of total bacterial and spike-in 16S rRNA gene copies, as well as for validation of accuracy of species- or genus-level identification.

As a further aspect in this context, using spike-in controls can also be applied for estimating absolute abundances of identified OTUs in complex microbiome datasets by spike-in-based calibration to microbial load (Stämmler et al. 2016). This normalization procedure circumvents common difficulties that may arise when interpreting compositional changes in the microbiome data (Gloor et al. 2017), since common normalization strategies and differential abundance analyses are often not applicable to compositional datasets (Weiss et al. 2017).

Due to methodological variations and high interlaboratory variability, standardization of microbiome analysis by the development of one generally accepted universal standard operating procedure is very idealistic. Therefore, the implementation of well-designed controls, definition of stringent specifications for their release, and the continuous monitoring of the analytical process is even more important.

19.5 From Association to Causation: Approaches to Study Functional Host-Microbiota Relationships

Metagenomics provide a powerful tool for the characterization of the human microbiome which enable the identification of disease-specific microbial signatures, as well as the possibility of microbiome monitoring after therapeutic interventions. Because of inherent methodological limitations of metagenome-based methods for community profiling as described in the previous sections, the development of new strategies for moving beyond correlations toward a mechanistic understanding of host-microbiota interactions and the identification of causal microorganisms or consortia is essential. In recent years, in vivo, in vitro, in silico, and ex vivo approaches to studying interactions between the host and microbiota have been developed. Furthermore, integration of additional "omics" data like metabolomics, whole shotgun metagenomics, metaproteomics, or metatranscriptomics may help to further refine disease associations derived from differential abundances to help identify potential mechanistic host-microbiota relationships. However, these methods often only generate hypotheses, which should be evaluated and validated by in vitro co-culture systems or in vivo animal models such as gnotobiotic and germ-free mice.

The extrapolation of the original Koch's postulates to commensals and the identification of beneficial bacteria have been suggested (Neville et al. 2018). This means that the microorganism has to be associated with host health, can be isolated in pure culture, should alleviate disease when introduced into a new host, and can be re-isolated from the new host. Therefore, cultivation-based techniques play an important role in the validation and investigation of mechanistic models of microbiota-host interactions. New advances in cultivation techniques by highly parallel testing of thousands of culture and co-culture conditions referred to as "culturomics" have enabled the cultivation and phenotypic characterization of previously unculturable microorganisms (Lagier et al. 2012).

Micro-fluidic-based high-throughput co-culture systems mimicking the gastrointestinal tract in vitro have been invented to investigate metabolic interactions between the microbiota and intestinal epithelial cells. The *simulated human intestinal microbial ecosystem (SHIME) reactor* (Molly et al. 1993) and the *human gut-on-chip* approach (Kim et al. 2012) are two examples of in vitro approaches. However, their application is complex, and results may be difficult to transfer from these strongly simplified models to humans. This also applies to other approaches. For example, microbial shifts associated with human disease from animal models to human are often nontransferable (Nguyen et al. 2015). However, germ-free or gnotobiotic animal models provide powerful tools for testing hypotheses derived from HTS-based association studies and advance our understanding of functional host-microbiota relationships.

In a pioneering study, a novel strategy referred to as "triangulation of microbe-phenotype relationships" was applied to identify bacteria which were causally related to an observed phenotype (Surana and Kasper 2017). The fecal microbiota of MMb mice, which are sensitive to the manifestation of fatal colitis after DSS induction, were compared to the microbiota of HMb mice, who rarely died and lost significantly less weight after DSS administration. A hybrid microbiota of these two groups was characterized by HTS, which suggested a transfer of microbiota between these phenotypes. Cohousing resulted in an intermediate phenotype. Comparisons of microbiome profiles at the shortest time point after phenotypic changes were detectable after cohousing with non-cohoused mice were used to refine differentially abundant taxa between the two groups. In this way, members of the *Lachnospiraceae* family were found to be significantly correlated to protection from colitis. In a culture approach, this protective effect could be confirmed by the identification of a novel bacterial species: *Clostridium immunis*. This species exhibited protection of colitis-related death after administration to colitis-prone mice. In general, this method should be applicable to other animal or human microbiome studies and impressively describes a promising tool for refining differentially abundant patterns from association studies to promote the identification of causal microorganisms.

19.6 Clinical Implementation

In summary, HTS-based techniques for the characterization of the human microbiota have the potential to be used in clinical practice for the identification of disease-specific signatures in patient microbiota, as well as for monitoring responses to therapeutic interventions (Fig. 19.3). The valid interpretation of microbial patterns in the context of a dynamically changing ecosystem inhabited by an individually shaped microbiota requires careful integration of clinical metadata with microbiome signatures. Potentially confounding factors such as medication, nutritional information, or methodological aspects such as the type of PCR primers used, or the variable 16S rRNA gene regions covered, all need to be taken into consideration when deriving medical reports from microbiome sequencing data. Also, the most current valid information from cross-sectional, case-control, and longitudinal clinical studies has to be included in curated meta-databases to take advantage of up-to-date advances in microbiome science in the interpretation of microbiota profiles.

In a future clinical setting, transitions of microbial communities from a stable "symbiotic" state to an unstable "dysbiotic" state, or vice versa, can be monitored by longitudinal HTS approaches

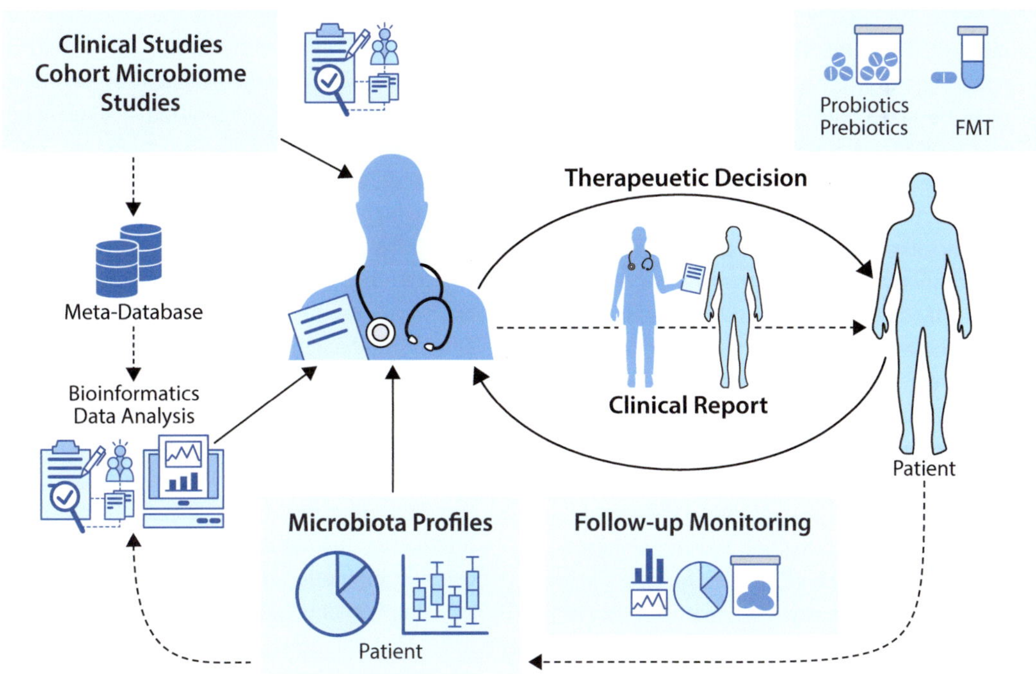

Fig. 19.3 Clinical implementation of HTS-based microbiome analyses and their use in monitoring therapeutic responses

through the assessment of follow-up samples. Therapeutic decisions to maintain a diverse and healthy microbiota can lead to fecal microbiota transplantation or the gavage of defined microbial consortia of beneficial microbes, as well as the dietary modulation by probiotics or prebiotics. These therapeutic interventions should be monitored by follow-up samples.

Controversy

Flooding the current scientific literature, a host of existing case-control studies have identified associations between differentially altered microbiota and a large number of diseases. Taking place on a purely descriptive level, direct causal relationships have only been shown in a small number of selected cases after interdisciplinary integration of sequencing and metabolite datasets, as well as by validation of findings in germ-free and gnotobiotic animal models. Overall, this constitutes a risk for over interpretation and drawing of erroneous and premature conclusions from association studies, which leads to the generation of incorrect diagnostic findings for poorly characterized disease patterns. Thus, the step toward the clinical application of HTS-based microbiome analysis must be taken with caution. Methodological bias leads to a lack of standardization, and the influence of environmental or host factors on the structural and functional composition of the microbiome is not yet well understood.

Without a doubt, major advances in the understanding of microbiota-associated pathologies, as well as promising treatment outcomes for hitherto therapeutically challenging diseases after application of fecal microbiota transfer, are understandably injecting new hopes in the field. Monitoring of responses to these therapeutic interventions is a promising starting point for the implementation of HTS techniques into clinical diagnostics. However, methods should not be implemented into the clinical diagnostic routine without meeting the requirements for methodological standardization and identification of possible risks.

History

Since the isolation and scientific characterization of the first bacterium from the human gastrointestinal tract by Theodor Escherich in 1886, characterization of the gastrointestinal microbiota has been dominated by adapted culture techniques, especially after the revolutionary inventions of anaerobic culture techniques by Robert Hungate. Even if laboratory cultivation techniques have been constantly refined, molecular-based methods like PCR, invention of “first-generation” DNA sequencing techniques by Frederik Sanger in the late 1970s, and the proposal of using 16S rRNA genes in bacterial phylogenetic by Carl Woese in 1990 have facilitated a breakthrough in the comprehensive molecular characterization of microorganisms. These methods have thus far shaped biological and medical science as well as the field of microbiological routine diagnostics, enabling both the detection and molecular characterization of human pathogens. The market launch of so-called second- or next-generation sequencing platforms by 454 Life Sciences in 2005 represented a quantum leap in molecular biology, which allowed culture-independent characterization of complex microbial communities for the first time. Since then, enormous progress has been made toward a better understanding of the mutualistic relationships between the human microbiota and its host and their role in a wide range of pathologies. Continuous multidisciplinary developments in HTS platforms, mass spectrometry for the characterization of host-interacting metabolites, as well as sophisticated computational tools for the integration of complex datasets is an ongoing process.

Notably, the advent of new fourth-generation nanopore-based sequencing platforms allows longer read length and finer resolution of microbiota profiles in the near future.

Highlights

- High-throughput sequencing has broken new ground in characterizing the human microbiota, independent of cultivation, and is finding a broad application in the medical research of host-microbiota associations in human pathologies.
- Methodological bias can be introduced during individual stages of the implemented laboratory protocols ranging from sampling, DNA extraction, PCR amplification, and library generation to high-throughput sequencing and bioinformatic analysis and is impeding the overall comparability of microbiota studies.
- The use of internal controls such as spike-in bacteria can support monitoring and standardization of HTS-based microbiome analyses.
- Clinical application of microbiome sequencing is aggravated by large inter-individual differences and multiple confounding environmental or host factors.
- Further validation studies and new methods are needed to advance from correlation to causation in microbiota-host interaction studies.

References

Bahl, M. I., Laursen, M. F., & Dalgaard, M. D. (2017). Genomic GC-content affects the accuracy of 16S rRNA gene sequencing based microbial profiling due to PCR bias. *Frontiers in Microbiology, 8*, 1–8.

Barb, J. J., Oler, A. J., Kim, H. S., Chalmers, N., Wallen, G. R., Cashion, A., et al. (2016). Development of an analysis pipeline characterizing multiple hypervariable regions of 16S rRNA using mock samples. *PLoS One, 11*, 1–18.

Benitez-Paez, A., Portune, K., & Sanz, Y. (2015). Species-level resolution of 16S rRNA gene amplicons sequenced through MinIONTM portable nanopore sequencer. *Gigascience, 5*, 4.

Burke, C. M., & Darling, A. E. (2016). A method for high precision sequencing of near full-length 16S rRNA genes on an Illumina MiSeq. *PeerJ, 4*, e2492.

Callahan, B. J., McMurdie, P. J., & Holmes, S. P. (2017). Exact sequence variants should replace operational taxonomic units in marker-gene data analysis. *The ISME Journal, 11*, 2639–2643.

Choo, J. M., Leong, L. E. X., & Rogers, G. B. (2015). Sample storage conditions significantly influence faecal microbiome profiles. *Scientific Reports, 5*, 16350.

Costea, P. I., Zeller, G., Sunagawa, S., Pelletier, E., Alberti, A., Levenez, F., et al. (2017). Towards standards for human fecal sample processing in metagenomic studies. Nat. *Biotechnology, 35*, 1069–1076.

Davido, B., Batista, R., Michelon, H., Lepainteur, M., Bouchand, F., Lepeule, R., et al. (2017). Is faecal microbiota transplantation an option to eradicate highly drug-resistant enteric bacteria carriage? *The Journal of Hospital Infection, 95*, 433–437.

Donaldson, G. P., Lee, S. M., & Mazmanian, S. K. (2015). Gut biogeography of the bacterial microbiota. *Nature Reviews. Microbiology, 14*, 20–32.

Duvallet, C., Gibbons, S. M., Gurry, T., Irizarry, R. A., & Alm, E. J. (2017). Meta-analysis of gut microbiome studies identifies diseasespecific and shared responses. *Nature Communications, 8*, 1784.

Forslund, K., Hildebrand, F., Nielsen, T., Falony, G., Le Chatelier, E., Sunagawa, S., et al. (2015). Disentangling type 2 diabetes and metformin treatment signatures in the human gut microbiota. *Nature, 528*, 262–266.

Fouhy, F., Clooney, A. G., Stanton, C., Claesson, M. J., & Cotter, P. D. (2016). 16S rRNA gene sequencing of mock microbial populations- impact of DNA extraction method, primer choice and sequencing platform. *BMC Microbiology, 16*, 123.

Franzosa, E. A., Morgan, X. C., Segata, N., Waldron, L., Reyes, J., Earl, A. M., et al. (2014). Relating the metatranscriptome and metagenome of the human gut. *Proceedings of the National Academy of Sciences of the United States of America, 111*, E2329–E2338.

Gagan, J., & Van Allen, E. M. (2015). Next-generation sequencing to guide cancer therapy. *Genome Medicine, 7*, 80.

Gloor, G. B., Macklaim, J. M., Pawlowsky-Glahn, V., & Egozcue, J. J. (2017). Microbiome datasets are compositional: And this is not optional. *Frontiers in Microbiology, 8*, 2224.

Gohl, D. M., Vangay, P., Garbe, J., MacLean, A., Hauge, A., Becker, A., et al. (2016). Systematic improvement of amplicon marker gene methods for increased accuracy in microbiome studies. *Nature Biotechnology, 34*, 942–949.

Hiergeist, A., Gläsner, J., Reischl, U., & Gessner, A. (2015). Analyses of intestinal microbiota: Culture versus sequencing. *ILAR Journal, 56*, 228–240.

Hiergeist, A., Reischl, U., Priority Program 1656 Intestinal Microbiota Consortium/Quality Assessment Participants, & Gessner, A. (2016). Multicenter quality assessment of 16S ribosomal DNA-sequencing for microbiome analyses reveals high inter-center variability. *International Journal of Medical Microbiology, 306*, 334–342.

Jones, M. B., Highlander, S. K., Anderson, E. L., Li, W., Dayrit, M., Klitgord, N., et al. (2015). Library preparation methodology can influence genomic and functional predictions in human microbiome research. *Proceedings of the National Academy of Sciences of the United States of America, 112*, 14024–14029.

Kearney, S. M., Gibbons, S. M., Poyet, M., Gurry, T., Bullock, K., Allegretti, J., et al. (2017). Endospores and other lysisresistant bacteria comprise a widely shared core community within the human microbiota. *BioRxiv*, 221713.

Kim, H. J., Huh, D., Hamilton, G., & Ingber, D. E. (2012). Human gut-on-a-chip inhabited by microbial flora that experiences intestinal peristalsis-like motions and flow. *Lab on a Chip, 12*, 2165.

Kim, D., Hofstaedter, C. E., Zhao, C., Mattei, L., Tanes, C., Clarke, E., et al. (2017). Optimizing methods and dodging pitfalls in microbiome research. *Microbiome, 5*, 52.

Klindworth, A., Pruesse, E., Schweer, T., Peplies, J., Quast, C., Horn, M., et al. (2013). Evaluation of general 16S ribosomal RNA gene PCR primers for classical and next-generation sequencing-based diversity studies. *Nucleic Acids Research, 41*, e1.

Koppel, N., Maini Rekdal, V., & Balskus, E. P. (2017). Chemical transformation of xenobiotics by the human gut microbiota. *Science, 356*, eaag2770.

Lagier, J. C., Armougom, F., Million, M., Hugon, P., Pagnier, I., Robert, C., et al. (2012). Microbial culturomics: Paradigm shift in the human gut microbiome study. *Clinical Microbiology and Infection, 18*, 1185–1193.

Lagkouvardos, I., Joseph, D., Kapfhammer, M., Giritli, S., Horn, M., Haller, D., et al. (2016). IMNGS: A comprehensive open resource of processed 16S rRNA microbial profiles for ecology and diversity studies. *Scientific Reports, 6*, 33721.

Leclercq, S., Mian, F. M., Stanisz, A. M., Bindels, L. B., Cambier, E., Ben-Amram, H., et al. (2017). Low-dose penicillin in early life induces long-term changes in murine gut microbiota, brain cytokines and behavior. *Nature Communications, 8*, 15062.

Lee, S. T. M., Kahn, S. A., Delmont, T. O., Shaiber, A., Esen, ö. C., Hubert, N. A., et al. (2017). Tracking microbial colonization in fecal microbiota transplantation experiments via genome-resolved metagenomics. *Microbiome, 5*.

Leinonen, R., Sugawara, H., & Shumway, M. (2011). The sequence read archive. *Nucleic Acids Research, 39*.

Maurice, C. F., Haiser, H. J., & Turnbaugh, P. J. (2013). Xenobiotics shape the physiology and gene expression of the active human gut microbiome. *Cell, 152*, 39–50.

Molly, K., Vande Woestyne, M., & Verstraete, W. (1993). Development of a 5-step multi-chamber reactor as a simulation of the human intestinal microbial ecosystem. *Applied Microbiology and Biotechnology, 39*, 254–258.

Neville, B. A., Forster, S. C., & Lawley, T. D. (2018). Commensal Koch's postulates: Establishing causation in human microbiota research. *Current Opinion in Microbiology, 42*, 47–52.

Nguyen, T. L. A., Vieira-Silva, S., Liston, A., & Raes, J. (2015). How informative is the mouse for human gut microbiota research? *Disease Models and Mechanisms, 8*, 1–16.

Nielsen, H. B., Almeida, M., Juncker, A. S., Rasmussen, S., Li, J., Sunagawa, S., et al. (2014). Identification and assembly of genomes and genetic elements in complex metagenomic samples without using reference genomes. *Nature Biotechnology, 32*, 822–828.

Raju, S., Ellonen, P., De Vos, W. M., Eriksson, J. G., Weiderpass, E., Rounge, T. B., et al. (2018). Reproducibility and repeatability of six high-throughput 16S rDNA sequencing protocols for microbiota profiling. *J Microbiol Methods, 147*, 76–86.

Santiago, A., Panda, S., Mengels, G., Martinez, X., Azpiroz, F., Dore, J., et al. (2014). Processing faecal samples: A step forward for standards in microbial community analysis. *BMC Microbiology, 14*, 112.

Schirmer, M., Ijaz, U. Z., D'Amore, R., Hall, N., Sloan, W. T., & Quince, C. (2015). Insight into biases and sequencing errors for amplicon sequencing with the Illumina MiSeq platform. *Nucleic Acids Research, 43*, e37.

Sinha, R., Abu-Ali, G., Vogtmann, E., Fodor, A.A., Ren, B., Amir, A., et al., Microbiome Quality Control Project Consortium. (2017). Assessment of variation in microbial community amplicon sequencing by the Microbiome Quality Control (MBQC) project consortium. *Nature Biotechnology, 35*, 1077–1086.

Song, S. J., Amir, A., Metcalf, J. L., Amato, K. R., Xu, Z. Z., Humphrey, G., et al. (2016). Preservation methods differ in fecal microbiome stability, affecting suitability for field studies. *mSystems, 1*, e00021-16.

Stämmler, F., Gläsner, J., Hiergeist, A., Holler, E., Weber, D., Oefner, P. J., et al. (2016). Adjusting microbiome profiles for differences in microbial load by spike-in bacteria. *Microbiome, 4*, 28.

Surana, N. K., & Kasper, D. L. (2017). Moving beyond microbiome-wide associations to causal microbe identification. *Nature, 552*(7684), 244–247.

Taur, Y., & Pamer, E. G. (2014). Harnessing microbiota to kill a pathogen: Fixing the microbiota to treat Clostridium difficile infections. *Nature Medicine, 20*, 246–247.

Thaiss, C. A., Levy, M., Korem, T., Dohnalová, L., Shapiro, H., Jaitin, D. A., et al. (2016). Microbiota diurnal rhythmicity programs host transcriptome oscillations. *Cell, 167*, 1495–1510.e12.

Velásquez-Mejía, E. P., de la Cuesta-Zuluaga, J., & Escobar, J. S. (2018). Impact of DNA extraction, sample dilution, and reagent contamination on 16S rRNA gene sequencing of human feces. *Applied Microbiology and Biotechnology, 102*, 403–411.

Wagner, J., Coupland, P., Browne, H. P., Lawley, T. D., Francis, S. C., & Parkhill, J. (2016). Evaluation of PacBio sequencing for full-length bacterial 16S rRNA gene classification. *BMC Microbiology, 16*, 1–17.

Walker, A. W., Martin, J. C., Scott, P., Parkhill, J., Flint, H. J., & Scott, K. P. (2015). 16S rRNA gene-based profiling of the human infant gut microbiota is strongly influenced by sample processing and PCR primer choice. *Microbiome, 3*, 26.

Wang, J., & Jia, H. (2016). Metagenome-wide association studies: Fine-mining the microbiome. *Nature Reviews. Microbiology, 14*, 508–522.

Weiss, S., Xu, Z. Z., Peddada, S., Amir, A., Bittinger, K., Gonzalez, A., et al. (2017). Normalization and microbial differential abundance strategies depend upon data characteristics. *Microbiome, 5*, 27.

Wilson, M. R., Suan, D., Duggins, A., Schubert, R. D., Khan, L. M., Sample, H. A., et al. (2017). A novel cause of chronic viral meningoencephalitis: Cache Valley virus. *Annals of Neurology, 82*, 105–114.

Yang, C., & Iwasaki, W. (2014). MetaMetaDB: A database and analytic system for investigating microbial habitability. *PLoS One, 9*, e87126.

Zoetendal, E. G., Von Wright, A., Vilpponen-Salmela, T., Ben-Amor, K., Akkermans, A. D. L., & De Vos, W. M. (2002). Mucosa-associated bacteria in the human gastrointestinal tract are uniformly distributed along the colon and differ from the community recovered from feces. *Applied and Environmental Microbiology, 68*, 3401–3407.

Fecal Transplantation

20

Franziska Schmidt and Britta Siegmund

Abstract

Fecal transplantation or fecal microbiota transplantation (FMT) represents a therapeutic approach that has been applied in early Chinese medicine for diarrhea and has only recently found the way into medicine. However, while it is an intriguing concept that a disease such as *Clostridium difficile*-associated colitis can be "cured" by the FMT with a transferred intestinal microbiota, it became at the same time apparent that several factors have to be considered. While the data for *C. difficile*-associated colitis are based on a placebo-controlled trial, the data for many other indications including inflammatory bowel diseases are less clear. Thus, there is the risk of transferring potential infectious disease as well as phenotypic properties such as obesity. Consequently, the donor screening has to be clearly defined. The present book chapter will summarize the development of the field over the last decade and will provide an outlook about possible innovations in the foreseeable future.

F. Schmidt · B. Siegmund (✉)
Medizinische Klinik für Gastroenterologie, Infektiologie, Rheumatologie, Charité – Universitätsmedizin Berlin, Campus Benjamin Franklin, Berlin, Germany
e-mail: britta.siegmund@charite.de

Transplantation has been subject of medical innovation for the last decades. However, the first thought of transplantation is mostly related to organ transplantation with all its opportunities and difficulties. Over the last decade, a novel "form" of transplantation has entered not only the medical field but even found entrance to medical guidelines, namely, fecal microbiota transplantation (FMT). One has to differ here two situations. In human, FMT from the donor into an already colonized host only redirects the colonization and composition of the intestinal microbiota, whereas in mice, the transfer is usually performed into germfree mice, where the transplant is in fact colonizing the host. Although, as outlined in the next paragraph, there were multiple reports in the literature in the past on the beneficial effects of FMT in patients, it never gained major attention in the medical community. However, the interest in the microbiota field did rise profoundly due to two developments: (1) the introduction of affordable high-throughput sequencing techniques and (2) the microbiome-disease association studies. The initial prominent one by Turnbaugh and colleagues was the transfer of feces from obese mice into germfree mice, which induced a significant weight increase in the germfree group (Turnbaugh et al. 2006). With this chapter, we aim to provide first a historic introduction to the field, followed by describing the data for pseudomembranous enterocolitis and ultimately

D. Haller (ed.), *The Gut Microbiome in Health and Disease*,
https://doi.org/10.1007/978-3-319-90545-7_20

novel areas of interest as well as fecal preparations and routes of application.

20.1 Historic View

When looking in depth in medical history, one has to admit that the field evolved as early as in the fourth century in China. At this time, a Chinese medicine doctor prescribed a suspension of human feces for patients with diarrhea. The suspension was administered orally. This strategy was considered a miracle since it rescued patients from death. This therapeutic strategy is first described in the Chinese handbook of emergency medicine, *Zhou Hou Bei Ji Fang* (Ge 2000; Zhang et al. 2012). In addition, in the sixteenth century, during the Ming dynasty, Li Shizhen prescribed feces in various forms (fresh suspension, dry feces, infant feces) as treatment against various gastrointestinal symptoms including severe diarrhea, vomiting, and constipation, all of which is included in the book of Chinese medicine *Ben Cao Gang Mu*. At this time it was labeled as "yellow soup" probably to achieve better acceptance (Li 2011; Zhang et al. 2012). In Europe there are early reports suggesting that this idea was probably introduced in veterinary medicine in the seventeenth century (Borody et al. 2004; Brandt et al. 2012). Next, during World War II, bacterial dysentery was treated with fresh and warm camel feces by German soldiers following recommendations of the Bedouins (Lewin 2001). Ultimately, the first "medical" FMT for pseudomembranous colitis was performed in 1958 by Dr. Eiseman in Denver, Colorado (Eiseman et al. 1958). The following decades this strategy entered medical therapy as third-line recommendation in *The Sanford Guide to Antimicrobial Therapy* (Gilbert et al. 2004). It was not until studies were performed during the last decade that FMT entered clinical medicine (Fig. 20.1). With this book chapter, the critical studies as well as the recent developments in the field will be summarized.

20.2 Classic Fecal Microbiota Transplantation in *Clostridium difficile* Colitis

Initially, *Clostridium difficile* infection (CDI), at this time named antibiotic-associated colitis, was established after the increasing use of antibiotics. Only in 1978, *C. difficile* could be identified as the causative pathogen (George et al. 1978a, b; Larson et al. 1978). The incidence of CDI has increased dramatically over the last decade, partly mediated by the emergence of hypervirulent strains (Surawicz and Alexander 2011). A retrospective review from Canada in 2000 indicated a fourfold rise in incidence since 1991, paralleled by a tenfold increase for individuals over 65 years of age (Miller et al. 2011) that was subsequently confirmed in other populations (Koo et al. 2014). The fecal-oral route of transmission in parallel to the resistance of the pathogen facilitates transmission in particular in hospitals; however this is beyond the scope of this chapter and is addressed in more detail elsewhere (Surawicz and Alexander 2011).

Of note, even before the identification of the pathogen, the first patient received a FMT for pseudomembranous colitis and could be cured (Eiseman et al. 1958). The term pseudomembranous colitis derives from the characteristic endoscopic finding of "pseudomembranes" by which the endoscopist can make the diagnosis by eye in the majority of cases (see Fig. 20.2).

In the mid-2000s, the field of microbiota experienced novel attention majorly driven by colonization experiments in mice that resulted in a change of phenotype (Turnbaugh et al. 2006). At the beginning of 2000, the first uncontrolled series of FMT in patients with CDI were reported (Aas et al. 2003). The concept is intriguing; antibiotic therapy and other intestinal diseases facilitate the emergence of CDI and result in a collapse of the microbiota composition and subsequently diarrhea and possibly a life-threatening condition. These case series indicated that by performing a FMT, intestinal functionality can be restored (Khoruts and Sadowsky 2011). The approach seemed to be

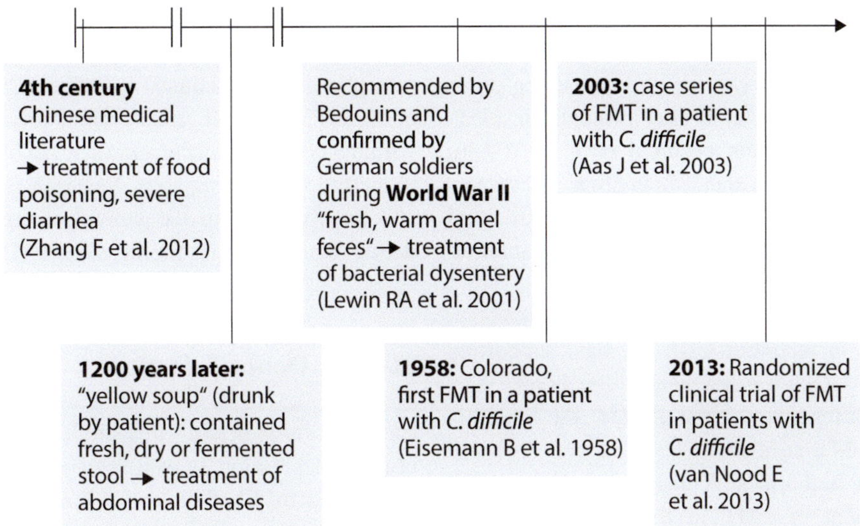

Fig. 20.1 Historic timeline of the development of fecal microbiota transplantation. *FMT* fecal microbiota transplantation; *C. difficile*, *Clostridium difficile*

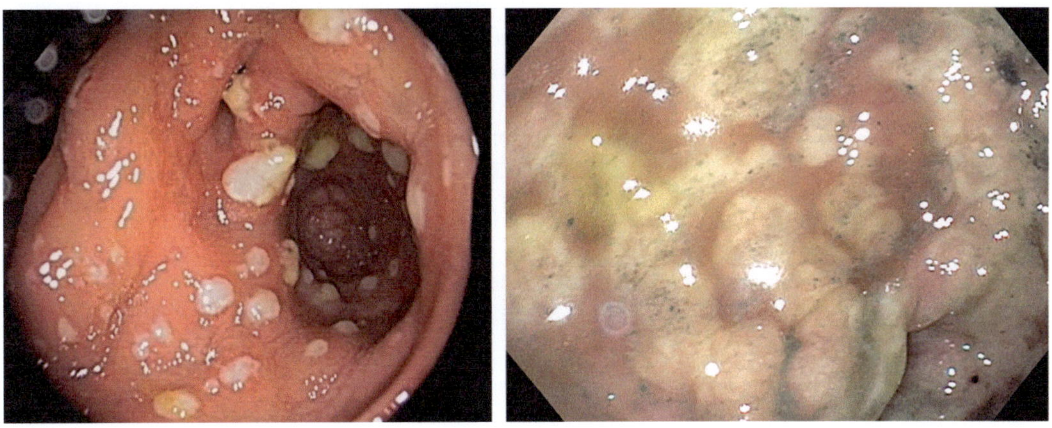

Fig. 20.2 Representative endoscopic picture of pseudomembranous colitis. Shown are representative endoscopic pictures with the characteristic white pseudomembranes. The left picture presents *Clostridium difficile* colitis with fewer pseudomembranes; the right picture represents full-blown disease. The endoscopic pictures were kindly provided by PD Dr. Christian Bojarski from our clinic

robust since it did not matter whether the FMT was performed via a nasoduodenal tube in the small intestine or endoscopically in the colon (Postigo and Kim 2012).

Finally, in 2013 the Amsterdam group presented data from the first controlled clinical trial in *The New England Journal of Medicine* (van Nood et al. 2013). The study included patients suffering from recurrent CDI and assigned them to one of the three groups:

1. An initial vancomycin regimen for 4 days, followed by bowel lavage and subsequent application of donor feces through a nasoduodenal tube
2. A standard vancomycin regimen for 14 days
3. A standard vancomycin regimen with bowel lavage

The primary endpoint was the resolution of diarrhea without a relapse after 10 weeks. The study had to be stopped after the interim analysis since 81% of group (1) had a resolution of CDI after a single infusion; two of the remaining three patients showed resolution after a second transplant. In contrast, resolution occurred only in 31% of patients in group (2) and 23% in group (3), respectively. Not surprisingly, resolution was associated with an increase in bacterial diversity comparable to healthy donors (van Nood et al. 2013).

While these data strongly support the indication for FMT in recurrent CDI, the data supporting refractory CDI are rather weak at this point.

From a regulatory point of view, the introduction of FMT seemed to be in a gray zone of regulatory authorities. A discussion of the initial development of FMT in the United States of America is reflected by two early publications in *Nature Biotechnology* (Ratner 2014; Smith et al. 2014). Today the United States represents one of the few countries with published guidelines for FMT and a central stool bank called OpenBiome; the respective forms are available at http://www.idsociety.org/FMT/. In Europe, until the beginning of 2017, the authoritative published guidelines and recommendations have been shared as expert opinions rather than expert-based consensus reports. This gap was filled by a recently published European consensus conference on microbiota transplantation in clinical practice (Cammarota et al. 2017). Several critical points are addressed within a consensus document that will be outlined in the following paragraphs. Nevertheless, there is still an open gap in the individual countries to define regulatory rules for the entire FMT process.

20.2.1 Donor Selection

Case reports indicating that properties including obesity, depression as well as cancer risk can be transferred by FMT underline the necessity that donor selection represents a critical step that warrants special selection. The consensus paper suggests a two-step process where the actual testing for risk factors is preceded by a simple interview (Cammarota et al. 2017). This interview addresses risk factors or known previous relevant infectious diseases as well as risk factors known to predispose for infectious diseases including use of illegal drugs, reception of blood products, body tattoo, healthcare worker, and many more (see Table 20.1). In the second step, the interview

Table 20.1 Step 1: Noninvasive donor evaluation [adapted from Cammarota et al. (2017)]

Area of question	Specific issues
Infectious diseases	– History of, exposure to: HIV, HBV, HCV, syphilis, HTLV I and II, malaria, trypanosomiasis, tuberculosis – Systemic infection not controlled at the time of donation – Illegal drugs, other risky behavior, recipient of transplant (ever) or blood product (<12 months), needle stick, body tattoo, piercing, earring, acupuncture (all <6 months), medical treatment in poorly hygienic conditions, risk of transmission of prions – Recent parasitosis or infection with GI involvement – Recent (<6 months) travel to tropical countries or countries at high risk of communicable disease or traveller's diarrhea – Recent (<6 months) vaccination with a live attenuated virus – Healthcare worker, individuals working with animals
GI, metabolic and neurological disorders	– History of chronic GI disorder (functional, inflammatory, malignant) or recent development of diarrhea, hematochezia – High risk of GI cancer, polyposis – History of neurological/neurodegenerative or psychiatric disorders – Body mass index > 25
Medication that can affect intestinal microbiota composition	– Exposure to antibiotics, immunosuppressants, chemotherapy (<3 months) – Chronic exposure to proton pump inhibitors

GI gastrointestinal tract; *HIV* human immunodeficiency virus; *HBV* hepatitis B virus; *HCV* hepatitis C virus; *HTLV* human T lymphotrophic virus

Table 20.2 Step 2: Donor testing, blood, and stool [adapted from Cammarota et al. (2017)]

Area of question	Specific issues
Blood	**Viruses**: Cytomegalovirus, Epstein-Barr virus, hepatitis A/B/C/E, HIV 1/-2, (specific situation: HTLV I/II) **Bacteria**: Syphilis (*Treponema pallidum*) **Parasites**: *Entamoeba histolytica*, (specific situation: *Strongyloides stercoralis*) **General**: Complete blood cell count (+ differential), C-reactive protein, ESR, albumin, creatinine, electrolytes, aminotransferases, bilirubin, gamma-glutamyltransferase, alkaline phosphatase
Stool	**Bacteria**: *Clostridium difficile*, enteric pathogens (*Salmonella, Shigella, Campylobacter, Escherichia coli* 0157 H7, *Yersinia*, vancomycin-resistant enterococci, methicillin-resistant *Staphylococcus aureus*, Gram-negative multidrug-resistant bacteria), (specific situations: *Vibrio cholerae*, *Listeria monocytogenes, Helicobacter pylori* fecal antigen) **Viruses**: Norovirus, (specific situation: Rotavirus) **Parasites**: *Giardia lamblia, Cryptosporidium parvum*, protozoa (including *Blastocystis hominis*) and helminths, (specific situations: *Isospora*, *Microsporidia*) **General**: Fecal occult blood testing, calprotectin

ESR erythrocyte sedimentation rate; *HIV* human immunodeficiency virus; *HTLV* human T lymphotrophic virus

should address the points emphasized in the first sentence of this paragraph that include history or known gastrointestinal disorders, increased risk for gastrointestinal cancer, and history of neurological, psychiatric disorders or obesity. Having successfully completed the interview, potential donors will enter the step of extensive blood and stool testing (overview provided in Table 20.2). This extensive list of testing immediately reveals the next problem for clinical practice, where the donor is sometimes required on short notice; however the described qualification process for donors represents a rather time-consuming process. This led to the evaluation whether or not frozen stool samples might equally work.

20.2.2 Preparation of Transplant

The evolving problem in clinical practice is that the stool should be used within 6 h. Although for CDI no difference could be detected between anaerobic and aerobic processing (Hamilton et al. 2012; Lee et al. 2016; Satokari et al. 2015), the preparation and storage should be as brief as possible. A minimum amount of 30 g of feces is required and should be suspended in saline and subsequently sieved. A dedicated space as well as protective measures for the preparing personnel should be available (Cammarota et al. 2017). To overcome the need of pretested available donors, a trial was conducted comparing frozen and fresh FMT in CDI (Lee et al. 2016). One hundred and fourteen patients were randomized to receive frozen and 118 to receive fresh FMT, respectively. Eight of the first and 17 patients of the second group did not complete the study. No statistical difference could be detected when comparing both groups. Clinical resolution was received in both groups in about 90%. The authors conclude that frozen FMT can equally be used in CDI.

20.2.3 Preparation of Recipient and Route of Application

There is only weak evidence for any preparatory protocol for recipients. Nevertheless, the European consensus protocol agreed on two steps: (1) Patients with recurrent CDI should be treated with vancomycin or fidaxomicin at least for 3 days before FMT. Antibiotics should be stopped 12–48 h before FMT. (2) The recipient should be prepared with bowel lavage by polyethylene glycol before FMT (Cammarota et al. 2017). Having prepared the recipient, the question arises which route should be chosen for FMT, the oral route into the small intestine or vice versa an endoscopic administration via colonoscope.

For CDI where in the majority of cases a single transplantation suffices, it seems that it does not matter whether to perform the FMT via colonoscopy where the transplant should be infused into the right colon or to place it in the upper gastrointestinal tract through the working channel of a gastroscope or through a nasogastric, nasojejunal, or gastrostomy tube. However, patients should be kept in a 45° upright position for 4 h after infusion from above in order to prevent aspiration. There is also the possibility to perform FMT via enema, which will be discussed below in particular for the indication ulcerative colitis. Nevertheless, this is sometimes difficult since the patients are required to hold the infused material for at least 30 min and to remain supine (Lee et al. 2016).

20.2.4 Mechanisms Behind and Novel Directions

What is protecting from CDI and how does FMT potentially work? *C. difficile* is producing the toxins *tcdA* and *tcdB*; both have been shown to inhibit epithelial Rho guanosine triphosphatases, thus inducing a barrier defect that facilitates colitis (Leffler and Lamont 2015). In addition, serum antitoxin antibodies protect against CDI. Hence mounting a serum antibody response during an initial episode of CDI has been associated with relative protection (Kelly 1996; Kyne et al. 2000, 2001). Probably most relevant with regard to the efficacy of FMT is that colonization with a non-toxigenic strain also conveys protection (Blossom and McDonald 2007; Sambol et al. 2002).

While FMT has been proven successful in CDI, the role of viral alterations has been neglected. This gap has been filled recently, by a study where the intestinal virome was analyzed by ultra-deep metagenomic sequencing on fecal samples of patients with CDI and healthy controls (Zuo et al. 2018). The CDI patients were partly treated with vancomycin or with FMT, and the enteric virome alterations were evaluated. Not surprisingly, the enteric virome shows distinct changes in patients with CDI. Treatment response in FMT was paralleled by a high colonization level of donor-derived *Caudovirales* taxa in the host. Hence, the authors conclude that *Caudovirales* bacteriophages may play a role in mediating therapeutic efficacy in CDI.

20.2.5 Summarizing Thoughts for CDI

Considering the complexity of the patient and donor selection as well as the procedure of performing FMT, the consensus statement concludes that referral centers should be established that are implemented in hospitals with appropriate expertise and facilities. A referral center should include a multidisciplinary team that consists of gastroenterologist, microbiologists, and infectious disease physicians (Cammarota et al. 2017).

20.3 Other Indications

20.3.1 Ulcerative Colitis

In order to consider FMT as a therapeutic option, one should anticipate changes in the microbiota composition in the disease of interest. There are numerous reports indicating that in particular the diversity of the intestinal microbiota decreases profoundly during the flares of inflammatory bowel diseases (IBD) (Sepehri et al. 2007; Walker et al. 2011). Thus the idea, in parallel to the FMT-induced obesity, that FMT might be able to induce remission appeared quite intriguing.

Currently, only a few placebo-controlled studies are available. Thus in the following, we will outline the complexity of the topic by focusing on the available controlled clinical studies for ulcerative colitis. An overview of the controlled trials is provided in Table 20.3. Two placebo-controlled studies, published head-to-head in 2015 in *Gastroenterology*, started the field (Moayyedi et al. 2015; Rossen et al. 2015). Both studies included patients with mild-to-moderate

Table 20.3 Fecal microbiota transfer in inflammatory bowel disease-controlled trials

Trial	Design	Patients	Donor strategy	Controls	Primary endpoint	Comments	References
Mild to moderate UC	Double-blind, randomized, FMT performed at the start of the study and 3 weeks later	$n =$ 23 FMT; $n =$ 25 control	Feces from donors (single donor), administration via nasoduodenal tube	Autologous fecal microbiota	Clinical remission (simple clinical colitis activity index scores ≤ 2) combined with ≥ 1-point decrease in the Mayo endoscopic score at week 12	– Primary endpoint missed – Diversity	Rossen et al. (2015)
Active UC	Double-blind, randomized controlled trial of FMT vs. placebo in active UC; FMT at start and after 3 weeks	$n =$ 38 FMT; $n =$ 37 placebo	FMT administered via enema (single donor)	Water enema	Remission of UC defined as a Mayo score ≤ 2 with an endoscopic Mayo score of 0 at week 7	– Donor B – Primary endpoint missed – Trial was stopped early for futility by the Data Monitoring and Safety Committee	Moayyedi et al. (2015)
Active UC (Mayo 4–6)	Multicenter, doubleblind, randomized (1:1), placebo-controlled trial; first FMT via colonoscopic infusion, followed by enemas 5 days per week for 8 weeks	$n =$ 42 FMT; $n =$ 43 placebo	FMT as multi-donor (3–7 unrelated donors)	Placebo	Steroid-free clinical remission with endoscopic remission or response (Mayo score ≤ 2, all subscores ≤ 1, ≥ 1-point reduction in endoscopic subscore) at week 8	– Multidonor concept – Frozen – Primary endpoint met – Quality of life	Paramsothy et al. (2017)
Active UC (Mayo 3–10 with an endoscopic Mayo subscore ≥ 2)	Multicenter randomized, doubleblind, placebocontrolled trial, first FMT via colonoscope followed by 2 enemas by day 7	$n =$ 38 FMT; $n =$ 35 control	FMT: anaerobically prepared donor stool (pooled from 3–4 donors) were stored frozen at 80 °C, thawed and then administered	Autologous FMT	Steroid-free remission (total Mayo score ≤ 2 with and endoscopic Mayo score of ≤ 1 at week 8)	– Anaerobic preparation – Multidonor – More efficient	Costello et al. (2017)

FMT fecal microbiota transfer; *UC* ulcerative colitis

ulcerative colitis. In the study by Rossen et al., 25 patients were assigned to each group, either to FMT or to autologous FMT, respectively. FMT was administered via duodenal tube at weeks 1 and 3. The primary endpoint consisted of clinical remission combined with endoscopic improvement (decrease in Mayo score of $\geq$ 1 point) at week 12 (Rossen et al. 2015). Although the study failed the primary endpoint, a tendency toward clinical improvement could be observed. Remarkably, patients that improved throughout the trial, independent from the assigned group, were characterized by an increased diversity of their intestinal microbiota, a hallmark for a "healthier" microbiota (Rossen et al. 2015). In the study by Moayyedi et al., 38 patients were assigned to the FMT group, while 37 were assigned to the placebo, in this case water, group (Moayyedi et al. 2015). All patients received 50 ml enemas once weekly for a total of 6 weeks. The primary endpoint in week 7 consisted of clinical remission (Mayo score $\leq$2) plus endoscopic remission (Mayo score =0). The combined endpoint was not reached; however in the secondary endpoint, clinical remission statistical significance was observed. However, more remarkable of this trial are the observations made with regard to the donor. The study was initiated in July 2012, and until April 2013, 50% of the patients in the FMT group received the transplant from donors A and B, respectively. During this time period, two patients went into remission, both of them receiving FMT from donor B. From April 2013 until November 2013, donor B was unavailable; hence patients received FMT from donors A, C, D, E, and F. However, only two reached remission (donors E and F). Starting from November 2013 until the end of study in June 2014, the board decided to choose the now again available donor B as single donor. During this time period, another five patients achieved remission. This strongly underlines that there is a significant difference in outcome depending on the donor. Unsolved until now is the identification of this "high-quality" donor ahead (Moayyedi et al. 2015). However, several groups worldwide are currently working on strategies of how to identify this unique type of donor.

To overcome this donor selection problem, a recent study applied a multi-donor strategy (Paramsothy et al. 2017). The study again included ulcerative colitis patients with mild-to-moderate disease. Patients were 1:1 randomized to receive either multi-donor FMT ($n = 41$) or placebo ($n = 40$). The first application was performed via colonoscopy followed by five enemas/week until week 8. The primary endpoint at week 8 was steroid-free clinical as well as endoscopic remission (Paramsothy et al. 2017). The primary endpoint in this study was reached, thus representing the first controlled trial indicating that this strategy can indeed be successful. Several points are remarkable; besides the multi-donor concept, the authors decided to use a rather intense protocol when compared to the previously published studies. However, although the study reached the primary endpoint, it is remarkable that the quality of life did not differ between the two groups possibly reflecting the high-intensity protocol. In addition, although a multi-donor concept was applied, secondary analysis revealed a donor that when included led to a better outcome (Siegmund 2017). Again, the studies proved in line with the study by Rossen et al. that remission was associated with an increase in microbial diversity (Paramsothy et al. 2017; Rossen et al. 2015).

The most recent study at this point continues with the multi-donor concept; each preparation included three to four donors (Costello et al. 2017). The technical difference here is that an anaerobic preparation was applied. Considering that ulcerative colitis has been shown to present with a lower abundance of anaerobic bacteria, it is rational to expect that anaerobic processing of samples would be relevant for FMT success in the treatment of this disease (Sokol et al. 2009).

Again the study included patients with mild-to-moderate ulcerative colitis and performed a 1:1 randomization, 38 patients receiving the anaerobic, multi-donor FMT and 35 receiving autologous FMT, respectively (Costello et al. 2017). The first FMT was applied via colonoscopy followed by two enemas until day 7. The primary endpoint was again steroid-free clinical and endoscopic remission in week 8. The primary endpoint reached clinical significance and hence the study might set a novel standard. While an anaerobic preparation in CDI did not result in a clinical significance, the effect in ulcerative colitis is remarkable. An additional open controlled trial of repeated FMT in ulcerative colitis patients provides evidence that donors with a high bacterial richness and a high abundance of *Akkermansia muciniphila*, unclassified *Ruminococcaceae*, and *Ruminococcus* spp. were more likely to induce remission, thus underlining that the taxonomic composition is a critical factor with regard to efficacy of FMT in ulcerative colitis patients (Kump et al. 2018).

Where do we stand now for ulcerative colitis? Based in particular on the recent studies, there is evidence that FMT might actually provide a novel form of maintenance therapy. In contrast to CDI, it appears to be very unlikely that an actual cure can be achieved by FMT once FMT is withdrawn. However, considering the gaps of knowledge, several issues should be filled before entering clinical routine. The ultimate solution obviously would be to identify the critical microbiota composition that can then be manufactured without the actual need of a donor and hence the associated risks. At this point this solution seems to be far away, and the multi-donor approach in combination with anaerobic preparation might present a transient though difficult solution.

20.3.2 Irritable Bowel Syndrome and Metabolic Syndrome

Irritable bowel syndrome (IBS) shares a number of symptoms with IBD including abdominal pain, bloating, and reduced quality of life. However, endoscopic signs of inflammation are lacking. Thus, a recent study aimed to compare FMT with placebo in patients with IBS. For this, a double-blind, randomized, placebo-controlled, parallel-group, single-center study was performed, including patients with a moderate-to-severe IBS. In this study FMT induced a significant symptom improvement; however larger studies are required for confirmation (Johnsen et al. 2018).

A recent trial aimed at investigating the role of intestinal microbiota on insulin sensitivity. Thus the effect of lean donor *versus* autologous FMT to male recipients with metabolic syndrome was performed. Eighteen weeks after FMT, no metabolic changes were observed; however, insulin sensitivity 6 weeks post FMT was significantly improved, and the microbiota composition altered (Kootte et al. 2017).

20.4 Potential Risks and Considerations

Most surprising for the medical community, FMT is extremely well accepted by patients. This includes some potential dangers, since a significant number of patients believe that, based on a few case reports, FMT might work for more diseases and underestimate associated risks. The most prominent example is the broadly discussed case report where a patient became obese after receiving FMT from an obese donor (Alang and Kelly 2015). Although other causes of weight increase could not be excluded, considering the above-cited work in mice, it seems possible that the weight increase was mediated by FMT (Ridaura et al. 2013). In a recent work the group expanded their data by showing that germfree mice colonized with microbiota from obese or lean monozygotic twins resembled the microbiota of the donor. In addition, the obesity-associated phenotype from one twin could be transferred to germfree mice (Ridaura et al. 2013). Other diseases that can potentially be transferred are summarized in "Controversy".

20.5 Novel Developments, Challenges, and Future Directions

The most demanding challenge for the upcoming years is the selection of the best donor. This seems to be more important for ulcerative colitis, while the donor selection did not matter (besides the standard selection process) for CDI. In addition to the above-discussed anaerobic preparation for ulcerative colitis patients, a recently published study contributes a novel flavor to the field. The study by Ott et al. performed a novel form of FMT in CDI. They applied a single-donor strategy; however the transplant was filtered through a sterile filter before transplantation. Thus, the transplant included bacterial metabolites, bacteriophages as well as viruses, but no viable bacteria (Ott et al. 2017). They chose five patients with complicated CDI and comorbidities, all of whom reached resolution of diarrhea after FMT. Similar to the above-discussed studies, microbial diversity was dramatically decreased in the recipients and recovered profoundly after transplantation. This raises the question whether it is required to transplant whole bacteria or whether the transplantation of the intestinal milieu, here reflected by the metabolites, is sufficient or not.

An overview of current indications for FMT, various FMT preparations, as well as routes of application is summarized in Fig. 20.3.

20.6 Conclusion

Even after a decade of a sudden increase in interest, the field "microbiota" can be described as an emerging field. While in the beginning the majority of publications were purely descriptive, the field is now focusing on functional consequences and mechanistic aspects. It is tempting to speculate where we will be in 5 years from now. From a medical point of view, the hope is that we will be able to define the mandatory composition of microbes or even only their metabolites and have them readily manufactured without the need of actual donors. Which indications besides recurrent *C. difficile* infection will be relevant remains to be answered over the next years by performing additional randomized controlled trials.

▶ **Controversy**

The use of microbiota transfer in less severe diseases than infections with *C. difficile* has to be analyzed carefully. The possible introduction of pathogens or induction of changes in phenotypes in the recipient are risks using FMT (Collins et al. 2013). Alang and Kelly reported a case of weight gain after FMT in a female patient getting stool from an obese donor. The authors cannot exclude other possible mechanisms for the increase in body weight in the patient (Alang and Kelly 2015). However, others demonstrated in mouse experiments that obesity could be transferred by the microbiota (Ridaura et al. 2013). The authors showed that germfree mice colonized with microbiota from obese or lean twins resembled the microbiota of the donor. Additionally, the obesity-associated phenotype from the human was transmissible to the germfree mice colonized with microbiota from the obese donor. Moreover, FMT can

Fecal microbiota transfer

Route of administration	**Orally** ("yellow soup", duodenal tube, capsules)		**Rectally** (endoscopically, enema)	
Strategies of preparation	"Yellow soup"	Filtrate	Anaerobic preparation	"Sterile" filtrate
Frequency of administration	Once	Multiple times (once/week up to daily)		
Indications	***C. difficile*** **colitis**	Ulcerative colitis		???

Fig. 20.3 Overview of indications, application routes, and transplant preparation

transmit the predisposition to develop colorectal cancer in mice (Couturier-Maillard et al. 2013). Not only phenotypic changes were influenced by the microbial transfer but also behavior (Bercik et al. 2011; Deltheil et al. 2008). Therefore the use of FMT to treat or improve symptoms in neurological diseases like Parkinson's disease, multiple sclerosis, or autism was investigated (Borody and Campbell 2012; Sandler et al. 2000). Further controlled studies have to proof the impact of FMT in neurological and psychiatric disorders.

History

The first notification of transplantation of feces in humans is going back to the fourth century. Chinese traditional medical literature described the treatment of food poisoning or severe diarrhea by stool transfer (Zhang et al. 2012). One thousand two-hundred years later, patients with abdominal diseases drank the so-called yellow soup (Ge 2000; Li 2011). This medication contained fresh, dry, or fermented stool. In the last century, treatment of bacterial dysentery with fresh and warm camel feces was recommended by Bedouins and performed by German soldiers during World War II (Lewin 2001). In 1958 the first fecal microbiota transplantation (FMT) in modern medicine was done in patients with pseudomembranous enterocolitis. Dr. Eiseman treated the patients with fecal enema and the disease ameliorated (Eiseman et al. 1958). In 2003 case series using FMT in patients with *Clostridium difficile* infection were published (Aas et al. 2003). Ten years later the first randomized clinical trial analyzed the treatment of patients with recurrent *C. difficile* infections with FMT (van Nood et al. 2013).

Highlights

- The process of transfer of stool from a healthy donor to a recipient is called fecal microbiota transplantation (FMT).
- FMT is an effective treatment of antibiotic-associated colitis caused by *C. difficile* infection.
- Patients with ulcerative colitis treated successfully with fecal transfer achieved remission with paralleled increase in gut microbial diversity.
- Donor selection, preparation of transplant and recipient, and the route of administration are critical factors for FMT.
- The selection of a suitable donor will be the challenge for the future. May the use of sterile fecal filtrate could overcome problems transferring pathogens.

References

Aas, J., Gessert, C. E., & Bakken, J. S. (2003). Recurrent Clostridium difficile colitis: Case series involving 18 patients treated with donor stool administered via a nasogastric tube. *Clinical Infectious Diseases, 36*, 580–585.

Alang, N., & Kelly, C. R. (2015). Weight gain after fecal microbiota transplantation. *Open Forum Infectious Diseases, 2*, ofv004.

Bercik, P., Denou, E., Collins, J., Jackson, W., Lu, J., Jury, J., et al. (2011). The intestinal microbiota affect central levels of brain-derived neurotropic factor and behavior in mice. *Gastroenterology, 141*, 599–609 609 e591–593.

Blossom, D. B., & McDonald, L. C. (2007). The challenges posed by reemerging Clostridium difficile infection. *Clinical Infectious Diseases, 45*, 222–227.

Borody, T. J., & Campbell, J. (2012). Fecal microbiota transplantation: Techniques, applications, and issues. *Gastroenterology Clinics of North America, 41*, 781–803.

Borody, T. J., Warren, E. F., Leis, S. M., Surace, R., Ashman, O., & Siarakas, S. (2004). Bacteriotherapy using fecal flora: Toying with human motions. *Journal of Clinical Gastroenterology, 38*, 475–483.

Brandt, L. J., Aroniadis, O. C., Mellow, M., Kanatzar, A., Kelly, C., Park, T., et al. (2012). Long-term follow-up of colonoscopic fecal microbiota transplant for recurrent Clostridium difficile infection. *The American Journal of Gastroenterology, 107*, 1079–1087.

Cammarota, G., Ianiro, G., Tilg, H., Rajilic-Stojanovic, M., Kump, P., Satokari, R., et al. (2017). European consensus conference on faecal microbiota transplantation in clinical practice. *Gut, 66*, 569–580.

Collins, S. M., Kassam, Z., & Bercik, P. (2013). The adoptive transfer of behavioral phenotype via the intestinal microbiota: Experimental evidence and clinical implications. *Current Opinion in Microbiology, 16*, 240–245.

Costello, S., Waters, O., Bryant, R., Katsikeros, R., Makanyanga, J., Schoeman, M., et al. (2017). OP36: Short duration, low intensity pooled faecal microbiota transplantation induces remission in patients with mild-moderately active ulcerative colitis: a randomized controlled trial. *JCC, 11S1*, S23.

Couturier-Maillard, A., Secher, T., Rehman, A., Normand, S., De Arcangelis, A., Haesler, R., et al. (2013). NOD2-mediated dysbiosis predisposes mice to transmissible colitis and colorectal cancer. *The Journal of Clinical Investigation, 123*, 700–711.

Deltheil, T., Guiard, B. P., Cerdan, J., David, D. J., Tanaka, K. F., Reperant, C., et al. (2008). Behavioral and serotonergic consequences of decreasing or increasing hippocampus brain-derived neurotrophic factor protein levels in mice. *Neuropharmacology, 55*, 1006–1014.

Eiseman, B., Silen, W., Bascom, G. S., & Kauvar, A. J. (1958). Fecal enema as an adjunct in the treatment of pseudomembranous enterocolitis. *Surgery, 44*, 854–859.

Ge, H. D. D. (2000). *Zhou Hou Bei Ji Fang*. Tianjin: Tianjin Science & Technology Press.

George, R. H., Symonds, J. M., Dimock, F., Brown, J. D., Arabi, Y., Shinagawa, N., et al. (1978a). Identification of Clostridium difficile as a cause of pseudomembranous colitis. *British Medical Journal, 1*, 695.

George, W. L., Sutter, V. L., Goldstein, E. J., Ludwig, S. L., & Finegold, S. M. (1978b). Aetiology of antimicrobial-agent-associated colitis. *Lancet, 1*, 802–803.

Gilbert, D. N., Moellering, R. C., Eliopoulos, G. M., & Sande, M. A. (2004). *The Sanford guide to antimicrobial therapy* (34th ed.). Sperryville, VA: Antimicromial Therapy, Inc..

Hamilton, M. J., Weingarden, A. R., Sadowsky, M. J., & Khoruts, A. (2012). Standardized frozen preparation for transplantation of fecal microbiota for recurrent Clostridium difficile infection. *The American Journal of Gastroenterology, 107*, 761–767.

Johnsen, P. H., Hilpusch, F., Cavanagh, J. P., Leikanger, I. S., Kolstad, C., Valle, P. C., et al. (2018). Faecal microbiota transplantation versus placebo for moderate-to-severe irritable bowel syndrome: a double-blind, randomised, placebo-controlled, parallel-group, single-centre trial. *The Lancet Gastroenterology and Hepatology, 3*, 17–24.

Kelly, C. P. (1996). Immune response to Clostridium difficile infection. *European Journal of Gastroenterology and Hepatology, 8*, 1048–1053.

Khoruts, A., & Sadowsky, M. J. (2011). Therapeutic transplantation of the distal gut microbiota. *Mucosal Immunology, 4*, 4–7.

Koo, H. L., Van, J. N., Zhao, M., Ye, X., Revell, P. A., Jiang, Z. D., et al. (2014). Real-time polymerase chain reaction detection of asymptomatic Clostridium difficile colonization and rising C. difficile-associated disease rates. *Infection Control and Hospital Epidemiology, 35*, 667–673.

Kootte, R. S., Levin, E., Salojarvi, J., Smits, L. P., Hartstra, A. V., Udayappan, S. D., et al. (2017). Improvement of insulin sensitivity after lean donor feces in metabolic syndrome is driven by baseline intestinal microbiota composition. *Cell Metabolism, 26*, 611–619 e616.

Kump, P., Wurm, P., Grochenig, H. P., Wenzl, H., Petritsch, W., Halwachs, B., et al. (2018). The taxonomic composition of the donor intestinal microbiota is a major factor influencing the efficacy of faecal microbiota transplantation in therapy refractory ulcerative colitis. *Alimentary Pharmacology and Therapeutics, 47*, 67–77.

Kyne, L., Warny, M., Qamar, A., & Kelly, C. P. (2000). Asymptomatic carriage of Clostridium difficile and serum levels of IgG antibody against toxin A. *The New England Journal of Medicine, 342*, 390–397.

Kyne, L., Warny, M., Qamar, A., & Kelly, C. P. (2001). Association between antibody response to toxin A and protection against recurrent Clostridium difficile diarrhoea. *Lancet, 357*, 189–193.

Larson, H. E., Price, A. B., Honour, P., & Borriello, S. P. (1978). Clostridium difficile and the aetiology of pseudomembranous colitis. *Lancet, 1*, 1063–1066.

Lee, C. H., Steiner, T., Petrof, E. O., Smieja, M., Roscoe, D., Nematallah, A., et al. (2016). Frozen vs fresh fecal microbiota transplantation and clinical resolution of diarrhea in patients with recurrent clostridium difficile infection: A randomized clinical trial. *JAMA, 315*, 142–149.

Leffler, D. A., & Lamont, J. T. (2015). Clostridium difficile infection. *The New England Journal of Medicine, 372*, 1539–1548.

Lewin, R. A. (2001). More on Merde. *Perspectives in Biology and Medicine, 44*, 594–607.

Li, S. M. D. (2011). *Ben Cao Gang Mu*. Bejing: Huaxia Press.

Miller, B. A., Chen, L. F., Sexton, D. J., & Anderson, D. J. (2011). Comparison of the burdens of hospital-onset, healthcare facility-associated Clostridium difficile Infection and of healthcare-associated infection due to methicillin-resistant Staphylococcus aureus in community hospitals. *Infection Control and Hospital Epidemiology, 32*, 387–390.

Moayyedi, P., Surette, M. G., Kim, P. T., Libertucci, J., Wolfe, M., Onischi, C., et al. (2015). Fecal microbiota transplantation induces remission in patients with active ulcerative colitis in a randomized controlled trial. *Gastroenterology, 149*, 102–109 e106.

Ott, S. J., Waetzig, G. H., Rehman, A., Moltzau-Anderson, J., Bharti, R., Grasis, J. A., et al. (2017). Efficacy of sterile fecal filtrate transfer for treating patients with Clostridium difficile Infection. *Gastroenterology, 152*, 799–811 e797.

Paramsothy, S., Kamm, M. A., Kaakoush, N. O., Walsh, A. J., van den Bogaerde, J., Samuel, D., et al. (2017). Multidonor intensive faecal microbiota transplantation for active ulcerative colitis: A randomised placebo-controlled trial. *Lancet, 389*, 1218–1228.

Postigo, R., & Kim, J. H. (2012). Colonoscopic versus nasogastric fecal transplantation for the treatment of Clostridium difficile infection: A review and pooled analysis. *Infection, 40*, 643–648.

Ratner, M. (2014). Fecal transplantation poses dilemma for FDA. *Nature Biotechnology, 32*, 401–402.

Ridaura, V. K., Faith, J. J., Rey, F. E., Cheng, J., Duncan, A. E., Kau, A. L., et al. (2013). Gut microbiota from twins discordant for obesity modulate metabolism in mice. *Science, 341*, 1241214.

Rossen, N. G., Fuentes, S., van der Spek, M. J., Tijssen, J. G., Hartman, J. H., Duflou, A., et al. (2015). Findings from a randomized controlled trial of fecal transplantation for patients with ulcerative colitis. *Gastroenterology, 149*, 110–118 e114.

Sambol, S. P., Merrigan, M. M., Tang, J. K., Johnson, S., & Gerding, D. N. (2002). Colonization for the prevention of Clostridium difficile disease in hamsters. *The Journal of Infectious Diseases, 186*, 1781–1789.

Sandler, R. H., Finegold, S. M., Bolte, E. R., Buchanan, C. P., Maxwell, A. P., Vaisanen, M. L., et al. (2000). Short-term benefit from oral vancomycin treatment of regressive-onset autism. *Journal of Child Neurology, 15*, 429–435.

Satokari, R., Mattila, E., Kainulainen, V., & Arkkila, P. E. (2015). Simple faecal preparation and efficacy of frozen inoculum in faecal microbiota transplantation for recurrent Clostridium difficile infection – An observational cohort study. *Alimentary Pharmacology and Therapeutics, 41*, 46–53.

Sepehri, S., Kotlowski, R., Bernstein, C. N., & Krause, D. O. (2007). Microbial diversity of inflamed and noninflamed gut biopsy tissues in inflammatory bowel disease. *Inflammatory Bowel Diseases, 13*, 675–683.

Siegmund, B. (2017). Is intensity the solution for FMT in ulcerative colitis? *Lancet, 389*, 1170–1172.

Smith, M., Kassam, Z., Edelstein, C., Burgess, J., & Alm, E. (2014). OpenBiome remains open to serve the medical community. *Nature Biotechnology, 32*, 867.

Sokol, H., Seksik, P., Furet, J. P., Firmesse, O., Nion-Larmurier, I., Beaugerie, L., et al. (2009). Low counts of Faecalibacterium prausnitzii in colitis microbiota. *Inflammatory Bowel Diseases, 15*, 1183–1189.

Surawicz, C. M., & Alexander, J. (2011). Treatment of refractory and recurrent Clostridium difficile infection. Nature reviews. *Gastroenterology and Hepatology, 8*, 330–339.

Turnbaugh, P. J., Ley, R. E., Mahowald, M. A., Magrini, V., Mardis, E. R., & Gordon, J. I. (2006). An obesity-associated gut microbiome with increased capacity for energy harvest. *Nature, 444*, 1027–1031.

van Nood, E., Vrieze, A., Nieuwdorp, M., Fuentes, S., Zoetendal, E. G., de Vos, W. M., et al. (2013). Duodenal infusion of donor feces for recurrent Clostridium difficile. *The New England Journal of Medicine, 368*, 407–415.

Walker, A. W., Sanderson, J. D., Churcher, C., Parkes, G. C., Hudspith, B. N., Rayment, N., et al. (2011). High-throughput clone library analysis of the mucosa-associated microbiota reveals dysbiosis and differences between inflamed and non-inflamed regions of the intestine in inflammatory bowel disease. *BMC Microbiology, 11*, 7.

Zhang, F., Luo, W., Shi, Y., Fan, Z., & Ji, G. (2012). Should we standardize the 1,700-year-old fecal microbiota transplantation? *The American Journal of Gastroenterology, 107*, 1755 author reply pp. 1755–1756.

Zuo, T., Wong, S. H., Lam, K., Lui, R., Cheung, K., Tang, W., et al. (2018, April). Bacteriophage transfer during faecal microbiota transplantation in Clostridium difficile infection is associated with treatment outcome. *Gut, 67*(4), 634–643.

Gnotobiology

21

Marijana Basic and André Bleich

Abstract

In this chapter we provide a brief overview on the historical development of gnotobiology, housing, and maintenance systems as well as procedures used today in the gnotobiotic facility/laboratory. The techniques and equipment that prompted the development of the gnotobiology field were developed more than half a century ago. However, the main principles of gnotobiotic work have remained unchanged over the years. The pioneers of gnotobiology were Nuttall and Thierfelder, who have rederived the first germ-free animals. However, groundbreaking advancements in the gnotobiology field were achieved in the mid-1900s by scientists gathered around Reyniers and Trexler at the LOBUND Institute. Since its beginning the main goals of gnotobiotic husbandry were to provide a nucleus of pathogen-free animals for the biomedical research and to elucidate the impact of microorganisms on their host health and physiology. However, to achieve these goals, the obstacles for long-term maintenance of germ-free animals needed to be overcome. The development of gnotobiotic equipment and prerequisites for long-term maintenance accompanied with methodological progress, creation of various mouse models and sequencing platforms, contributed to the rise of gnotobiotic research in the last decade. Today, gnotobiology represents a powerful platform for unraveling the mechanisms underlying the complex nature of host-microbiota interactions and probe the function of individual microbes in health and disease. Germ-free mice can be utilized to unravel the functionality of individual murine or human bacterial species, microbial consortia, or human fecal transplants in health and disease, under highly defined conditions. Thereby, gnotobiology can reveal crucial genetic, microbial, and environmental determinants underlying host-microbiota interactions.

Gnotobiology comprises the field of studies concerned with the rearing and maintaining animals or plants that are free of all microorganisms (germ-free) or associated only with known taxa or complex microbial communities. This field emerged as a discipline when scientists began studying the mutualistic, commensal, and parasitic relationship between microorganisms and their mammalian hosts. The term gnotobiology is derived from the Greek "gnotos" meaning well known and "bios" meaning life. A gnotobiotic animal is an animal in which all life forms have been completely

M. Basic · A. Bleich (✉)
Institute for Laboratory Animal Science, Hannover Medical School, Hannover, Germany
e-mail: basic.marijana@mh-hannover.de; bleich.andre@mh-hannover.de

D. Haller (ed.), *The Gut Microbiome in Health and Disease*,
https://doi.org/10.1007/978-3-319-90545-7_21

defined, thus this term includes both germ-free and animals associated with known taxa. Today, gnotobiotic animal models have become a potent tool to advance our understanding of the host-microbiome interactions. These models provide an excellent platform to characterize the functional role of single bacterial species and minimal bacterial consortia in the regulation of host physiology and disease pathogenesis. However, the rederivation, maintenance, and monitoring of gnotobiotic rodents require vigilance, intensive labor, and a high budget. Therefore, we would like to start this journey with the words of Thomas D. Luckey: "The essence of germ-free research is isolation. Isolation must be attained mechanically, proven scientifically, and understood philosophically" (Luckey 1963).

21.1 Biological Barriers

Today, laboratory animals, especially rodents, are housed in different biological barriers, which are designed to prevent animal contact with particular microbes. The level of barrier is defined by the exclusion list of certain microbes and by the methods used to preserve the status of the barrier. The germ-free barrier is the strictest biological barrier in animal facilities. Germ-free or *axenic* (from Greek "a" meaning without and "xenos" meaning stranger) animals are free of all foreign life forms apart from itself including bacteria, viruses, fungi, protozoa, and other saprophytic or parasitic life forms. However, indigenous viruses or other still uncharacterized life forms that have been integrated into the host genome and cannot be removed are considered to be an indivisible part of the animal. The specified pathogen-free (SPF) barrier is a common barrier in biomedical research institutions. The SPF animals are free from particular microbes, mostly bacteria, viruses, and parasites, that might induce disease or alter the course of the animal experiment. To exclude these pathogens, SPF animals are regularly monitored in accordance with recommendations for hygienic monitoring such as the recommendation list of the Federation of Laboratory Animal Science Associations (FELASA) (Mähler et al. 2014). However, the composition of the indigenous microbiota in these mice is unknown. On the other hand, the conventional housing represents the lowest barrier level, and it is outdated in modern animal facilities. The microbiological status of conventional animals is completely unknown. These animals are colonized with a non-defined microbiota and also not monitored for the presence of specified pathogens.

21.2 Historical Perspective

The necessary gnotobiotic techniques and equipment were developed more than 50 years ago (Gustafsson 1959; Reyniers and Sacksteder 1958; Trexler and Reynolds 1957; van der Waaij and Andreas 1971). The development of gnotobiotic techniques was propelled by two conflicting hypothesis. In 1885, Louis Pasteur postulated that animal life cannot exist without symbiotic or commensal bacteria. One year later Marceli Nencki proposed that animals without bacteria would live longer and be healthier due to absence of repulsive microbiological by-products in the gut (Rahija 2007). Therefore, to resolve this debate, the necessary equipment needed to be established.

Nuttall and Thierfelder are considered to be the pioneers of gnotobiology and germ-free research. In 1985 they delivered first germ-free guinea pigs by cesarean section into the sterile bell jar. They have kept them germ-free on sterile milk for 8 days. In a subsequent experiment, they fed the germ-free guinea pigs with sterile vegetables up to 13 days. However, they have observed that germ-free animals gained less weight than those carrying complex microbiota, demonstrating that the gut bacteria are necessary for the digestion of the vegetable diet (Nuttall and Thierfelder 1897).

Furthermore, in 1912 Kuester designed the first isolator prototype, in which germ-free animals were isolated from exposure to microbes (Küster 1915). In this isolator prototype, Kuester kept goats germ-free for over a month. These results partially opposed to Pasteur's hypothesis, as it was shown that life without bacteria was possible in several species for a limited time. However, it is important to note that these animals had to be fed with an unnatural diet fortified with vitamins, which are normally provided by the gut microbiota, to survive long term (Rahija 2007).

Groundbreaking work was then performed by the Laboratories of Bacteriology at the University of Notre Dame (LOBUND) group with Reyniers and Trexler. They have established large-scale gnotobiology laboratories. First isolators were designed as large stainless steel devices and were heavy, expensive, and difficult to handle. Additionally, these chambers were not efficient in housing of large gnotobiotic colonies. One of these systems was developed by Reyniers, which was designed as an autoclave with security glass windows. The second model of this isolator type was constructed by Gustafsson. His isolation chamber was designed as a steam sterilizer as well, but it also had a dip tank lock. Until 1957 the stainless steel isolators and steam sterilization were the backbone of the gnotobiotic technology. Subsequently, Trexler designed and introduced the first plastic polyvinyl chloride (PVC) flexible film isolator that is commonly used today (Trexler and Reynolds 1957). Furthermore, in 1960 Trexler attempted to make an entire room germ-free. The staff entering this room was wearing full body PVC suits. The room was entered by going through the germicidal shower followed by immersion in a dunk tank. However, even in spite of these strict precautions, contamination with human microbiota occurred. Therefore, this idea of germ-free room was abandoned. However, the designs and procedures developed for the first germ-free rooms were the precursor for the rodent barrier in facilities specialized for breeding and maintenance of SPF rodent colonies (Rahija 2007).

Germ-free research has developed independently at three institutions. In 1928 Glimstedt started the gnotobiotic program in Sweden at the University of Lund. In parallel, Reyniers founded the LOBUND Institute, which attracted top scientists such as P. C. Trexler and J. Pleasants, and became the world's leader in germ-free research. The third program was established by Miyakawa 15 years later in Japan at the Nagoya University (Rahija 2007). These institutes focused on the development of the equipment that would provide the physical barrier and sterile environment needed to grow germ-free animals for long-term studies. Moreover, their later studies were concentrated on the biological characterization of animals reared in germ-free environment. During these times many different species were rederived germ-free including chickens, rabbits, and nonhuman primates. The standard method of germ-free rederivation was cesarean section and hand nurturing. The rearing of germ-free rodents was difficult due to their small neonatal size, which made hand rearing challenging. First hand rearing of germ-free rats was reported by Gustafsson and shortly after by Reyniers (Gustafsson 1946; Reyniers et al. 1946). Several years later germ-free mice have been also successfully hand raised (Pleasants 1959). However, once the colony of germ-free mice was established, it was possible to maintain it for several generations. Hence, methods as hysterectomy and cross-fostering of neonates by germ-free foster mothers substituted hand feeding and made the germ-free rodents maintenance easier. Reproducing colonies of germ-free rodents were available for distribution to the scientific community since the late 1950s at the LOBUND Institute and the University of Lund (Pleasants 1959). However, it is important to note that the long-term maintenance of germ-free rodent colonies was initially unsuccessful due to growth retardation and premature death caused by nutritional deficiencies. This encouraged studies to identify vital proportion of dietary components to maintain animals under germ-free conditions. First studies have shown that in the absence of the gut bacteria, no adequate levels of vitamin K could be synthetized. Furthermore, it was shown that the sterilization process also degrades vitamins, and therefore the diet for germ-free rodents needs to be enriched with these components. Fortifying the rodent diet with supplements reduced the mortality in germ-free rodent colonies (Rahija 2007). Overcoming these obstacles in maintenance of germ-free colonies advanced the studies about anatomic and pathophysiologic changes associated with germ-free environment.

A second major achievement was the application of the gnotobiotic technology for the derivation and maintenance of SPF animals for biomedical research, as in that time, many mice were infected with pathogens (Foster 1959; Weisbroth et al. 1999). First seminars educating animal suppliers and scientific community about gnotobiotic technology and its use for developing and maintaining SPF nucleus colonies were given already in early 1960s (Trexler 1961).

Today, gnotobiotic animals emerged as a valuable tool to elucidate the function and the role of the intestinal microbiota for homeostasis as well as in the development of pathological conditions and disorders. Interest in studying host-microbiota interactions has increased since many human chronic disorders have been associated with the composition and changes in the intestinal microbiota, such as type 1 diabetes, metabolic syndrome, obesity, autoimmune arthritis, inflammatory bowel disease (IBD), irritable bowel syndrome (IBS), and liver disease (Becker et al. 2015; Bleich and Fox 2015; Bleich and Hansen 2012; Brenner et al. 2015; Hormannsperger et al. 2015; Keubler et al. 2015; Sartor 2008; Ussar et al. 2015). Germ-free mice and those colonized with minimal bacterial consortia can be utilized to analyze the impact of single or defined communities on the host mucosal immune system, for example, to unravel the mechanisms leading to physiological and/or pathological inflammation (Eun et al. 2014; Hansen et al. 2014; Hormannsperger et al. 2015; Kohashi et al. 1985; Steck et al. 2011). The strategy of using minimal bacterial consortia is allowing reduction of microbial complexity and causal analysis of a wide range of physiological and pathological events. Furthermore, the detailed dissection of microbiota-host interactions reveals important genetic, microbial, and environmental factors that contribute to the development of pathology in these models and promote the discovery of new therapeutic targets.

21.3 Housing and Maintenance of Germ-Free or Gnotobiotic Animals

21.3.1 Isolators

For long-term maintenance in sterile conditions, gnotobiotic animals are housed in positive pressure isolators. Isolators are designed as closed systems and thus decrease the risk of contamination. The isolator chamber creates the sterile workspace that is required to produce and maintain germ-free and gnotobiotic mice. Therefore, the isolator chamber needs to be constructed from materials that can form an impermeable physical barrier (Rahija 2007; Trexler 1983).

The basic components of an isolator are isolation chamber, air supply, air inlet filter, air outlet filter, transfer port and arm-length gloves (Fig. 21.1a, b). Isolators can be constructed as rigid, semirigid, or flexible film isolators made of plastic or stainless steel (Rahija 2007). Commonly used today are those made of plastic, e.g., flexible film isolators. These can be made of a great variety of plastic material such as PVC, polyethylene, or nylons (Trexler 1983). However, the most used material is clear PVC. The isolator chambers can be positioned on tables that can be adjusted to the proper height to facilitate handling and provide a surface to support the isolator.

The air inlet and outlet units contain protective filter such as high-efficiency particulate air (HEPA) filter. Isolators are maintained under positive air pressure by central ventilation systems or decentral blower unit. Due to positive pressure, the isolator chamber of flexible film isolator appears as a "bubble." Positive pressure is used to prevent entry of airborne contaminants through any puncture, and it is critical when rearing germ-free or gnotobiotic animals. Additionally, the flexible film isolators might contain a rigid frame, which supports the shape of the chamber.

Manipulation of animals and supplies in the isolator is carried out through arm-length gloves. The gloves are the most vulnerable part of the isolator and a common source of isolator contamination. It is recommended to inspect them for

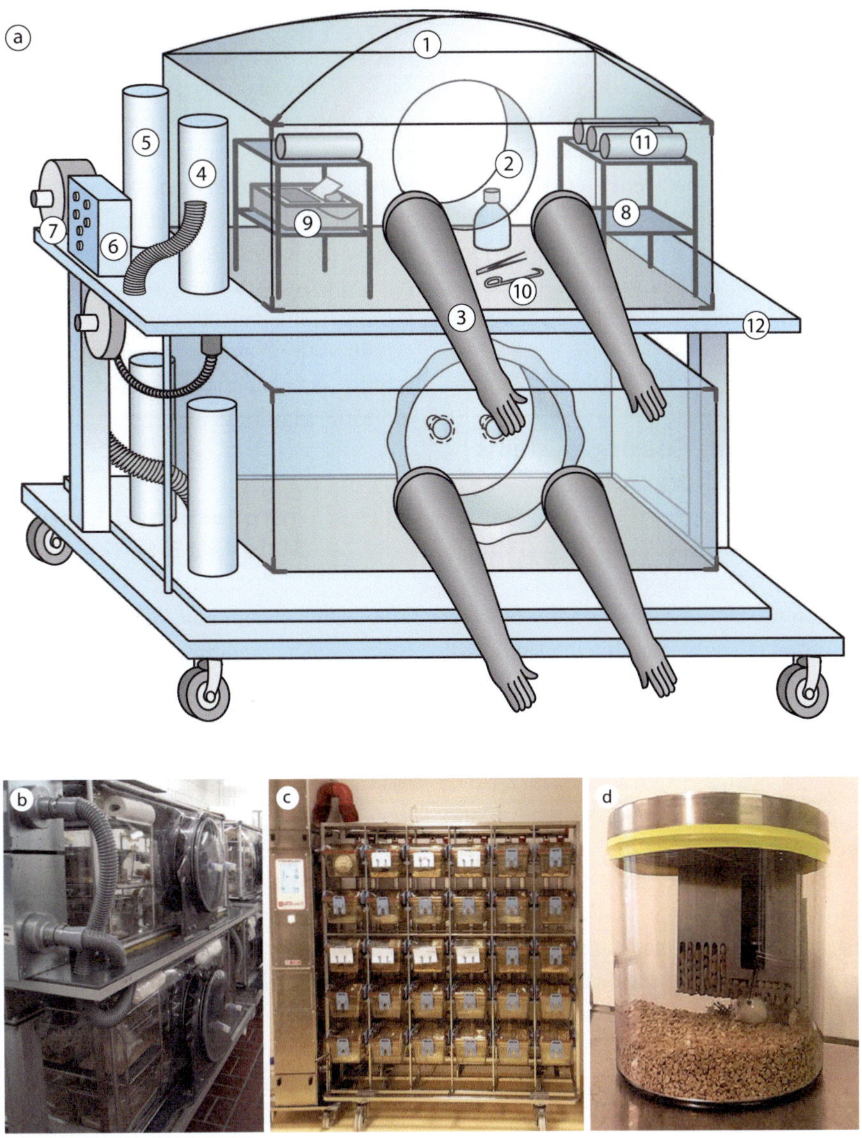

Fig. 21.1 Housing systems for germ-free and gnotobiotic rodents. (**a**) Scheme of the double-door port isolator (1-isolator chamber, 2-transfer port, 3-arm-lenght gloves, 4-air inlet filter, 5-air outlet filter, 6-console, 7-motor, 8-cage rack, 9-cage, 10-11-supplies for isolator husbandry, 12-adjustable table), (**b**) Isolator unit at Hannover Medical School, (**c**) Isocage system for short-term maintenance, (**d**) Gnotocage

holes, cracks, and excessive wear before every use. The isolator gloves can be made from nitrile rubber, latex, or polyurethane and should be changed in pairs to reduce the risk of irregular wear (Rahija 2007). Additionally, to protect them from inside, wearing of cloth gloves is recommended. For occupational health and safety reasons, glove material has to be compatible with the disinfectant under use.

The transfer port is crucial, as it represents the transition area between the isolator chamber and the room environment. It is mostly made of plastic or stainless steel cylinder, which can vary in size (Rahija 2007). The port functions as a portal for loading and sterilizing the surface of materials and supplies when entering the isolator. The other key role of the port is to maintain the physical barrier to prevent contamination of the chamber.

Depending on the type, the transfer port can be fitted with a single- or double-door. The double-door port is sealed with PVC caps on both sides (Fig. 21.1a, b). Sealing caps of the transfer port are equipped with small openings through which disinfectant such as peracetic acid can be introduced to decontaminate materials before they enter the isolator chamber. Single-door-type ports, which are otherwise easier to handle, require a transfer isolator for all exchange of material.

21.3.2 Short-Term Housing Systems for Gnotobiotic Animals

Each individual experiment has to be performed in one isolator, as microorganisms, once introduced into the isolator intentionally or accidentally as contaminants, rapidly spread to all cages. Therefore, each experimental group requires a separate isolator. Furthermore, these isolators must be completely disassembled and sterilized before next use (Vowles et al. 2016). As these procedures are time-consuming and expensive, alternative housing options were developed for gnotobiotic short-term studies. Gnotobiotic animals can be housed shortly in microisolator cages such as static gnotocages (e.g., Han-gnotocage) or individually ventilated cage (IVC)-type isocages (Fig. 21.1c, d) and handled under class II biosafety cabinet (Nicklas et al. 2015). The risk of contamination in these conditions is much higher than in isolators (Hecht et al. 2014; Hedrich and Nicklas 2012). However, this is a very efficient way to perform gnotobiotic experiments, as access and handling of the animals is much easier. The gnotocages are designed as autoclavable plastic containers and sealed with a metal lid filled with HEPA filter material (Fig. 21.1d). Additionally, gnotocages can also be utilized for shipment of germ-free and gnotobiotic animals to other research institutions. Isocages such as Tecniplast Isocage P is the new cage technology developed initially for biosafety levels 3 and 4. This cage technology represents a completely closed system, in which each cage is essentially a mini isolator (Fig. 21.1c). Air flows through HEPA filter into the cage. When the cage is removed from the rack, it automatically seals and prevents any airflow into the cage. As each cage is a separate hygienic unit, multiple different gnotobiotic experimental groups can be run simultaneously, without risk of cross-contamination between the cages. However, to ensure the quality of the experiments performed in these systems, the gnotobiotic status must be controlled on regular basis. Additionally, sterile handling of the cages requires extensive training and similar effort like working with large isolators.

21.3.3 Introduction of Supply Materials into the Isolator

The isolator contains cages with the animals, food, water, bedding, and all other supplies needed for husbandry or experimental procedures such as forceps to handle mice or a hook to pull material from the supply unit (cylinder or chamber) into the isolator chamber. Isolators can be loaded using a transport cylinder that is autoclaved with all supplies, which need to be imported into the isolator. Supply cylinders are usually used to sterilize food, bedding, nesting material, or other supplies needed in the isolator. The cylinder is made out of stainless steel and wrapped with filter (HEPA filter) to exclude microorganisms and allow for steam penetration during autoclaving (Rahija 2007). The cylinder must be packed loosely to allow circulation of air and steam. The open end is covered with polyethylene terephthalate polyester plastic film, after the cylinder has been loaded with supplies, to seal the cylinder during sterilization in the autoclave (Rahija 2007). After sterilization has been verified (see below), the cylinder can be attached to the transport port and materials can be imported into the isolator (Fig. 21.2a–c). The cylinder is attached to the transfer port with a flexible plastic transfer sleeve. The inside of the sleeve and the port is disinfected/sterilized using a highly effective chemical, e.g., by spraying it with buffered peracetic acid solution to create the sterile portal between the inside of the isolator

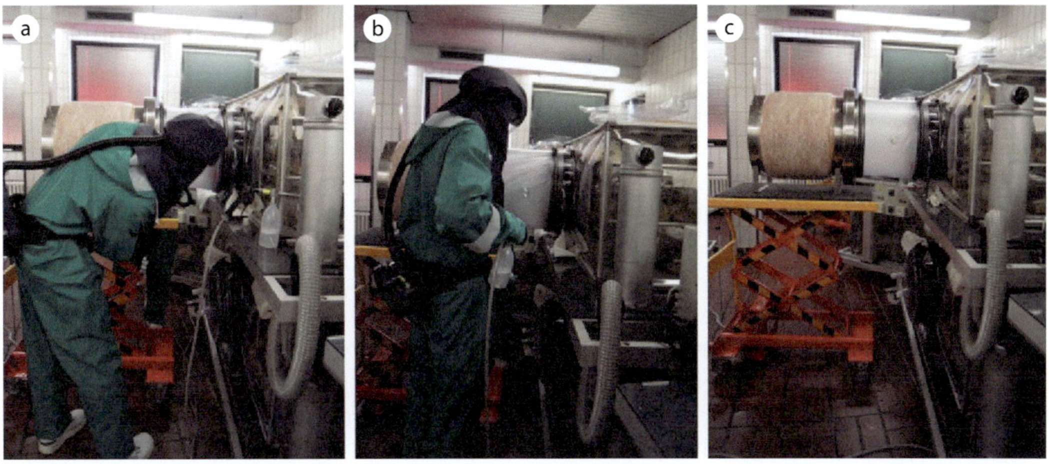

Fig. 21.2 Supply introduction into the isolator. (**a**) Docking of the supply cylinder onto isolator using a transfer sleeve, (**b**) Sterilization of the transfer sleeve using peracetic acid and creating the sterile portal for supply introduction, (**c**) The supply cylinder is docked on the isolator through transfer sleeve and supplies are ready to be introduced into isolator

and the supplies within the cylinder. During this procedure, the personnel needs to wear safety equipment like gas masks and overalls. After disinfection, the isolator internal door and the external seal on the transport cylinder are opened, and sterile materials are brought into the isolator.

21.3.4 Sterilization Methods

Prevention of contamination is essential in all aspects of germ-free husbandry. Therefore, the germ-free isolators and equipment in use need to be strictly monitored for sterility. Very important is to control and validate the sterilization methods and to inspect equipment regularly for damage or leaks. Several different methods can be used to sterilize goods such as autoclaving, irradiation, or gas sterilization, e.g., using ethylene oxide. As not all supplies can be autoclaved within the transfer cylinder before they are introduced into the isolator, surfaces of otherwise sterilized materials as well as the interior of newly assembled isolators need to be chemically sterilized. These chemical disinfectants need to be highly effective to ensure the germ-free state. Therefore, the appropriate method of sterilization depends on the type of material and how the material is transferred into the isolator (Nicklas et al. 2015; Vowles et al. 2016).

An autoclave is the most important piece of equipment used to sterilize organic materials such as food, bedding, and nesting but also water or other supplies. Most frequently autoclaving is performed for 30 min at 134 °C (Nicklas et al. 2015). However, the exact autoclaving parameters for particular materials need to be validated separately by experienced staff. Furthermore, optimal autoclaving environment is reached by packing food and bedding into small perforated bags (Nicklas et al. 2015). The autoclaving cycles can be validated by controlling the autoclave monitoring system and using bioindicators. Bioindicators are standard preparation of spore-forming heat-resistant organisms such as *Geobacillus stearothermophilus* (Rahija 2007). The growth of the bioindicator indicates that the supply cylinder is not sterile and that the sterilization process has failed.

The interior of all newly assembled isolators must be sterilized by using chemical sterilization agents before use. Preferred method is 2% buffered peracetic acid solution (Nicklas et al. 2015; Vowles et al. 2016). Alternatively, for this purpose other sterilizing agents such as chlorine dioxide or hydrogen peroxide can be used. The advantage of peracetic acid is that it can be used for the sterilization of heat sensitive materials such as PVC. Additionally, it is highly efficient at low concentrations and low temperatures and effective in both liquid and vapor phase. Peracetic acid is fast acting and reliably inactivates extremely resistant bacteria including spores as well as fungal spores. However, it does not penetrate parasite cysts and arthropod eggs. The decomposition products are acetic acid, hydrogen peroxide, oxygen, and water. However, the staff working with the peracetic acid should be protected from direct contact because it is an irritant and can damage the mucosal surfaces.

Irradiation can be used for sterilization of food. For gnotobiotic husbandry vacuum-sealed radiation-sterilized packages of food and bedding are commercially available (Rahija 2007). High doses of 50 kGy should ensure inactivation of relatively resistant murine pathogens like parvoviruses, but this might not apply to radio-resistant bacteria (Nicklas et al. 2015; Reuter et al. 2011). To control the risk of introducing bacterial contamination via irradiated food, all diet batches are sampled and tested for bacterial contamination. In addition to the standard microbiological tests, a dedicated isolator for verifying sterility of the irradiated food batches should be established, as an additional control method.

21.4 Microbiological Monitoring

Laboratory animals housed in SPF barriers and colonized with endogenous microbiota are usually monitored in accordance with the FELASA recommendations. These animals are regularly tested for the presence of microorganisms that might affect animal health and modify the experimental outcome (Mähler et al. 2014; Nicklas et al. 2015). The tested infectious agents are primarily species-specific viruses, bacteria, or parasites, which are rarely found as contaminants in gnotobiotic animals (Nicklas et al. 2015). Therefore, environmental samples have been traditionally more relevant in gnotobiotic monitoring. Contaminants that are most likely introduced into isolators include microorganisms able to stay viable under environmental conditions such as spore-forming bacteria, micrococci, or fungi. Therefore, analyses of pathogenic bacteria such as mycobacteria or viruses are less significant in the hygienic monitoring of gnotobiotic colonies, as it is unlikely that these pathogens will be introduced to an isolator (Nicklas et al. 2015). However, some microbes can be transmitted during the rederivation process. Therefore, it is important to verify the germ-free status of mice immediately after rederivation, and testing should include infectious agents based on the FELASA recommendation list.

For microbiological monitoring of germ-free and gnotobiotic colonies, samples obtained from animals, remaining food, and bedding as well as swabs from various positions are tested. To ensure the gnotobiotic status, animals and materials are tested regularly. The goal of microbiological monitoring in germ-free animals is to demonstrate the absence of all foreign life forms. On the other hand, hygienic monitoring of gnotobiotic animals carrying defined bacterial consortia seeks the confirmation of the presence of particular organisms but also the exclusion of all other organisms. For a successful microbiological monitoring of the gnotobiotic colonies, it is recommended to use different techniques as all methods have disadvantages. Furthermore, it is crucial that all procedures included in the monitoring process are done under sterile conditions.

From the isolators, swabs samples are mostly taken before animals are introduced. After that, bedding, remaining food and fresh fecal samples are recommended to be tested by culture every 4 weeks (Nicklas et al. 2015). Furthermore, animals are tested by a full necropsy every 3–6 months (Nicklas et al. 2015). However, it is important to adapt the frequency of testing based on the number of animal or supply transfers into the isolator. Diagnostic procedures include microscopic examination of gastrointestinal content for visual assessment of bacterial morphology and microbial culture of feces and various organs using media, temperature, and time schemes adapted for the increased spectrum of bacteria posing a risk in the gnotobiotic environment. The molecular diagnostics includes culture-independent methods such as quantitative polymerase chain reaction (qPCR) or 16S rRNA gene sequencing. This applies in particular to defined colonized animals, as culture-based approaches provide nonconclusive information about the status of the animals. Viral contaminations can be detected indirectly, via serological assessment. However, it is important to note that individual bacteria-like particles can be present in low numbers in the feces of germ-free animals. These particles can originate from dead bacteria or plant debris from food or bedding and thus must be distinguished from contaminations with living bacteria (Vowles et al. 2016). The detailed guidelines for the hygienic monitoring of the gnotobiotic facilities can be found elsewhere (Nicklas et al. 2015; Vowles et al. 2016).

21.5 Rederivation of Germ-Free Rodents

Rederivation refers to the process of creating germ-free animals from a colony that is not germ-free. Propagation of germ-free rodents is usually achieved by maintaining axenic colonies that are bred in germ-free isolators. However, if new germ-free strains are required, they must be rederived. Today, due to increased interest in microbiome research, the demand for new germ-free lines, including both classical inbred and genetically manipulated mutant strains, has resulted in more frequent request for rederivation of mice. Germ-free rodent strains can be rederived in two ways: by hysterectomy/cesarean section and fostering or by embryo transfer.

First germ-free rodents were derived by aseptic hysterectomy (surgical removal of uterus) into isolators where the neonates were hand raised (Pleasants 1959). However, once a germ-free rodent colony was established, it became possible to derive new germ-free strains by hysterectomy into a germ-free isolator and fostering the offspring on a germ-free dam.

Hysterectomy or cesarean section method requires precise timings of pregnancy of both, donor and foster dams. The foster dams need to be mated to ensure the fostering of rederived neonates. Ideally, the foster mothers give birth 1–2 days before the planned hysterectomy. Likewise, the mating of the donor dams needs to be carefully planned to ensure that offspring are viable and able to survive the stress of surgical delivery, transfer into isolators, and fostering. Namely, the pups are mature enough to survive the transfer only few hours before birth. However, the natural birth should not yet been started, as the rupture of the amniotic sac is an entry point for microorganisms. The next critical step in planning the hysterectomy rederivation is the gestation period of the donor and the "foster" strain. Depending on the strain background, the gestation period can vary between 19 and 21 days. Therefore, the best method of timing pregnancy is to pair mice overnight and check for vaginal plugs every morning thereafter.

During the hysterectomy procedure, the gravid uterus is removed and passed into the isolator through a germicidal dip tank such as 4% iodine bath (some techniques describe the transfer of the evolved pups through the dip tank). It is important that the transfer of pups into the isolator occurs rapidly to ensure their viability and that the aseptic technique is ensured during this process. Once inside, the uterus is opened and neonates are removed. Subsequently, they are cleaned and gently dried utilizing sterile cotton swabs and stimulated to breathe. Thorough cleaning is required to prevent cannibalism and rejection by the surrogate mother. Cross-fostering to germ-free dams is successful within the first 5–7 days post-delivery (Vowles et al. 2016). However, the disadvantage of this method is the higher rate of neonatal death because of immaturity, complications during the transfer into the isolator or maternal neglect. Furthermore, the hysterectomy rederivation procedure increases the risk of isolator contamination. Thus, it is recommended to have a separate isolator only for rederivation procedures. Additionally, hysterectomy does not eliminate microorganisms that contaminate the fetus after uterine implantation. Viruses and bacteria such as LCMV (lymphocytic choriomeningitis virus), LDV (lactate dehydrogenase elevating virus), *Pasteurella pneumotropica*, and *Mycoplasma* spp. can be vertically transmitted in mice (Baker 1998; Blackmore and Casillo 1972). Hence, microbiological monitoring should be performed on the mouse colony before rederivation procedure to confirm the absence of these organisms. If these organisms are present in the donor colony, the rederivation by embryo transfer should be considered.

The second method utilized to rederive germ-free animals is the embryo transfer method. This method is commonly used today to reconstitute rodent lines that have been cryopreserved or to derive lines free of opportunistic organisms and pathogens (Reetz et al. 1988). The advantage of this method is the circumvention of the vertical transmission and the mother neglect. Generally it is performed by implanting two-cell embryos into the oviduct of the pseudopregnant female. Embryos can be derived either directly from the donor female or after in vitro fertilization (IVF). However, as embryos allow stricter washing than sperm, IVF may harbor a higher risk of transmitting concomitantly excreted microbes (Janus et al. 2009). For direct derivation of embryos, donor females are usually superovulated by intraperitoneal administration of 5 international unit (IU) pregnant mare's serum gonadotropin (PMSG) followed by 5 IU human chorionic gonadotropin (hCG) 48 h later and mated (Gates 1956). Successful mating is confirmed the next morning by the presence of a visible vaginal plug. Only these females are sacrificed for embryo collection on day 2 of pregnancy. Two-cell embryos are collected by flushing the oviduct, then washed several times with phosphate buffered medium, and prepared for the transfer (Sarvari et al. 2013; Whittingham 1971). The morphological integrity of the embryos is confirmed using a stereomicroscope.

The recipient germ-free females are paired with vasectomized germ-free males 1 day after the donor mating and monitored for the presence of the vaginal plug. Only females with the vaginal plug are selected for the embryo transfer procedure. These females are removed from the isolator into a sterile cage, anesthetized, and embryos are transferred by a surgical procedure under germ-free conditions into the oviduct of the females. Embryo transfer method is described in detail elsewhere (Dorsch 2012). Shortly, a transverse incision across the lumbar area is made, and the cavity is opened with a small transverse incision in the area of the ovary. The ovary and oviduct are excised carefully, and the embryos are transferred into the oviduct. The ovary and oviduct are then redeposited in the abdominal cavity. After the transfer, females are placed back into the isolator until delivery, which is expected approximately 20–21 days later.

To prevent the contamination of the germ-free recipient females, the embryo transfer needs to be performed under complete asepsis and requires the setup of a surgical platform under germ-free conditions like dedicated clean benches with connectors for isolators (Vowles et al. 2016).

21.6 Physiology of Germ-Free Animals

Germ-free animals have adapted anatomically and physiologically to an environment and life without microbiological organisms. In this section only the major differences will be mentioned with a special regard to the changes observed within the gastrointestinal tract.

Major anatomical difference of germ-free animals to those with indigenous microbiota is the enlarged cecum. Germ-free mice have a 5 times larger cecal volume than mice housed in an SPF barrier (Fig. 21.3a). Due to the enlarged cecum size, cecal volvulus (torsion), ischemia, and cecal obstruction may occur and lead to the death of germ-free mice (Trexler 1983). The enlarged cecum is a consequence of altered osmolarity of the intestinal content due to not-degraded mucopolysaccharides, which bind sodium, and intestinal atonia that is caused by accumulation of muscle depressant substances normally degraded by bacteria (Wostmann 1981).

Furthermore, varieties of mucosal parameters are decreased in germ-free animals such as intestinal epithelial renewal cycle, enzyme production, and bowel motility. Germ-free animals have a much thinner intestinal wall than mice from SPF barrier (Fig. 21.3b, c). In the absence of bacteria, the milieu of the intestine is rather aerobic than

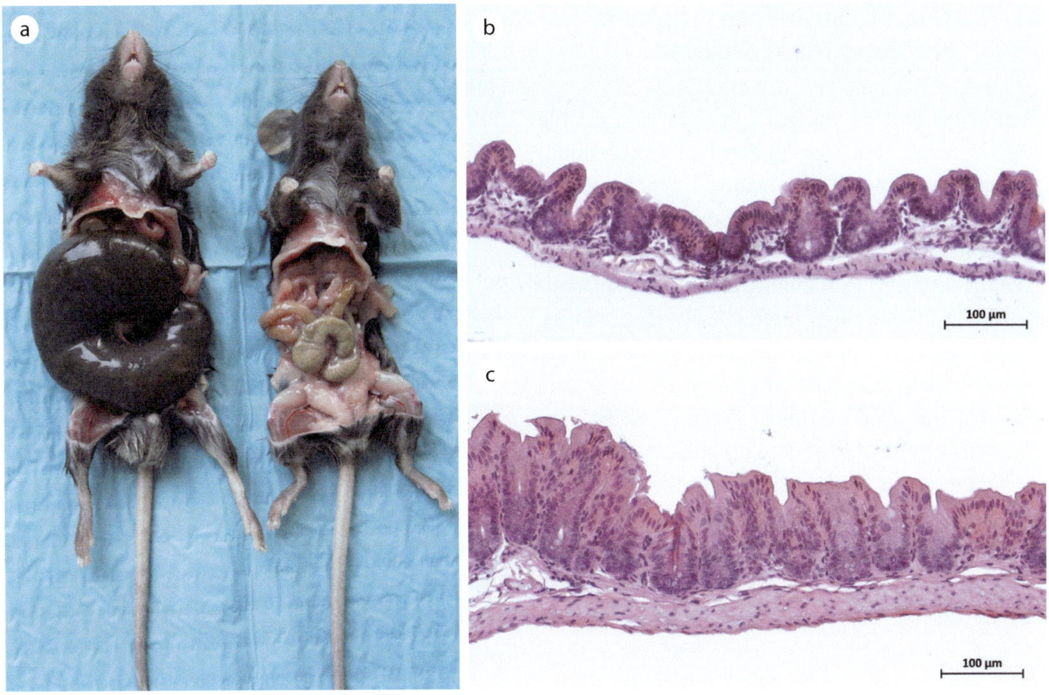

Fig. 21.3 Characteristics of germ-free mice. (**a**) Cecal size comparison between germ-free mouse (left) and SPF housed mice (right), (**b–c**) Cecal wall thickness of the mouse housed in the strict SPF barrier (**b**) and germfree mouse (**c**)

strictly anaerobic. Moreover, their fecal pellets exhibit increased water content and are softer than those of mice from the SPF barrier. Furthermore, germ-free animals have altered bile acid patterns in the intestine. During their circulation through the intestine, bile acids are exposed to microbial enzymes, which modify their structure and biological activity. Thus, the gut of germ-free animals contains mainly nonmodified (conjugated and primary) bile acids (Wostmann 1981).

The second prominent feature of germ-free animals is underdeveloped immune system caused by the decreased antigenic stimulation due to absence of live microbiota. Microbial colonization was shown to shape host immunity by stimulating the formation of organized lymphoid tissues and the function of immune cells (Belkaid and Hand 2014; Hooper et al. 2012; Round and Mazmanian 2009; Zhao and Elson 2018). The lymph nodes of germ-free mice are smaller, and the levels of circulating leukocytes and serum immunoglobulins are much lower than those from mice colonized with enteric microbiota (Coates 1975; Wostmann 1981). Furthermore, the gut-associated lymphoid tissue, intestinal T cells, especially CD4+ cells, and intestinal IgA-producing cells are greatly reduced in these mice as well. Additionally, the spleens of germ-free mice have relatively few germinal centers and poorly formed T and B cell zones (Macpherson et al. 2008; Rhee et al. 2004; Round and Mazmanian 2009).

Further anatomical adaptations are smaller heart, lung, and liver. The cardiac input of germ-free animals is one third lower (Wostmann 1981). Furthermore, the germ-free condition affects reproduction, even though germ-free mice grow and reproduce similar to mice raised under conventional conditions. The diestrus period is prolonged in germ-free females, which reduces estrus frequency and therefore copulation and implantation rate (Shimizu et al. 1998). Moreover, the typical ammonia odors are absent in the germ-free housing, as no bacterial ureases are present to hydrolyze urea. Therefore, the smell of ammonia could be an initial indicator of bacterial contamination (Nicklas et al. 2015).

21.7 Use of Germ-Free Animals for Functional Studies

Germ-free animal models provide a valuable experimental tool for functional microbiota-related research. Colonization of germ-free animals with a single microorganism or complex microbiota such as human feces or defined bacterial consortia allows standardized and causal analysis of microbe-host interactions (Fig. 21.4). Some of the anatomical and physiological abnormalities developed due to germ-free state can be reversed within several weeks by colonizing germ-free mice with intestinal bacteria (Schaedler et al. 1965b). However, the time point of bacterial association of an ex-germ-free host is critical for this reversal, as some phenotypes can only be reestablished if microbial colonization occurs during the early life. Multiple studies have shown that microbial exposure and microbiota establishment during the postnatal period have a profound impact on the future immune responses and are critical for the establishment of immune homeostasis. Hansen and colleagues have shown that the delayed microbial colonization of ex-germ-free animals at 3 weeks of age shifts the immune system toward a pro-inflammatory state (Hansen et al. 2012). Furthermore, the microbial exposure of germ-free mice during the neonatal period prevented the accumulation of mucosal invariant natural killer T cells and related pathology in mouse IBD and asthma models (Olszak et al. 2012). Thus, this critical time frame is described as the "neonatal window of opportunity" (Hansen et al. 2013).

Several methods can be used for associating germ-free animals with microorganisms. A complex microbiota can be transferred to germ-free

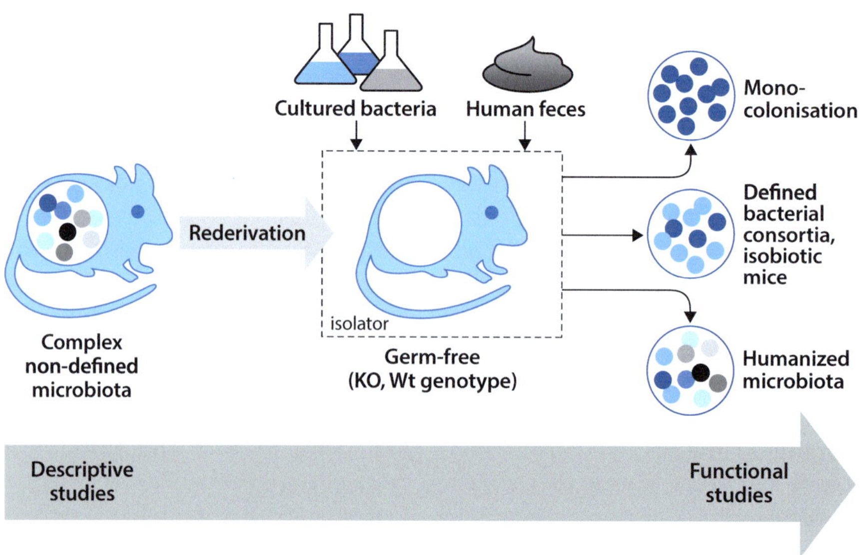

Fig. 21.4 Use of germ-free animals in functional studies. Germ-free animals are rederived from animals colonized by a complex non-defined microbiota. The new rederived strains (specific mutant strains or wild-type strains) represent a valuable tool for functional microbiome research, as these models are fully naïve in respect of microbial colonization. Germ-free animals are free of all life forms except their own cells. Therefore, these mice can be colonized with cultured microorganisms or complex microbiota such as human feces. Colonization of germ-free animal models with a single microorganism (monocolonization) contributes to the understanding and unraveling mechanisms of host-microbe and microbe-host interaction. Furthermore, defined microbial consortia can be established in germ-free mice (isobiotic mice) by colonizing them with particular cultured mixtures of microorganisms. Complex microbiota can be established in germ-free mice by inoculating human or rodent feces. Colonization of germ-free mice with human feces results in establishing human microbiota in a mouse model generating so called "humanized gnotobiotic mice". In summary, all listed approaches generate highly standardized models, which allow causal analysis of microbe-host interactions

mice by cohousing them together with conventional donor mice over several weeks. Furthermore, germ-free mice can be directly inoculated with fecal material or cultured microorganisms by oral gavage. An additional method is to inoculate the drinking water with pure culture, fresh feces, or cecum content and in this way transfer the microorganism into germ-free animals. This only works well for oxygen-tolerant organisms.

Schaedler was a pioneer in developing defined microbial cocktails that were used to associate germ-free animals (Schaedler et al. 1965b). These bacterial cocktails were composed of bacteria isolated from the normal mouse gut including *Lactobacillus* spp., *Enterococcus* spp., *Enterobacter* spp., *Bacteroides* spp., and *Clostridium* spp. (Schaedler et al. 1965a) and were used to reduce the mortality rate of germ-free mice placed into normal biological barriers. Schaedler and Dubos showed that the majority of the cultivable bacteria in the gastrointestinal tract were facultative anaerobes or obligate anaerobes, which make 99% of the bacterial microbiota in the gastrointestinal tract. Schaedler eventually developed a cocktail that was containing eight bacterial species, which was called the Schaedler flora (Schaedler et al. 1965b). This microbial population was the first defined mouse-derived microbial community. However, due to presence of facultative anaerobes, this minimal bacterial consortium was refined by Orcutt (Orcutt et al. 1987). The new bacteria present in the consortium represented the major constituents of the mouse indigenous microbiota and was used by major commercial rodent vendors (Wymore Brand et al. 2015). This minimal consortium was called "Altered Schaedler Flora" and is still in use worldwide, although individual components are not available in public strain collections.

Recently, one new bacterial minimal consortium called "Oligo-Mouse-Microbiota" was developed (Brugiroux et al. 2016). It consists of 12 strains, which provide partial colonization resistance against enteric infections. Additionally, colonies of humanized mice using defined bacterial consortia from human stool have been established, e.g., two simplified human microbial consortia (SIHUMI): one harbors human-derived IBD-related enteric bacteria (Wohlgemuth et al. 2011), and in the other species selection was based on the numerical importance and fermentative abilities in the gut (Becker et al. 2011). Colonization of germ-free mice with minimal bacterial consortia reduces the intestinal ecosystem complexity and allows better understanding and analyses of host-microbiota interactions.

Due to great variation of microbial composition in different facilities, the application of the minimal bacterial microbiota might contribute to the standardized analysis of host-microbe interactions. Furthermore, they might enable to better define phenotypes in mouse models by designing model specific microbiomes.

Besides, germ-free mice can be associated with human fecal material (Collins et al. 2015; Turnbaugh et al. 2009). This enables studying microbiota functions in human diseases. Several human disease phenotypes, including obesity (Ley et al. 2006; Ridaura et al. 2013; Turnbaugh et al. 2009), irritable bowel syndrome (Crouzet et al. 2013), and resistance to cancer therapy (Routy et al. 2018), can be recapitulated this way in mice and open the door for future mechanistic studies.

21.8 Concluding Remarks

The maintenance of germ-free and gnotobiotic animals requires very well-trained staff, and it is effortful and expensive. Germ-free animals are anatomically and physiologically different than those raised in the regular environment. These animals have become a powerful tool for analyzing the impact of the microbiome on the host physiology as well as microbiota-driven disease mechanisms.

Association of germ-free mice with single or defined minimal bacterial consortia enables to analyze protective or detrimental effects of particular microbes as well as to study microbial ecologies in a natural host. Therefore, gnotobiotic research provides tools essential for detailed analysis of host-microbe interactions under standardized conditions.

Controversy

Treating animals with broad-spectrum antibiotics or antibiotic cocktails including ampicillin, ciprofloxacin, metronidazole, streptomycin or vancomycin significantly reduces the load of the majority of bacterial species in the gut, but this treatment cannot completely deplete them (Schubert et al. 2015). This way, bacterial communities are suppressed but not eliminated. This approach is often used as an alternative to the germ-free animals. However, several limitations of antibiotic-treated mice need to be considered. Application of antibiotics can change gut function and metabolism, lead to overgrowth of antibiotic-resistant bacterial species and favor antibiotic resistances (Lundberg et al. 2016). Additionally, gut microbiota is crucial for immune system priming and thus the exposure to microorganisms before antibiotic depletion can have permanent or long-lasting effects on the host physiology, overriding antibiotic therapy. Furthermore, antibiotics do not target other members of the enteric microbiota such as fungi, bacteriophages and eukaryotic viruses, which also may contribute to the intestinal homeostasis or dysbiosis (Lundberg et al. 2016). Therefore, these models have a limited spectrum of applications and may only be a starting experiment to demonstrate if a particular disorder is microbiota-dependent or not. However, germ-free animals are also critically discussed due to their underdeveloped immune system and lack of immune system priming in early age, which is necessary for normal development of immune response (Hansen et al. 2012). This can be circumvented by establishing gnotobiotic colonies, which are colonized with defined and stabile microbial consortia (Becker et al. 2011; Brugiroux et al. 2016; Schaedler et al. 1965b). These colonies could then provide standardized animal models for research of host-microbiota interaction, in which the immune system was primed and normally developed since early age, but all organisms present are fully determined.

Highlights

- Germ-free animals differ in their anatomical and physiological characteristics to the animals carrying indigenous microbiota
- The microbiological status of gnotobiotic animals is fully defined
- Husbandry of gnotobiotic animals is labor intensive, costly and requires well trained staff
- Gnotobiotic animals are an effective tool for elucidating the effect of single or multiple microorganism on the host physiology or pathology under highly standardized conditions

References

Baker, D. G. (1998). Natural pathogens of laboratory mice, rats, and rabbits and their effects on research. *Clinical Microbiology Reviews, 11*, 231–266.

Becker, N., Kunath, J., Loh, G., & Blaut, M. (2011). Human intestinal microbiota: Characterization of a simplified and stable gnotobiotic rat model. *Gut Microbes, 2*, 25–33.

Becker, C., Neurath, M. F., & Wirtz, S. (2015). The intestinal microbiota in inflammatory bowel disease. *ILAR Journal/National Research Council, Institute of Laboratory Animal Resources, 56*, 192–204.

Belkaid, Y., & Hand, T. W. (2014). Role of the microbiota in immunity and inflammation. *Cell, 157*, 121–141.

Blackmore, D. K., & Casillo, S. (1972). Experimental investigation of uterine infections of mice due to Pasteurella pneumotropica. *Journal of Comparative Pathology, 82*, 471–475.

Bleich, A., & Fox, J. G. (2015). The mammalian microbiome and its importance in laboratory animal research. *ILAR Journal/National Research Council, Institute of Laboratory Animal Resources, 56*, 153–158.

Bleich, A., & Hansen, A. K. (2012). Time to include the gut microbiota in the hygienic standardisation of laboratory rodents. *Comparative Immunology, Microbiology and Infectious Diseases, 35*, 81–92.

Brenner, D. A., Paik, Y. H., & Schnabl, B. (2015). Role of gut microbiota in liver disease. *Journal of Clinical Gastroenterology, 49(Suppl 1)*, S25–S27.

Brugiroux, S., Beutler, M., Pfann, C., Garzetti, D., Ruscheweyh, H. J., Ring, D., et al. (2016). Genome-guided design of a defined mouse microbiota that confers colonization resistance against Salmonella enterica serovar Typhimurium. *Nature Microbiology, 2*, 16215.

Coates, M. E. (1975). Gnotobiotic animals in research: Their uses and limitations. *Laboratory Animals, 9*, 275–282.

Collins, J., Auchtung, J. M., Schaefer, L., Eaton, K. A., & Britton, R. A. (2015). Humanized microbiota mice as a model of recurrent Clostridium difficile disease. *Microbiome, 3*, 35.

Crouzet, L., Gaultier, E., Del'Homme, C., Cartier, C., Delmas, E., Dapoigny, M., et al. (2013). The hypersensitivity to colonic distension of IBS patients can be transferred to rats through their fecal microbiota. *Neurogastroenterology and Motility, 25*, e272–e282.

Dorsch, M. (2012). Cryopreservation of preimplantation embryos and gametes, and associated methods. In H. J. Hedrich (Ed.), *The laboratory mouse*. Amsterdam: Elsevier.

Eun, C. S., Mishima, Y., Wohlgemuth, S., Liu, B., Bower, M., Carroll, I. M., et al. (2014). Induction of bacterial antigen-specific colitis by a simplified human microbiota consortium in gnotobiotic interleukin-10-/- mice. *Infection and Immunity, 82*, 2239–2246.

Foster, H. L. (1959). Housing of disease-free vertebrates. *Annals of the New York Academy of Sciences, 78*, 80–88.

Gates, A. H. (1956). Viability and developmental capacity of eggs from immature mice treated with gonadotrophins. *Nature, 177*, 754–755.

Gustafsson, B. (1946). Germ-free rearing of rats. *Acta Anatomica, 2*, 376–391.

Gustafsson, B. E. (1959). Lightweight stainless steel systems for rearing germfree animals. *Annals of the New York Academy of Sciences, 78*, 17–28.

Hansen, C. H., Nielsen, D. S., Kverka, M., Zakostelska, Z., Klimesova, K., Hudcovic, T., et al. (2012). Patterns of early gut colonization shape future immune responses of the host. *PLoS One, 7*, e34043.

Hansen, C. H., Metzdorff, S. B., & Hansen, A. K. (2013). Customizing laboratory mice by modifying gut microbiota and host immunity in an early "window of opportunity". *Gut Microbes, 4*, 241–245.

Hansen, A. K., Hansen, C. H., Krych, L., & Nielsen, D. S. (2014). Impact of the gut microbiota on rodent models of human disease. *World Journal of Gastroenterology, 20*, 17727–17736.

Hecht, G., Bar-Nathan, C., Milite, G., Alon, I., Moshe, Y., Greenfeld, L., et al. (2014). A simple cage-autonomous method for the maintenance of the barrier status of germ-free mice during experimentation. *Laboratory Animals, 48*, 292–297.

Hedrich, H. J., & Nicklas, W. (2012). Housing and maintenance. In H. J. Hedrich (Ed.), *The laboratory mouse* (pp. 521–546). Academic Press: Oxford.

Hooper, L. V., Littman, D. R., & Macpherson, A. J. (2012). Interactions between the microbiota and the immune system. *Science, 336*, 1268–1273.

Hormannsperger, G., Schaubeck, M., & Haller, D. (2015). Intestinal microbiota in animal models of inflammatory diseases. *ILAR Journal/National Research Council, Institute of Laboratory Animal Resources, 56*, 179–191.

Janus, L. M., Smoczek, A., Hedrich, H. J., & Bleich, A. (2009). Risk assessment of minute virus of mice transmission during rederivation: Detection in reproductive organs, gametes, and embryos of mice after in vivo infection. *Biology of Reproduction, 81*, 1010–1015.

Keubler, L. M., Buettner, M., Hager, C., & Bleich, A. (2015). A multihit model: Colitis lessons from the Interleukin-10-deficient Mouse. *Inflammatory Bowel Diseases, 21*, 1967–1975.

Kohashi, O., Kohashi, Y., Takahashi, T., Ozawa, A., & Shigematsu, N. (1985). Reverse effect of gram-positive bacteria vs. gram-negative bacteria on adjuvant-induced arthritis in germfree rats. *Microbiology and Immunology, 29*, 487–497.

Küster, E. (1915). Die keimfreie Zuchtung von Säugetieren. In E. Abderhalden (Ed.) *Handbuch der biochemischen Arbeitsmethoden*, Berlin, pp. 311–323; 419–436.

Ley, R. E., Turnbaugh, P. J., Klein, S., & Gordon, J. I. (2006). Microbial ecology: Human gut microbes associated with obesity. *Nature, 444*, 1022–1023.

Luckey, T. D. (1963). *Germfree life and gnotobiology*. New York: Academic Press.

Lundberg, R., Toft, M. F., August, B., Hansen, A. K., & Hansen, C. H. (2016). Antibiotic-treated versus germ-free rodents for microbiota transplantation studies. *Gut Microbes, 7*, 68–74.

Macpherson, A. J., McCoy, K. D., Johansen, F. E., & Brandtzaeg, P. (2008). The immune geography of IgA induction and function. *Mucosal Immunology, 1*, 11–22.

Mähler, M., Berard, M., Feinstein, R., Gallagher, A., Illgen-Wilcke, B., Pritchett-Corning, K., et al. (2014). FELASA recommendations for the health monitoring of mouse, rat, hamster, guinea pig and rabbit colonies in breeding and experimental units. *Laboratory Animals, 48*, 178–192.

Nicklas, W., Keubler, L., & Bleich, A. (2015). Maintaining and monitoring the defined microbiota status of gnotobiotic rodents. *ILAR Journal, 56*, 241–249.

Nuttall, G. H. F., & Thierfelder, H. (1897). Tierisches Leben ohne Bakterien im Verdauungskanal. *Zeitschrift für Physiologische Chemie, 23*, 231–235.

Olszak, T., An, D., Zeissig, S., Vera, M. P., Richter, J., Franke, A., et al. (2012). Microbial exposure during early life has persistent effects on natural killer T cell function. *Science, 336*, 489–493.

Orcutt, R. P., Gianni, F. J., & Judge, R. J. (1987). Development of an "altered Schaedler flora" for NCI gnotobiotic rodents. *Microecology and Therapy, 17*, 59.

Pleasants, J. R. (1959). Rearing germfree cesarean-born rats, mice, and rabbits through weaning. *Annals of the New York Academy of Sciences, 78*, 116–126.

Rahija, R. J. (2007). Gnotobiotics. In J. G. Fox, M. T. Davidson, C. E. Newcomer, F. W. Quimby, & A. L. Smith (Eds.), *The mouse in biomedical research: Normative biology, husbandry, and models* (pp. 218–232). Elsevier.

Reetz, I. C., Wullenweber-Schmidt, M., Kraft, V., & Hedrich, H. J. (1988). Rederivation of inbred strains of mice by means of embryo transfer. *Laboratory Animal Science, 38*, 696–701.

Reuter, J. D., Livingston, R., & Leblanc, M. (2011). Management strategies for controlling endemic and seasonal mouse parvovirus infection in a barrier facility. *Laboratory Animal, 40*, 145–152.

Reyniers, J. A., & Sacksteder, M. R. (1958). Apparatus and method for shipping germ-free and disease-free animals via public transportation. *Applied Microbiology, 6*, 146–152.

Reyniers, J. A., Trexler, P. C., & Ervin, R. F. (1946). Rearing germ-free albino rats. *Lobund Reports*, 1–84.

Rhee, K. J., Sethupathi, P., Driks, A., Lanning, D. K., & Knight, K. L. (2004). Role of commensal bacteria in development of gut-associated lymphoid tissues and preimmune antibody repertoire. *Journal of Immunology, 172*, 1118–1124.

Ridaura, V. K., Faith, J. J., Rey, F. E., Cheng, J., Duncan, A. E., Kau, A. L., et al. (2013). Gut microbiota from twins discordant for obesity modulate metabolism in mice. *Science, 341*, 1241214.

Round, J. L., & Mazmanian, S. K. (2009). The gut microbiota shapes intestinal immune responses during health and disease. *Nature Reviews. Immunology, 9*, 313–323.

Routy, B., Le Chatelier, E., Derosa, L., Duong, C. P. M., Alou, M. T., Daillere, R., et al. (2018). Gut microbiome influences efficacy of PD-1-based immunotherapy against epithelial tumors. *Science, 359*, 91–97.

Sartor, R. B. (2008). Microbial influences in inflammatory bowel diseases. *Gastroenterology, 134*, 577–594.

Sarvari, A., Naderi, M. M., Sadeghi, M. R., & Akhondi, M. M. (2013). A technique for facile and precise transfer of mouse embryos. *Avicenna Journal of Medical Biotechnology, 5*, 62–65.

Schaedler, R. W., Dubos, R., & Costello, R. (1965a). The development of the bacterial flora in the gastrointestinal tract of mice. *The Journal of Experimental Medicine, 122*, 59–66.

Schaedler, R. W., Dubs, R., & Costello, R. (1965b). Association of germfree mice with bacteria isolated from normal mice. *The Journal of Experimental Medicine, 122*, 77–82.

Schubert, A. M., Sinani, H., & Schloss, P. D. (2015). Antibiotic-Induced Alterations of the Murine Gut Microbiota and Subsequent Effects on Colonization Resistance against Clostridium difficile. *mBio, 6*, e00974.

Shimizu, K., Muranaka, Y., Fujimura, R., Ishida, H., Tazume, S., & Shimamura, T. (1998). Normalization of reproductive function in germfree mice following bacterial contamination. *Experimental Animals, 47*, 151–158.

Steck, N., Hoffmann, M., Sava, I. G., Kim, S. C., Hahne, H., Tonkonogy, S. L., et al. (2011). Enterococcus faecalis metalloprotease compromises epithelial barrier and contributes to intestinal inflammation. *Gastroenterology, 141*, 959–971.

Trexler, P. C. (1961). The gnotobiote-review and future. *Bio-Medical Purview, 1*, 47–58.

Trexler, P. C. (1983). Gnotobiotics. In H. L. Forster & J. G. Fox (Eds.), *The mose in biomedical research* (pp. 1–15). New York: Academic Press.

Trexler, P. C., & Reynolds, L. I. (1957). Flexible film apparatus for the rearing and use of germfree animals. *Applied Microbiology, 5*, 406–412.

Turnbaugh, P. J., Ridaura, V. K., Faith, J. J., Rey, F. E., Knight, R., & Gordon, J. I. (2009). The effect of diet on the human gut microbiome: A metagenomic analysis in humanized gnotobiotic mice. *Science Translational Medicine, 1*, 6ra14.

Ussar, S., Griffin, N. W., Bezy, O., Fujisaka, S., Vienberg, S., Softic, S., et al. (2015). Interactions between gut microbiota, host genetics and diet modulate the predisposition to obesity and metabolic syndrome. *Cell Metabolism, 22*, 516–530.

van der Waaij, D., & Andreas, A. H. (1971). Prevention of airborne contamination and cross-contamination in germ-free mice by laminar flow. *The Journal of Hygiene, 69*, 83–89.

Vowles, C. J., Anderson, N. E., & Eaton, K. A. (2016). *Gnotobiotic mouse technology an illustrated guide*. Boca Raton: CRC Press.

Weisbroth, S. H., Geistfeld, J., Weisbroth, S. P., Williams, B., Feldman, S. H., Linke, M. J., et al. (1999). Latent Pneumocystis carinii infection in commercial rat colonies: Comparison of inductive immunosuppressants plus histopathology, PCR, and serology as detection methods. *Journal of Clinical Microbiology, 37*, 1441–1446.

Whittingham, D. G. (1971). Culture of mouse ova. *Journal of Reproduction and Fertility. Supplement, 14*, 7–21.

Wohlgemuth, S., Bower, M., Gulati, A., & Sartor, R. B. (2011). Simplified human microbiota – A humanized gnotobiotic rodent model to study complex microbe-host interactions in ileal Crohn's disease. *Inflammatory Bowel Disease, 17(Suppl 2), S75*.

Wostmann, B. S. (1981). The germfree animal in nutritional studies. *Annual Review of Nutrition, 1*, 257–279.

Wymore Brand, M., Wannemuehler, M. J., Phillips, G. J., Proctor, A., Overstreet, A. M., Jergens, A. E., et al. (2015). The altered schaedler flora: continued applications of a defined murine microbial community. *ILAR Journal/National Research Council, Institute of Laboratory Animal Resources, 56*, 169–178.

Zhao, Q., & Elson, C. O. (2018). Adaptive immune education by gut microbiota antigens. *Immunology, 154* (1), 28–37.

MIX
Papier aus verantwortungsvollen Quellen
Paper from responsible sources
FSC® C105338

If you have any concerns about our products,
you can contact us on
ProductSafety@springernature.com

In case Publisher is established outside the EU,
the EU authorized representative is:
Springer Nature Customer Service Center GmbH
Europaplatz 3, 69115 Heidelberg, Germany

Printed by Libri Plureos GmbH
in Hamburg, Germany